Essentials of Probability & Statistics for Engineers & Scientists

Ronald E. Walpole
Roanoke College

Raymond H. Myers
Virginia Tech

Sharon L. Myers
Radford University

Keying Ye
University of Texas at San Antonio

PEARSON

Boston Columbus Indianapolis New York San Francisco Upper Saddle River
Amsterdam Cape Town Dubai London Madrid Milan Munich Paris Montréal Toronto
Delhi Mexico City São Paulo Sydney Hong Kong Seoul Singapore Taipei Tokyo

Editor in Chief: *Deirdre Lynch*
Acquisitions Editor: *Christopher Cummings*
Executive Content Editor: *Christine O'Brien*
Sponsoring Editor: *Christina Lepre*
Associate Content Editor: *Dana Bettez*
Editorial Assistant: *Sonia Ashraf*
Senior Managing Editor: *Karen Wernholm*
Senior Production Project Manager: *Tracy Patruno*
Associate Director of Design: USHE North and West, *Andrea Nix*
Cover Designer: *Heather Scott*
Digital Assets Manager: *Marianne Groth*
Associate Media Producer: *Jean Choe*
Marketing Manager: *Erin Lane*
Marketing Assistant: *Kathleen DeChavez*
Senior Author Support/Technology Specialist: *Joe Vetere*
Rights and Permissions Advisor: *Michael Joyce*
Procurement Manager: *Evelyn Beaton*
Procurement Specialist: *Debbie Rossi*
Production Coordination: *Lifland et al., Bookmakers*
Composition: *Keying Ye*
Cover image: *Marjory Dressler/Dressler Photo-Graphics*

Library of Congress Cataloging-in-Publication Data

Essentials of probability & statistics for engineers & scientists/Ronald E. Walpole ... [et al.].
  p. cm.
   Shorter version of: Probability and statistics for engineers and scientists. c2011.
   ISBN 0-321-78373-5
   1. Engineering—Statistical methods. 2. Probabilities. I. Walpole, Ronald E. II. Probability and statistics for engineers and scientists. III. Title: Essentials of probability and statistics for engineers and scientists.
  TA340.P738 2013
  620.001'5192—dc22

                       2011007277

1 2 3 4 5 6 7 8 9 10—EB—15 14 13 12 11

www.pearsonhighered.com

ISBN 10: 0-321-78373-5
ISBN 13: 978-0-321-78373-8

Contents

3   Some Probability Distributions 101

4   Sampling Distributions and Data Descriptions 157

Preface

General Approach and Mathematical Level

This text was designed for a one-semester course that covers the essential topics needed for a fundamental understanding of basic statistics and its applications in the fields of engineering and the sciences. A balance between theory and application is maintained throughout the text. Coverage of analytical tools in statistics is enhanced with the use of calculus when discussion centers on rules and concepts in probability. Students using this text should have the equivalent of the completion of one semester of differential and integral calculus. Linear algebra would be helpful but not necessary if the instructor chooses not to include Section 7.11 on multiple linear regression using matrix algebra.

Class projects and case studies are presented throughout the text to give the student a deeper understanding of real-world usage of statistics. Class projects provide the opportunity for students to work alone or in groups to gather their own experimental data and draw inferences using the data. In some cases, the work conducted by the student involves a problem whose solution will illustrate the meaning of a concept and/or will provide an empirical understanding of an important statistical result. Case studies provide commentary to give the student a clear understanding in the context of a practical situation. The comments we affectionately call "Pot Holes" at the end of each chapter present the big picture and show how the chapters relate to one another. They also provide warnings about the possible misuse of statistical techniques presented in the chapter. A large number of exercises are available to challenge the student. These exercises deal with real-life scientific and engineering applications. The many data sets associated with the exercises are available for download from the website http://www.pearsonhighered.com/mathstatsresources.

Content and Course Planning

This textbook contains nine chapters. The first two chapters introduce the notion of random variables and their properties, including their role in characterizing data sets. Fundamental to this discussion is the distinction, in a practical sense, between populations and samples.

In Chapter 3, both discrete and continuous random variables are illustrated with examples. The binomial, Poisson, hypergeometric, and other useful discrete distributions are discussed. In addition, continuous distributions include the normal, gamma, and exponential. In all cases, real-life scenarios are given to reveal how these distributions are used in practical engineering problems.

The material on specific distributions in Chapter 3 is followed in Chapter 4 by practical topics such as random sampling and the types of descriptive statistics that convey the center of location and variability of a sample. Examples involving the sample mean and sample variance are included. Following the introduction of central tendency and variability is a substantial amount of material dealing with the importance of sampling distributions. Real-life illustrations highlight how sampling distributions are used in basic statistical inference. Central Limit type methodology is accompanied by the mechanics and purpose behind the use of the normal, Student t, χ^2, and f distributions, as well as examples that illustrate their use. Students are exposed to methodology that will be brought out again in later chapters in the discussions of estimation and hypothesis testing. This fundamental methodology is accompanied by illustration of certain important graphical methods, such as stem-and-leaf and box-and-whisker plots. Chapter 4 presents the first of several case studies involving real data.

Chapters 5 and 6 complement each other, providing a foundation for the solution of practical problems in which estimation and hypothesis testing are employed. Statistical inference involving a single mean and two means, as well as one and two proportions, is covered. Confidence intervals are displayed and thoroughly discussed; prediction intervals and tolerance intervals are touched upon. Problems with paired observations are covered in detail.

In Chapter 7, the basics of simple linear regression (SLR) and multiple linear regression (MLR) are covered in a depth suitable for a one-semester course. Chapters 8 and 9 use a similar approach to expose students to the standard methodology associated with analysis of variance (ANOVA). Although regression and ANOVA are challenging topics, the clarity of presentation, along with case studies, class projects, examples, and exercises, allows students to gain an understanding of the essentials of both.

In the discussion of rules and concepts in probability, the coverage of analytical tools is enhanced through the use of calculus. Though the material on multiple linear regression in Chapter 7 covers the essential methodology, students are not burdened with the level of matrix algebra and relevant manipulations that they would confront in a text designed for a two-semester course.

Computer Software

Case studies, beginning in Chapter 4, feature computer printout and graphical material generated using both SAS® and MINITAB®. The inclusion of the computer reflects our belief that students should have the experience of reading and interpreting computer printout and graphics, even if the software in the text is not that which is used by the instructor. Exposure to more than one type of software can broaden the experience base for the student. There is no reason to believe that the software used in the course will be that which the student will be called upon to use in a professional setting.

Supplements

Instructor's Solutions Manual. This resource contains worked-out solutions to all text exercises and is available for download from Pearson's Instructor Resource Center at www.pearsonhighered.com/irc.

Student's Solutions Manual. ISBN-10: 0-321-78399-9; ISBN-13: 978-0-321-78399-8. This resource contains complete solutions to selected exercises. It is available for purchase from MyPearsonStore at www.mypearsonstore.com, or ask your local representative for value pack options.

PowerPoint® *Lecture Slides.* These slides include most of the figures and tables from the text. Slides are available for download from Pearson's Instructor Resource Center at www.pearsonhighered.com/irc.

Looking for more comprehensive coverage for a two-semester course? See the more comprehensive book *Probability and Statistics for Engineers and Scientists*, 9th edition, by Walpole, Myers, Myers, and Ye (ISBN-10: 0-321-62911-6; ISBN-13: 978-0-321-62911-1).

Acknowledgments

We are indebted to those colleagues who provided many helpful suggestions for this text. They are David Groggel, *Miami University*; Lance Hemlow, *Raritan Valley Community College*; Ying Ji, *University of Texas at San Antonio*; Thomas Kline, *University of Northern Iowa*; Sheila Lawrence, *Rutgers University*; Luis Moreno, *Broome County Community College*; Donald Waldman, *University of Colorado—Boulder*; and Marlene Will, *Spalding University*. We would also like to thank Delray Schultz, *Millersville University*, and Keith Friedman, *University of Texas — Austin*, for ensuring the accuracy of this text.

We would like to thank the editorial and production services provided by numerous people from Pearson/Prentice Hall, especially the editor in chief Deirdre Lynch, acquisitions editor Chris Cummings, sponsoring editor Christina Lepre, associate content editor Dana Bettez, editorial assistant Sonia Ashraf, production project manager Tracy Patruno, and copyeditor Sally Lifland. We thank the Virginia Tech Statistical Consulting Center, which was the source of many real-life data sets.

R.H.M.
S.L.M.
K.Y.

This book is dedicated to

Billy and Julie
R.H.M. and S.L.M.

Limin, Carolyn and Emily
K.Y.

Chapter 1

Introduction to Statistics and Probability

1.1 Overview: Statistical Inference, Samples, Populations, and the Role of Probability

Beginning in the 1980s and continuing into the 21st century, a great deal of attention has been focused on *improvement of quality* in American industry. Much has been said and written about the Japanese "industrial miracle," which began in the middle of the 20th century. The Japanese were able to succeed where we and other countries had failed—namely, to create an atmosphere that allows the production of high-quality products. Much of the success of the Japanese has been attributed to the use of *statistical methods* and statistical thinking among management personnel.

Use of Scientific Data

The use of statistical methods in manufacturing, development of food products, computer software, energy sources, pharmaceuticals, and many other areas involves the gathering of information or **scientific data**. Of course, the gathering of data is nothing new. It has been done for well over a thousand years. Data have been collected, summarized, reported, and stored for perusal. However, there is a profound distinction between collection of scientific information and **inferential statistics**. It is the latter that has received rightful attention in recent decades.

The offspring of inferential statistics has been a large "toolbox" of statistical methods employed by statistical practitioners. These statistical methods are designed to contribute to the process of making scientific judgments in the face of **uncertainty** and **variation**. The product density of a particular material from a manufacturing process will not always be the same. Indeed, if the process involved is a batch process rather than continuous, there will be not only variation in material density among the batches that come off the line (batch-to-batch variation), but also within-batch variation. Statistical methods are used to analyze data from a process such as this one in order to gain more sense of where in the process changes may be made to improve the **quality** of the process. In this process, qual-

1

ity may well be defined in relation to closeness to a target density value in harmony with *what portion of the time* this closeness criterion is met. An engineer may be concerned with a specific instrument that is used to measure sulfur monoxide in the air during pollution studies. If the engineer has doubts about the effectiveness of the instrument, there are two **sources of variation** that must be dealt with. The first is the variation in sulfur monoxide values that are found at the same locale on the same day. The second is the variation between values observed and the **true** amount of sulfur monoxide that is in the air at the time. If either of these two sources of variation is exceedingly large (according to some standard set by the engineer), the instrument may need to be replaced. In a biomedical study of a new drug that reduces hypertension, 85% of patients experienced relief, while it is generally recognized that the current drug, or "old" drug, brings relief to 80% of patients that have chronic hypertension. However, the new drug is more expensive to make and may result in certain side effects. Should the new drug be adopted? This is a problem that is encountered (often with much more complexity) frequently by pharmaceutical firms in conjunction with the FDA (Federal Drug Administration). Again, the consideration of variation needs to be taken into account. The "85%" value is based on a certain number of patients chosen for the study. Perhaps if the study were repeated with new patients the observed number of "successes" would be 75%! It is the natural variation from study to study that must be taken into account in the decision process. Clearly this variation is important, since variation from patient to patient is endemic to the problem.

Variability in Scientific Data

In the problems discussed above the statistical methods used involve dealing with variability, and in each case the variability to be studied is that encountered in scientific data. If the observed product density in the process were always the same and were always on target, there would be no need for statistical methods. If the device for measuring sulfur monoxide always gives the same value and the value is accurate (i.e., it is correct), no statistical analysis is needed. If there were no patient-to-patient variability inherent in the response to the drug (i.e., it either always brings relief or not), life would be simple for scientists in the pharmaceutical firms and FDA and no statistician would be needed in the decision process. Statistics researchers have produced an enormous number of analytical methods that allow for analysis of data from systems like those described above. This reflects the true nature of the science that we call inferential statistics, namely, using techniques that allow us to go beyond merely reporting data to drawing conclusions (or inferences) about the scientific system. Statisticians make use of fundamental laws of probability and statistical inference to draw conclusions about scientific systems. Information is gathered in the form of **samples**, or collections of **observations**. The process of sampling will be introduced in this chapter, and the discussion continues throughout the entire book.

Samples are collected from **populations**, which are collections of all individuals or individual items of a particular type. At times a population signifies a scientific system. For example, a manufacturer of computer boards may wish to eliminate defects. A sampling process may involve collecting information on 50 computer boards sampled randomly from the process. Here, the population is all computer boards manufactured by the firm over a specific period of time. If an improvement is made in the computer board process and a second sample of boards is collected, any conclusions drawn regarding the effectiveness of the change in process should extend to the entire population of computer boards produced under the "improved process." In a drug experiment, a sample of patients is taken and each is given a specific drug to reduce blood pressure. The interest is focused on drawing conclusions about the population of those who suffer from hypertension.

Often, it is very important to collect scientific data in a systematic way, with planning being high on the agenda. At times the planning is, by necessity, quite limited. We often focus only on certain properties or characteristics of the items or objects in the population. Each characteristic has particular engineering or, say, biological importance to the "customer," the scientist or engineer who seeks to learn about the population. For example, in one of the illustrations above the quality of the process had to do with the product density of the output of a process. An engineer may need to study the effect of process conditions, temperature, humidity, amount of a particular ingredient, and so on. He or she can systematically move these **factors** to whatever levels are suggested according to whatever prescription or **experimental design** is desired. However, a forest scientist who is interested in a study of factors that influence wood density in a certain kind of tree cannot necessarily design an experiment. This case may require an **observational study** in which data are collected in the field but **factor levels** can not be preselected. Both of these types of studies lend themselves to methods of statistical inference. In the former, the quality of the inferences will depend on proper planning of the experiment. In the latter, the scientist is at the mercy of what can be gathered. For example, it is sad if an agronomist is interested in studying the effect of rainfall on plant yield and the data are gathered during a drought.

The importance of statistical thinking by managers and the use of statistical inference by scientific personnel is widely acknowledged. Research scientists gain much from scientific data. Data provide understanding of scientific phenomena. Product and process engineers learn a great deal in their off-line efforts to improve the process. They also gain valuable insight by gathering production data (on-line monitoring) on a regular basis. This allows them to determine necessary modifications in order to keep the process at a desired level of quality.

There are times when a scientific practitioner wishes only to gain some sort of summary of a set of data represented in the sample. In other words, inferential statistics is not required. Rather, a set of single-number statistics or **descriptive statistics** is helpful. These numbers give a sense of center of the location of the data, variability in the data, and the general nature of the distribution of observations in the sample. Though no specific statistical methods leading to **statistical inference** are incorporated, much can be learned. At times, descriptive statistics are accompanied by graphics. Modern statistical software packages allow for computation of **means**, **medians**, **standard deviations**, and other single-number statistics as well as production of graphs that show a "footprint" of the nature of the sample, including histograms, stem-and-leaf plots, scatter plots, dot plots, and box plots.

The Role of Probability

From this chapter to Chapter 3, we deal with fundamental notions of probability. A thorough grounding in these concepts allows the reader to have a better understanding of statistical inference. Without some formalism of probability theory, the student cannot appreciate the true interpretation from data analysis through modern statistical methods. It is quite natural to study probability prior to studying statistical inference. Elements of probability allow us to quantify the strength or "confidence" in our conclusions. In this sense, concepts in probability form a major component that supplements statistical methods and helps us gauge the strength of the statistical inference. The discipline of probability, then, provides the transition between descriptive statistics and inferential methods. Elements of probability allow the conclusion to be put into the language that the science or engineering practitioners require. An example follows that will enable the reader to understand the notion of a *P*-value, which often provides the "bottom line" in the interpretation of results from the use of statistical methods.

Example 1.1: Suppose that an engineer encounters data from a manufacturing process in which 100 items are sampled and 10 are found to be defective. It is expected and anticipated that occasionally there will be defective items. Obviously these 100 items represent the sample. However, it has been determined that in the long run, the company can only tolerate 5% defective in the process. Now, the elements of probability allow the engineer to determine how conclusive the sample information is regarding the nature of the process. In this case, the **population** conceptually represents all possible items from the process. Suppose we learn that *if the process is acceptable*, that is, if it does produce items no more than 5% of which are defective, there is a probability of 0.0282 of obtaining 10 or more defective items in a random sample of 100 items from the process. This small probability suggests that the process does, indeed, have a long-run rate of defective items that exceeds 5%. In other words, under the condition of an acceptable process, the sample information obtained would rarely occur. However, it did occur! Clearly, though, it would occur with a much higher probability if the process defective rate exceeded 5% by a significant amount.

From this example it becomes clear that the elements of probability aid in the translation of sample information into something conclusive or inconclusive about the scientific system. In fact, what was learned likely is alarming information to the engineer or manager. Statistical methods, which we will actually detail in Chapter 6, produced a *P*-value of 0.0282. The result suggests that the process **very likely is not acceptable**. The concept of a ***P*-value** is dealt with at length in succeeding chapters. The example that follows provides a second illustration.

Example 1.2: Often the nature of the scientific study will dictate the role that probability and deductive reasoning play in statistical inference. Exercise 5.28 on page 221 provides data associated with a study conducted at Virginia Tech on the development of a relationship between the roots of trees and the action of a fungus. Minerals are transferred from the fungus to the trees and sugars from the trees to the fungus. Two samples of 10 northern red oak seedlings were planted in a greenhouse, one containing seedlings treated with nitrogen and the other containing seedlings with no nitrogen. All other environmental conditions were held constant. All seedlings contained the fungus *Pisolithus tinctorus*. More details are supplied in Chapter 5. The stem weights in grams were recorded after the end of 140 days. The data are given in Table 1.1.

Table 1.1: Data Set for Example 1.2

No Nitrogen	Nitrogen
0.32	0.26
0.53	0.43
0.28	0.47
0.37	0.49
0.47	0.52
0.43	0.75
0.36	0.79
0.42	0.86
0.38	0.62
0.43	0.46

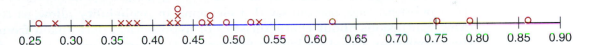

Figure 1.1: A dot plot of stem weight data.

In this example there are two samples from two **separate populations**. The purpose of the experiment is to determine if the use of nitrogen has an influence on the growth of the roots. The study is a comparative study (i.e., we seek to compare the two populations with regard to a certain important characteristic). It is instructive to plot the data as shown in the dot plot of Figure 1.1. The ∘ values represent the "nitrogen" data and the × values represent the "no-nitrogen" data.

Notice that the general appearance of the data might suggest to the reader that, on average, the use of nitrogen increases the stem weight. Four nitrogen observations are considerably larger than any of the no-nitrogen observations. Most of the no-nitrogen observations appear to be below the center of the data. The appearance of the data set would seem to indicate that nitrogen is effective. But how can this be quantified? How can all of the apparent visual evidence be summarized in some sense? As in the preceding example, the fundamentals of probability can be used. The conclusions may be summarized in a probability statement or P-value. We will not show here the statistical inference that produces the summary probability. As in Example 1.1, these methods will be discussed in Chapter 6. The issue revolves around the "probability that data like these could be observed" *given that nitrogen has no effect*, in other words, given that both samples were generated from the same population. Suppose that this probability is small, say 0.03. That would certainly be strong evidence that the use of nitrogen does indeed influence (apparently increases) average stem weight of the red oak seedlings.

How Do Probability and Statistical Inference Work Together?

It is important for the reader to understand the clear distinction between the discipline of probability, a science in its own right, and the discipline of inferential statistics. As we have already indicated, the use or application of concepts in probability allows real-life interpretation of the results of statistical inference. As a result, it can be said that statistical inference makes use of concepts in probability. One can glean from the two examples above that the sample information is made available to the analyst and, with the aid of statistical methods and elements of probability, conclusions are drawn about some feature of the population (the process does not appear to be acceptable in Example 1.1, and nitrogen does appear to influence average stem weights in Example 1.2). Thus for a statistical problem, **the sample along with inferential statistics allows us to draw conclusions about the population, with inferential statistics making clear use of elements of probability**. This reasoning is *inductive* in nature. Now as we move into Section 1.4 and beyond, the reader will note that, unlike what we do in our two examples here, we will not focus on solving statistical problems. Many examples will be given in which no sample is involved. There will be a population clearly described with all features of the population known. Then questions of importance will focus on the nature of data that might hypothetically be drawn from the population. Thus, one can say that **elements in probability allow us to draw conclusions about characteristics of hypothetical data taken from the population, based on known features of the population**. This type of reasoning is *deductive* in nature. Figure 1.2 shows the fundamental relationship between probability and inferential statistics.

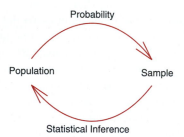

Figure 1.2: Fundamental relationship between probability and inferential statistics.

Now, in the grand scheme of things, which is more important, the field of probability or the field of statistics? They are both very important and clearly are complementary. The only certainty concerning the pedagogy of the two disciplines lies in the fact that if statistics is to be taught at more than merely a "cookbook" level, then the discipline of probability must be taught first. This rule stems from the fact that nothing can be learned about a population from a sample until the analyst learns the rudiments of uncertainty in that sample. For example, consider Example 1.1. The question centers around whether or not the population, defined by the process, is no more than 5% defective. In other words, the conjecture is that **on the average** 5 out of 100 items are defective. Now, the sample contains 100 items and 10 are defective. Does this support the conjecture or refute it? On the surface it would appear to be a refutation of the conjecture because 10 out of 100 seem to be "a bit much." But without elements of probability, how do we know? Only through the study of material in future chapters will we learn the conditions under which the process is acceptable (5% defective). The probability of obtaining 10 or more defective items in a sample of 100 is 0.0282.

We have given two examples where the elements of probability provide a summary that the scientist or engineer can use as evidence on which to build a decision. The bridge between the data and the conclusion is, of course, based on foundations of statistical inference, distribution theory, and sampling distributions discussed in future chapters.

1.2 Sampling Procedures; Collection of Data

In Section 1.1 we discussed very briefly the notion of sampling and the sampling process. While sampling appears to be a simple concept, the complexity of the questions that must be answered about the population or populations necessitates that the sampling process be very complex at times. While the notion of sampling is discussed in a technical way in Chapter 4, we shall endeavor here to give some common-sense notions of sampling. This is a natural transition to a discussion of the concept of variability.

Simple Random Sampling

The importance of proper sampling revolves around the degree of confidence with which the analyst is able to answer the questions being asked. Let us assume that only a single population exists in the problem. Recall that in Example 1.2 two populations were involved. **Simple random sampling** implies that any particular sample of a specified *sample size* has the same chance of being selected as any other sample of the same size. The term **sample size** simply means the number of elements in the sample. Obviously, a table of random numbers can be utilized in sample selection in many instances. The virtue of simple random sampling is that it aids in the elimination of the problem of having the sample reflect a different (possibly more confined) population than the one about which inferences need to be made. For example, a sample is to be chosen to answer certain questions regarding political preferences in a certain state in the United States. The sample involves the choice of, say, 1000 families, and a survey is to be conducted. Now, suppose it turns out that random sampling is not used. Rather, all or nearly all of the 1000 families chosen live in an urban setting. It is believed that political preferences in rural areas differ from those in urban areas. In other words, the sample drawn actually confined the population and thus the inferences need to be confined to the "limited population," and in this case confining may be undesirable. If, indeed, the inferences need to be made about the state as a whole, the sample of size 1000 described here is often referred to as a **biased sample**.

As we hinted earlier, simple random sampling is not always appropriate. Which alternative approach is used depends on the complexity of the problem. Often, for example, the sampling units are not homogeneous and naturally divide themselves into nonoverlapping groups that are homogeneous. These groups are called *strata,* and a procedure called *stratified random sampling* involves random selection of a sample *within* each stratum. The purpose is to be sure that each of the strata is neither over- nor underrepresented. For example, suppose a sample survey is conducted in order to gather preliminary opinions regarding a bond referendum that is being considered in a certain city. The city is subdivided into several ethnic groups which represent natural strata. In order not to disregard or overrepresent any group, separate random samples of families could be chosen from each group.

Experimental Design

The concept of randomness or random assignment plays a huge role in the area of **experimental design**, which was introduced very briefly in Section 1.1 and is an important staple in almost any area of engineering or experimental science. This will also be discussed at length in Chapter 8. However, it is instructive to give a brief presentation here in the context of random sampling. A set of so-called **treatments** or **treatment combinations** becomes the populations to be studied or compared in some sense. An example is the nitrogen versus no-nitrogen treatments in Example 1.2. Another simple example would be placebo versus active drug, or in a corrosion fatigue study we might have treatment combinations that involve specimens that are coated or uncoated as well as conditions of low or high humidity to which the specimens are exposed. In fact, there are four treatment or factor combinations (i.e., 4 populations), and many scientific questions may be asked and answered through statistical and inferential methods. Consider first the situation in Example 1.2. There are 20 diseased seedlings involved in the experiment. It is easy to see from the data themselves that the seedlings are different from each other. Within the nitrogen group (or the no-nitrogen group) there is considerable **variability** in the stem weights. This variability is due to what is generally called the **experimental unit**. This is a very important concept in inferential statistics, in fact one whose description will not end in this chapter. The nature of the variability is very important. If it is too large, stemming from a condition of excessive nonhomogeneity in experimental units, the variability will "wash out" any detectable difference between the two populations. Recall that in this case that did not occur.

The dot plot in Figure 1.1 and *P*-value indicated a clear distinction between these two conditions. What role do those experimental units play in the data-taking process itself? The common-sense and, indeed, quite standard approach is to assign the 20 seedlings or experimental units **randomly to the two treatments or conditions**. In the drug study, we may decide to use a total of 200 available patients, patients that clearly will be different in some sense. They are the experimental units. However, they all may have the same chronic condition for which the drug is a potential treatment. Then in a so-called **completely randomized design**, 100 patients are assigned randomly to the placebo and 100 to the active drug. Again, it is these experimental units within a group or treatment that produce the variability in data results (i.e., variability in the measured result), say blood pressure, or whatever drug efficacy value is important. In the corrosion fatigue study, the experimental units are the specimens that are the subjects of the corrosion.

Why Assign Experimental Units Randomly?

What is the possible negative impact of not randomly assigning experimental units to the treatments or treatment combinations? This is seen most clearly in the case of the drug study. Among the characteristics of the patients that produce variability in the results are age, gender, and weight. Suppose merely by chance the placebo group contains a sample of people that are predominately heavier than those in the treatment group. Perhaps heavier individuals have a tendency to have a higher blood pressure. This clearly biases the result, and indeed, any result obtained through the application of statistical inference may have little to do with the drug and more to do with differences in weights among the two samples of patients.

We should emphasize the attachment of importance to the term **variability**. Excessive variability among experimental units "camouflages" scientific findings. In future sections, we attempt to characterize and quantify measures of variability. In sections that follow, we introduce and discuss specific quantities that can be computed in samples; the quantities give a sense of the nature of the sample with respect to center of location of the data and variability in the data. A discussion of several of these single-number measures serves to provide a preview of what statistical information will be important components of the statistical methods that are used in future chapters. These measures that help characterize the nature of the data set fall into the category of **descriptive statistics**. This material is a prelude to a brief presentation of pictorial and graphical methods that go even further in characterization of the data set. The reader should understand that the statistical methods illustrated here will be used throughout the text. In order to offer the reader a clearer picture of what is involved in experimental design studies, we offer Example 1.3.

Example 1.3: A corrosion study was made in order to determine whether coating an aluminum metal with a corrosion retardation substance reduced the amount of corrosion. The coating is a protectant that is advertised to minimize fatigue damage in this type of material. Also of interest is the influence of humidity on the amount of corrosion. A corrosion measurement can be expressed in thousands of cycles to failure. Two levels of coating, no coating and chemical corrosion coating, were used. In addition, the two relative humidity levels are 20% relative humidity and 80% relative humidity.

The experiment involves four treatment combinations that are listed in the table that follows. There are eight experimental units used, and they are aluminum specimens prepared; two are assigned randomly to each of the four treatment combinations. The data are presented in Table 1.2.

The corrosion data are averages of two specimens. A plot of the averages is pictured in Figure 1.3. A relatively large value of cycles to failure represents a small amount of corrosion. As one might expect, an increase in humidity appears to make the corrosion worse. The use of the chemical corrosion coating procedure appears to reduce corrosion.

In this experimental design illustration, the engineer has systematically selected the four treatment combinations. In order to connect this situation to concepts with which the reader has been exposed to this point, it should be assumed that the

Table 1.2: Data for Example 1.3

Coating	Humidity	Average Corrosion in Thousands of Cycles to Failure
Uncoated	20%	975
	80%	350
Chemical Corrosion	20%	1750
	80%	1550

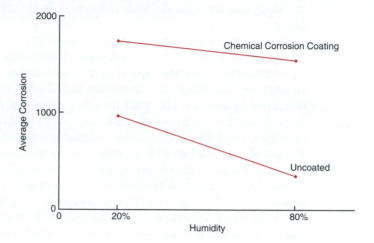

Figure 1.3: Corrosion results for Example 1.3.

conditions representing the four treatment combinations are four separate populations and that the two corrosion values observed for each population are important pieces of information. The importance of the average in capturing and summarizing certain features in the population will be highlighted in Section 4.2. While we might draw conclusions about the role of humidity and the impact of coating the specimens from the figure, we cannot truly evaluate the results from an analytical point of view without taking into account the *variability around* the average. Again, as we indicated earlier, if the two corrosion values for each treatment combination are close together, the picture in Figure 1.3 may be an accurate depiction. But if each corrosion value in the figure is an average of two values that are widely dispersed, then this variability may, indeed, truly "wash away" any information that appears to come through when one observes averages only. The foregoing example illustrates these concepts:

(1) random assignment of treatment combinations (coating, humidity) to experimental units (specimens)

(2) the use of sample averages (average corrosion values) in summarizing sample information

(3) the need for consideration of measures of variability in the analysis of any sample or sets of samples

1.3 Discrete and Continuous Data

Statistical inference through the analysis of observational studies or designed experiments is used in many scientific areas. The data gathered may be **discrete** or **continuous**, depending on the area of application. For example, a chemical engineer may be interested in conducting an experiment that will lead to conditions where yield is maximized. Here, of course, the yield may be in percent or grams/pound, measured on a continuum. On the other hand, a toxicologist conducting a combination drug experiment may encounter data that are binary in nature (i.e., the patient either responds or does not).

Great distinctions are made between discrete and continuous data in the probability theory that allow us to draw statistical inferences. Often applications of statistical inference are found when the data are *count data*. For example, an engineer may be interested in studying the number of radioactive particles passing through a counter in, say, 1 millisecond. Personnel responsible for the efficiency of a port facility may be interested in the properties of the number of oil tankers arriving each day at a certain port city. In Chapter 3, several distinct scenarios, leading to different ways of handling data, are discussed for situations with count data.

Special attention even at this early stage of the textbook should be paid to some details associated with binary data. Applications requiring statistical analysis of binary data are voluminous. Often the measure that is used in the analysis is the *sample proportion*. Obviously the binary situation involves two categories. If there are n units involved in the data and x is defined as the number that fall into category 1, then $n - x$ fall into category 2. Thus, x/n is the sample proportion in category 1, and $1 - x/n$ is the sample proportion in category 2. In the biomedical application, 50 patients may represent the sample units, and if 20 out of 50 experienced an improvement in a stomach ailment (common to all 50) after all were given the drug, then $\frac{20}{50} = 0.4$ is the sample proportion for which the drug was a success and $1 - 0.4 = 0.6$ is the sample proportion for which the drug was not successful. Actually the basic numerical measurement for binary data is generally denoted by either 0 or 1. For example, in our medical example, a successful result is denoted by a 1 and a nonsuccess by a 0. As a result, the sample proportion is actually a sample mean of the ones and zeros. For the successful category,

$$\frac{x_1 + x_2 + \cdots + x_{50}}{50} = \frac{1 + 1 + 0 + \cdots + 0 + 1}{50} = \frac{20}{50} = 0.4.$$

1.4 Probability: Sample Space and Events

Sample Space

In the study of statistics, we are concerned basically with the presentation and interpretation of **chance outcomes** that occur in a planned study or scientific investigation. For example, we may record the number of accidents that occur monthly at the intersection of Driftwood Lane and Royal Oak Drive, hoping to justify the installation of a traffic light; we might classify items coming off an assembly line as "defective" or "nondefective"; or we may be interested in the volume of gas released in a chemical reaction when the concentration of an acid is varied. Hence, the statistician is often dealing with either numerical data, representing counts or measurements, or **categorical data**, which can be classified according to some criterion.

We shall refer to any recording of information, whether it be numerical or categorical, as an **observation**. Thus, the numbers 2, 0, 1, and 2, representing the number of accidents that occurred for each month from January through April during the past year at the intersection of Driftwood Lane and Royal Oak Drive, constitute a set of observations. Similarly, the categorical data *N, D, N, N*, and *D*, representing the items found to be defective or nondefective when five items are inspected, are recorded as observations.

Statisticians use the word **experiment** to describe any process that generates a set of data. A simple example of a statistical experiment is the tossing of a coin. In this experiment, there are only two possible outcomes, heads or tails. Another experiment might be the launching of a missile and observing of its velocity at specified times. The opinions of voters concerning a new sales tax can also be considered as observations of an experiment. We are particularly interested in the observations obtained by repeating the experiment several times. In most cases, the outcomes will depend on chance and, therefore, cannot be predicted with certainty. If a chemist runs an analysis several times under the same conditions, he or she will obtain different measurements, indicating an element of chance in the experimental procedure. Even when a coin is tossed repeatedly, we cannot be certain that a given toss will result in a head. However, we know the entire set of possibilities for each toss.

Definition 1.1: | The set of all possible outcomes of a statistical experiment is called the **sample space** and is represented by the symbol *S*.

Each outcome in a sample space is called an **element** or a **member** of the sample space, or simply a **sample point**. If the sample space has a finite number of elements, we may *list* the members separated by commas and enclosed in braces. Thus, the sample space *S*, of possible outcomes when a coin is flipped, may be written

$$S = \{H, T\},$$

where *H* and *T* correspond to heads and tails, respectively.

Example 1.4: | Consider the experiment of tossing a die. If we are interested in the number that shows on the top face, the sample space is

$$S_1 = \{1, 2, 3, 4, 5, 6\}.$$

If we are interested only in whether the number is even or odd, the sample space is simply

$$S_2 = \{\text{even, odd}\}.$$

Example 1.4 illustrates the fact that more than one sample space can be used to describe the outcomes of an experiment. In this case, S_1 provides more information than S_2. If we know which element in S_1 occurs, we can tell which outcome in S_2 occurs; however, a knowledge of what happens in S_2 is of little help in determining which element in S_1 occurs. In general, it is desirable to use the sample space that gives the most information concerning the outcomes of the experiment. In some experiments, it is helpful to list the elements of the sample space systematically by means of a **tree diagram**.

Example 1.5: Suppose that three items are selected at random from a manufacturing process. Each item is inspected and classified defective, D, or nondefective, N. To list the elements of the sample space providing the most information, we construct the tree diagram of Figure 1.4. Now, the various paths along the branches of the tree give the distinct sample points. Starting with the first path, we get the sample point DDD, indicating the possibility that all three items inspected are defective. As we proceed along the other paths, we see that the sample space is

$$S = \{DDD, \; DDN, \; DND, \; DNN, \; NDD, \; NDN, \; NND, \; NNN\}.$$

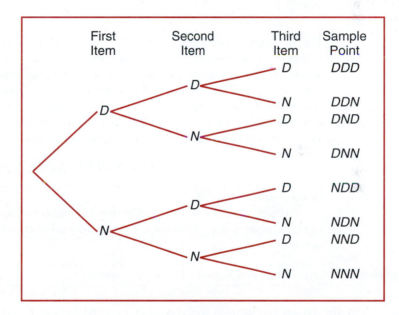

Figure 1.4: Tree diagram for Example 1.5.

Sample spaces with a large or infinite number of sample points are best described by a **statement** or **rule method**. For example, if the possible outcomes of an experiment are the set of cities in the world with a population over 1 million, our sample space is written

$$S = \{x \mid x \text{ is a city with a population over 1 million}\},$$

which reads "S is the set of all x such that x is a city with a population over 1 million." The vertical bar is read "such that." Similarly, if S is the set of all points

(x, y) on the boundary or the interior of a circle of radius 2 with center at the origin, we write the **rule**

$$S = \{(x, y) \mid x^2 + y^2 \le 4\}.$$

Whether we describe the sample space by the rule method or by listing the elements will depend on the specific problem at hand. The rule method has practical advantages, particularly for many experiments where listing becomes a tedious chore.

Consider the situation of Example 1.5 in which items from a manufacturing process are either D, defective, or N, nondefective. There are many important statistical procedures called sampling plans that determine whether or not a "lot" of items is considered satisfactory. One such plan involves sampling until k defectives are observed. Suppose the experiment is to sample items randomly until one defective item is observed. The sample space for this case is

$$S = \{D, ND, NND, NNND, \dots\}.$$

Events

For any given experiment, we may be interested in the occurrence of certain **events** rather than in the occurrence of a specific element in the sample space. For instance, we may be interested in the event A that the outcome when a die is tossed is divisible by 3. This will occur if the outcome is an element of the subset $A = \{3, 6\}$ of the sample space S_1 in Example 1.4. As a further illustration, we may be interested in the event B that the number of defectives is greater than 1 in Example 1.5. This will occur if the outcome is an element of the subset

$$B = \{DDN, DND, NDD, DDD\}$$

of the sample space S.

To each event we assign a collection of sample points, which constitute a subset of the sample space. That subset represents all of the elements for which the event is true.

Definition 1.2: | An **event** is a subset of a sample space.

Example 1.6: | Given the sample space $S = \{t \mid t \ge 0\}$, where t is the life in years of a certain electronic component, then the event A that the component fails before the end of the fifth year is the subset $A = \{t \mid 0 \le t < 5\}$. ⬛

It is conceivable that an event may be a subset that includes the entire sample space S or a subset of S called the **null set** and denoted by the symbol ϕ, which contains no elements at all. For instance, if we let A be the event of detecting a microscopic organism with the naked eye in a biological experiment, then $A = \phi$. Also, if

$$B = \{x \mid x \text{ is an even factor of 7}\},$$

then B must be the null set, since the only possible factors of 7 are the odd numbers 1 and 7.

Consider an experiment where the smoking habits of the employees of a manufacturing firm are recorded. A possible sample space might classify an individual as a nonsmoker, a light smoker, a moderate smoker, or a heavy smoker. Let the subset of smokers be some event. Then all the nonsmokers correspond to a different event, also a subset of S, which is called the **complement** of the set of smokers.

Definition 1.3: The **complement** of an event A with respect to S is the subset of all elements of S that are not in A. We denote the complement of A by the symbol A'.

Example 1.7: Let R be the event that a red card is selected from an ordinary deck of 52 playing cards, and let S be the entire deck. Then R' is the event that the card selected from the deck is not a red card but a black card.

Example 1.8: Consider the sample space

$$S = \{\text{book, cell phone, mp3, paper, stationery, laptop}\}.$$

Let $A = \{\text{book, stationery, laptop, paper}\}$. Then the complement of A is $A' = \{\text{cell phone, mp3}\}$.

We now consider certain operations with events that will result in the formation of new events. These new events will be subsets of the same sample space as the given events. Suppose that A and B are two events associated with an experiment. In other words, A and B are subsets of the same sample space S. For example, in the tossing of a die we might let A be the event that an even number occurs and B the event that a number greater than 3 shows. Then the subsets $A = \{2, 4, 6\}$ and $B = \{4, 5, 6\}$ are subsets of the same sample space

$$S = \{1, 2, 3, 4, 5, 6\}.$$

Note that *both* A and B will occur on a given toss if the outcome is an element of the subset $\{4, 6\}$, which is just the **intersection** of A and B.

Definition 1.4: The **intersection** of two events A and B, denoted by the symbol $A \cap B$, is the event containing all elements that are common to A and B.

Example 1.9: Let E be the event that a person selected at random in a classroom is majoring in engineering, and let F be the event that the person is female. Then $E \cap F$ is the event of all female engineering students in the classroom.

Example 1.10: Let $V = \{a, e, i, o, u\}$ and $C = \{l, r, s, t\}$; then it follows that $V \cap C = \phi$. That is, V and C have no elements in common and, therefore, cannot both simultaneously occur.

For certain statistical experiments it is by no means unusual to define two events, A and B, that cannot both occur simultaneously. The events A and B are then said to be **mutually exclusive**. Stated more formally, we have the following definition:

Definition 1.5: Two events A and B are **mutually exclusive**, or **disjoint**, if $A \cap B = \phi$, that is, if A and B have no elements in common.

Example 1.11: A cable television company offers programs on eight different channels, three of which are affiliated with ABC, two with NBC, and one with CBS. The other two are an educational channel and the ESPN sports channel. Suppose that a person subscribing to this service turns on a television set without first selecting the channel. Let A be the event that the program belongs to the NBC network and B the event that it belongs to the CBS network. Since a television program cannot belong to more than one network, the events A and B have no programs in common. Therefore, the intersection $A \cap B$ contains no programs, and consequently the events A and B are mutually exclusive. ⌐

Often one is interested in the occurrence of at least one of two events associated with an experiment. Thus, in the die-tossing experiment, if

$$A = \{2, 4, 6\} \text{ and } B = \{4, 5, 6\},$$

we might be interested in either A or B occurring or both A and B occurring. Such an event, called the **union** of A and B, will occur if the outcome is an element of the subset $\{2, 4, 5, 6\}$.

Definition 1.6: The **union** of the two events A and B, denoted by the symbol $A \cup B$, is the event containing all the elements that belong to A or B or both.

Example 1.12: Let $A = \{a, b, c\}$ and $B = \{b, c, d, e\}$; then $A \cup B = \{a, b, c, d, e\}$. ⌐

Example 1.13: Let P be the event that an employee selected at random from an oil drilling company smokes cigarettes. Let Q be the event that the employee selected drinks alcoholic beverages. Then the event $P \cup Q$ is the set of all employees who either drink or smoke or do both. ⌐

Example 1.14: If $M = \{x \mid 3 < x < 9\}$ and $N = \{y \mid 5 < y < 12\}$, then

$$M \cup N = \{z \mid 3 < z < 12\}. ⌐$$

The relationship between events and the corresponding sample space can be illustrated graphically by means of **Venn diagrams**. In a Venn diagram we let the sample space be a rectangle and represent events by circles drawn inside the rectangle. Thus, in Figure 1.5, we see that

$$A \cap B = \text{ regions 1 and 2,}$$
$$B \cap C = \text{ regions 1 and 3,}$$
$$A \cup C = \text{ regions 1, 2, 3, 4, 5, and 7,}$$
$$B' \cap A = \text{ regions 4 and 7,}$$
$$A \cap B \cap C = \text{ region 1,}$$
$$(A \cup B) \cap C' = \text{ regions 2, 6, and 7,}$$

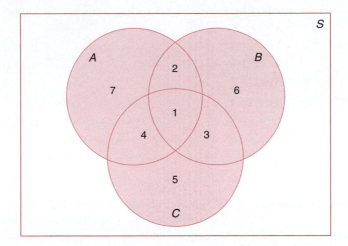

Figure 1.5: Events represented by various regions.

and so forth.

In Figure 1.6, we see that events A, B, and C are all subsets of the sample space S. It is also clear that event B is a subset of event A; event $B \cap C$ has no elements and hence B and C are mutually exclusive; event $A \cap C$ has at least one element; and event $A \cup B = A$. Figure 1.6 might, therefore, depict a situation where we select a card at random from an ordinary deck of 52 playing cards and observe whether the following events occur:

A: the card is red,

B: the card is the jack, queen, or king of diamonds,

C: the card is an ace.

Clearly, the event $A \cap C$ consists of only the two red aces.

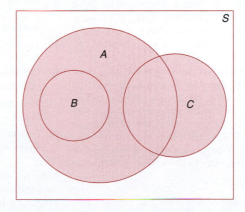

Figure 1.6: Events of the sample space S.

Several results that follow from the foregoing definitions, which may easily be verified by means of Venn diagrams, are as follows:

1. $A \cap \phi = \phi$.

2. $A \cup \phi = A$.

3. $A \cap A' = \phi$.

4. $A \cup A' = S$.

5. $S' = \phi$.

6. $\phi' = S$.

7. $(A')' = A$.

8. $(A \cap B)' = A' \cup B'$.

9. $(A \cup B)' = A' \cap B'$.

Exercises

1.1 List the elements of each of the following sample spaces:

(a) the set of integers between 1 and 50 divisible by 8;

(b) the set $S = \{x \mid x^2 + 4x - 5 = 0\}$;

(c) the set of outcomes when a coin is tossed until a tail or three heads appear;

(d) the set $S = \{x \mid x \text{ is a continent}\}$;

(e) the set $S = \{x \mid 2x - 4 \geq 0 \text{ and } x < 1\}$.

1.2 Use the rule method to describe the sample space S consisting of all points in the first quadrant inside a circle of radius 3 with center at the origin.

1.3 Which of the following events are equal?

(a) $A = \{1, 3\}$;

(b) $B = \{x \mid x \text{ is a number on a die}\}$;

(c) $C = \{x \mid x^2 - 4x + 3 = 0\}$;

(d) $D = \{x \mid x \text{ is the number of heads when six coins are tossed}\}$.

1.4 Two jurors are selected from 4 alternates to serve at a murder trial. Using the notation $A_1 A_3$, for example, to denote the simple event that alternates 1 and 3 are selected, list the 6 elements of the sample space S.

1.5 An experiment consists of tossing a die and then flipping a coin once if the number on the die is even. If the number on the die is odd, the coin is flipped twice. Using the notation $4H$, for example, to denote the outcome that the die comes up 4 and then the coin comes up heads, and $3HT$ to denote the outcome that the die comes up 3 followed by a head and then a tail on the coin, construct a tree diagram to show the 18 elements of the sample space S.

1.6 For the sample space of Exercise 1.5,

(a) list the elements corresponding to the event A that a number less than 3 occurs on the die;

(b) list the elements corresponding to the event B that

two tails occur;

(c) list the elements corresponding to the event A';

(d) list the elements corresponding to the event $A' \cap B$;

(e) list the elements corresponding to the event $A \cup B$.

1.7 The resumés of two male applicants for a college teaching position in chemistry are placed in the same file as the resumés of two female applicants. Two positions become available, and the first, at the rank of assistant professor, is filled by selecting one of the four applicants at random. The second position, at the rank of instructor, is then filled by selecting at random one of the remaining three applicants. Using the notation $M_2 F_1$, for example, to denote the simple event that the first position is filled by the second male applicant and the second position is then filled by the first female applicant,

(a) list the elements of a sample space S;

(b) list the elements of S corresponding to event A that the position of assistant professor is filled by a male applicant;

(c) list the elements of S corresponding to event B that exactly one of the two positions is filled by a male applicant;

(d) list the elements of S corresponding to event C that neither position is filled by a male applicant;

(e) list the elements of S corresponding to the event $A \cap B$;

(f) list the elements of S corresponding to the event $A \cup C$;

(g) construct a Venn diagram to illustrate the intersections and unions of the events A, B, and C.

1.8 An engineering firm is hired to determine if certain waterways in Virginia are safe for fishing. Samples are taken from three rivers.

(a) List the elements of a sample space S, using the letters F for safe to fish and N for not safe to fish.

(b) List the elements of S corresponding to event E

that at least two of the rivers are safe for fishing.

(c) Define an event that has as its elements the points

$$\{FFF, NFF, FFN, NFN\}.$$

1.9 Construct a Venn diagram to illustrate the possible intersections and unions for the following events relative to the sample space consisting of all automobiles made in the United States.

F: Four door, S: Sun roof, P: Power steering.

1.10 Exercise and diet are being studied as possible substitutes for medication to lower blood pressure. Three groups of subjects will be used to study the effect of exercise. Group 1 is sedentary, while group 2 walks and group 3 swims for 1 hour a day. Half of each of the three exercise groups will be on a salt-free diet. An additional group of subjects will not exercise or restrict their salt, but will take the standard medication. Use Z for sedentary, W for walker, S for swimmer, Y for salt, N for no salt, M for medication, and F for medication free.

(a) Show all of the elements of the sample space S.

(b) Given that A is the set of nonmedicated subjects and B is the set of walkers, list the elements of $A \cup B$.

(c) List the elements of $A \cap B$.

1.11 If $S = \{0, 1, 2, 3, 4, 5, 6, 7, 8, 9\}$ and $A = \{0, 2, 4, 6, 8\}$, $B = \{1, 3, 5, 7, 9\}$, $C = \{2, 3, 4, 5\}$, and $D = \{1, 6, 7\}$, list the elements of the sets corresponding to the following events:

(a) $A \cup C$;

(b) $A \cap B$;

(c) C';

(d) $(C' \cap D) \cup B$;

(e) $(S \cap C)'$;

(f) $A \cap C \cap D'$.

1.12 If $S = \{x \mid 0 < x < 12\}$, $M = \{x \mid 1 < x < 9\}$, and $N = \{x \mid 0 < x < 5\}$, find

(a) $M \cup N$;

(b) $M \cap N$;

(c) $M' \cap N'$.

1.13 Let A, B, and C be events relative to the sample space S. Using Venn diagrams, shade the areas representing the following events:

(a) $(A \cap B)'$;

(b) $(A \cup B)'$;

(c) $(A \cap C) \cup B$.

1.14 Which of the following pairs of events are mutually exclusive?

(a) A golfer scoring the lowest 18-hole round in a 72-hole tournament and losing the tournament.

(b) A poker player getting a flush (all cards in the same suit) and 3 of a kind on the same 5-card hand.

(c) A mother giving birth to a baby girl and a set of twin daughters on the same day.

(d) A chess player losing the last game and winning the match.

1.15 Suppose that a family is leaving on a summer vacation in their camper and that M is the event that they will experience mechanical problems, T is the event that they will receive a ticket for committing a traffic violation, and V is the event that they will arrive at a campsite with no vacancies. Referring to the Venn diagram of Figure 1.7, state in words the events represented by the following regions:

(a) region 5;

(b) region 3;

(c) regions 1 and 2 together;

(d) regions 4 and 7 together;

(e) regions 3, 6, 7, and 8 together.

1.16 Referring to Exercise 1.15 and the Venn diagram of Figure 1.7, list the numbers of the regions that represent the following events:

(a) The family will experience no mechanical problems and will not receive a ticket for a traffic violation but will arrive at a campsite with no vacancies.

(b) The family will experience both mechanical problems and trouble in locating a campsite with a vacancy but will not receive a ticket for a traffic violation.

(c) The family will either have mechanical trouble or arrive at a campsite with no vacancies but will not receive a ticket for a traffic violation.

(d) The family will not arrive at a campsite with no vacancies.

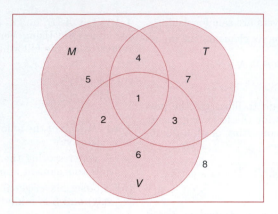

Figure 1.7: Venn diagram for Exercises 1.15 and 1.16.

1.5 Counting Sample Points

Frequently, we are interested in a sample space that contains as elements all possible orders or arrangements of a group of objects. For example, we may want to know how many different arrangements are possible for sitting 6 people around a table, or we may ask how many different orders are possible for drawing 2 lottery tickets from a total of 20. The different arrangements are called **permutations**.

Definition 1.7: | A **permutation** is an arrangement of all or part of a set of objects.

Consider the three letters a, b, and c. The possible permutations are abc, acb, bac, bca, cab, and cba. Thus, we see that there are 6 distinct arrangements.

Theorem 1.1: | The number of permutations of n objects is $n!$.

The number of permutations of the four letters a, b, c, and d will be $4! = 24$. Now consider the number of permutations that are possible by taking two letters at a time from four. These would be ab, ac, ad, ba, bc, bd, ca, cb, cd, da, db, and dc. Consider that we have two positions to fill, with $n_1 = 4$ choices for the first and then $n_2 = 3$ choices for the second, for a total of

$$n_1 n_2 = (4)(3) = 12$$

permutations. In general, n distinct objects taken r at a time can be arranged in

$$n(n-1)(n-2)\cdots(n-r+1)$$

ways. We represent this product by the symbol

$$_nP_r = \frac{n!}{(n-r)!}.$$

As a result, we have the theorem that follows.

Theorem 1.2: The number of permutations of n distinct objects taken r at a time is

$$_nP_r = \frac{n!}{(n-r)!}.$$

Example 1.15: In one year, three awards (research, teaching, and service) will be given to a class of 25 graduate students in a statistics department. If each student can receive at most one award, how many possible selections are there?

Solution: Since the awards are distinguishable, it is a permutation problem. The total number of sample points is

$$_{25}P_3 = \frac{25!}{(25-3)!} = \frac{25!}{22!} = (25)(24)(23) = 13{,}800.$$

Example 1.16: A president and a treasurer are to be chosen from a student club consisting of 50 people. How many different choices of officers are possible if

(a) there are no restrictions;

(b) A will serve only if he is president;

(c) B and C will serve together or not at all;

(d) D and E will not serve together?

Solution: (a) The total number of choices of officers, without any restrictions, is

$$_{50}P_2 = \frac{50!}{48!} = (50)(49) = 2450.$$

(b) Since A will serve only if he is president, we have two situations here: (i) A is selected as the president, which yields 49 possible outcomes for the treasurer's position, or (ii) officers are selected from the remaining 49 people without A, which has the number of choices $_{49}P_2 = (49)(48) = 2352$. Therefore, the total number of choices is $49 + 2352 = 2401$.

(c) The number of selections when B and C serve together is 2. The number of selections when both B and C are not chosen is $_{48}P_2 = 2256$. Therefore, the total number of choices in this situation is $2 + 2256 = 2258$.

(d) The number of selections when D serves as an officer but not E is $(2)(48) = 96$, where 2 is the number of positions D can take and 48 is the number of selections of the other officer from the remaining people in the club except E. The number of selections when E serves as an officer but not D is also $(2)(48) = 96$. The number of selections when both D and E are not chosen is $_{48}P_2 = 2256$. Therefore, the total number of choices is $(2)(96) + 2256 = 2448$. This problem also has another short solution: Since D and E can only serve together in 2 ways, the answer is $2450 - 2 = 2448$.

Permutations that occur by arranging objects in a circle are called **circular permutations**. Two circular permutations are not considered different unless corresponding objects in the two arrangements are preceded or followed by a different object as we proceed in a clockwise direction. For example, if 4 people are playing bridge, we do not have a new permutation if they all move one position in a clockwise direction. By considering one person in a fixed position and arranging the other three in 3! ways, we find that there are 6 distinct arrangements for the bridge game.

Theorem 1.3: The number of permutations of n objects arranged in a circle is $(n-1)!$.

So far we have considered permutations of distinct objects. That is, all the objects were completely different or distinguishable. Obviously, if the letters b and c are both equal to x, then the 6 permutations of the letters a, b, and c become axx, axx, xax, xax, xxa, and xxa, of which only 3 are distinct. Therefore, with 3 letters, 2 being the same, we have $3!/2! = 3$ distinct permutations. With 4 different letters a, b, c, and d, we have 24 distinct permutations. If we let $a = b = x$ and $c = d = y$, we can list only the following distinct permutations: $xxyy$, $xyxy$, $yxxy$, $yyxx$, $xyyx$, and $yxyx$. Thus, we have $4!/(2!\,2!) = 6$ distinct permutations.

Theorem 1.4: The number of distinct permutations of n things of which n_1 are of one kind, n_2 of a second kind, ..., n_k of a kth kind is

$$\frac{n!}{n_1!n_2!\cdots n_k!}.$$

Example 1.17: In a college football training session, the defensive coordinator needs to have 10 players standing in a row. Among these 10 players, there are 1 freshman, 2 sophomores, 4 juniors, and 3 seniors. How many different ways can they be arranged in a row if only their class level will be distinguished?

Solution: Directly using Theorem 1.4, we find that the total number of arrangements is

$$\frac{10!}{1!\,2!\,4!\,3!} = 12{,}600.$$

Often we are concerned with the number of ways of partitioning a set of n objects into r subsets called **cells**. A partition has been achieved if the intersection of every possible pair of the r subsets is the empty set ϕ and if the union of all subsets gives the original set. The order of the elements within a cell is of no importance. Consider the set $\{a,\,e,\,i,\,o,\,u\}$. The possible partitions into two cells in which the first cell contains 4 elements and the second cell 1 element are

$$\{(a,e,i,o),(u)\}, \{(a,i,o,u),(e)\}, \{(e,i,o,u),(a)\}, \{(a,e,o,u),(i)\}, \{(a,e,i,u),(o)\}.$$

We see that there are 5 ways to partition a set of 4 elements into two subsets, or cells, containing 4 elements in the first cell and 1 element in the second.

The number of partitions for this illustration is denoted by the symbol

$$\binom{5}{4,\,1} = \frac{5!}{4!\,1!} = 5,$$

where the top number represents the total number of elements and the bottom numbers represent the number of elements going into each cell. We state this more generally in Theorem 1.5.

Theorem 1.5: The number of ways of partitioning a set of n objects into r cells with n_1 elements in the first cell, n_2 elements in the second, and so forth, is

$$\binom{n}{n_1, n_2, \ldots, n_r} = \frac{n!}{n_1! n_2! \cdots n_r!},$$

where $n_1 + n_2 + \cdots + n_r = n$.

Example 1.18: In how many ways can 7 graduate students be assigned to 1 triple and 2 double hotel rooms during a conference?

Solution: The total number of possible partitions would be

$$\binom{7}{3, 2, 2} = \frac{7!}{3! \, 2! \, 2!} = 210.$$

In many problems, we are interested in the number of ways of selecting r objects from n without regard to order. These selections are called **combinations**. A combination is actually a partition with two cells, the one cell containing the r objects selected and the other cell containing the $(n-r)$ objects that are left. The number of such combinations, denoted by

$$\binom{n}{r, n-r}, \text{ is usually shortened to } \binom{n}{r},$$

since the number of elements in the second cell must be $n - r$.

Theorem 1.6: The number of combinations of n distinct objects taken r at a time is

$$\binom{n}{r} = \frac{n!}{r!(n-r)!}.$$

Example 1.19: A young boy asks his mother to get 5 Game-Boy^{TM} cartridges from his collection of 10 arcade and 5 sports game cartridges. How many ways are there that his mother can get 3 arcade and 2 sports game cartridges?

Solution: The number of ways of selecting 3 cartridges from 10 is

$$\binom{10}{3} = \frac{10!}{3! \, (10-3)!} = 120.$$

The number of ways of selecting 2 cartridges from 5 is

$$\binom{5}{2} = \frac{5!}{2! \, 3!} = 10.$$

Hence for the total we have $(120)(10) = 1200$ ways.

Example 1.20: How many different letter arrangements can be made from the letters in the word *STATISTICS*?

Solution: Using the same argument as in the discussion for Theorem 1.6, in this example we can actually apply Theorem 1.5 to obtain

$$\binom{10}{3,3,2,1,1} = \frac{10!}{3!\ 3!\ 2!\ 1!\ 1!} = 50,400.$$

Here we have 10 total letters, with 2 letters (S, T) appearing 3 times each, letter I appearing twice, and letters A and C appearing once each. Or this result can be obtained directly by using Theorem 1.4. ⏺

Exercises

1.17 Registrants at a large convention are offered 6 sightseeing tours on each of 3 days. In how many ways can a person arrange to go on a sightseeing tour planned by this convention?

1.18 In a medical study, patients are classified in 8 ways according to whether they have blood type AB^+, AB^-, A^+, A^-, B^+, B^-, O^+, or O^-, and also according to whether their blood pressure is low, normal, or high. Find the number of ways in which a patient can be classified.

1.19 Students at a private liberal arts college are classified as being freshmen, sophomores, juniors, or seniors, and also according to whether they are male or female. Find the total number of possible classifications for the students of that college.

1.20 A California study concluded that following 7 simple health rules can extend a man's life by 11 years on the average and a woman's life by 7 years. These 7 rules are as follows: no smoking, get regular exercise, use alcohol only in moderation, get 7 to 8 hours of sleep, maintain proper weight, eat breakfast, and do not eat between meals. In how many ways can a person adopt 5 of these rules to follow

(a) if the person presently violates all 7 rules?

(b) if the person never drinks and always eats breakfast?

1.21 A developer of a new subdivision offers a prospective home buyer a choice of 4 designs, 3 different heating systems, a garage or carport, and a patio or screened porch. How many different plans are available to this buyer?

1.22 A drug for the relief of asthma can be purchased from 5 different manufacturers in liquid, tablet, or capsule form, all of which come in regular and extra strength. How many different ways can a doctor prescribe the drug for a patient suffering from asthma?

1.23 In a fuel economy study, each of 3 race cars is tested using 5 different brands of gasoline at 7 test sites located in different regions of the country. If 2 drivers are used in the study, and test runs are made once under each distinct set of conditions, how many test runs are needed?

1.24 In how many different ways can a true-false test consisting of 9 questions be answered?

1.25 A witness to a hit-and-run accident told the police that the license number contained the letters RLH followed by 3 digits, the first of which was a 5. If the witness cannot recall the last 2 digits, but is certain that all 3 digits are different, find the maximum number of automobile registrations that the police may have to check.

1.26 (a) In how many ways can 6 people be lined up to get on a bus?

(b) If 3 specific persons, among 6, insist on following each other, how many ways are possible?

(c) If 2 specific persons, among 6, refuse to follow each other, how many ways are possible?

1.27 A contractor wishes to build 9 houses, each different in design. In how many ways can he place these houses on a street if 6 lots are on one side of the street and 3 lots are on the opposite side?

1.28 (a) How many distinct permutations can be made from the letters of the word *COLUMNS*?

(b) How many of these permutations start with the letter *M*?

1.29 In how many ways can 4 boys and 5 girls sit in a row if the boys and girls must alternate?

1.30 (a) How many three-digit numbers can be formed from the digits 0, 1, 2, 3, 4, 5, and 6 if each digit can be used only once?
(b) How many of these are odd numbers?
(c) How many are greater than 330?

1.31 In a regional spelling bee, the 8 finalists consist of 3 boys and 5 girls. Find the number of sample points in the sample space *S* for the number of possible orders at the conclusion of the contest for
(a) all 8 finalists;
(b) the first 3 positions.

1.32 Four married couples have bought 8 seats in the same row for a concert. In how many different ways can they be seated

(a) with no restrictions?
(b) if each couple is to sit together?
(c) if all the men sit together to the right of all the women?

1.33 Find the number of ways that 6 teachers can be assigned to 4 sections of an introductory psychology course if no teacher is assigned to more than one section.

1.34 Three lottery tickets for first, second, and third prizes are drawn from a group of 40 tickets. Find the number of sample points in *S* for awarding the 3 prizes if each contestant holds only 1 ticket.

1.35 In how many ways can 5 different trees be planted in a circle?

1.36 In how many ways can 3 oaks, 4 pines, and 2 maples be arranged along a property line if one does not distinguish among trees of the same kind?

1.37 How many ways are there that no two students will have the same birth date in a class of size 60?

1.6 Probability of an Event

Perhaps it was humankind's unquenchable thirst for gambling that led to the early development of probability theory. In an effort to increase their winnings, gamblers called upon mathematicians to provide optimum strategies for various games of chance. Some of the mathematicians providing these strategies were Pascal, Leibniz, Fermat, and James Bernoulli. As a result of this development of probability theory, statistical inference, with all its predictions and generalizations, has branched out far beyond games of chance to encompass many other fields associated with chance occurrences, such as politics, business, weather forecasting, and scientific research. For these predictions and generalizations to be reasonably accurate, an understanding of basic probability theory is essential.

What do we mean when we make the statement "John will probably win the tennis match," or "I have a fifty-fifty chance of getting an even number when a die is tossed," or "The university is not likely to win the football game tonight," or "Most of our graduating class will likely be married within 3 years"? In each case, we are expressing an outcome of which we are not certain, but owing to past information or from an understanding of the structure of the experiment, we have some degree of confidence in the validity of the statement.

Throughout the remainder of this chapter, we consider only those experiments for which the sample space contains a finite number of elements. The likelihood of the occurrence of an event resulting from such a statistical experiment is evaluated by means of a set of real numbers, called **weights** or **probabilities**, ranging from 0 to 1. To every point in the sample space we assign a probability such that the sum of all probabilities is 1. If we have reason to believe that a certain sample point is quite likely to occur when the experiment is conducted, the probability assigned should be close to 1. On the other hand, a probability closer to 0 is assigned to a sample point that is not likely to occur. In many experiments, such as tossing a fair coin or a balanced die, all the sample points have the same chance of occurring and are assigned equal probabilities. For points outside the sample space, that is, for simple events that cannot possibly occur, we assign a probability of 0.

To find the probability of an event A, we sum all the probabilities assigned to the sample points in A. This sum is called the **probability** of A and is denoted by $P(A)$.

Definition 1.8:

The **probability** of an event A is the sum of the weights of all sample points in A. Therefore,

$$0 \leq P(A) \leq 1, \quad P(\phi) = 0, \quad \text{and} \quad P(S) = 1.$$

Furthermore, if A_1, A_2, A_3, ... is a sequence of mutually exclusive events, then

$$P(A_1 \cup A_2 \cup A_3 \cup \cdots) = P(A_1) + P(A_2) + P(A_3) + \cdots.$$

Example 1.21: A coin is tossed twice. What is the probability that at least 1 head occurs?

Solution: The sample space for this experiment is

$$S = \{HH, HT, TH, TT\}.$$

If the coin is balanced, each of these outcomes is equally likely to occur. Therefore, we assign a probability of ω to each sample point. Then $4\omega = 1$, or $\omega = 1/4$. If A represents the event of at least 1 head occurring, then

$$A = \{HH, HT, TH\} \text{ and } P(A) = \frac{1}{4} + \frac{1}{4} + \frac{1}{4} = \frac{3}{4}.$$

Example 1.22: A die is loaded in such a way that an even number is twice as likely to occur as an odd number. If E is the event that a number less than 4 occurs on a single toss of the die, find $P(E)$.

Solution: The sample space is $S = \{1, 2, 3, 4, 5, 6\}$. We assign a probability of w to each odd number and a probability of $2w$ to each even number. Since the sum of the probabilities must be 1, we have $9w = 1$, or $w = 1/9$. Hence, probabilities of $1/9$ and $2/9$ are assigned to each odd and even number, respectively. Therefore,

$$E = \{1, 2, 3\} \text{ and } P(E) = \frac{1}{9} + \frac{2}{9} + \frac{1}{9} = \frac{4}{9}.$$

If the sample space for an experiment contains N elements, all of which are equally likely to occur, we assign a probability equal to $1/N$ to each of the N points. The probability of any event A containing n of these N sample points is then the ratio of the number of elements in A to the number of elements in S.

> **Definition 1.9:** If an experiment can result in any one of N different equally likely outcomes, and if exactly n of these outcomes correspond to event A, then the probability of event A is
>
> $$P(A) = \frac{n}{N}.$$

Example 1.23: A statistics class for engineers consists of 25 industrial, 10 mechanical, 10 electrical, and 8 civil engineering students. If a person is randomly selected by the instructor to answer a question, find the probability that the student chosen is (a) an industrial engineering major and (b) a civil engineering or an electrical engineering major.

Solution: Denote by I, M, E, and C the students majoring in industrial, mechanical, electrical, and civil engineering, respectively. The total number of students in the class is 53, all of whom are equally likely to be selected.

(a) Since 25 of the 53 students are majoring in industrial engineering, the probability of event I, selecting an industrial engineering major at random, is

$$P(I) = \frac{25}{53}.$$

(b) Since 18 of the 53 students are civil or electrical engineering majors, it follows that

$$P(C \cup E) = \frac{18}{53}.$$

1.7 Additive Rules

Often it is easiest to calculate the probability of some event from known probabilities of other events. This may well be true if the event in question can be represented as the union of two other events or as the complement of some event. Several important laws that frequently simplify the computation of probabilities follow. The first, called the **additive rule**, applies to unions of events.

> **Theorem 1.7:** If A and B are two events, then
>
> $$P(A \cup B) = P(A) + P(B) - P(A \cap B).$$

Proof: Consider the Venn diagram in Figure 1.8. The $P(A \cup B)$ is the sum of the probabilities of the sample points in $A \cup B$. Now $P(A) + P(B)$ is the sum of all the probabilities in A plus the sum of all the probabilities in B. Therefore, we have added the probabilities in $(A \cap B)$ twice. Since these probabilities add up to $P(A \cap B)$, we must subtract this probability once to obtain the sum of the probabilities in $A \cup B$.

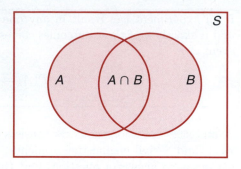

Figure 1.8: Additive rule of probability.

Corollary 1.1: If A and B are mutually exclusive, then

$$P(A \cup B) = P(A) + P(B).$$

Corollary 1.1 is an immediate result of Theorem 1.7, since if A and B are mutually exclusive, $A \cap B = 0$ and then $P(A \cap B) = P(\phi) = 0$. In general, we can write Corollary 1.2.

Corollary 1.2: If $A_1,\ A_2, \ldots, A_n$ are mutually exclusive, then

$$P(A_1 \cup A_2 \cup \cdots \cup A_n) = P(A_1) + P(A_2) + \cdots + P(A_n).$$

A collection of events $\{A_1, A_2, \ldots, A_n\}$ of a sample space S is called a **partition** of S if $A_1, A_2, \ldots, A_n$ are mutually exclusive and $A_1 \cup A_2 \cup \cdots \cup A_n = S$. Thus, we have

Corollary 1.3: If $A_1,\ A_2, \ldots, A_n$ is a partition of sample space S, then

$$P(A_1 \cup A_2 \cup \cdots \cup A_n) = P(A_1) + P(A_2) + \cdots + P(A_n) = P(S) = 1.$$

As one might expect, Theorem 1.7 extends in an analogous fashion.

Theorem 1.8: For three events A, B, and C,

$$\begin{aligned}
P(A \cup B \cup C) = {} & P(A) + P(B) + P(C) \\
& - P(A \cap B) - P(A \cap C) - P(B \cap C) + P(A \cap B \cap C).
\end{aligned}$$

Example 1.24: John is going to graduate from an industrial engineering department in a university by the end of the semester. After being interviewed at two companies he likes, he assesses that his probability of getting an offer from company A is 0.8, and his probability of getting an offer from company B is 0.6. If he believes that the probability that he will get offers from both companies is 0.5, what is the probability that he will get at least one offer from these two companies?

***Solution*:** Using the additive rule, we have

$$P(A \cup B) = P(A) + P(B) - P(A \cap B) = 0.8 + 0.6 - 0.5 = 0.9.$$

Example 1.25: What is the probability of getting a total of 7 or 11 when a pair of fair dice is tossed?

***Solution*:** Let A be the event that 7 occurs and B the event that 11 comes up. Now, a total of 7 occurs for 6 of the 36 sample points, and a total of 11 occurs for only 2 of the sample points. Since all sample points are equally likely, we have $P(A) = 1/6$ and $P(B) = 1/18$. The events A and B are mutually exclusive, since a total of 7 and 11 cannot both occur on the same toss. Therefore,

$$P(A \cup B) = P(A) + P(B) = \frac{1}{6} + \frac{1}{18} = \frac{2}{9}.$$

This result could also have been obtained by counting the total number of points for the event $A \cup B$, namely 8, and writing

$$P(A \cup B) = \frac{n}{N} = \frac{8}{36} = \frac{2}{9}.$$

Theorem 1.7 and its three corollaries should help the reader gain more insight into probability and its interpretation. Corollaries 1.1 and 1.2 suggest the very intuitive result dealing with the probability of occurrence of at least one of a number of events, no two of which can occur simultaneously. The probability that at least one occurs is the sum of the probabilities of occurrence of the individual events. The third corollary simply states that the highest value of a probability (unity) is assigned to the entire sample space S.

Example 1.26: If the probabilities are, respectively, 0.09, 0.15, 0.21, and 0.23 that a person purchasing a new automobile will choose the color green, white, red, or blue, what is the probability that a given buyer will purchase a new automobile that comes in one of those colors?

***Solution*:** Let G, W, R, and B be the events that a buyer selects, respectively, a green, white, red, or blue automobile. Since these four events are mutually exclusive, the probability is

$$P(G \cup W \cup R \cup B) = P(G) + P(W) + P(R) + P(B)$$
$$= 0.09 + 0.15 + 0.21 + 0.23 = 0.68.$$

Often it is more difficult to calculate the probability that an event occurs than it is to calculate the probability that the event does not occur. Should this be the case for some event A, we simply find $P(A')$ first and then, using Theorem 1.9, find $P(A)$ by subtraction.

Theorem 1.9: If A and A' are complementary events, then

$$P(A) + P(A') = 1.$$

Proof: Since $A \cup A' = S$ and the sets A and A' are disjoint,

$$1 = P(S) = P(A \cup A') = P(A) + P(A').$$

Example 1.27: If the probabilities that an automobile mechanic will service 3, 4, 5, 6, 7, or 8 or more cars on any given workday are, respectively, 0.12, 0.19, 0.28, 0.24, 0.10, and 0.07, what is the probability that he will service at least 5 cars on his next day at work?

Solution: Let E be the event that at least 5 cars are serviced. Now, $P(E) = 1 - P(E')$, where E' is the event that fewer than 5 cars are serviced. Since

$$P(E') = 0.12 + 0.19 = 0.31,$$

it follows from Theorem 1.9 that

$$P(E) = 1 - 0.31 = 0.69.$$

Example 1.28: Suppose the manufacturer's specifications for the length of a certain type of computer cable are 2000 ± 10 millimeters. In this industry, it is known that short cable is just as likely to be defective (not meeting specifications) as long cable. That is, the probability of randomly producing a cable with length exceeding 2010 millimeters is equal to the probability of producing a cable with length smaller than 1990 millimeters. The probability that the production procedure meets specifications is known to be 0.99.

(a) What is the probability that a cable selected randomly is too long?

(b) What is the probability that a randomly selected cable is longer than 1990 millimeters?

Solution: Let M be the event that a cable meets specifications. Let S and L be the events that the cable is too short and too long, respectively. Then

(a) $P(M) = 0.99$ and $P(S) = P(L) = (1 - 0.99)/2 = 0.005$.

(b) Denoting by X the length of a randomly selected cable, we have

$$P(1990 \leq X \leq 2010) = P(M) = 0.99.$$

Since $P(X \geq 2010) = P(L) = 0.005$,

$$P(X \geq 1990) = P(M) + P(L) = 0.995.$$

This also can be solved by using Theorem 1.9:

$$P(X \geq 1990) + P(X < 1990) = 1.$$

Thus, $P(X \geq 1990) = 1 - P(S) = 1 - 0.005 = 0.995$.

Exercises

1.38 Suppose that in a college senior class of 500 students it is found that 210 smoke, 258 drink alcoholic beverages, 216 eat between meals, 122 smoke and drink alcoholic beverages, 83 eat between meals and drink alcoholic beverages, 97 smoke and eat between meals, and 52 engage in all three of these bad health practices. If a member of this senior class is selected at random, find the probability that the student

(a) smokes but does not drink alcoholic beverages;

(b) eats between meals and drinks alcoholic beverages but does not smoke;

(c) neither smokes nor eats between meals.

1.39 Find the errors in each of the following statements:

(a) The probabilities that an automobile salesperson will sell 0, 1, 2, or 3 cars on any given day in February are, respectively, 0.19, 0.38, 0.29, and 0.15.

(b) The probability that it will rain tomorrow is 0.40, and the probability that it will not rain tomorrow is 0.52.

(c) The probabilities that a printer will make 0, 1, 2, 3, or 4 or more mistakes in setting a document are, respectively, 0.19, 0.34, −0.25, 0.43, and 0.29.

(d) On a single draw from a deck of playing cards, the probability of selecting a heart is 1/4, the probability of selecting a black card is 1/2, and the probability of selecting both a heart and a black card is 1/8.

1.40 An automobile manufacturer is concerned about a possible recall of its best-selling four-door sedan. If there were a recall, there is a probability of 0.25 of a defect in the brake system, 0.18 of a defect in the transmission, 0.17 of a defect in the fuel system, and 0.40 of a defect in some other area.

(a) What is the probability that the defect is in the brakes or the fueling system if the probability of defects in both systems simultaneously is 0.15?

(b) What is the probability that there are no defects in either the brakes or the fueling system?

1.41 The probability that an American industry will locate in Shanghai, China, is 0.7, the probability that it will locate in Beijing, China, is 0.4, and the probability that it will locate in either Shanghai or Beijing or both is 0.8. What is the probability that the industry will locate

(a) in both cities?

(b) in neither city?

1.42 From past experience, a stockbroker believes that under present economic conditions a customer will invest in tax-free bonds with a probability of 0.6, will invest in mutual funds with a probability of 0.3, and will invest in both tax-free bonds and mutual funds with a probability of 0.15. At this time, find the probability that a customer will invest

(a) in either tax-free bonds or mutual funds;

(b) in neither tax-free bonds nor mutual funds.

1.43 A box contains 500 envelopes, of which 75 contain $100 in cash, 150 contain $25, and 275 contain $10. An envelope may be purchased for $25. What is the sample space for the different amounts of money? Assign probabilities to the sample points and then find the probability that the first envelope purchased contains less than $100.

1.44 If 3 books are picked at random from a shelf containing 5 novels, 3 books of poems, and a dictionary, what is the probability that

(a) the dictionary is selected?

(b) 2 novels and 1 book of poems are selected?

1.45 In a high school graduating class of 100 students, 54 studied mathematics, 69 studied history, and 35 studied both mathematics and history. If one of these students is selected at random, find the probability that

(a) the student took mathematics or history;

(b) the student did not take either of these subjects;

(c) the student took history but not mathematics.

1.46 Dom's Pizza Company uses taste testing and statistical analysis of the data prior to marketing any new product. Consider a study involving three types of crusts (thin, thin with garlic and oregano, and thin with bits of cheese). Dom's is also studying three sauces (standard, a new sauce with more garlic, and a new sauce with fresh basil).

(a) How many combinations of crust and sauce are involved?

(b) What is the probability that a judge will get a plain thin crust with a standard sauce for his first taste test?

1.47 According to *Consumer Digest* (July/August 1996), the probable location of personal computers (PCs) in the home is as follows:

Adult bedroom: 0.03
Child bedroom: 0.15
Other bedroom: 0.14
Office or den: 0.40
Other rooms: 0.28

(a) What is the probability that a PC is in a bedroom?

(b) What is the probability that it is not in a bedroom?

(c) Suppose a household is selected at random from households with a PC; in what room would you expect to find a PC?

1.48 Interest centers around the life of an electronic component. Suppose it is known that the probability that the component survives for more than 6000 hours is 0.42. Suppose also that the probability that the component survives *no longer than* 4000 hours is 0.04.

(a) What is the probability that the life of the component is less than or equal to 6000 hours?

(b) What is the probability that the life is greater than 4000 hours?

1.49 Consider the situation of Exercise 1.48. Let A be the event that the component fails a particular test and B be the event that the component displays strain but does not actually fail. Event A occurs with probability 0.20, and event B occurs with probability 0.35.

(a) What is the probability that the component does not fail the test?

(b) What is the probability that the component works perfectly well (i.e., neither displays strain nor fails the test)?

(c) What is the probability that the component either fails or shows strain in the test?

1.50 Factory workers are constantly encouraged to practice zero tolerance when it comes to accidents in factories. Accidents can occur because the working environment or conditions themselves are unsafe. On the other hand, accidents can occur due to carelessness or so-called human error. In addition, the worker's shift, 7:00 A.M.–3:00 P.M. (day shift), 3:00 P.M.–11:00 P.M. (evening shift), or 11:00 P.M.–7:00 A.M. (graveyard shift), may be a factor. During the last year, 300 accidents have occurred. The percentages of the accidents for the condition combinations are as follows:

Shift	Unsafe Conditions	Human Error
Day	5%	32%
Evening	6%	25%
Graveyard	2%	30%

If an accident report is selected randomly from the 300 reports,

(a) what is the probability that the accident occurred on the graveyard shift?

(b) what is the probability that the accident occurred due to human error?

(c) what is the probability that the accident occurred due to unsafe conditions?

(d) what is the probability that the accident occurred on either the evening or the graveyard shift?

1.51 Consider the situation of Example 1.27 on page 30.

(a) What is the probability that no more than 4 cars will be serviced by the mechanic?

(b) What is the probability that he will service fewer than 8 cars?

(c) What is the probability that he will service either 3 or 4 cars?

1.52 Interest centers around the nature of an oven purchased at a particular department store. It can be either a gas or an electric oven. Consider the decisions made by six distinct customers.

(a) Suppose that the probability is 0.40 that at most two of these individuals purchase an electric oven. What is the probability that at least three purchase the electric oven?

(b) Suppose it is known that the probability that all six purchase the electric oven is 0.007 while 0.104 is the probability that all six purchase the gas oven. What is the probability that at least one of each type is purchased?

1.53 It is common in many industrial areas to use a filling machine to fill boxes full of product. This occurs in the food industry as well as other areas in which the product is used in the home (for example, detergent). These machines are not perfect, and indeed they may A, fill to specification, B, underfill, and C, overfill. Generally, the practice of underfilling is that which one hopes to avoid. Let $P(B) = 0.001$ while $P(A) = 0.990$.

(a) Give $P(C)$.

(b) What is the probability that the machine does not underfill?

(c) What is the probability that the machine either overfills or underfills?

1.54 Consider the situation of Exercise 1.53. Suppose 50,000 boxes of detergent are produced per week and suppose also that those underfilled are "sent back," with customers requesting reimbursement of the purchase price. Suppose also that the cost of production is known to be $4.00 per box while the purchase price is $4.50 per box.

(a) What is the weekly profit under the condition of no defective boxes?

(b) What is the loss in profit expected due to under-filling?

1.55 As the situation of Exercise 1.53 might suggest, statistical procedures are often used for control of quality (i.e., industrial quality control). At times, the *weight* of a product is an important variable to control. Specifications are given for the weight of a certain packaged product, and a package is rejected if it is either too light or too heavy. Historical data suggest that 0.95 is the probability that the product meets weight specifications whereas 0.002 is the probability that the product is too light. For each single packaged product, the manufacturer invests $20.00 in production and the purchase price for the consumer is $25.00.

(a) What is the probability that a package chosen randomly from the production line is too heavy?

(b) For each 10,000 packages sold, what profit is received by the manufacturer if all packages meet weight specification?

(c) Assuming that all defective packages are rejected and rendered worthless, how much is the profit reduced on 10,000 packages due to failure to meet weight specification?

1.56 Prove that

$$P(A' \cap B') = 1 + P(A \cap B) - P(A) - P(B).$$

1.8 Conditional Probability, Independence, and the Product Rule

One very important concept in probability theory is conditional probability. In some applications, the practitioner is interested in the probability structure under certain restrictions. For instance, in epidemiology, rather than studying the chance that a person from the general population has diabetes, it might be of more interest to know this probability for a distinct group such as Asian women in the age range of 35 to 50 or Hispanic men in the age range of 40 to 60. This type of probability is called a conditional probability.

Conditional Probability

The probability of an event B occurring when it is known that some event A has occurred is called a **conditional probability** and is denoted by $P(B|A)$. The symbol $P(B|A)$ is usually read "the probability that B occurs given that A occurs" or simply "the probability of B, given A."

Consider the event B of getting a perfect square when a die is tossed. The die is constructed so that the even numbers are twice as likely to occur as the odd numbers. Based on the sample space $S = \{1, 2, 3, 4, 5, 6\}$, with probabilities of $1/9$ and $2/9$ assigned, respectively, to the odd and even numbers, the probability of B occurring is $1/3$. Now suppose that it is known that the toss of the die resulted in a number greater than 3. We are now dealing with a reduced sample space $A = \{4, 5, 6\}$, which is a subset of S. To find the probability that B occurs, relative to the space A, we must first assign new probabilities to the elements of A proportional to their original probabilities such that their sum is 1. Assigning a probability of w to the odd number in A and a probability of $2w$ to the two even numbers, we have $5w = 1$, or $w = 1/5$. Relative to the space A, we find that B contains the single element 4. Denoting this event by the symbol $B|A$, we write $B|A = \{4\}$, and hence

$$P(B|A) = \frac{2}{5}.$$

This example illustrates that events may have different probabilities when considered relative to different sample spaces.

We can also write

$$P(B|A) = \frac{2}{5} = \frac{2/9}{5/9} = \frac{P(A \cap B)}{P(A)},$$

where $P(A \cap B)$ and $P(A)$ are found from the original sample space S. In other words, a conditional probability relative to a subspace A of S may be calculated directly from the probabilities assigned to the elements of the original sample space S.

Definition 1.10: | The conditional probability of B, given A, denoted by $P(B|A)$, is defined by

$$P(B|A) = \frac{P(A \cap B)}{P(A)}, \quad \text{provided} \quad P(A) > 0.$$

As an additional illustration, suppose that our sample space S is the population of adults in a small town who have completed the requirements for a college degree. We shall categorize them according to gender and employment status. The data are given in Table 1.3.

Table 1.3: Categorization of the Adults in a Small Town

	Employed	Unemployed	Total
Male	460	40	500
Female	140	260	400
Total	600	300	900

One of these individuals is to be selected at random for a tour throughout the country to publicize the advantages of establishing new industries in the town. We shall be concerned with the following events:

M: a man is chosen,

E: the one chosen is employed.

Using the reduced sample space E, we find that

$$P(M|E) = \frac{460}{600} = \frac{23}{30}.$$

Let $n(A)$ denote the number of elements in any set A. Using this notation, since each adult has an equal chance of being selected, we can write

$$P(M|E) = \frac{n(E \cap M)}{n(E)} = \frac{n(E \cap M)/n(S)}{n(E)/n(S)} = \frac{P(E \cap M)}{P(E)},$$

where $P(E \cap M)$ and $P(E)$ are found from the original sample space S. To verify this result, note that

$$P(E) = \frac{600}{900} = \frac{2}{3} \quad \text{and} \quad P(E \cap M) = \frac{460}{900} = \frac{23}{45}.$$

Hence,

$$P(M|E) = \frac{23/45}{2/3} = \frac{23}{30},$$

as before.

Example 1.29: The probability that a regularly scheduled flight departs on time is $P(D) = 0.83$; the probability that it arrives on time is $P(A) = 0.82$; and the probability that it departs and arrives on time is $P(D \cap A) = 0.78$. Find the probability that a plane (a) arrives on time, given that it departed on time, and (b) departed on time, given that it has arrived on time.

Solution: Using Definition 1.10, we have the following.

(a) The probability that a plane arrives on time, given that it departed on time, is

$$P(A|D) = \frac{P(D \cap A)}{P(D)} = \frac{0.78}{0.83} = 0.94.$$

(b) The probability that a plane departed on time, given that it has arrived on time, is

$$P(D|A) = \frac{P(D \cap A)}{P(A)} = \frac{0.78}{0.82} = 0.95.$$

The notion of conditional probability provides the capability of reevaluating the idea of probability of an event in light of additional information, that is, when it is known that another event has occurred. The probability $P(A|B)$ is an updating of $P(A)$ based on the knowledge that event B has occurred. In Example 1.29, it is important to know the probability that the flight arrives on time. One is given the information that the flight did not depart on time. Armed with this additional information, one can calculate the more pertinent probability $P(A|D')$, that is, the probability that it arrives on time, given that it did not depart on time. In many situations, the conclusions drawn from observing the more important conditional probability change the picture entirely. In this example, the computation of $P(A|D')$ is

$$P(A|D') = \frac{P(A \cap D')}{P(D')} = \frac{0.82 - 0.78}{0.17} = 0.24.$$

As a result, the probability of an on-time arrival is diminished severely in the presence of the additional information.

Example 1.30: The concept of conditional probability has countless uses in both industrial and biomedical applications. Consider an industrial process in the textile industry in which strips of a particular type of cloth are being produced. These strips can be defective in two ways, length and nature of texture. For the case of the latter, the process of identification is very complicated. It is known from historical information on the process that 10% of strips fail the length test, 5% fail the texture test, and only 0.8% fail both tests. If a strip is selected randomly from the process and a quick measurement identifies it as failing the length test, what is the probability that it is texture defective?

Solution: Consider the events

$$L: \text{ length defective,} \qquad T: \text{ texture defective.}$$

Given that the strip is length defective, the probability that this strip is texture defective is given by

$$P(T|L) = \frac{P(T \cap L)}{P(L)} = \frac{0.008}{0.1} = 0.08.$$

Thus, knowing the conditional probability provides considerably more information than merely knowing $P(T)$. ⌐

Independent Events

In the die-tossing experiment discussed on page 33, we note that $P(B|A) = 2/5$ whereas $P(B) = 1/3$. That is, $P(B|A) \neq P(B)$, indicating that B depends on A. Now consider an experiment in which 2 cards are drawn in succession from an ordinary deck, with replacement. The events are defined as

   A: the first card is an ace,

   B: the second card is a spade.

Since the first card is replaced, our sample space for both the first and the second draw consists of 52 cards, containing 4 aces and 13 spades. Hence,

$$P(B|A) = \frac{13}{52} = \frac{1}{4} \quad \text{and} \quad P(B) = \frac{13}{52} = \frac{1}{4}.$$

That is, $P(B|A) = P(B)$. When this is true, the events A and B are said to be **independent**.

   Although conditional probability allows for an alteration of the probability of an event in the light of additional material, it also enables us to understand better the very important concept of **independence** or, in the present context, independent events. In the airport illustration in Example 1.29, $P(A|D)$ differs from $P(A)$. This suggests that the occurrence of D influenced A, and this is certainly expected in this illustration. However, consider the situation where we have events A and B and

$$P(A|B) = P(A).$$

In other words, the occurrence of B had no impact on the chance of occurrence of A. Here the occurrence of A is independent of the occurrence of B. The importance of the concept of independence cannot be overemphasized. It plays a vital role in material in virtually all chapters in this book and in all areas of applied statistics.

Definition 1.11: Two events A and B are **independent** if and only if

$$P(B|A) = P(B) \quad \text{or} \quad P(A|B) = P(A),$$

assuming the existences of the conditional probabilities. Otherwise, A and B are **dependent**.

The condition $P(B|A) = P(B)$ implies that $P(A|B) = P(A)$, and conversely. For the card-drawing experiments, where we showed that $P(B|A) = P(B) = 1/4$, we also can see that $P(A|B) = P(A) = 1/13$.

The Product Rule, or the Multiplicative Rule

Multiplying the formula in Definition 1.10 by $P(A)$, we obtain the following important **multiplicative rule** (or **product rule**), which enables us to calculate the probability that two events will both occur.

Theorem 1.10: If in an experiment the events A and B can both occur, then

$$P(A \cap B) = P(A)P(B|A), \text{ provided } P(A) > 0.$$

Thus, the probability that both A and B occur is equal to the probability that A occurs multiplied by the conditional probability that B occurs, given that A occurs. Since the events $A \cap B$ and $B \cap A$ are equivalent, it follows from Theorem 1.10 that we can also write

$$P(A \cap B) = P(B \cap A) = P(B)P(A|B).$$

In other words, it does not matter which event is referred to as A and which event is referred to as B.

Example 1.31: Suppose that we have a fuse box containing 20 fuses, of which 5 are defective. If 2 fuses are selected at random and removed from the box in succession without replacing the first, what is the probability that both fuses are defective?

Solution: We shall let A be the event that the first fuse is defective and B the event that the second fuse is defective; then we interpret $A \cap B$ as the event that A occurs and then B occurs after A has occurred. The probability of first removing a defective fuse is $1/4$; then the probability of removing a second defective fuse from the remaining 4 is $4/19$. Hence,

$$P(A \cap B) = \left(\frac{1}{4}\right)\left(\frac{4}{19}\right) = \frac{1}{19}.$$

If, in Example 1.31, the first fuse is replaced and the fuses thoroughly rearranged before the second is removed, then the probability of a defective fuse on the second selection is still $1/4$; that is, $P(B|A) = P(B)$ and the events A and B are independent. When this is true, we can substitute $P(B)$ for $P(B|A)$ in Theorem 1.10 to obtain the following special multiplicative rule.

Theorem 1.11: | Two events A and B are independent if and only if

$$P(A \cap B) = P(A)P(B).$$

Therefore, to obtain the probability that two independent events will both occur, we simply find the product of their individual probabilities.

Example 1.32: A small town has one fire engine and one ambulance available for emergencies. The probability that the fire engine is available when needed is 0.98, and the probability that the ambulance is available when called is 0.92. In the event of an injury resulting from a burning building, find the probability that both the ambulance and the fire engine will be available, assuming they operate independently.

Solution: Let A and B represent the respective events that the fire engine and the ambulance are available. Then

$$P(A \cap B) = P(A)P(B) = (0.98)(0.92) = 0.9016.$$

Example 1.33: An electrical system consists of four components as illustrated in Figure 1.9. The system works if components A and B work and either of the components C or D works. The reliability (probability of working) of each component is also shown in Figure 1.9. Find the probability that (a) the entire system works and (b) component C does not work, given that the entire system works. Assume that the four components work independently.

Solution: In this configuration of the system, A, B, and the subsystem C and D constitute a serial circuit system, whereas the subsystem C and D itself is a parallel circuit system.

(a) Clearly the probability that the entire system works can be calculated as follows:

$$\begin{aligned}
P[A \cap B \cap (C \cup D)] &= P(A)P(B)P(C \cup D) = P(A)P(B)[1 - P(C' \cap D')] \\
&= P(A)P(B)[1 - P(C')P(D')] \\
&= (0.9)(0.9)[1 - (1 - 0.8)(1 - 0.8)] = 0.7776.
\end{aligned}$$

The equalities above hold because of the independence among the four components.

(b) To calculate the conditional probability in this case, notice that

$$\begin{aligned}
P &= \frac{P(\text{the system works but } C \text{ does not work})}{P(\text{the system works})} \\
&= \frac{P(A \cap B \cap C' \cap D)}{P(\text{the system works})} = \frac{(0.9)(0.9)(1 - 0.8)(0.8)}{0.7776} = 0.1667.
\end{aligned}$$

The multiplicative rule can be extended to more than two-event situations.

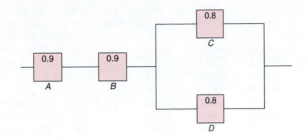

Figure 1.9: An electrical system for Example 1.33.

Theorem 1.12: If, in an experiment, the events $A_1, A_2, \ldots, A_k$ can occur, then

$$P(A_1 \cap A_2 \cap \cdots \cap A_k)$$
$$= P(A_1)P(A_2|A_1)P(A_3|A_1 \cap A_2) \cdots P(A_k|A_1 \cap A_2 \cap \cdots \cap A_{k-1}).$$

If the events $A_1, A_2, \ldots, A_k$ are independent, then

$$P(A_1 \cap A_2 \cap \cdots \cap A_k) = P(A_1)P(A_2) \cdots P(A_k).$$

The property of independence stated in Theorem 1.11 can be extended to deal with more than two events. Consider, for example, the case of three events A, B, and C. It is not sufficient to only have that $P(A \cap B \cap C) = P(A)P(B)P(C)$ as a definition of independence among the three. Suppose $A = B$ and $C = \phi$, the null set. Although $A \cap B \cap C = \phi$, which results in $P(A \cap B \cap C) = 0 = P(A)P(B)P(C)$, events A and B are not independent. Hence, we have the following definition.

Definition 1.12: A collection of events $\mathcal{A} = \{A_1, \ldots, A_n\}$ are mutually independent if for any subset of $\mathcal{A}$, $A_{i_1}, \ldots, A_{i_k}$, for $k \leq n$, we have

$$P(A_{i_1} \cap \cdots \cap A_{i_k}) = P(A_{i_1}) \cdots P(A_{i_k}).$$

Exercises

1.57 If R is the event that a convict committed armed robbery and D is the event that the convict sold drugs, state in words what probabilities are expressed by

(a) $P(R|D)$;

(b) $P(D'|R)$;

(c) $P(R'|D')$.

1.58 In an experiment to study the relationship of hypertension and smoking habits, the following data are collected for 180 individuals:

	Nonsmokers	Moderate Smokers	Heavy Smokers
H	21	36	30
NH	48	26	19

where H and NH in the table stand for *Hypertension* and *Nonhypertension*, respectively. If one of these individuals is selected at random, find the probability that the person is

(a) experiencing hypertension, given that the person is a heavy smoker;

(b) a nonsmoker, given that the person is experiencing no hypertension.

1.59 In *USA Today* (Sept. 5, 1996), the results of a survey involving the use of sleepwear while traveling were listed as follows:

	Male	Female	Total
Underwear	0.220	0.024	0.244
Nightgown	0.002	0.180	0.182
Nothing	0.160	0.018	0.178
Pajamas	0.102	0.073	0.175
T-shirt	0.046	0.088	0.134
Other	0.084	0.003	0.087

(a) What is the probability that a traveler is a female who sleeps in the nude?

(b) What is the probability that a traveler is male?

(c) Assuming the traveler is male, what is the probability that he sleeps in pajamas?

(d) What is the probability that a traveler is male if the traveler sleeps in pajamas or a T-shirt?

1.60 A manufacturer of a flu vaccine is concerned about the quality of its flu serum. Batches of serum are processed by three different departments having rejection rates of 0.10, 0.08, and 0.12, respectively. The inspections by the three departments are sequential and independent.

(a) What is the probability that a batch of serum survives the first departmental inspection but is rejected by the second department?

(b) What is the probability that a batch of serum is rejected by the third department?

1.61 The probability that a vehicle entering the Luray Caverns has Canadian license plates is 0.12; the probability that it is a camper is 0.28; and the probability that it is a camper with Canadian license plates is 0.09. What is the probability that

(a) a camper entering the Luray Caverns has Canadian license plates?

(b) a vehicle with Canadian license plates entering the Luray Caverns is a camper?

(c) a vehicle entering the Luray Caverns does not have Canadian plates or is not a camper?

1.62 For married couples living in a certain suburb, the probability that the husband will vote on a bond referendum is 0.21, the probability that the wife will vote on the referendum is 0.28, and the probability that both the husband and the wife will vote is 0.15. What is the probability that

(a) at least one member of a married couple will vote?

(b) a wife will vote, given that her husband will vote?

(c) a husband will vote, given that his wife will not vote?

1.63 The probability that a doctor correctly diagnoses a particular illness is 0.7. Given that the doctor makes an incorrect diagnosis, the probability that the patient files a lawsuit is 0.9. What is the probability that the doctor makes an incorrect diagnosis and the patient sues?

1.64 The probability that an automobile being filled with gasoline also needs an oil change is 0.25; the probability that it needs a new oil filter is 0.40; and the probability that both the oil and the filter need changing is 0.14.

(a) If the oil has to be changed, what is the probability that a new oil filter is needed?

(b) If a new oil filter is needed, what is the probability that the oil has to be changed?

1.65 In 1970, 11% of Americans completed four years of college; 43% of them were women. In 1990, 22% of Americans completed four years of college; 53% of them were women (*Time*, Jan. 19, 1996).

(a) Given that a person completed four years of college in 1970, what is the probability that the person was a woman?

(b) What is the probability that a woman finished four years of college in 1990?

(c) What is the probability that a man had not finished college in 1990?

1.66 Before the distribution of certain statistical software, every fourth compact disk (CD) is tested for accuracy. The testing process consists of running four independent programs and checking the results. The failure rates for the four testing programs are, respectively, 0.01, 0.03, 0.02, and 0.01.

(a) What is the probability that a CD was tested and failed any test?

(b) Given that a CD was tested, what is the probability that it failed program 2 or 3?

(c) In a sample of 100, how many CDs would you expect to be rejected?

(d) Given that a CD was defective, what is the probability that it was tested?

1.67 A town has two fire engines operating independently. The probability that a specific engine is available when needed is 0.96.

(a) What is the probability that neither is available when needed?

(b) What is the probability that a fire engine is available when needed?

1.68 Pollution of the rivers in the United States has been a problem for many years. Consider the following events:

   A: the river is polluted,

   B: a sample of water tested detects pollution,

   C: fishing is permitted.

Assume $P(A) = 0.3$, $P(B|A) = 0.75$, $P(B|A') = 0.20$, $P(C|A \cap B) = 0.20$, $P(C|A' \cap B) = 0.15$, $P(C|A \cap B') = 0.80$, and $P(C|A' \cap B') = 0.90$.

(a) Find $P(A \cap B \cap C)$.

(b) Find $P(B' \cap C)$.

(c) Find $P(C)$.

(d) Find the probability that the river is polluted, given that fishing is permitted and the sample tested did not detect pollution.

1.69 A circuit system is given in Figure 1.10. Assume the components fail independently.

(a) What is the probability that the entire system works?

(b) Given that the system works, what is the probability that component A is not working?

1.70 Suppose the diagram of an electrical system is as given in Figure 1.11. What is the probability that the system works? Assume the components fail independently.

1.71 In the situation of Exercise 1.69, it is known that the system does not work. What is the probability that component A also does not work?

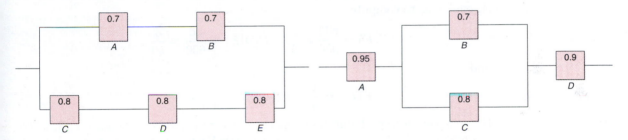

Figure 1.10: Diagram for Exercise 1.69.   Figure 1.11: Diagram for Exercise 1.70.

1.9 Bayes' Rule

Bayesian statistics is a collection of tools that is used in a special form of statistical inference which applies in the analysis of experimental data in many practical situations in science and engineering. Bayes' rule is one of the most important rules in probability theory.

Total Probability

Let us now return to the illustration of Section 1.8, where an individual is being selected at random from the adults of a small town to tour the country and publicize the advantages of establishing new industries in the town. Suppose that we are now given the additional information that 36 of those employed and 12 of those unemployed are members of the Rotary Club. We wish to find the probability of the event A that the individual selected is a member of the Rotary Club. Referring to Figure 1.12, we can write A as the union of the two mutually exclusive events $E \cap A$ and $E' \cap A$. Hence, $A = (E \cap A) \cup (E' \cap A)$, and by Corollary 1.1 of Theorem 1.7, and then Theorem 1.10, we can write

$$P(A) = P[(E \cap A) \cup (E' \cap A)] = P(E \cap A) + P(E' \cap A)$$
$$= P(E)P(A|E) + P(E')P(A|E').$$

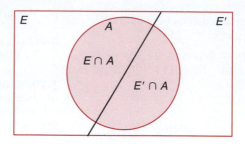

Figure 1.12: Venn diagram for the events A, E, and E'.

The data of Section 1.8, together with the additional data given above for the set A, enable us to compute

$$P(E) = \frac{600}{900} = \frac{2}{3}, \quad P(A|E) = \frac{36}{600} = \frac{3}{50},$$

and

$$P(E') = \frac{1}{3}, \quad P(A|E') = \frac{12}{300} = \frac{1}{25}.$$

If we display these probabilities by means of the tree diagram of Figure 1.13, where the first branch yields the probability $P(E)P(A|E)$ and the second branch yields the probability $P(E')P(A|E')$, it follows that

$$P(A) = \left(\frac{2}{3}\right)\left(\frac{3}{50}\right) + \left(\frac{1}{3}\right)\left(\frac{1}{25}\right) = \frac{4}{75}.$$

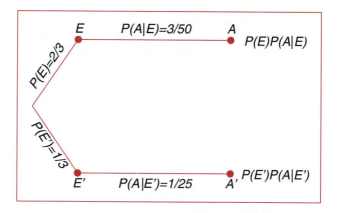

Figure 1.13: Tree diagram for the data on page 34, using additional information given above.

A generalization of the foregoing illustration to the case where the sample space is partitioned into k subsets is covered by the following theorem, sometimes called the **theorem of total probability** or the **rule of elimination**.

Theorem 1.13: If the events $B_1, B_2, \ldots, B_k$ constitute a partition of the sample space S such that $P(B_i) \neq 0$ for $i = 1, 2, \ldots, k$, then for any event A of S,

$$P(A) = \sum_{i=1}^{k} P(B_i \cap A) = \sum_{i=1}^{k} P(B_i)P(A|B_i).$$

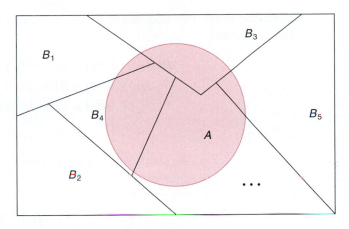

Figure 1.14: Partitioning the sample space S.

Proof: Consider the Venn diagram of Figure 1.14. The event A is seen to be the union of the mutually exclusive events

$$B_1 \cap A, \ B_2 \cap A, \ \ldots, \ B_k \cap A;$$

that is,

$$A = (B_1 \cap A) \cup (B_2 \cap A) \cup \cdots \cup (B_k \cap A).$$

Using Corollary 1.2 of Theorem 1.7 and Theorem 1.10, we have

$$P(A) = P[(B_1 \cap A) \cup (B_2 \cap A) \cup \cdots \cup (B_k \cap A)]$$
$$= P(B_1 \cap A) + P(B_2 \cap A) + \cdots + P(B_k \cap A)$$
$$= \sum_{i=1}^{k} P(B_i \cap A)$$
$$= \sum_{i=1}^{k} P(B_i)P(A|B_i).$$

Example 1.34: In a certain assembly plant, three machines, B_1, B_2, and B_3, make 30%, 45%, and 25%, respectively, of the products. It is known from past experience that 2%, 3%, and 2% of the products made by each machine, respectively, are defective. Now, suppose that a finished product is randomly selected. What is the probability that it is defective?

Solution: Consider the following events:

A: the product is defective,

B_1: the product is made by machine B_1,

B_2: the product is made by machine B_2,

B_3: the product is made by machine B_3.

Applying the rule of elimination, we can write

$$P(A) = P(B_1)P(A|B_1) + P(B_2)P(A|B_2) + P(B_3)P(A|B_3).$$

Referring to the tree diagram of Figure 1.15, we find that the three branches give the probabilities

$$P(B_1)P(A|B_1) = (0.3)(0.02) = 0.006,$$
$$P(B_2)P(A|B_2) = (0.45)(0.03) = 0.0135,$$
$$P(B_3)P(A|B_3) = (0.25)(0.02) = 0.005,$$

and hence

$$P(A) = 0.006 + 0.0135 + 0.005 = 0.0245.$$

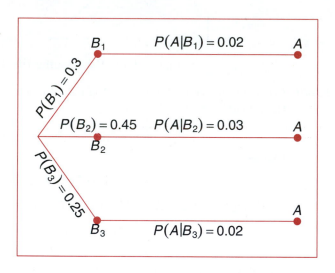

Figure 1.15: Tree diagram for Example 1.34.

Bayes' Rule

Instead of asking for $P(A)$ in Example 1.34, by the rule of elimination, suppose that we now consider the problem of finding the conditional probability $P(B_i|A)$. In other words, suppose that a product was randomly selected and it is defective. What is the probability that this product was made by machine B_i? Questions of this type can be answered by using the following theorem, called **Bayes' rule**:

Theorem 1.14: (Bayes' Rule) If the events $B_1, B_2, \ldots, B_k$ constitute a partition of the sample space S such that $P(B_i) \neq 0$ for $i = 1, 2, \ldots, k$, then for any event A in S such that $P(A) \neq 0$,

$$P(B_r|A) = \frac{P(B_r \cap A)}{\sum_{i=1}^{k} P(B_i \cap A)} = \frac{P(B_r)P(A|B_r)}{\sum_{i=1}^{k} P(B_i)P(A|B_i)} \quad \text{for } r = 1, 2, \ldots, k.$$

Proof: By the definition of conditional probability,

$$P(B_r|A) = \frac{P(B_r \cap A)}{P(A)},$$

and then using Theorem 1.13 in the denominator, we have

$$P(B_r|A) = \frac{P(B_r \cap A)}{\sum_{i=1}^{k} P(B_i \cap A)} = \frac{P(B_r)P(A|B_r)}{\sum_{i=1}^{k} P(B_i)P(A|B_i)},$$

which completes the proof.

Example 1.35: With reference to Example 1.34, if a product was chosen randomly and found to be defective, what is the probability that it was made by machine B_3?

Solution: Using Bayes' rule to write

$$P(B_3|A) = \frac{P(B_3)P(A|B_3)}{P(B_1)P(A|B_1) + P(B_2)P(A|B_2) + P(B_3)P(A|B_3)},$$

and then substituting the probabilities calculated in Example 1.34, we have

$$P(B_3|A) = \frac{0.005}{0.006 + 0.0135 + 0.005} = \frac{0.005}{0.0245} = \frac{10}{49}.$$

In view of the fact that a defective product was selected, this result suggests that it probably was not made by machine B_3.

Example 1.36: A manufacturing firm employs three analytical plans for the design and development of a particular product. For cost reasons, all three are used at varying times. In fact, plans 1, 2, and 3 are used for 30%, 20%, and 50% of the products, respectively. The defect rate is different for the three procedures as follows:

$$P(D|P_1) = 0.01, \qquad P(D|P_2) = 0.03, \qquad P(D|P_3) = 0.02,$$

where $P(D|P_j)$ is the probability of a defective product, given plan j. If a random product was observed and found to be defective, which plan was most likely used and thus responsible?

Solution: From the statement of the problem

$$P(P_1) = 0.30, \quad P(P_2) = 0.20, \quad \text{and} \quad P(P_3) = 0.50,$$

we must find $P(P_j|D)$ for $j = 1, 2, 3$. Bayes' rule (Theorem 1.14) shows

$$P(P_1|D) = \frac{P(P_1)P(D|P_1)}{P(P_1)P(D|P_1) + P(P_2)P(D|P_2) + P(P_3)P(D|P_3)}$$

$$= \frac{(0.30)(0.01)}{(0.3)(0.01) + (0.20)(0.03) + (0.50)(0.02)} = \frac{0.003}{0.019} = 0.158.$$

Similarly,

$$P(P_2|D) = \frac{(0.03)(0.20)}{0.019} = 0.316 \text{ and } P(P_3|D) = \frac{(0.02)(0.50)}{0.019} = 0.526.$$

The conditional probability of a defect given plan 3 is the largest of the three; thus a defective for a random product is most likely the result of the use of plan 3.

Exercises

1.72 Police plan to enforce speed limits by using radar traps at four different locations within the city limits. The radar traps at each of the locations L_1, L_2, L_3, and L_4 will be operated 40%, 30%, 20%, and 30% of the time. If a person who is speeding on her way to work has probabilities of 0.2, 0.1, 0.5, and 0.2, respectively, of passing through these locations, what is the probability that she will receive a speeding ticket?

1.73 In a certain region of the country it is known from past experience that the probability of selecting an adult over 40 years of age with cancer is 0.05. If the probability of a doctor correctly diagnosing a person with cancer as having the disease is 0.78 and the probability of incorrectly diagnosing a person without cancer as having the disease is 0.06, what is the probability that an adult over 40 years of age is diagnosed as having cancer?

1.74 If the person in Exercise 1.72 received a speeding ticket on her way to work, what is the probability that she passed through the radar trap located at L_2?

1.75 Referring to Exercise 1.73, what is the probability that a person diagnosed as having cancer actually has the disease?

1.76 A regional telephone company operates three identical relay stations at different locations. During a one-year period, the number of malfunctions reported by each station and the causes are shown below.

Station	A	B	C
Problems with electricity supplied	2	1	1
Computer malfunction	4	3	2
Malfunctioning electrical equipment	5	4	2
Caused by other human errors	7	7	5

Suppose that a malfunction was reported and it was found to be caused by other human errors. What is the probability that it came from station C?

1.77 Suppose that the four inspectors at a film factory are supposed to stamp the expiration date on each package of film at the end of the assembly line. John, who stamps 20% of the packages, fails to stamp the expiration date once in every 200 packages; Tom, who stamps 60% of the packages, fails to stamp the expiration date once in every 100 packages; Jeff, who stamps 15% of the packages, fails to stamp the expiration date once in every 90 packages; and Pat, who stamps 5% of the packages, fails to stamp the expiration date once in every 200 packages. If a customer complains that her package of film does not show the expiration date, what is the probability that it was inspected by John?

1.78 Denote by A, B, and C the events that a grand prize is behind doors A, B, and C, respectively. Suppose you randomly picked a door, say A. The game host opened a door, say B, and showed there was no prize behind it. Now the host offers you the option of either staying at the door that you picked (A) or switching to the remaining unopened door (C). Use probability to explain whether you should switch or not.

1.79 A paint-store chain produces and sells latex and semigloss paint. Based on long-range sales, the probability that a customer will purchase latex paint is 0.75. Of those that purchase latex paint, 60% also purchase rollers. But only 30% of semigloss paint buyers purchase rollers. A randomly selected buyer purchases a roller and a can of paint. What is the probability that the paint is latex?

Review Exercises

1.80 A truth serum has the property that 90% of the guilty suspects are properly judged while, of course, 10% of the guilty suspects are improperly found innocent. On the other hand, innocent suspects are misjudged 1% of the time. If the suspect was selected from a group of suspects of which only 5% have ever committed a crime, and the serum indicates that he is guilty, what is the probability that he is innocent?

1.81 An allergist claims that 50% of the patients she tests are allergic to some type of weed. What is the probability that

(a) exactly 3 of her next 4 patients are allergic to weeds?

(b) none of her next 4 patients is allergic to weeds?

1.82 By comparing appropriate regions of Venn diagrams, verify that

(a) $(A \cap B) \cup (A \cap B') = A$;

(b) $A' \cap (B' \cup C) = (A' \cap B') \cup (A' \cap C)$.

1.83 The probabilities that a service station will pump gas into 0, 1, 2, 3, 4, or 5 or more cars during a certain 30-minute period are 0.03, 0.18, 0.24, 0.28, 0.10, and 0.17, respectively. Find the probability that in this 30-minute period

(a) more than 2 cars receive gas;

(b) at most 4 cars receive gas;

(c) 4 or more cars receive gas.

1.84 A large industrial firm uses three local motels to provide overnight accommodations for its clients. From past experience it is known that 20% of the clients are assigned rooms at the Ramada Inn, 50% at the Sheraton, and 30% at the Lakeview Motor Lodge. If the plumbing is faulty in 5% of the rooms at the Ramada Inn, in 4% of the rooms at the Sheraton, and in 8% of the rooms at the Lakeview Motor Lodge, what is the probability that

(a) a client will be assigned a room with faulty plumbing?

(b) a person with a room having faulty plumbing was assigned accommodations at the Lakeview Motor Lodge?

1.85 The probability that a patient recovers from a delicate heart operation is 0.8. What is the probability that

(a) exactly 2 of the next 3 patients who have this operation survive?

(b) all of the next 3 patients who have this operation survive?

1.86 In a certain federal prison, it is known that 2/3 of the inmates are under 25 years of age. It is also known that 3/5 of the inmates are male and that 5/8 of the inmates are female or 25 years of age or older. What is the probability that a prisoner selected at random from this prison is female and at least 25 years old?

1.87 A shipment of 12 television sets contains 3 defective sets. In how many ways can a hotel purchase 5 of these sets and receive at least 2 of the defective sets?

1.88 A certain federal agency employs three consulting firms (A, B, and C) with probabilities 0.40, 0.35, and 0.25, respectively. From past experience it is known that the probabilities of cost overruns for the firms are 0.05, 0.03, and 0.15, respectively. Suppose a cost overrun is experienced by the agency.

(a) What is the probability that the consulting firm involved is company C?

(b) What is the probability that it is company A?

1.89 A manufacturer is studying the effects of cooking temperature, cooking time, and type of cooking oil on making potato chips. Three different temperatures, 4 different cooking times, and 3 different oils are to be used.

(a) What is the total number of combinations to be studied?

(b) How many combinations will be used for each type of oil?

(c) Discuss why permutations are not an issue in this exercise.

1.90 Consider the situation in Exercise 1.89, and suppose that the manufacturer can try only two combinations in a day.

(a) What is the probability that any given set of two runs is chosen?

(b) What is the probability that the highest temperature is used in either of these two combinations?

1.91 A certain form of cancer is known to be found in women over 60 with probability 0.07. A blood test exists for the detection of the disease, but the test is not infallible. In fact, it is known that 10% of the time the test gives a false negative (i.e., the test incorrectly gives a negative result) and 5% of the time the test gives a false positive (i.e., incorrectly gives a positive result). If a woman over 60 is known to have taken the test and received a favorable (i.e., negative) result, what is the probability that she has the disease?

1.92 A producer of a certain type of electronic component ships to suppliers in lots of twenty. Suppose that 60% of all such lots contain no defective components, 30% contain one defective component, and 10% contain two defective components. A lot is picked, two components from the lot are randomly selected and tested, and neither is defective.

(a) What is the probability that zero defective components exist in the lot?

(b) What is the probability that one defective exists in the lot?

(c) What is the probability that two defectives exist in the lot?

1.93 A construction company employs two sales engineers. Engineer 1 does the work of estimating cost for 70% of jobs bid by the company. Engineer 2 does the work for 30% of jobs bid by the company. It is known that the error rate for engineer 1 is such that 0.02 is the probability of an error when he does the work, whereas the probability of an error in the work of engineer 2 is 0.04. Suppose a bid arrives and a serious error occurs in estimating cost. Which engineer would you guess did the work? Explain and show all work.

1.94 In the field of quality control, the science of statistics is often used to determine if a process is "out of control." Suppose the process is, indeed, out of control and 20% of items produced are defective.

(a) If three items arrive off the process line in succession, what is the probability that all three are defective?

(b) If four items arrive in succession, what is the probability that three are defective?

1.95 An industrial plant is conducting a study to determine how quickly injured workers are back on the job following injury. Records show that 10% of all injured workers are admitted to the hospital for treatment and 15% are back on the job the next day. In addition, studies show that 2% are both admitted for hospital treatment and back on the job the next day.

If a worker is injured, what is the probability that the worker will either be admitted to a hospital or be back on the job the next day or both?

1.96 A firm is accustomed to training operators who do certain tasks on a production line. Those operators who attend the training course are known to be able to meet their production quotas 90% of the time. New operators who do not take the training course only meet their quotas 65% of the time. Fifty percent of new operators attend the course. Given that a new operator meets her production quota, what is the probability that she attended the program?

1.97 During bad economic times, industrial workers are dismissed and are often replaced by machines. The history of 100 workers whose loss of employment is attributable to technological advances is reviewed. For each of these individuals, it is determined if he or she was given an alternative job within the same company, found a job with another company in the same field, found a job in a new field, or has been unemployed for 1 year. In addition, the union status of each worker is recorded. The following table summarizes the results.

	Union	Nonunion
Same Company	40	15
New Company (same field)	13	10
New Field	4	11
Unemployed	2	5

(a) If the selected worker found a job with a new company in the same field, what is the probability that the worker is a union member?

(b) If the worker is a union member, what is the probability that the worker has been unemployed for a year?

1.98 Group Project: Give each student a bag of chocolate M&Ms. Divide the students into groups of 5 or 6. Calculate the relative frequency distribution for color of M&Ms for each group.

(a) What is your estimated probability of randomly picking a yellow? a red?

(b) Redo the calculations for the whole classroom. Did the estimates change?

(c) Do you believe there is an equal number of each color in a process batch? Discuss.

Chapter 2

Random Variables, Distributions, and Expectations

2.1 Concept of a Random Variable

Statistics is concerned with making inferences about populations and population characteristics. Experiments are conducted with results that are subject to chance. The testing of a number of electronic components is an example of a **statistical experiment**, a term that is used to describe any process by which several chance observations are generated. It is often important to allocate a numerical description to the outcome. For example, the sample space giving a detailed description of each possible outcome when three electronic components are tested may be written

$$S = \{NNN, NND, NDN, DNN, NDD, DND, DDN, DDD\},$$

where N denotes nondefective and D denotes defective. One is naturally concerned with the number of defectives that occur. Thus, each point in the sample space will be *assigned a numerical value* of 0, 1, 2, or 3. These values are, of course, random quantities *determined by the outcome of the experiment*. They may be viewed as values assumed by the *random variable X*, the number of defective items when three electronic components are tested.

Definition 2.1: A **random variable** is a variable that associates a real number with each element in the sample space.

We shall use a capital letter, say X, to denote a random variable and its corresponding small letter, x in this case, for one of its values. In the electronic component testing illustration above, we notice that the random variable X assumes the value 2 for all elements in the subset

$$E = \{DDN, DND, NDD\}$$

of the sample space S. That is, each possible value of X represents an event that is a subset of the sample space for the given experiment.

49

Example 2.1: Two balls are drawn in succession without replacement from an urn containing 4 red balls and 3 black balls. The possible outcomes and the values y of the random variable Y, where Y is the number of red balls, are

Sample Space	y
RR	2
RB	1
BR	1
BB	0

Example 2.2: A stockroom clerk returns three safety helmets at random to three steel mill employees who had previously checked them. If Smith, Jones, and Brown, in that order, receive one of the three hats, list the sample space for the possible orders of returning the helmets, and find the value m of the random variable M that represents the number of correct matches.

Solution: If S, J, and B stand for Smith's, Jones's, and Brown's helmets, respectively, then the possible arrangements in which the helmets may be returned and the number of correct matches are

Sample Space	m
SJB	3
SBJ	1
BJS	1
JSB	1
JBS	0
BSJ	0

In each of the two preceding examples, the sample space contains a finite number of elements. On the other hand, when a die is thrown until a 5 occurs, we obtain a sample space with an unending sequence of elements,

$$S = \{F, NF, NNF, NNNF, \dots\},$$

where F and N represent, respectively, the occurrence and nonoccurrence of a 5. But even in this experiment, the number of elements can be equated to the number of whole numbers so that there is a first element, a second element, a third element, and so on, and in this sense can be counted.

Definition 2.2: If a sample space contains a finite number of possibilities or an unending sequence with as many elements as there are whole numbers, it is called a **discrete sample space**.

When the random variable is categorical in nature, it is often called a *dummy* variable. A good illustration is the case in which the random variable is binary in nature, as shown in the following example.

Example 2.3: Consider the simple experiment in which components are arriving from the production line and they are stipulated to be defective or not defective. Define the random variable X by

$$X = \begin{cases} 1, & \text{if the component is defective,} \\ 0, & \text{if the component is not defective.} \end{cases}$$

Clearly the assignment of 1 or 0 is arbitrary though quite convenient. This will become clear in later chapters. The random variable for which 0 and 1 are chosen to describe the two possible values is called a **Bernoulli random variable**. ⌐

Further illustrations of random variables appear in the following examples.

Example 2.4: Statisticians use **sampling plans** to either accept or reject batches or lots of material. Suppose one of these sampling plans involves sampling independently 10 items from a lot of 100 items in which 12 are defective.

Let X be the random variable defined as the number of items found defective in the sample of 10. In this case, the random variable takes on the values $0, 1, 2, \ldots, 9, 10$. ⌐

Example 2.5: Suppose a sampling plan involves sampling items from a process until a defective is observed. The evaluation of the process will depend on how many consecutive nondefective items are observed. In that regard, let X be a random variable defined by the number of items observed before a defective is found. With N a nondefective and D a defective, the outcomes in the sample space are D given $X = 1$, ND given $X = 2$, NND given $X = 3$, and so on. ⌐

Example 2.6: Interest centers around the proportion of people who respond to a certain mail order solicitation. Let X be that proportion. X is a random variable that takes on all values x for which $0 \leq x \leq 1$. ⌐

Example 2.7: Let X be the random variable defined by the waiting time, in hours, between successive speeders spotted by a radar unit. The random variable X takes on all values x for which $x \geq 0$. ⌐

The outcomes of some statistical experiments may be neither finite nor countable. Such is the case, for example, when one conducts an investigation measuring the distances that a certain make of automobile will travel over a prescribed test course on 5 liters of gasoline. Assuming distance to be a variable measured to any degree of accuracy, then clearly we have an infinite number of possible distances in the sample space that cannot be equated to the number of whole numbers. Or, if one were to record the length of time for a chemical reaction to take place, once again the possible time intervals making up our sample space would be infinite in number and uncountable. We see now that all sample spaces need not be discrete.

Definition 2.3: If a sample space contains an infinite number of possibilities equal to the number of points on a line segment, it is called a **continuous sample space**.

A random variable is called a **discrete random variable** if its set of possible outcomes is countable. The random variables in Examples 2.1 to 2.5 are discrete random variables. But a random variable whose set of possible values is an entire interval of real numbers is not discrete. When a random variable can take on values on a continuous scale, it is called a **continuous random variable**. Often the possible values of a continuous random variable are precisely the same values that are contained in the continuous sample space. Obviously, the random variables described in Examples 2.6 and 2.7 are continuous random variables.

In most practical problems, continuous random variables represent *measured* data, such as all possible heights, weights, temperatures, distances, or life periods, whereas discrete random variables represent *count* data, such as the number of defectives in a sample of k items or the number of highway fatalities per year in a given state. Note that the random variables Y and M of Examples 2.1 and 2.2 both represent count data, Y the number of red balls and M the number of correct hat matches.

2.2 Discrete Probability Distributions

A discrete random variable assumes each of its values with a certain probability. In the case of tossing a coin three times, the variable X, representing the number of heads, assumes the value 2 with probability 3/8, since 3 of the 8 equally likely sample points result in two heads and one tail. If one assumes equal weights for the simple events in Example 2.2, the probability that no employee gets back the right helmet, that is, the probability that M assumes the value 0, is 1/3. The possible values m of M and their probabilities are

m	0	1	3
$P(M = m)$	$\frac{1}{3}$	$\frac{1}{2}$	$\frac{1}{6}$

Note that the values of m exhaust all possible cases and hence the probabilities add to 1.

Frequently, it is convenient to represent all the probabilities of the values of a random variable X by a formula. Such a formula would necessarily be a function of the numerical values x that we shall denote by $f(x)$, $g(x)$, $r(x)$, and so forth. Therefore, we write $f(x) = P(X = x)$; that is, $f(3) = P(X = 3)$. The set of ordered pairs $(x, f(x))$ is called the **probability mass function**, **probability function**, or **probability distribution** of the discrete random variable X.

Definition 2.4: The set of ordered pairs $(x, f(x))$ is a **probability mass function**, **probability function**, or **probability distribution** of the discrete random variable X if, for each possible outcome x,

1. $f(x) \geq 0$,

2. $\sum_x f(x) = 1$,

3. $P(X = x) = f(x)$.

Example 2.8: A shipment of 20 similar laptop computers to a retail outlet contains 3 that are defective. If a school makes a random purchase of 2 of these computers, find the probability distribution for the number of defectives.

Solution: Let X be a random variable whose values x are the possible numbers of defective computers purchased by the school. Then x can only take the numbers 0, 1, and 2. Now

$$f(0) = P(X = 0) = \frac{\binom{3}{0}\binom{17}{2}}{\binom{20}{2}} = \frac{68}{95}, \quad f(1) = P(X = 1) = \frac{\binom{3}{1}\binom{17}{1}}{\binom{20}{2}} = \frac{51}{190},$$

$$f(2) = P(X = 2) = \frac{\binom{3}{2}\binom{17}{0}}{\binom{20}{2}} = \frac{3}{190}.$$

Thus, the probability distribution of X is

x	0	1	2
$f(x)$	$\frac{68}{95}$	$\frac{51}{190}$	$\frac{3}{190}$

Example 2.9: If a car agency sells 50% of its inventory of a certain foreign car equipped with side airbags, find a formula for the probability distribution of the number of cars with side airbags among the next 4 cars sold by the agency.

Solution: Since the probability of selling an automobile with side airbags is 0.5, the $2^4 = 16$ points in the sample space are equally likely to occur. Therefore, the denominator for all probabilities, and also for our function, is 16. To obtain the number of ways of selling 3 cars with side airbags, we need to consider the number of ways of partitioning 4 outcomes into two cells, with 3 cars with side airbags assigned to one cell and the model without side airbags assigned to the other. This can be done in $\binom{4}{3} = 4$ ways. In general, the event of selling x models with side airbags and $4 - x$ models without side airbags can occur in $\binom{4}{x}$ ways, where x can be 0, 1, 2, 3, or 4. Thus, the probability distribution $f(x) = P(X = x)$ is

$$f(x) = \frac{1}{16}\binom{4}{x}, \quad \text{for } x = 0, 1, 2, 3, 4.$$

There are many problems where we may wish to compute the probability that the observed value of a random variable X will be less than or equal to some real number x. Writing $F(x) = P(X \leq x)$ for every real number x, we define $F(x)$ to be the **cumulative distribution function** of the random variable X.

Definition 2.5: The **cumulative distribution function** $F(x)$ of a discrete random variable X with probability distribution $f(x)$ is

$$F(x) = P(X \leq x) = \sum_{t \leq x} f(t), \quad \text{for } -\infty < x < \infty.$$

For the random variable M, the number of correct matches in Example 2.2, we have

$$F(2) = P(M \leq 2) = f(0) + f(1) = \frac{1}{3} + \frac{1}{2} = \frac{5}{6}.$$

The cumulative distribution function of M is

$$F(m) = \begin{cases} 0, & \text{for } m < 0, \\ \frac{1}{3}, & \text{for } 0 \leq m < 1, \\ \frac{5}{6}, & \text{for } 1 \leq m < 3, \\ 1, & \text{for } m \geq 3. \end{cases}$$

One should pay particular notice to the fact that the cumulative distribution function is a monotone nondecreasing function defined not only for the values assumed by the given random variable but for all real numbers.

Example 2.10: Find the cumulative distribution function of the random variable X in Example 2.9. Using $F(x)$, verify that $f(2) = 3/8$.

Solution: Direct calculations of the probability distribution of Example 2.9 give $f(0) = 1/16$, $f(1) = 1/4$, $f(2) = 3/8$, $f(3) = 1/4$, and $f(4) = 1/16$. Therefore,

$$F(0) = f(0) = \frac{1}{16},$$

$$F(1) = f(0) + f(1) = \frac{5}{16},$$

$$F(2) = f(0) + f(1) + f(2) = \frac{11}{16},$$

$$F(3) = f(0) + f(1) + f(2) + f(3) = \frac{15}{16},$$

$$F(4) = f(0) + f(1) + f(2) + f(3) + f(4) = 1.$$

Hence,

$$F(x) = \begin{cases} 0, & \text{for } x < 0, \\ \frac{1}{16}, & \text{for } 0 \le x < 1, \\ \frac{5}{16}, & \text{for } 1 \le x < 2, \\ \frac{11}{16}, & \text{for } 2 \le x < 3, \\ \frac{15}{16}, & \text{for } 3 \le x < 4, \\ 1 & \text{for } x \ge 4. \end{cases}$$

Now

$$f(2) = F(2) - F(1) = \frac{11}{16} - \frac{5}{16} = \frac{3}{8}.$$

It is often helpful to look at a probability distribution in graphic form. One might plot the points $(x, f(x))$ of Example 2.9 to obtain Figure 2.1. By joining the points to the x axis either with a dashed or with a solid line, we obtain a probability mass function plot. Figure 2.1 makes it easy to see what values of X are most likely to occur, and it also indicates a perfectly symmetric situation in this case.

Instead of plotting the points $(x, f(x))$, we more frequently construct rectangles, as in Figure 2.2. Here the rectangles are constructed so that their bases of equal width are centered at each value x and their heights are equal to the corresponding probabilities given by $f(x)$. The bases are constructed so as to leave no space between the rectangles. Figure 2.2 is called a **probability histogram**.

Since each base in Figure 2.2 has unit width, $P(X = x)$ is equal to the area of the rectangle centered at x. Even if the bases were not of unit width, we could adjust the heights of the rectangles to give areas that would still equal the probabilities of X assuming any of its values x. This concept of using areas to represent

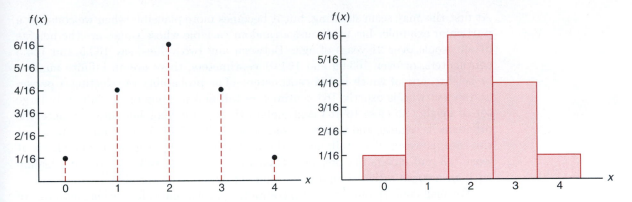

Figure 2.1: Probability mass function plot.   Figure 2.2: Probability histogram.

probabilities is necessary for our consideration of the probability distribution of a continuous random variable.

The graph of the cumulative distribution function of Example 2.10, which appears as a step function in Figure 2.3, is obtained by plotting the points $(x, F(x))$.

Certain probability distributions are applicable to more than one physical situation. The probability distribution of Example 2.10, for example, also applies to the random variable Y, where Y is the number of heads when a coin is tossed 4 times, or to the random variable W, where W is the number of red cards that occur when 4 cards are drawn at random from a deck in succession with each card replaced and the deck shuffled before the next drawing. Special discrete distributions that can be applied to many different experimental situations will be considered in Chapter 3.

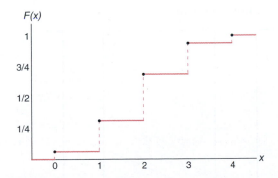

Figure 2.3: Discrete cumulative distribution function.

2.3   Continuous Probability Distributions

A continuous random variable has a probability of 0 of assuming *exactly* any of its values. Consequently, its probability distribution cannot be given in tabular form.

At first this may seem startling, but it becomes more plausible when we consider a particular example. Let us discuss a random variable whose values are the heights of all people over 21 years of age. Between any two values, say 163.5 and 164.5 centimeters, or even 163.99 and 164.01 centimeters, there are an infinite number of heights, one of which is 164 centimeters. The probability of selecting a person at random who is exactly 164 centimeters tall and not one of the infinitely large set of heights so close to 164 centimeters that you cannot humanly measure the difference is remote, and thus we assign a probability of 0 to the event. This is not the case, however, if we talk about the probability of selecting a person who is at least 163 centimeters but not more than 165 centimeters tall. Now we are dealing with an interval rather than a point value of our random variable.

We shall concern ourselves with computing probabilities for various intervals of continuous random variables such as $P(a < X < b)$, $P(W \geq c)$, and so forth. Note that when X is continuous,

$$P(a < X \leq b) = P(a < X < b) + P(X = b) = P(a < X < b).$$

That is, it does not matter whether we include an endpoint of the interval or not. This is not true, though, when X is discrete.

Although the probability distribution of a continuous random variable cannot be presented in tabular form, it can be stated as a formula. Such a formula would necessarily be a function of the numerical values of the continuous random variable X and as such will be represented by the functional notation $f(x)$. In dealing with continuous variables, $f(x)$ is usually called the **probability density function**, or simply the **density function**, of X. Since X is defined over a continuous sample space, it is possible for $f(x)$ to have a finite number of discontinuities. However, most density functions that have practical applications in the analysis of statistical data are continuous and their graphs may take any of several forms, some of which are shown in Figure 2.4. Because areas will be used to represent probabilities and probabilities are positive numerical values, the density function must lie entirely above the x axis.

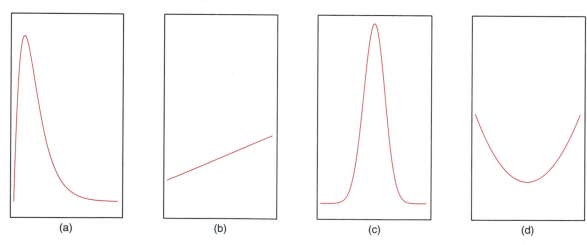

Figure 2.4: Typical density functions.

A probability density function is constructed so that the area under its curve bounded by the x axis is equal to 1 when computed over the range of X for which $f(x)$ is defined. Should this range of X be a finite interval, it is always possible to extend the interval to include the entire set of real numbers by defining $f(x)$ to be zero at all points in the extended portions of the interval. In Figure 2.5, the probability that X assumes a value between a and b is equal to the shaded area under the density function between the ordinates at $x = a$ and $x = b$, and from integral calculus is given by

$$P(a < X < b) = \int_a^b f(x)\, dx.$$

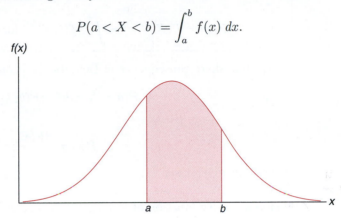

Figure 2.5: $P(a < X < b)$.

Definition 2.6: The function $f(x)$ is a **probability density function** (pdf) for the continuous random variable X, defined over the set of real numbers, if

1. $f(x) \geq 0$, for all $x \in R$,

2. $\int_{-\infty}^{\infty} f(x)\, dx = 1$,

3. $P(a < X < b) = \int_a^b f(x)\, dx$.

Example 2.11: Suppose that the error in the reaction temperature, in °C, for a controlled laboratory experiment is a continuous random variable X having the probability density function

$$f(x) = \begin{cases} \frac{x^2}{3}, & -1 < x < 2, \\ 0, & \text{elsewhere.} \end{cases}$$

(a) Verify that $f(x)$ is a density function.

(b) Find $P(0 < X \leq 1)$.

Solution: We use Definition 2.6.

(a) Obviously, $f(x) \geq 0$. To verify condition 2 in Definition 2.6, we have

$$\int_{-\infty}^{\infty} f(x)\, dx = \int_{-1}^{2} \frac{x^2}{3}\, dx = \left. \frac{x^3}{9} \right|_{-1}^{2} = \frac{8}{9} + \frac{1}{9} = 1.$$

(b) Using formula 3 in Definition 2.6, we obtain

$$P(0 < X \le 1) = \int_0^1 \frac{x^2}{3}\,dx = \frac{x^3}{9}\Big|_0^1 = \frac{1}{9}.$$

Definition 2.7: The **cumulative distribution function** $F(x)$ of a continuous random variable X with density function $f(x)$ is

$$F(x) = P(X \le x) = \int_{-\infty}^{x} f(t)\,dt, \quad \text{for } -\infty < x < \infty.$$

As an immediate consequence of Definition 2.7, one can write the two results

$$P(a < X < b) = F(b) - F(a)$$

and

$$f(x) = \frac{dF(x)}{dx},$$

if the derivative exists.

Example 2.12: For the density function of Example 2.11, find $F(x)$, and use it to evaluate $P(0 < X \le 1)$.

Solution: For $-1 < x < 2$,

$$F(x) = \int_{-\infty}^{x} f(t)\,dt = \int_{-1}^{x} \frac{t^2}{3}\,dt = \frac{t^3}{9}\Big|_{-1}^{x} = \frac{x^3 + 1}{9}.$$

Therefore,

$$F(x) = \begin{cases} 0, & x < -1, \\ \frac{x^3+1}{9}, & -1 \le x < 2, \\ 1, & x \ge 2. \end{cases}$$

The cumulative distribution function $F(x)$ is expressed in Figure 2.6. Now

$$P(0 < X \le 1) = F(1) - F(0) = \frac{2}{9} - \frac{1}{9} = \frac{1}{9},$$

which agrees with the result obtained by using the density function in Example 2.11.

Example 2.13: The Department of Energy (DOE) puts projects out on bid and generally estimates what a reasonable bid should be. Call the estimate b. The DOE has determined that the density function of the winning (low) bid is

$$f(y) = \begin{cases} \frac{5}{8b}, & \frac{2}{5}b \le y \le 2b, \\ 0, & \text{elsewhere}. \end{cases}$$

Find $F(y)$ and use it to determine the probability that the winning bid is less than the DOE's preliminary estimate b.

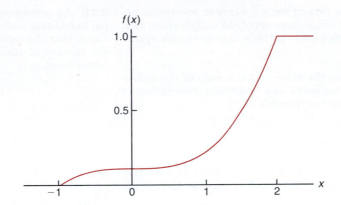

Figure 2.6: Continuous cumulative distribution function.

Solution: For $2b/5 \leq y \leq 2b$,

$$F(y) = \int_{2b/5}^{y} \frac{5}{8b} \, dy = \frac{5t}{8b} \bigg|_{2b/5}^{y} = \frac{5y}{8b} - \frac{1}{4}.$$

Thus,

$$F(y) = \begin{cases} 0, & y < \frac{2}{5}b, \\ \frac{5y}{8b} - \frac{1}{4}, & \frac{2}{5}b \leq y < 2b, \\ 1, & y \geq 2b. \end{cases}$$

To determine the probability that the winning bid is less than the preliminary bid estimate b, we have

$$P(Y \leq b) = F(b) = \frac{5}{8} - \frac{1}{4} = \frac{3}{8}.$$

Exercises

2.1 Classify the following random variables as discrete or continuous:

X: the number of automobile accidents per year in Virginia.

Y: the length of time to play 18 holes of golf.

M: the amount of milk produced yearly by a particular cow.

N: the number of eggs laid each month by a hen.

P: the number of building permits issued each month in a certain city.

Q: the weight of grain produced per acre.

2.2 An overseas shipment of 5 foreign automobiles contains 2 that have slight paint blemishes. If an agency receives 3 of these automobiles at random, list the elements of the sample space S, using the letters B and N for blemished and nonblemished, respectively; then to each sample point assign a value x of the random variable X representing the number of automobiles with paint blemishes purchased by the agency.

2.3 Let W be a random variable giving the number of heads minus the number of tails in three tosses of a coin. List the elements of the sample space S for the three tosses of the coin and to each sample point assign a value w of W.

2.4 A coin is flipped until 3 heads in succession occur. List only those elements of the sample space that require 6 or less tosses. Is this a discrete sample space? Explain.

2.5 Determine the value c so that each of the following functions can serve as a probability distribution of the discrete random variable X:

(a) $f(x) = c(x^2 + 4)$, for $x = 0, 1, 2, 3$;

(b) $f(x) = c\binom{2}{x}\binom{3}{3-x}$, for $x = 0, 1, 2$.

2.6 The shelf life, in days, for bottles of a certain prescribed medicine is a random variable having the density function

$$f(x) = \begin{cases} \frac{20,000}{(x+100)^3}, & x > 0, \\ 0, & \text{elsewhere.} \end{cases}$$

Find the probability that a bottle of this medicine will have a shell life of

(a) at least 200 days;

(b) anywhere from 80 to 120 days.

2.7 The total number of hours, measured in units of 100 hours, that a family runs a vacuum cleaner over a period of one year is a continuous random variable X that has the density function

$$f(x) = \begin{cases} x, & 0 < x < 1, \\ 2 - x, & 1 \leq x < 2, \\ 0, & \text{elsewhere.} \end{cases}$$

Find the probability that over a period of one year, a family runs their vacuum cleaner

(a) less than 120 hours;

(b) between 50 and 100 hours.

2.8 The proportion of people who respond to a certain mail-order solicitation is a continuous random variable X that has the density function

$$f(x) = \begin{cases} \frac{2(x+2)}{5}, & 0 < x < 1, \\ 0, & \text{elsewhere.} \end{cases}$$

(a) Show that $P(0 < X < 1) = 1$.

(b) Find the probability that more than 1/4 but fewer than 1/2 of the people contacted will respond to this type of solicitation.

2.9 A shipment of 7 television sets contains 2 defective sets. A hotel makes a random purchase of 3 of the sets. If x is the number of defective sets purchased by the hotel, find the probability distribution of X. Express the results graphically as a probability histogram.

2.10 An investment firm offers its customers municipal bonds that mature after varying numbers of years. Given that the cumulative distribution function of T, the number of years to maturity for a randomly selected bond, is

$$F(t) = \begin{cases} 0, & t < 1, \\ \frac{1}{4}, & 1 \leq t < 3, \\ \frac{1}{2}, & 3 \leq t < 5, \\ \frac{3}{4}, & 5 \leq t < 7, \\ 1, & t \geq 7, \end{cases}$$

find

(a) $P(T = 5)$;

(b) $P(T > 3)$;

(c) $P(1.4 < T < 6)$;

(d) $P(T \leq 5 \mid T \geq 2)$.

2.11 The probability distribution of X, the number of imperfections per 10 meters of a synthetic fabric in continuous rolls of uniform width, is given by

x	0	1	2	3	4
$f(x)$	0.41	0.37	0.16	0.05	0.01

Construct the cumulative distribution function of X.

2.12 The waiting time, in hours, between successive speeders spotted by a radar unit is a continuous random variable with cumulative distribution function

$$F(x) = \begin{cases} 0, & x < 0, \\ 1 - e^{-8x}, & x \geq 0. \end{cases}$$

Find the probability of waiting less than 12 minutes between successive speeders

(a) using the cumulative distribution function of X;

(b) using the probability density function of X.

2.13 Find the cumulative distribution function of the random variable X representing the number of defectives in Exercise 2.9. Then using $F(x)$, find

(a) $P(X = 1)$;

(b) $P(0 < X \leq 2)$.

2.14 Construct a graph of the cumulative distribution function of Exercise 2.13.

2.15 Consider the density function

$$f(x) = \begin{cases} k\sqrt{x}, & 0 < x < 1, \\ 0, & \text{elsewhere.} \end{cases}$$

(a) Evaluate k.

(b) Find $F(x)$ and use it to evaluate

$$P(0.3 < X < 0.6).$$

2.16 Three cards are drawn in succession from a deck without replacement. Find the probability distribution for the number of spades.

2.17 From a box containing 4 dimes and 2 nickels, 3 coins are selected at random without replacement. Find the probability distribution for the total T of the 3 coins. Express the probability distribution graphically as a probability histogram.

2.18 Find the probability distribution for the number of jazz CDs when 4 CDs are selected at random from a collection consisting of 5 jazz CDs, 2 classical CDs, and 3 rock CDs. Express your results by means of a formula.

2.19 The time to failure in hours of an important piece of electronic equipment used in a manufactured DVD player has the density function

$$f(x) = \begin{cases} \frac{1}{2000}\exp(-x/2000), & x \geq 0, \\ 0, & x < 0. \end{cases}$$

(a) Find $F(x)$.

(b) Determine the probability that the component (and thus the DVD player) lasts more than 1000 hours before the component needs to be replaced.

(c) Determine the probability that the component fails before 2000 hours.

2.20 A cereal manufacturer is aware that the weight of the product in the box varies slightly from box to box. In fact, considerable historical data have allowed the determination of the density function that describes the probability structure for the weight (in ounces). Letting X be the random variable weight, in ounces, the density function can be described as

$$f(x) = \begin{cases} \frac{2}{5}, & 23.75 \leq x \leq 26.25, \\ 0, & \text{elsewhere.} \end{cases}$$

(a) Verify that this is a valid density function.

(b) Determine the probability that the weight is smaller than 24 ounces.

(c) The company desires that the weight exceeding 26 ounces be an extremely rare occurrence. What is the probability that this rare occurrence does actually occur?

2.21 An important factor in solid missile fuel is the particle size distribution. Significant problems occur if the particle sizes are too large. From production data in the past, it has been determined that the particle size (in micrometers) distribution is characterized by

$$f(x) = \begin{cases} 3x^{-4}, & x > 1, \\ 0, & \text{elsewhere.} \end{cases}$$

(a) Verify that this is a valid density function.

(b) Evaluate $F(x)$.

(c) What is the probability that a random particle from the manufactured fuel exceeds 4 micrometers?

2.22 Measurements of scientific systems are always subject to variation, some more than others. There are many structures for measurement error, and statisticians spend a great deal of time modeling these errors. Suppose the measurement error X of a certain physical quantity is decided by the density function

$$f(x) = \begin{cases} k(3 - x^2), & -1 \leq x \leq 1, \\ 0, & \text{elsewhere.} \end{cases}$$

(a) Determine k that renders $f(x)$ a valid density function.

(b) Find the probability that a random error in measurement is less than $1/2$.

(c) For this particular measurement, it is undesirable if the *magnitude* of the error (i.e., $|x|$) exceeds 0.8. What is the probability that this occurs?

2.23 Based on extensive testing, it is determined by the manufacturer of a washing machine that the time Y (in years) before a major repair is required is characterized by the probability density function

$$f(y) = \begin{cases} \frac{1}{4}e^{-y/4}, & y \geq 0, \\ 0, & \text{elsewhere.} \end{cases}$$

(a) Critics would certainly consider the product a bargain if it is unlikely to require a major repair before the sixth year. Comment on this by determining $P(Y > 6)$.

(b) What is the probability that a major repair occurs in the first year?

2.24 The proportion of the budget for a certain type of industrial company that is allotted to environmental and pollution control is coming under scrutiny. A data collection project determines that the distribution of these proportions is given by

$$f(y) = \begin{cases} 5(1 - y)^4, & 0 \leq y \leq 1, \\ 0, & \text{elsewhere.} \end{cases}$$

(a) Verify that the above is a valid density function.

(b) What is the probability that a company chosen at random expends less than 10% of its budget on environmental and pollution controls?

(c) What is the probability that a company selected at random spends more than 50% of its budget on environmental and pollution controls?

2.25 Suppose a certain type of small data processing firm is so specialized that some have difficulty making a profit in their first year of operation. The probability density function that characterizes the proportion Y that make a profit is given by

$$f(y) = \begin{cases} ky^4(1-y)^3, & 0 \le y \le 1, \\ 0, & \text{elsewhere.} \end{cases}$$

(a) What is the value of k that renders the above a valid density function?

(b) Find the probability that at most 50% of the firms make a profit in the first year.

(c) Find the probability that at least 80% of the firms make a profit in the first year.

2.26 Magnetron tubes are produced on an automated assembly line. A sampling plan is used periodically to assess quality of the lengths of the tubes. This measurement is subject to uncertainty. It is thought that the probability that a random tube meets length specification is 0.99. A sampling plan is used in which the lengths of 5 random tubes are measured.

(a) Show that the probability function of Y, the number out of 5 that meet length specification, is given by the following discrete probability function:

$$f(y) = \frac{5!}{y!(5-y)!}(0.99)^y(0.01)^{5-y},$$

for $y = 0, 1, 2, 3, 4, 5$.

(b) Suppose random selections are made off the line and 3 are outside specifications. Use $f(y)$ above either to support or to refute the conjecture that the probability is 0.99 that a single tube meets specifications.

2.27 Suppose it is known from large amounts of historical data that X, the number of cars that arrive at a specific intersection during a 20-second time period, is characterized by the following discrete probability function:

$$f(x) = e^{-6}\frac{6^x}{x!}, \quad \text{for } x = 0, 1, 2, \ldots.$$

(a) Find the probability that in a specific 20-second time period, more than 8 cars arrive at the intersection.

(b) Find the probability that only 2 cars arrive.

2.28 On a laboratory assignment, if the equipment is working, the density function of the observed outcome, X, is

$$f(x) = \begin{cases} 2(1-x), & 0 < x < 1, \\ 0, & \text{otherwise.} \end{cases}$$

(a) Calculate $P(X \le 1/3)$.

(b) What is the probability that X will exceed 0.5?

(c) Given that $X \ge 0.5$, what is the probability that X will be less than 0.75?

2.4   Joint Probability Distributions

Our study of random variables and their probability distributions in the preceding sections was restricted to one-dimensional sample spaces, in that we recorded outcomes of an experiment as values assumed by a single random variable. There will be situations, however, where we may find it desirable to record the simultaneous outcomes of several random variables. For example, we might measure the amount of precipitate P and volume V of gas released from a controlled chemical experiment, giving rise to a two-dimensional sample space consisting of the outcomes (p, v), or we might be interested in the hardness H and tensile strength T of cold-drawn copper, resulting in the outcomes (h, t). In a study to determine the likelihood of success in college based on high school data, we might use a three-dimensional sample space and record for each individual his or her aptitude test score, high school class rank, and grade-point average at the end of freshman year in college.

If X and Y are two discrete random variables, the probability distribution for their simultaneous occurrence can be represented by a function with values $f(x, y)$ for any pair of values (x, y) within the range of the random variables X and Y. It is customary to refer to this function as the **joint probability distribution** of

X and Y.

Hence, in the discrete case,

$$f(x, y) = P(X = x, Y = y);$$

that is, the values $f(x, y)$ give the probability that outcomes x and y occur at the same time. For example, if an 18-wheeler is to have its tires serviced and X represents the number of miles these tires have been driven and Y represents the number of tires that need to be replaced, then $f(30000, 5)$ is the probability that the tires are used over 30,000 miles and the truck needs 5 new tires.

Definition 2.8: The function $f(x, y)$ is a **joint probability distribution** or **probability mass function** of the discrete random variables X and Y if

1. $f(x, y) \geq 0$, for all (x, y),

2. $\sum_x \sum_y f(x, y) = 1$,

3. $P(X = x, Y = y) = f(x, y)$.

For any region A in the xy plane, $P[(X, Y) \in A] = \sum \sum_A f(x, y)$.

Example 2.14: Two ballpoint pens are selected at random from a box that contains 3 blue pens, 2 red pens, and 3 green pens. If X is the number of blue pens selected and Y is the number of red pens selected, find

(a) the joint probability function $f(x, y)$,

(b) $P[(X, Y) \in A]$, where A is the region $\{(x, y) | x + y \leq 1\}$.

Solution: The possible pairs of values (x, y) are $(0, 0)$, $(0, 1)$, $(1, 0)$, $(1, 1)$, $(0, 2)$, and $(2, 0)$.

(a) Now, $f(0, 1)$, for example, represents the probability that a red and a green pen are selected. The total number of equally likely ways of selecting any 2 pens from the 8 is $\binom{8}{2} = 28$. The number of ways of selecting 1 red from 2 red pens and 1 green from 3 green pens is $\binom{2}{1}\binom{3}{1} = 6$. Hence, $f(0, 1) = 6/28 = 3/14$. Similar calculations yield the probabilities for the other cases, which are presented in Table 2.1. Note that the probabilities sum to 1. In Chapter 3, it will become clear that the joint probability distribution of Table 2.1 can be represented by the formula

$$f(x, y) = \frac{\binom{3}{x}\binom{2}{y}\binom{3}{2-x-y}}{\binom{8}{2}},$$

for $x = 0, 1, 2$; $y = 0, 1, 2$; and $0 \leq x + y \leq 2$.

(b) The probability that (X, Y) fall in the region A is

$$P[(X, Y) \in A] = P(X + Y \leq 1) = f(0, 0) + f(0, 1) + f(1, 0)$$
$$= \frac{3}{28} + \frac{3}{14} + \frac{9}{28} = \frac{9}{14}.$$

Table 2.1: Joint Probability Distribution for Example 2.14

	$f(x,y)$	x 0	x 1	x 2	Row Totals
y	0	$\frac{3}{28}$	$\frac{9}{28}$	$\frac{3}{28}$	$\frac{15}{28}$
	1	$\frac{3}{14}$	$\frac{3}{14}$	0	$\frac{3}{7}$
	2	$\frac{1}{28}$	0	0	$\frac{1}{28}$
Column Totals		$\frac{5}{14}$	$\frac{15}{28}$	$\frac{3}{28}$	1

When X and Y are continuous random variables, the **joint density function** $f(x, y)$ is a surface lying above the xy plane, and $P[(X, Y) \in A]$, where A is any region in the xy plane, is equal to the volume of the right cylinder bounded by the base A and the surface.

Definition 2.9: The function $f(x, y)$ is a **joint probability density function** of the continuous random variables X and Y if

1. $f(x, y) \geq 0$, for all (x, y),

2. $\int_{-\infty}^{\infty} \int_{-\infty}^{\infty} f(x, y) \, dx \, dy = 1$,

3. $P[(X, Y) \in A] = \int \int_A f(x, y) \, dx \, dy$, for any region A in the xy plane.

Example 2.15: A privately owned business operates both a drive-in facility and a walk-in facility. On a randomly selected day, let X and Y, respectively, be the proportions of the time that the drive-in and the walk-in facilities are in use, and suppose that the joint density function of these random variables is

$$f(x, y) = \begin{cases} \frac{2}{5}(2x + 3y), & 0 \leq x \leq 1, 0 \leq y \leq 1, \\ 0, & \text{elsewhere.} \end{cases}$$

(a) Verify condition 2 of Definition 2.9.

(b) Find $P[(X, Y) \in A]$, where $A = \{(x, y) \mid 0 < x < \frac{1}{2}, \frac{1}{4} < y < \frac{1}{2}\}$.

Solution: (a) The integration of $f(x, y)$ over the whole region is

$$\int_{-\infty}^{\infty} \int_{-\infty}^{\infty} f(x, y) \, dx \, dy = \int_0^1 \int_0^1 \frac{2}{5}(2x + 3y) \, dx \, dy$$

$$= \int_0^1 \left(\frac{2x^2}{5} + \frac{6xy}{5} \right) \Big|_{x=0}^{x=1} dy$$

$$= \int_0^1 \left(\frac{2}{5} + \frac{6y}{5} \right) dy = \left(\frac{2y}{5} + \frac{3y^2}{5} \right) \Big|_0^1 = \frac{2}{5} + \frac{3}{5} = 1.$$

(b) To calculate the probability, we use

$$P[(X, Y) \in A] = P\left(0 < X < \frac{1}{2}, \frac{1}{4} < Y < \frac{1}{2}\right)$$

$$= \int_{1/4}^{1/2} \int_0^{1/2} \frac{2}{5}(2x + 3y) \, dx \, dy$$

$$= \int_{1/4}^{1/2} \left(\frac{2x^2}{5} + \frac{6xy}{5}\right)\Big|_{x=0}^{x=1/2} \, dy = \int_{1/4}^{1/2} \left(\frac{1}{10} + \frac{3y}{5}\right) dy$$

$$= \left(\frac{y}{10} + \frac{3y^2}{10}\right)\Big|_{1/4}^{1/2}$$

$$= \frac{1}{10}\left[\left(\frac{1}{2} + \frac{3}{4}\right) - \left(\frac{1}{4} + \frac{3}{16}\right)\right] = \frac{13}{160}.$$

Given the joint probability distribution $f(x, y)$ of the discrete random variables X and Y, the probability distribution $g(x)$ of X alone is obtained by summing $f(x, y)$ over all the values of Y at each value of x. Similarly, the probability distribution $h(y)$ of Y alone is obtained by summing $f(x, y)$ over the values of X. We define $g(x)$ and $h(y)$ to be the **marginal distributions** of X and Y, respectively. When X and Y are continuous random variables, summations are replaced by integrals. We can now make the following general definition.

Definition 2.10: The **marginal distributions** of X alone and of Y alone are

$$g(x) = \sum_y f(x, y) \quad \text{and} \quad h(y) = \sum_x f(x, y)$$

for the discrete case, and

$$g(x) = \int_{-\infty}^{\infty} f(x, y) \, dy \quad \text{and} \quad h(y) = \int_{-\infty}^{\infty} f(x, y) \, dx$$

for the continuous case.

The term *marginal* is used here because, in the discrete case, the values of $g(x)$ and $h(y)$ are just the marginal totals of the respective columns and rows when the values of $f(x, y)$ are displayed in a rectangular table.

Example 2.16: Show that the column and row totals of Table 2.1 give the marginal distribution of X alone and of Y alone.

Solution: For the random variable X, we see that

$$g(0) = f(0, 0) + f(0, 1) + f(0, 2) = \frac{3}{28} + \frac{3}{14} + \frac{1}{28} = \frac{5}{14},$$

$$g(1) = f(1, 0) + f(1, 1) + f(1, 2) = \frac{9}{28} + \frac{3}{14} + 0 = \frac{15}{28},$$

and

$$g(2) = f(2, 0) + f(2, 1) + f(2, 2) = \frac{3}{28} + 0 + 0 = \frac{3}{28},$$

which are just the column totals of Table 2.1. In a similar manner we could show that the values of $h(y)$ are given by the row totals. In tabular form, these marginal distributions may be written as follows:

x	0	1	2
$g(x)$	$\frac{5}{14}$	$\frac{15}{28}$	$\frac{3}{28}$

y	0	1	2
$h(y)$	$\frac{15}{28}$	$\frac{3}{7}$	$\frac{1}{28}$

Example 2.17: Find $g(x)$ and $h(y)$ for the joint density function of Example 2.15.

Solution: By definition,

$$g(x) = \int_{-\infty}^{\infty} f(x, y) \, dy = \int_{0}^{1} \frac{2}{5}(2x + 3y) \, dy = \left(\frac{4xy}{5} + \frac{6y^2}{10} \right) \Bigg|_{y=0}^{y=1} = \frac{4x + 3}{5},$$

for $0 \le x \le 1$, and $g(x) = 0$ elsewhere. Similarly,

$$h(y) = \int_{-\infty}^{\infty} f(x, y) \, dx = \int_{0}^{1} \frac{2}{5}(2x + 3y) \, dx = \frac{2(1 + 3y)}{5},$$

for $0 \le y \le 1$, and $h(y) = 0$ elsewhere.

The fact that the marginal distributions $g(x)$ and $h(y)$ are indeed the probability distributions of the individual variables X and Y alone can be verified by showing that the conditions of Definition 2.4 or Definition 2.6 are satisfied. For example, in the continuous case

$$\int_{-\infty}^{\infty} g(x) \, dx = \int_{-\infty}^{\infty} \int_{-\infty}^{\infty} f(x, y) \, dy \, dx = 1,$$

and

$$P(a < X < b) = P(a < X < b, -\infty < Y < \infty)$$
$$= \int_{a}^{b} \int_{-\infty}^{\infty} f(x, y) \, dy \, dx = \int_{a}^{b} g(x) \, dx.$$

In Section 2.1, we stated that the value x of the random variable X represents an event that is a subset of the sample space. If we use the definition of conditional probability as stated in Chapter 1,

$$P(B|A) = \frac{P(A \cap B)}{P(A)}, \text{ provided } P(A) > 0,$$

where A and B are now the events defined by $X = x$ and $Y = y$, respectively, then

$$P(Y = y \mid X = x) = \frac{P(X = x, Y = y)}{P(X = x)} = \frac{f(x, y)}{g(x)}, \text{ provided } g(x) > 0,$$

where X and Y are discrete random variables.

It is not difficult to show that the function $f(x, y)/g(x)$, which is strictly a function of y with x fixed, satisfies all the conditions of a probability distribution. This is also true when $f(x, y)$ and $g(x)$ are the joint density and marginal distribution, respectively, of continuous random variables. As a result, it is extremely important that we make use of the special type of distribution of the form $f(x, y)/g(x)$ in order to be able to effectively compute conditional probabilities. This type of distribution is called a **conditional probability distribution**; the formal definition follows.

Definition 2.11:
> Let X and Y be two random variables, discrete or continuous. The **conditional distribution** of the random variable Y given that $X = x$ is
>
> $$f(y|x) = \frac{f(x, y)}{g(x)}, \quad \text{provided } g(x) > 0.$$
>
> Similarly, the conditional distribution of X given that $Y = y$ is
>
> $$f(x|y) = \frac{f(x, y)}{h(y)}, \quad \text{provided } h(y) > 0.$$

If we wish to find the probability that the discrete random variable X falls between a and b when it is known that the discrete variable $Y = y$, we evaluate

$$P(a < X < b \mid Y = y) = \sum_{a < x < b} f(x|y),$$

where the summation extends over all values of X between a and b. When X and Y are continuous, we evaluate

$$P(a < X < b \mid Y = y) = \int_a^b f(x|y)\, dx.$$

Example 2.18: Referring to Example 2.14, find the conditional distribution of X, given that $Y = 1$, and use it to determine $P(X = 0 \mid Y = 1)$.

Solution: We need to find $f(x|y)$, where $y = 1$. First, we find that

$$h(1) = \sum_{x=0}^{2} f(x, 1) = \frac{3}{14} + \frac{3}{14} + 0 = \frac{3}{7}.$$

Now

$$f(x|1) = \frac{f(x, 1)}{h(1)} = \left(\frac{7}{3}\right) f(x, 1), \quad x = 0, 1, 2.$$

Therefore,

$$f(0|1) = \left(\frac{7}{3}\right) f(0, 1) = \left(\frac{7}{3}\right)\left(\frac{3}{14}\right) = \frac{1}{2}, \quad f(1|1) = \left(\frac{7}{3}\right) f(1, 1) = \left(\frac{7}{3}\right)\left(\frac{3}{14}\right) = \frac{1}{2},$$

$$f(2|1) = \left(\frac{7}{3}\right) f(2, 1) = \left(\frac{7}{3}\right) (0) = 0,$$

and the conditional distribution of X, given that $Y = 1$, is

$$
\begin{array}{c|ccc}
x & 0 & 1 & 2 \\
\hline
f(x|1) & \frac{1}{2} & \frac{1}{2} & 0
\end{array}
$$

Finally,

$$
P(X = 0 \mid Y = 1) = f(0|1) = \frac{1}{2}.
$$

Therefore, if it is known that 1 of the 2 pens selected is red, we have a probability equal to 1/2 that the other pen is not blue.

Example 2.19: The joint density for the random variables (X, Y), where X is the temperature change and Y is the proportion of the spectrum that shifts for a certain atomic particle, is

$$
f(x, y) = \begin{cases} 10xy^2, & 0 < x < y < 1, \\ 0, & \text{elsewhere.} \end{cases}
$$

(a) Find the marginal densities $g(x)$, $h(y)$, and the conditional density $f(y|x)$.

(b) Find the probability that the spectrum shifts more than half of the total observations, given that the temperature is increased by 0.25 unit.

Solution: (a) By definition,

$$
g(x) = \int_{-\infty}^{\infty} f(x, y)\, dy = \int_{x}^{1} 10xy^2\, dy
$$

$$
= \left. \frac{10}{3} xy^3 \right|_{y=x}^{y=1} = \frac{10}{3} x(1 - x^3), \ 0 < x < 1,
$$

$$
h(y) = \int_{-\infty}^{\infty} f(x, y)\, dx = \int_{0}^{y} 10xy^2\, dx = \left. 5x^2 y^2 \right|_{x=0}^{x=y} = 5y^4, \ 0 < y < 1.
$$

Now

$$
f(y|x) = \frac{f(x, y)}{g(x)} = \frac{10xy^2}{\frac{10}{3} x(1 - x^3)} = \frac{3y^2}{1 - x^3}, \ 0 < x < y < 1.
$$

(b) Therefore,

$$
P\left(Y > \frac{1}{2} \ \middle| \ X = 0.25\right) = \int_{1/2}^{1} f(y \mid x = 0.25)\, dy = \int_{1/2}^{1} \frac{3y^2}{1 - 0.25^3}\, dy = \frac{8}{9}.
$$

Example 2.20: Given the joint density function

$$
f(x, y) = \begin{cases} \frac{x(1+3y^2)}{4}, & 0 < x < 2, \ 0 < y < 1, \\ 0, & \text{elsewhere,} \end{cases}
$$

find $g(x)$, $h(y)$, $f(x|y)$, and evaluate $P(\frac{1}{4} < X < \frac{1}{2} \mid Y = \frac{1}{3})$.

Solution: By the definition of the marginal density, for $0 < x < 2$,

$$
g(x) = \int_{-\infty}^{\infty} f(x, y)\, dy = \int_{0}^{1} \frac{x(1 + 3y^2)}{4}\, dy
$$

$$
= \left. \left(\frac{xy}{4} + \frac{xy^3}{4} \right) \right|_{y=0}^{y=1} = \frac{x}{2},
$$

and for $0 < y < 1$,

$$h(y) = \int_{-\infty}^{\infty} f(x,y)\ dx = \int_0^2 \frac{x(1+3y^2)}{4}\,dx$$
$$= \left. \left(\frac{x^2}{8} + \frac{3x^2y^2}{8} \right) \right|_{x=0}^{x=2} = \frac{1+3y^2}{2}.$$

Therefore, using the conditional density definition, for $0 < x < 2$,

$$f(x|y) = \frac{f(x,y)}{h(y)} = \frac{x(1+3y^2)/4}{(1+3y^2)/2} = \frac{x}{2},$$

and

$$P\left(\frac{1}{4} < X < \frac{1}{2}\ \middle|\ Y = \frac{1}{3} \right) = \int_{1/4}^{1/2} \frac{x}{2}\,dx = \frac{3}{64}.$$

Statistical Independence

If $f(x|y)$ does not depend on y, as is the case for Example 2.20, then $f(x|y) = g(x)$ and $f(x,y) = g(x)h(y)$. The proof follows by substituting

$$f(x,y) = f(x|y)h(y)$$

into the marginal distribution of X. That is,

$$g(x) = \int_{-\infty}^{\infty} f(x,y)\ dy = \int_{-\infty}^{\infty} f(x|y)h(y)\ dy.$$

If $f(x|y)$ does not depend on y, we may write

$$g(x) = f(x|y) \int_{-\infty}^{\infty} h(y)\ dy.$$

Now

$$\int_{-\infty}^{\infty} h(y)\ dy = 1,$$

since $h(y)$ is the probability density function of Y. Therefore,

$$g(x) = f(x|y) \quad \text{and then} \quad f(x,y) = g(x)h(y).$$

It should make sense to the reader that if $f(x|y)$ does not depend on y, then of course the outcome of the random variable Y has no impact on the outcome of the random variable X. In other words, we say that X and Y are independent random variables. We now offer the following formal definition of statistical independence.

Definition 2.12: Let X and Y be two random variables, discrete or continuous, with joint probability distribution $f(x, y)$ and marginal distributions $g(x)$ and $h(y)$, respectively. The random variables X and Y are said to be **statistically independent** if and only if

$$f(x, y) = g(x)h(y)$$

for all (x, y) within their range.

The continuous random variables of Example 2.20 are statistically independent, since the product of the two marginal distributions gives the joint density function. This is obviously not the case, however, for the continuous variables of Example 2.19. Checking for statistical independence of discrete random variables requires a more thorough investigation, since it is possible to have the product of the marginal distributions equal to the joint probability distribution for some but not all combinations of (x, y). If you can find any point (x, y) for which $f(x, y)$ is defined such that $f(x, y) \neq g(x)h(y)$, the discrete variables X and Y are not statistically independent.

Example 2.21: Show that the random variables of Example 2.14 are not statistically independent.

Proof: Let us consider the point $(0, 1)$. From Table 2.1 we find the three probabilities $f(0, 1)$, $g(0)$, and $h(1)$ to be

$$f(0, 1) = \frac{3}{14},$$

$$g(0) = \sum_{y=0}^{2} f(0, y) = \frac{3}{28} + \frac{3}{14} + \frac{1}{28} = \frac{5}{14},$$

$$h(1) = \sum_{x=0}^{2} f(x, 1) = \frac{3}{14} + \frac{3}{14} + 0 = \frac{3}{7}.$$

Clearly,

$$f(0, 1) \neq g(0)h(1),$$

and therefore X and Y are not statistically independent.

All the preceding definitions concerning two random variables can be generalized to the case of n random variables. Let $f(x_1, x_2, \ldots, x_n)$ be the joint probability function of the random variables $X_1, X_2, \ldots, X_n$. The marginal distribution of X_1, for example, is

$$g(x_1) = \sum_{x_2} \cdots \sum_{x_n} f(x_1, x_2, \ldots, x_n)$$

for the discrete case, and

$$g(x_1) = \int_{-\infty}^{\infty} \cdots \int_{-\infty}^{\infty} f(x_1, x_2, \ldots, x_n) \, dx_2 \, dx_3 \cdots dx_n$$

for the continuous case. We can now obtain **joint marginal distributions** such as $g(x_1, x_2)$, where

$$g(x_1, x_2) = \begin{cases} \sum_{x_3} \cdots \sum_{x_n} f(x_1, x_2, \ldots, x_n) & \text{(discrete case)}, \\ \int_{-\infty}^{\infty} \cdots \int_{-\infty}^{\infty} f(x_1, x_2, \ldots, x_n) \, dx_3 \, dx_4 \cdots dx_n & \text{(continuous case)}. \end{cases}$$

We could consider numerous conditional distributions. For example, the **joint conditional distribution** of X_1, X_2, and X_3, given that $X_4 = x_4, X_5 = x_5, \ldots, X_n = x_n$, is written

$$f(x_1, x_2, x_3 \mid x_4, x_5, \ldots, x_n) = \frac{f(x_1, x_2, \ldots, x_n)}{g(x_4, x_5, \ldots, x_n)},$$

where $g(x_4, x_5, \ldots, x_n)$ is the joint marginal distribution of the random variables $X_4, X_5, \ldots, X_n$.

A generalization of Definition 2.12 leads to the following definition for the mutual statistical independence of the variables $X_1, X_2, \ldots, X_n$.

Definition 2.13: Let $X_1, X_2, \ldots, X_n$ be n random variables, discrete or continuous, with joint probability distribution $f(x_1, x_2, \ldots, x_n)$ and marginal distribution $f_1(x_1), f_2(x_2), \ldots, f_n(x_n)$, respectively. The random variables $X_1, X_2, \ldots, X_n$ are said to be mutually **statistically independent** if and only if

$$f(x_1, x_2, \ldots, x_n) = f_1(x_1)f_2(x_2) \cdots f_n(x_n)$$

for all $(x_1, x_2, \ldots, x_n)$ within their range.

Example 2.22: Suppose that the shelf life, in years, of a certain perishable food product packaged in cardboard containers is a random variable whose probability density function is given by

$$f(x) = \begin{cases} e^{-x}, & x > 0, \\ 0, & \text{elsewhere.} \end{cases}$$

Let X_1, X_2, and X_3 represent the shelf lives for three of these containers selected independently and find $P(X_1 < 2, 1 < X_2 < 3, X_3 > 2)$.

Solution: Since the containers were selected independently, we can assume that the random variables X_1, X_2, and X_3 are statistically independent, having the joint probability density

$$f(x_1, x_2, x_3) = f(x_1)f(x_2)f(x_3) = e^{-x_1}e^{-x_2}e^{-x_3} = e^{-x_1-x_2-x_3},$$

for $x_1 > 0$, $x_2 > 0$, $x_3 > 0$, and $f(x_1, x_2, x_3) = 0$ elsewhere. Hence

$$P(X_1 < 2, 1 < X_2 < 3, X_3 > 2) = \int_2^{\infty} \int_1^3 \int_0^2 e^{-x_1-x_2-x_3} \, dx_1 \, dx_2 \, dx_3$$
$$= (1 - e^{-2})(e^{-1} - e^{-3})e^{-2} = 0.0372.$$

What Are Important Characteristics of Probability Distributions and Where Do They Come From?

This is an important point in the text to provide the reader with a transition into the next three chapters. We have given illustrations in both examples and exercises of practical scientific and engineering situations in which probability distributions and their properties are used to solve important problems. These probability distributions, either discrete or continuous, were introduced through phrases like "it is known that" or "suppose that" or even in some cases "historical evidence suggests that." These are situations in which the nature of the distribution and even a good estimate of the probability structure can be determined through historical data, data from long-term studies, or even large amounts of planned data. However, not all probability functions and probability density functions are derived from large amounts of historical data. There are a substantial number of situations in which the nature of the scientific scenario suggests a distribution type. For example, when independent repeated observations are binary in nature (e.g., defective or not, survive or not, allergic or not) with value 0 or 1, the distribution covering this situation is called the **binomial distribution** and the probability function is known and will be demonstrated in its generality in Chapter 3.

A second part of this transition to material in future chapters deals with the notion of **population parameters** or **distributional parameters**. We will discuss later in this chapter the notions of a **mean** and **variance** and provide a vision for the concepts in the context of a population. Indeed, the population mean and variance are easily found from the probability function for the discrete case or the probability density function for the continuous case. These parameters and their importance in the solution of many types of real-world problems will provide much of the material in Chapters 4 through 9.

Exercises

2.29 Determine the values of c so that the following functions represent joint probability distributions of the random variables X and Y:

(a) $f(x, y) = cxy$, for $x = 1, 2, 3$; $y = 1, 2, 3$;

(b) $f(x, y) = c|x - y|$, for $x = -2, 0, 2$; $y = -2, 3$.

2.30 If the joint probability distribution of X and Y is given by

$$f(x, y) = \frac{x + y}{30}, \quad \text{for } x = 0, 1, 2, 3; \ y = 0, 1, 2,$$

find

(a) $P(X \le 2, Y = 1)$;

(b) $P(X > 2, Y \le 1)$;

(c) $P(X > Y)$;

(d) $P(X + Y = 4)$.

(e) Find the marginal distribution of X; of Y.

2.31 From a sack of fruit containing 3 oranges, 2 apples, and 3 bananas, a random sample of 4 pieces of fruit is selected. If X is the number of oranges and Y is the number of apples in the sample, find

(a) the joint probability distribution of X and Y;

(b) $P[(X, Y) \in A]$, where A is the region that is given by $\{(x, y) \mid x + y \le 2\}$;

(c) $P(Y = 0 \mid X = 2)$;

(d) the conditional distribution of y, given $X = 2$.

2.32 A fast-food restaurant operates both a drive-through facility and a walk-in facility. On a randomly selected day, let X and Y, respectively, be the proportions of the time that the drive-through and walk-in facilities are in use, and suppose that the joint density function of these random variables is

$$f(x, y) = \begin{cases} \frac{2}{3}(x + 2y), & 0 \le x \le 1, \ 0 \le y \le 1, \\ 0, & \text{elsewhere.} \end{cases}$$

(a) Find the marginal density of X.
(b) Find the marginal density of Y.
(c) Find the probability that the drive-through facility is busy less than one-half of the time.

2.33 A candy company distributes boxes of chocolates with a mixture of creams, toffees, and cordials. Suppose that the weight of each box is 1 kilogram, but the individual weights of the creams, toffees, and cordials vary from box to box. For a randomly selected box, let X and Y represent the weights of the creams and the toffees, respectively, and suppose that the joint density function of these variables is

$$f(x,y) = \begin{cases} 24xy, & 0 \le x \le 1,\ 0 \le y \le 1,\ x+y \le 1, \\ 0, & \text{elsewhere.} \end{cases}$$

(a) Find the probability that in a given box the cordials account for more than 1/2 of the weight.
(b) Find the marginal density for the weight of the creams.
(c) Find the probability that the weight of the toffees in a box is less than 1/8 of a kilogram if it is known that creams constitute 3/4 of the weight.

2.34 Let X and Y denote the lengths of life, in years, of two components in an electronic system. If the joint density function of these variables is

$$f(x,y) = \begin{cases} e^{-(x+y)}, & x > 0,\ y > 0, \\ 0, & \text{elsewhere,} \end{cases}$$

find $P(0 < X < 1 \mid Y = 2)$.

2.35 Let X denote the reaction time, in seconds, to a certain stimulus and Y denote the temperature (°F) at which a certain reaction starts to take place. Suppose that two random variables X and Y have the joint density

$$f(x,y) = \begin{cases} 4xy, & 0 < x < 1,\ 0 < y < 1, \\ 0, & \text{elsewhere.} \end{cases}$$

Find
(a) $P(0 \le X \le \frac{1}{2}$ and $\frac{1}{4} \le Y \le \frac{1}{2})$;
(b) $P(X < Y)$.

2.36 Each rear tire on an experimental airplane is supposed to be filled to a pressure of 40 pounds per square inch (psi). Let X denote the actual air pressure for the right tire and Y denote the actual air pressure for the left tire. Suppose that X and Y are random variables with the joint density function

$$f(x,y) = \begin{cases} k(x^2+y^2), & 30 \le x < 50,\ 30 \le y < 50, \\ 0, & \text{elsewhere.} \end{cases}$$

(a) Find k.
(b) Find $P(30 \le X \le 40$ and $40 \le Y < 50)$.
(c) Find the probability that both tires are underfilled.

2.37 Let X denote the diameter of an armored electric cable and Y denote the diameter of the ceramic mold that makes the cable. Both X and Y are scaled so that they range between 0 and 1. Suppose that X and Y have the joint density

$$f(x,y) = \begin{cases} \frac{1}{y}, & 0 < x < y < 1, \\ 0, & \text{elsewhere.} \end{cases}$$

Find $P(X + Y > 1/2)$.

2.38 The amount of kerosene, in thousands of liters, in a tank at the beginning of any day is a random amount Y from which a random amount X is sold during that day. Suppose that the tank is not resupplied during the day so that $x \le y$, and assume that the joint density function of these variables is

$$f(x,y) = \begin{cases} 2, & 0 < x \le y < 1, \\ 0, & \text{elsewhere.} \end{cases}$$

(a) Determine if X and Y are independent.
(b) Find $P(1/4 < X < 1/2 \mid Y = 3/4)$.

2.39 Let X denote the number of times a certain numerical control machine will malfunction: 1, 2, or 3 times on any given day. Let Y denote the number of times a technician is called on an emergency call. Their joint probability distribution is given as

		x		
	$f(x,y)$	1	2	3
	1	0.05	0.05	0.10
y	3	0.05	0.10	0.35
	5	0.00	0.20	0.10

(a) Evaluate the marginal distribution of X.
(b) Evaluate the marginal distribution of Y.
(c) Find $P(Y = 3 \mid X = 2)$.

2.40 Suppose that X and Y have the following joint probability distribution:

		x	
	$f(x,y)$	2	4
	1	0.10	0.15
y	3	0.20	0.30
	5	0.10	0.15

(a) Find the marginal distribution of X.
(b) Find the marginal distribution of Y.

2.41 Given the joint density function

$$f(x,y) = \begin{cases} \frac{6-x-y}{8}, & 0 < x < 2, \ 2 < y < 4, \\ 0, & \text{elsewhere,} \end{cases}$$

find $P(1 < Y < 3 \mid X = 1)$.

2.42 A coin is tossed twice. Let Z denote the number of heads on the first toss and W the total number of heads on the 2 tosses. If the coin is unbalanced and a head has a 40% chance of occurring, find
(a) the joint probability distribution of W and Z;
(b) the marginal distribution of W;
(c) the marginal distribution of Z;
(d) the probability that at least 1 head occurs.

2.43 Determine whether the two random variables of Exercise 2.40 are dependent or independent.

2.44 Determine whether the two random variables of Exercise 2.39 are dependent or independent.

2.45 Let X, Y, and Z have the joint probability density function

$$f(x,y,z) = \begin{cases} kxy^2z, & 0 < x,y < 1, \ 0 < z < 2, \\ 0, & \text{elsewhere.} \end{cases}$$

(a) Find k.

(b) Find $P(X < \frac{1}{4}, Y > \frac{1}{2}, 1 < Z < 2)$.

2.46 The joint density function of the random variables X and Y is

$$f(x,y) = \begin{cases} 6x, & 0 < x < 1, \ 0 < y < 1 - x, \\ 0, & \text{elsewhere.} \end{cases}$$

(a) Show that X and Y are not independent.
(b) Find $P(X > 0.3 \mid Y = 0.5)$.

2.47 Determine whether the two random variables of Exercise 2.35 are dependent or independent.

2.48 The joint probability density function of the random variables X, Y, and Z is

$$f(x,y,z) = \begin{cases} \frac{4xyz^2}{9}, & 0 < x,y < 1, \ 0 < z < 3, \\ 0, & \text{elsewhere.} \end{cases}$$

Find
(a) the joint marginal density function of Y and Z;
(b) the marginal density of Y;
(c) $P(\frac{1}{4} < X < \frac{1}{2}, \ Y > \frac{1}{3}, \ 1 < Z < 2)$;
(d) $P(0 < X < \frac{1}{2} \mid Y = \frac{1}{4}, \ Z = 2)$.

2.49 Determine whether the two random variables of Exercise 2.36 are dependent or independent.

2.5   Mean of a Random Variable

We can refer to the **population mean of the random variable X** or the **mean of the probability distribution of X** and write it as μ_x, or simply as μ when it is clear to which random variable we refer. It is also common among statisticians to refer to this mean as the mathematical expectation, or the expected value of the random variable X, and denote it as $E(X)$.

Assuming that one fair coin was tossed twice, we find that the sample space for our experiment is

$$S = \{HH, HT, TH, TT\}.$$

Denote by X the number of heads. Since the 4 sample points are all equally likely, it follows that

$$P(X = 0) = P(TT) = \frac{1}{4}, \quad P(X = 1) = P(TH) + P(HT) = \frac{1}{2},$$

and

$$P(X = 2) = P(HH) = \frac{1}{4},$$

where a typical element, say TH, indicates that the first toss resulted in a tail followed by a head on the second toss. Now, these probabilities are just the relative frequencies for the given events in the long run. Therefore,

$$\mu = E(X) = (0)\left(\frac{1}{4}\right) + (1)\left(\frac{1}{2}\right) + (2)\left(\frac{1}{4}\right) = 1.$$

This result means that a person who tosses 2 coins over and over again will, on the average, get 1 head per toss.

Definition 2.14: Let X be a random variable with probability distribution $f(x)$. The **mean**, or **expected value**, of X is

$$\mu = E(X) = \sum_x x f(x)$$

if X is discrete, and

$$\mu = E(X) = \int_{-\infty}^{\infty} x f(x)\ dx$$

if X is continuous.

Example 2.23: A lot containing 7 components is sampled by a quality inspector; the lot contains 4 good components and 3 defective components. A sample of 3 is taken by the inspector. Find the expected value of the number of good components in this sample.

Solution: Let X represent the number of good components in the sample. The probability distribution of X is

$$f(x) = \frac{\binom{4}{x}\binom{3}{3-x}}{\binom{7}{3}}, \qquad x = 0, 1, 2, 3.$$

Simple calculations yield $f(0) = 1/35$, $f(1) = 12/35$, $f(2) = 18/35$, and $f(3) = 4/35$. Therefore,

$$\mu = E(X) = (0)\left(\frac{1}{35}\right) + (1)\left(\frac{12}{35}\right) + (2)\left(\frac{18}{35}\right) + (3)\left(\frac{4}{35}\right) = \frac{12}{7} = 1.7.$$

Thus, if a sample of size 3 is selected at random over and over again from a lot of 4 good components and 3 defective components, it will contain, on average, 1.7 good components.

Example 2.24: A salesperson for a medical device company has two appointments on a given day. At the first appointment, he believes that he has a 70% chance to make the deal, from which he can earn $1000 commission if successful. On the other hand, he thinks he only has a 40% chance to make the deal at the second appointment, from which, if successful, he can make $1500. What is his expected commission based on his own probability belief? Assume that the appointment results are independent of each other.

Solution: First, we know that the salesperson, for the two appointments, can have 4 possible commission totals: $0, $1000, $1500, and $2500. We then need to calculate their associated probabilities. By independence, we obtain

$$f(\$0) = (1 - 0.7)(1 - 0.4) = 0.18, \quad f(\$2500) = (0.7)(0.4) = 0.28,$$
$$f(\$1000) = (0.7)(1 - 0.4) = 0.42, \text{ and } f(\$1500) = (1 - 0.7)(0.4) = 0.12.$$

Therefore, the expected commission for the salesperson is

$$E(X) = (\$0)(0.18) + (\$1000)(0.42) + (\$1500)(0.12) + (\$2500)(0.28)$$
$$= \$1300.$$

Examples 2.23 and 2.24 are designed to allow the reader to gain some insight into what we mean by the expected value of a random variable. In both cases the random variables are discrete. We follow with an example involving a continuous random variable, where an engineer is interested in the *mean life* of a certain type of electronic device. This is an illustration of a *time to failure* problem that occurs often in practice. The expected value of the life of a device is an important parameter for its evaluation.

Example 2.25: Let X be the random variable that denotes the life in hours of a certain electronic device. The probability density function is

$$f(x) = \begin{cases} \frac{20,000}{x^3}, & x > 100, \\ 0, & \text{elsewhere.} \end{cases}$$

Find the expected life of this type of device.

Solution: Using Definition 2.14, we have

$$\mu = E(X) = \int_{100}^{\infty} x \frac{20,000}{x^3} \, dx = \int_{100}^{\infty} \frac{20,000}{x^2} \, dx = 200.$$

Therefore, we can expect this type of device to last, *on average*, 200 hours.

Now let us consider a new random variable $g(X)$, which depends on X; that is, each value of $g(X)$ is determined by the value of X. For instance, $g(X)$ might be X^2 or $3X - 1$, and whenever X assumes the value 2, $g(X)$ assumes the value $g(2)$. In particular, if X is a discrete random variable with probability distribution $f(x)$, for $x = -1, 0, 1, 2$, and $g(X) = X^2$, then

$$P[g(X) = 0] = P(X = 0) = f(0),$$
$$P[g(X) = 1] = P(X = -1) + P(X = 1) = f(-1) + f(1),$$
$$P[g(X) = 4] = P(X = 2) = f(2),$$

and so the probability distribution of $g(X)$ may be written

$g(x)$	0	1	4
$P[g(X) = g(x)]$	$f(0)$	$f(-1) + f(1)$	$f(2)$

By the definition of the expected value of a random variable, we obtain

$$\mu_{g(X)} = E[g(x)] = 0f(0) + 1[f(-1) + f(1)] + 4f(2)$$
$$= (-1)^2 f(-1) + (0)^2 f(0) + (1)^2 f(1) + (2)^2 f(2) = \sum_x g(x)f(x).$$

This result is generalized in Theorem 2.1 for both discrete and continuous random variables.

Theorem 2.1: Let X be a random variable with probability distribution $f(x)$. The expected value of the random variable $g(X)$ is

$$\mu_{g(X)} = E[g(X)] = \sum_x g(x)f(x)$$

if X is discrete, and

$$\mu_{g(X)} = E[g(X)] = \int_{-\infty}^{\infty} g(x)f(x)\ dx$$

if X is continuous.

Example 2.26: Suppose that the number of cars X that pass through a car wash between 4:00 P.M. and 5:00 P.M. on any sunny Friday has the following probability distribution:

x	4	5	6	7	8	9
$P(X = x)$	$\frac{1}{12}$	$\frac{1}{12}$	$\frac{1}{4}$	$\frac{1}{4}$	$\frac{1}{6}$	$\frac{1}{6}$

Let $g(X) = 2X - 1$ represent the amount of money, in dollars, paid to the attendant by the manager. Find the attendant's expected earnings for this particular time period.

Solution: By Theorem 2.1, the attendant can expect to receive

$$E[g(X)] = E(2X - 1) = \sum_{x=4}^{9} (2x - 1)f(x)$$

$$= (7)\left(\frac{1}{12}\right) + (9)\left(\frac{1}{12}\right) + (11)\left(\frac{1}{4}\right) + (13)\left(\frac{1}{4}\right)$$

$$+ (15)\left(\frac{1}{6}\right) + (17)\left(\frac{1}{6}\right) = \$12.67.$$

Example 2.27: Let X be a random variable with density function

$$f(x) = \begin{cases} \frac{x^2}{3}, & -1 < x < 2, \\ 0, & \text{elsewhere.} \end{cases}$$

Find the expected value of $g(X) = 4X + 3$.

Solution: By Theorem 2.1, we have

$$E(4X + 3) = \int_{-1}^{2} \frac{(4x + 3)x^2}{3}\ dx = \frac{1}{3}\int_{-1}^{2} (4x^3 + 3x^2)\ dx = 8.$$

We shall now extend our concept of mathematical expectation to the case of two random variables X and Y with joint probability distribution $f(x, y)$.

Definition 2.15: Let X and Y be random variables with joint probability distribution $f(x, y)$. The mean, or expected value, of the random variable $g(X, Y)$ is

$$\mu_{g(X,Y)} = E[g(X,Y)] = \sum_x \sum_y g(x,y)f(x,y)$$

if X and Y are discrete, and

$$\mu_{g(X,Y)} = E[g(X,Y)] = \int_{-\infty}^{\infty} \int_{-\infty}^{\infty} g(x,y)f(x,y)\ dx\ dy$$

if X and Y are continuous.

Generalization of Definition 2.15 for the calculation of mathematical expectations of functions of several random variables is straightforward.

Example 2.28: Let X and Y be the random variables with joint probability distribution indicated in Table 2.1 on page 64. Find the expected value of $g(X, Y) = XY$. The table is reprinted here for convenience.

	$f(x,y)$	x = 0	x = 1	x = 2	Row Totals
	0	3/28	9/28	3/28	15/28
y 1		3/14	3/14	0	3/7
2		1/28	0	0	1/28
Column Totals		5/14	15/28	3/28	1

Solution: By Definition 2.15, we write

$$E(XY) = \sum_{x=0}^{2} \sum_{y=0}^{2} xy f(x,y)$$

$$= (0)(0)f(0,0) + (0)(1)f(0,1)$$
$$\quad + (1)(0)f(1,0) + (1)(1)f(1,1) + (2)(0)f(2,0)$$

$$= f(1,1) = \frac{3}{14}.$$

Example 2.29: Find $E(Y/X)$ for the density function

$$f(x,y) = \begin{cases} \frac{x(1+3y^2)}{4}, & 0 < x < 2,\ 0 < y < 1, \\ 0, & \text{elsewhere.} \end{cases}$$

Solution: We have

$$E\left(\frac{Y}{X}\right) = \int_0^1 \int_0^2 \frac{y(1+3y^2)}{4}\ dx\ dy = \int_0^1 \frac{y+3y^3}{2}\ dy = \frac{5}{8}.$$

Note that if $g(X, Y) = X$ in Definition 2.15, we have

$$E(X) = \begin{cases} \sum_x \sum_y x f(x,y) = \sum_x x g(x) & \text{(discrete case)}, \\ \int_{-\infty}^{\infty} \int_{-\infty}^{\infty} x f(x,y) \, dy \, dx = \int_{-\infty}^{\infty} x g(x) \, dx & \text{(continuous case)}, \end{cases}$$

where $g(x)$ is the marginal distribution of X. Therefore, in calculating $E(X)$ over a two-dimensional space, one may use either the joint probability distribution of X and Y or the marginal distribution of X. Similarly, we define

$$E(Y) = \begin{cases} \sum_y \sum_x y f(x,y) = \sum_y y h(y) & \text{(discrete case)}, \\ \int_{-\infty}^{\infty} \int_{-\infty}^{\infty} y f(x,y) \, dx \, dy = \int_{-\infty}^{\infty} y h(y) \, dy & \text{(continuous case)}, \end{cases}$$

where $h(y)$ is the marginal distribution of the random variable Y.

Exercises

2.50 The probability distribution of the discrete random variable X is

$$f(x) = \binom{3}{x} \left(\frac{1}{4}\right)^x \left(\frac{3}{4}\right)^{3-x}, \quad x = 0, 1, 2, 3.$$

Find the mean of X.

2.51 The probability distribution of X, the number of imperfections per 10 meters of a synthetic fabric in continuous rolls of uniform width, is given in Exercise 2.11 on page 60 as

x	0	1	2	3	4
$f(x)$	0.41	0.37	0.16	0.05	0.01

Find the average number of imperfections per 10 meters of this fabric.

2.52 A coin is biased such that a head is three times as likely to occur as a tail. Find the expected number of tails when this coin is tossed twice.

2.53 Find the mean of the random variable T representing the total of the three coins in Exercise 2.17 on page 61.

2.54 In a gambling game, a woman is paid $3 if she draws a jack or a queen and $5 if she draws a king or an ace from an ordinary deck of 52 playing cards. If she draws any other card, she loses. How much should she pay to play if the game is fair?

2.55 By investing in a particular stock, a person can make a profit in one year of $4000 with probability 0.3 or take a loss of $1000 with probability 0.7. What is this person's expected gain?

2.56 Suppose that an antique jewelry dealer is interested in purchasing a gold necklace for which the probabilities are $0.22, 0.36, 0.28$, and 0.14, respectively, that she will be able to sell it for a profit of $250, sell it for a profit of $150, break even, or sell it for a loss of $150. What is her expected profit?

2.57 The density function of coded measurements of the pitch diameter of threads of a fitting is

$$f(x) = \begin{cases} \frac{4}{\pi(1+x^2)}, & 0 < x < 1, \\ 0, & \text{elsewhere.} \end{cases}$$

Find the expected value of X.

2.58 Two tire-quality experts examine stacks of tires and assign a quality rating to each tire on a 3-point scale. Let X denote the rating given by expert A and Y denote the rating given by B. The following table gives the joint distribution for X and Y.

$f(x,y)$		y 1	2	3
	1	0.10	0.05	0.02
x	2	0.10	0.35	0.05
	3	0.03	0.10	0.20

Find μ_X and μ_Y.

2.59 The density function of the continuous random variable X, the total number of hours, in units of 100 hours, that a family runs a vacuum cleaner over a period of one year, is given in Exercise 2.7 on page 60 as

$$f(x) = \begin{cases} x, & 0 < x < 1, \\ 2 - x, & 1 \le x < 2, \\ 0, & \text{elsewhere.} \end{cases}$$

Find the average number of hours per year that families run their vacuum cleaners.

2.60 If a dealer's profit, in units of $5000, on a new automobile can be looked upon as a random variable X having the density function

$$f(x) = \begin{cases} 2(1-x), & 0 < x < 1, \\ 0, & \text{elsewhere,} \end{cases}$$

find the average profit per automobile.

2.61 Assume that two random variables (X, Y) are uniformly distributed on a circle with radius a. Then the joint probability density function is

$$f(x, y) = \begin{cases} \frac{1}{\pi a^2}, & x^2 + y^2 \le a^2, \\ 0, & \text{otherwise.} \end{cases}$$

Find μ_X, the expected value of X.

2.62 Find the proportion X of individuals who can be expected to respond to a certain mail-order solicitation if X has the density function

$$f(x) = \begin{cases} \frac{2(x+2)}{5}, & 0 < x < 1, \\ 0, & \text{elsewhere.} \end{cases}$$

2.63 Let X be a random variable with the following probability distribution:

x	-3	6	9
$f(x)$	1/6	1/2	1/3

Find $\mu_{g(X)}$, where $g(X) = (2X + 1)^2$.

2.64 Suppose that you are inspecting a lot of 1000 light bulbs, among which 20 are defectives. You choose two light bulbs randomly from the lot without replacement. Let

$$X_1 = \begin{cases} 1, & \text{if the 1st light bulb is defective,} \\ 0, & \text{otherwise,} \end{cases}$$

$$X_2 = \begin{cases} 1, & \text{if the 2nd light bulb is defective,} \\ 0, & \text{otherwise.} \end{cases}$$

Find the probability that at least one light bulb chosen is defective. [*Hint*: Compute $P(X_1 + X_2 = 1)$.]

2.65 A large industrial firm purchases several new word processors at the end of each year, the exact number depending on the frequency of repairs in the previous year. Suppose that the number of word processors, X, purchased each year has the following probability distribution:

x	0	1	2	3
$f(x)$	1/10	3/10	2/5	1/5

If the cost of the desired model is $1200 per unit and at the end of the year a refund of $50X^2$ dollars will be issued, how much can this firm expect to spend on new word processors during this year?

2.66 The hospitalization period, in days, for patients following treatment for a certain type of kidney disorder is a random variable $Y = X + 4$, where X has the density function

$$f(x) = \begin{cases} \frac{32}{(x+4)^3}, & x > 0, \\ 0, & \text{elsewhere.} \end{cases}$$

Find the average number of days that a person is hospitalized following treatment for this disorder.

2.67 Suppose that X and Y have the following joint probability function:

		x	
$f(x,y)$		2	4
	1	0.10	0.15
y 3		0.20	0.30
	5	0.10	0.15

(a) Find the expected value of $g(X, Y) = XY^2$.
(b) Find μ_X and μ_Y.

2.68 Referring to the random variables whose joint probability distribution is given in Exercise 2.31 on page 72,
(a) find $E(X^2Y - 2XY)$;
(b) find $\mu_X - \mu_Y$.

2.69 In Exercise 2.19 on page 61, a density function is given for the time to failure of an important component of a DVD player. Find the mean number of hours to failure of the component and thus the DVD player.

2.70 Let X and Y be random variables with joint density function

$$f(x, y) = \begin{cases} 4xy, & 0 < x, \ y < 1, \\ 0, & \text{elsewhere.} \end{cases}$$

Find the expected value of $Z = \sqrt{X^2 + Y^2}$.

2.71 Exercise 2.21 on page 61 dealt with an important particle size distribution characterized by

$$f(x) = \begin{cases} 3x^{-4}, & x > 1, \\ 0, & \text{elsewhere.} \end{cases}$$

(a) Plot the density function.

(b) Give the mean particle size.

2.72 Consider the information in Exercise 2.20 on page 61. The problem deals with the weight in ounces of the product in a cereal box, with

$$f(x) = \begin{cases} \frac{2}{5}, & 23.75 \le x \le 26.25, \\ 0, & \text{elsewhere.} \end{cases}$$

(a) Plot the density function.

(b) Compute the expected value, or mean weight, in ounces.

(c) Are you surprised at your answer in (b)? Explain why or why not.

2.73 Consider Exercise 2.24 on page 61.

(a) What is the mean proportion of the budget allocated to environmental and pollution control?

(b) What is the probability that a company selected at random will have allocated to environmental and pollution control a proportion that exceeds the population mean given in (a)?

2.74 In Exercise 2.23 on page 61, the distribution of times before a major repair of a washing machine was given as

$$f(y) = \begin{cases} \frac{1}{4}e^{-y/4}, & y \ge 0, \\ 0, & \text{elsewhere.} \end{cases}$$

What is the population mean of the times to repair?

2.75 In Exercise 2.11 on page 60, the distribution of the number of imperfections per 10 meters of synthetic fabric is given by

x	0	1	2	3	4
$f(x)$	0.41	0.37	0.16	0.05	0.01

(a) Plot the probability function.

(b) Find the expected number of imperfections, $E(X) = \mu$.

(c) Find $E(X^2)$.

2.6 Variance and Covariance of Random Variables

The mean, or expected value, of a random variable X is of special importance in statistics because it describes where the probability distribution is centered. By itself, however, the mean does not give an adequate description of the shape of the distribution. We also need to characterize the variability in the distribution. In Figure 2.7, we have the histograms of two discrete probability distributions that have the same mean, $\mu = 2$, but differ considerably in variability, or the dispersion of their observations about the mean.

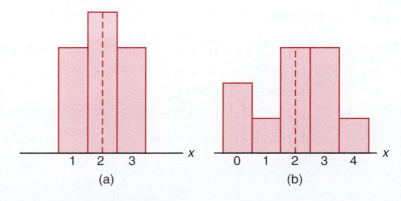

Figure 2.7: Distributions with equal means and unequal dispersions.

The most important measure of variability of a random variable X is obtained by applying Theorem 2.1 with $g(X) = (X - \mu)^2$. The quantity is referred to as the **variance of the random variable X** or the **variance of the probability distribution of X** and is denoted by $\mathrm{Var}(X)$ or the symbol σ_X^2, or simply by σ^2 when it is clear to which random variable we refer.

Definition 2.16: Let X be a random variable with probability distribution $f(x)$ and mean μ. The **variance** of X is

$$\sigma^2 = E[(X - \mu)^2] = \sum_x (x - \mu)^2 f(x), \qquad \text{if } X \text{ is discrete, and}$$

$$\sigma^2 = E[(X - \mu)^2] = \int_{-\infty}^{\infty} (x - \mu)^2 f(x)\, dx, \qquad \text{if } X \text{ is continuous.}$$

The positive square root of the variance, σ, is the **standard deviation** of X.

The quantity $x - \mu$ in Definition 2.16 is called the **deviation of an observation from its mean**. Since the deviations are squared and then averaged, σ^2 will be much smaller for a set of x values that are close to μ than it will be for a set of values that vary considerably from μ.

Example 2.30: Let the random variable X represent the number of automobiles that are used for official business purposes on any given workday. The probability distribution for company A [Figure 2.7(a)] is

x	1	2	3
$f(x)$	0.3	0.4	0.3

and that for company B [Figure 2.7(b)] is

x	0	1	2	3	4
$f(x)$	0.2	0.1	0.3	0.3	0.1

Show that the variance of the probability distribution for company B is greater than that for company A.

Solution: For company A, we find that

$$\mu_A = E(X) = (1)(0.3) + (2)(0.4) + (3)(0.3) = 2.0,$$

and then

$$\sigma_A^2 = \sum_{x=1}^{3}(x - 2)^2 = (1 - 2)^2(0.3) + (2 - 2)^2(0.4) + (3 - 2)^2(0.3) = 0.6.$$

For company B, we have

$$\mu_B = E(X) = (0)(0.2) + (1)(0.1) + (2)(0.3) + (3)(0.3) + (4)(0.1) = 2.0,$$

and then

$$\sigma_B^2 = \sum_{x=0}^{4}(x - 2)^2 f(x)$$

$$= (0 - 2)^2(0.2) + (1 - 2)^2(0.1) + (2 - 2)^2(0.3)$$

$$+ (3 - 2)^2(0.3) + (4 - 2)^2(0.1) = 1.6.$$

Clearly, the variance of the number of automobiles that are used for official business purposes is greater for company B than for company A.

An alternative and preferred formula for finding σ^2, which often simplifies the calculations, is stated in the following theorem and its proof is left to the reader.

Theorem 2.2: The variance of a random variable X is

$$\sigma^2 = E(X^2) - \mu^2.$$

Example 2.31: Let the random variable X represent the number of defective parts for a machine when 3 parts are sampled from a production line and tested. The following is the probability distribution of X.

x	0	1	2	3
$f(x)$	0.51	0.38	0.10	0.01

Using Theorem 2.2, calculate σ^2.

Solution: First, we compute

$$\mu = (0)(0.51) + (1)(0.38) + (2)(0.10) + (3)(0.01) = 0.61.$$

Now,

$$E(X^2) = (0)(0.51) + (1)(0.38) + (4)(0.10) + (9)(0.01) = 0.87.$$

Therefore,

$$\sigma^2 = 0.87 - (0.61)^2 = 0.4979.$$

Example 2.32: The weekly demand for bottled water, in thousands of liters, from a local chain of efficiency stores is a continuous random variable X having the probability density

$$f(x) = \begin{cases} 2(x-1), & 1 < x < 2, \\ 0, & \text{elsewhere.} \end{cases}$$

Find the mean and variance of X.

Solution: Calculating $E(X)$ and $E(X^2)$, we have

$$\mu = E(X) = 2 \int_1^2 x(x-1) \, dx = \frac{5}{3}$$

and

$$E(X^2) = 2 \int_1^2 x^2(x-1) \, dx = \frac{17}{6}.$$

Therefore,

$$\sigma^2 = \frac{17}{6} - \left(\frac{5}{3}\right)^2 = \frac{1}{18}.$$

At this point, the variance or standard deviation has meaning only when we compare two or more distributions that have the same units of measurement.

Therefore, we could compare the variances of the distributions of contents, measured in liters, of bottles of orange juice from two companies, and the larger value would indicate the company whose product was more variable or less uniform. It would not be meaningful to compare the variance of a distribution of heights to the variance of a distribution of aptitude scores.

We shall now extend our concept of the variance of a random variable X to include random variables related to X. For the random variable $g(X)$, the variance is denoted by $\sigma^2_{g(X)}$ and is calculated by means of the following theorem.

Theorem 2.3: Let X be a random variable with probability distribution $f(x)$. The variance of the random variable $g(X)$ is

$$\sigma^2_{g(X)} = E\{[g(X) - \mu_{g(X)}]^2\} = \sum_x [g(x) - \mu_{g(X)}]^2 f(x)$$

if X is discrete, and

$$\sigma^2_{g(X)} = E\{[g(X) - \mu_{g(X)}]^2\} = \int_{-\infty}^{\infty} [g(x) - \mu_{g(X)}]^2 f(x)\ dx$$

if X is continuous.

Proof: Since $g(X)$ is itself a random variable with mean $\mu_{g(X)}$ and probability distribution $f(x)$, as indicated in Theorem 2.1, the result follows directly from Definition 2.16 that

$$\sigma^2_{g(X)} = E\{[g(X) - \mu_{g(X)}]\}.$$

Now, applying Theorem 2.1 again to the random variable $[g(X) - \mu_{g(X)}]^2$ completes the proof. ◼

Example 2.33: Calculate the variance of $g(X) = 2X + 3$, where X is a random variable with probability distribution

x	0	1	2	3
$f(x)$	$\frac{1}{4}$	$\frac{1}{8}$	$\frac{1}{2}$	$\frac{1}{8}$

Solution: First, we find the mean of the random variable $2X + 3$. According to Theorem 2.1,

$$\mu_{2X+3} = E(2X + 3) = \sum_{x=0}^{3} (2x + 3)f(x) = 6.$$

Now, using Theorem 2.3, we have

$$\sigma^2_{2X+3} = E\{[(2X + 3) - \mu_{2x+3}]^2\} = E[(2X + 3 - 6)^2]$$

$$= E(4X^2 - 12X + 9) = \sum_{x=0}^{3} (4x^2 - 12x + 9)f(x) = 4.$$ ◼

Example 2.34: Let X be a random variable having the density function given in Example 2.27 on page 77. Find the variance of the random variable $g(X) = 4X + 3$.

Solution: In Example 2.27, we found that $\mu_{4X+3} = 8$. Now, using Theorem 2.3,

$$\sigma_{4X+3}^2 = E\{[(4X+3) - 8]^2\} = E[(4X - 5)^2]$$

$$= \int_{-1}^{2} (4x - 5)^2 \frac{x^2}{3} \, dx = \frac{1}{3} \int_{-1}^{2} (16x^4 - 40x^3 + 25x^2) \, dx = \frac{51}{5}.$$

If $g(X, Y) = (X - \mu_X)(Y - \mu_Y)$, where $\mu_X = E(X)$ and $\mu_Y = E(Y)$, Definition 2.15 yields an expected value called the **covariance** of X and Y, which we denote by σ_{XY} or $\text{Cov}(X, Y)$.

Definition 2.17: Let X and Y be random variables with joint probability distribution $f(x, y)$. The covariance of X and Y is

$$\sigma_{XY} = E[(X - \mu_X)(Y - \mu_Y)] = \sum_x \sum_y (x - \mu_X)(y - \mu_y) f(x, y)$$

if X and Y are discrete, and

$$\sigma_{XY} = E[(X - \mu_X)(Y - \mu_Y)] = \int_{-\infty}^{\infty} \int_{-\infty}^{\infty} (x - \mu_X)(y - \mu_y) f(x, y) \, dx \, dy$$

if X and Y are continuous.

The covariance between two random variables is a measure of the nature of the association between the two. If large values of X often result in large values of Y or small values of X result in small values of Y, positive $X - \mu_X$ will often result in positive $Y - \mu_Y$ and negative $X - \mu_X$ will often result in negative $Y - \mu_Y$. Thus, the product $(X - \mu_X)(Y - \mu_Y)$ will tend to be positive. On the other hand, if large X values often result in small Y values, the product $(X - \mu_X)(Y - \mu_Y)$ will tend to be negative. The *sign* of the covariance indicates whether the relationship between two dependent random variables is positive or negative. When X and Y are statistically independent, it can be shown that the covariance is zero (see Corollary 2.5). The converse, however, is not generally true. Two variables may have zero covariance and still not be statistically independent. Note that the covariance only describes the *linear* relationship between two random variables. Therefore, if a covariance between X and Y is zero, X and Y may have a nonlinear relationship, which means that they are not necessarily independent.

The alternative and preferred formula for σ_{XY} is stated by Theorem 2.4 and the proof of the theorem is left to the reader.

Theorem 2.4: The covariance of two random variables X and Y with means μ_X and μ_Y, respectively, is given by

$$\sigma_{XY} = E(XY) - \mu_X \mu_Y.$$

Example 2.35: Example 2.14 on page 63 describes a situation involving the number of blue pens X and the number of red pens Y selected from a box. Two pens are selected at random from a box, and the following is the joint probability distribution:

	$f(x,y)$	0	1	2	$h(y)$
			x		
	0	3/28	9/28	3/28	15/28
y	1	3/14	3/14	0	3/7
	2	1/28	0	0	1/28
	$g(x)$	5/14	15/28	3/28	1

Find the covariance of X and Y.

Solution: From Example 2.28, we see that $E(XY) = 3/14$. Now

$$\mu_X = \sum_{x=0}^{2} xg(x) = (0)\left(\frac{5}{14}\right) + (1)\left(\frac{15}{28}\right) + (2)\left(\frac{3}{28}\right) = \frac{3}{4},$$

and

$$\mu_Y = \sum_{y=0}^{2} yh(y) = (0)\left(\frac{15}{28}\right) + (1)\left(\frac{3}{7}\right) + (2)\left(\frac{1}{28}\right) = \frac{1}{2}.$$

Therefore,

$$\sigma_{XY} = E(XY) - \mu_X\mu_Y = \frac{3}{14} - \left(\frac{3}{4}\right)\left(\frac{1}{2}\right) = -\frac{9}{56}.$$

Example 2.36: The fraction X of male runners and the fraction Y of female runners who compete in marathon races are described by the joint density function

$$f(x,y) = \begin{cases} 8xy, & 0 \leq y \leq x \leq 1, \\ 0, & \text{elsewhere.} \end{cases}$$

Find the covariance of X and Y.

Solution: We first compute the marginal density functions. They are

$$g(x) = \begin{cases} 4x^3, & 0 \leq x \leq 1, \\ 0, & \text{elsewhere,} \end{cases}$$

and

$$h(y) = \begin{cases} 4y(1 - y^2), & 0 \leq y \leq 1, \\ 0, & \text{elsewhere.} \end{cases}$$

From these marginal density functions, we compute

$$\mu_X = E(X) = \int_0^1 4x^4 \, dx = \frac{4}{5} \text{ and } \mu_Y = \int_0^1 4y^2(1 - y^2) \, dy = \frac{8}{15}.$$

From the joint density function given above, we have

$$E(XY) = \int_0^1 \int_y^1 8x^2y^2 \, dx \, dy = \frac{4}{9}.$$

Then

$$\sigma_{XY} = E(XY) - \mu_X\mu_Y = \frac{4}{9} - \left(\frac{4}{5}\right)\left(\frac{8}{15}\right) = \frac{4}{225}.$$

Although the covariance between two random variables does provide information regarding the nature of the linear relationship, the magnitude of σ_{XY} *does not indicate anything regarding the strength of the relationship*, since σ_{XY} is not scale-free. Its magnitude will depend on the units used to measure both X and Y. There is a scale-free version of the covariance called the **correlation coefficient** that is used widely in statistics.

Definition 2.18: Let X and Y be random variables with covariance σ_{XY} and standard deviations σ_X and σ_Y, respectively. The correlation coefficient of X and Y is

$$\rho_{XY} = \frac{\sigma_{XY}}{\sigma_X\sigma_Y}.$$

It should be clear to the reader that ρ_{XY} is free of the units of X and Y. The correlation coefficient satisfies the inequality $-1 \le \rho_{XY} \le 1$. It assumes a value of zero when $\sigma_{XY} = 0$. Where there is an exact linear dependency, say $Y \equiv a + bX$, $\rho_{XY} = 1$ if $b > 0$ and $\rho_{XY} = -1$ if $b < 0$. (See Exercise 2.86.) The correlation coefficient is the subject of more discussion in Chapter 7, where we deal with linear regression.

Example 2.37: Find the correlation coefficient between X and Y in Example 2.35.
Solution: Since

$$E(X^2) = (0^2)\left(\frac{5}{14}\right) + (1^2)\left(\frac{15}{28}\right) + (2^2)\left(\frac{3}{28}\right) = \frac{27}{28}$$

and

$$E(Y^2) = (0^2)\left(\frac{15}{28}\right) + (1^2)\left(\frac{3}{7}\right) + (2^2)\left(\frac{1}{28}\right) = \frac{4}{7},$$

we obtain

$$\sigma_X^2 = \frac{27}{28} - \left(\frac{3}{4}\right)^2 = \frac{45}{112} \text{ and } \sigma_Y^2 = \frac{4}{7} - \left(\frac{1}{2}\right)^2 = \frac{9}{28}.$$

Therefore, the correlation coefficient between X and Y is

$$\rho_{XY} = \frac{\sigma_{XY}}{\sigma_X\sigma_Y} = \frac{-9/56}{\sqrt{(45/112)(9/28)}} = -\frac{1}{\sqrt{5}}.$$

Example 2.38: Find the correlation coefficient of X and Y in Example 2.36.
Solution: Because

$$E(X^2) = \int_0^1 4x^5\,dx = \frac{2}{3} \text{ and } E(Y^2) = \int_0^1 4y^3(1-y^2)\,dy = 1 - \frac{2}{3} = \frac{1}{3},$$

we conclude that

$$\sigma_X^2 = \frac{2}{3} - \left(\frac{4}{5}\right)^2 = \frac{2}{75} \text{ and } \sigma_Y^2 = \frac{1}{3} - \left(\frac{8}{15}\right)^2 = \frac{11}{225}.$$

Hence,

$$\rho_{XY} = \frac{4/225}{\sqrt{(2/75)(11/225)}} = \frac{4}{\sqrt{66}}.$$

Note that although the covariance in Example 2.37 is larger in magnitude (disregarding the sign) than that in Example 2.38, the relationship of the magnitudes of the correlation coefficients in these two examples is just the reverse. This is evidence that we cannot look at the magnitude of the covariance to decide on how strong the relationship is.

Exercises

2.76 Use Definition 2.16 on page 82 to find the variance of the random variable X of Exercise 2.55 on page 79.

2.77 The random variable X, representing the number of errors per 100 lines of software code, has the following probability distribution:

x	2	3	4	5	6
$f(x)$	0.01	0.25	0.4	0.3	0.04

Using Theorem 2.2 on page 83, find the variance of X.

2.78 Suppose that the probabilities are 0.4, 0.3, 0.2, and 0.1, respectively, that 0, 1, 2, or 3 power failures will strike a certain subdivision in any given year. Find the mean and variance of the random variable X representing the number of power failures striking this subdivision.

2.79 A dealer's profit, in units of \$5000, on a new automobile is a random variable X having the density function given in Exercise 2.60 on page 80. Find the variance of X.

2.80 The proportion of people who respond to a certain mail-order solicitation is a random variable X having the density function given in Exercise 2.62 on page 80. Find the variance of X.

2.81 The total number of hours, in units of 100 hours, that a family runs a vacuum cleaner over a period of one year is a random variable X having the density function given in Exercise 2.59 on page 79. Find the variance of X.

2.82 Referring to Exercise 2.62 on page 80, find $\sigma^2_{g(X)}$ for the function $g(X) = 3X^2 + 4$.

2.83 The length of time, in minutes, for an airplane to obtain clearance for takeoff at a certain airport is a random variable $Y = 3X - 2$, where X has the density function

$$f(x) = \begin{cases} \frac{1}{4}e^{-x/4}, & x > 0 \\ 0, & \text{elsewhere.} \end{cases}$$

Find the mean and variance of the random variable Y.

2.84 Find the covariance of the random variables X and Y of Exercise 2.36 on page 73.

2.85 For the random variables X and Y whose joint density function is given in Exercise 2.32 on page 72, find the covariance.

2.86 Given a random variable X, with standard deviation σ_X, and a random variable $Y = a + bX$, show that if $b < 0$, the correlation coefficient $\rho_{XY} = -1$, and if $b > 0$, $\rho_{XY} = 1$.

2.87 Consider the situation in Exercise 2.75 on page 81. The distribution of the number of imperfections per 10 meters of synthetic fabric is given by

x	0	1	2	3	4
$f(x)$	0.41	0.37	0.16	0.05	0.01

Find the variance and standard deviation of the number of imperfections.

2.88 For a laboratory assignment, if the equipment is working, the density function of the observed outcome X is

$$f(x) = \begin{cases} 2(1 - x), & 0 < x < 1, \\ 0, & \text{otherwise.} \end{cases}$$

Find the variance and standard deviation of X.

2.89 For the random variables X and Y in Exercise 2.31 on page 72, determine the correlation coefficient between X and Y.

2.90 Random variables X and Y follow a joint distribution

$$f(x, y) = \begin{cases} 2, & 0 < x \le y < 1, \\ 0, & \text{otherwise.} \end{cases}$$

Determine the correlation coefficient between X and Y.

2.7 Means and Variances of Linear Combinations of Random Variables

We now develop some useful properties that will simplify the calculations of means and variances of random variables that appear in later chapters. These properties will permit us to deal with expectations in terms of other parameters that are either known or easily computed. All the results that we present here are valid for both discrete and continuous random variables. Proofs are given only for the continuous case. We begin with a theorem and two corollaries that should be, intuitively, reasonable to the reader.

Theorem 2.5: If a and b are constants, then

$$E(aX + b) = aE(X) + b.$$

Proof: By the definition of expected value,

$$E(aX + b) = \int_{-\infty}^{\infty} (ax + b)f(x)\ dx = a \int_{-\infty}^{\infty} xf(x)\ dx + b \int_{-\infty}^{\infty} f(x)\ dx.$$

The first integral on the right is $E(X)$ and the second integral equals 1. Therefore, we have

$$E(aX + b) = aE(X) + b.$$

Corollary 2.1: Setting $a = 0$, we see that $E(b) = b$.

Corollary 2.2: Setting $b = 0$, we see that $E(aX) = aE(X)$.

Example 2.39: Applying Theorem 2.5 to the discrete random variable $h(X) = 2X - 1$, rework Example 2.26 on page 77.

Solution: According to Theorem 2.5, we can write

$$E(2X - 1) = 2E(X) - 1.$$

Now

$$\mu = E(X) = \sum_{x=4}^{9} xf(x)$$

$$= (4)\left(\frac{1}{12}\right) + (5)\left(\frac{1}{12}\right) + (6)\left(\frac{1}{4}\right) + (7)\left(\frac{1}{4}\right) + (8)\left(\frac{1}{6}\right) + (9)\left(\frac{1}{6}\right) = \frac{41}{6}.$$

Therefore,

$$\mu_{2X-1} = (2)\left(\frac{41}{6}\right) - 1 = \$12.67,$$

as before.

For the addition or subtraction of two functions of the random variable X, we have the following theorem to compute its mean. The proof of the theorem is left to the reader.

Theorem 2.6: The expected value of the sum or difference of two or more functions of a random variable X is the sum or difference of the expected values of the functions. That is,

$$E[g(X) \pm h(X)] = E[g(X)] \pm E[h(X)].$$

Example 2.40: Let X be a random variable with probability distribution as follows:

x	0	1	2	3
$f(x)$	$\frac{1}{3}$	$\frac{1}{2}$	0	$\frac{1}{6}$

Find the expected value of $Y = (X - 1)^2$.

Solution: Applying Theorem 2.6 to the function $Y = (X - 1)^2$, we can write

$$E[(X - 1)^2] = E(X^2 - 2X + 1) = E(X^2) - 2E(X) + E(1).$$

From Corollary 2.1, $E(1) = 1$, and by direct computation,

$$E(X) = (0)\left(\frac{1}{3}\right) + (1)\left(\frac{1}{2}\right) + (2)(0) + (3)\left(\frac{1}{6}\right) = 1 \text{ and}$$

$$E(X^2) = (0)\left(\frac{1}{3}\right) + (1)\left(\frac{1}{2}\right) + (4)(0) + (9)\left(\frac{1}{6}\right) = 2.$$

Hence,

$$E[(X - 1)^2] = 2 - (2)(1) + 1 = 1.$$

Example 2.41: The weekly demand for a certain drink, in thousands of liters, at a chain of convenience stores is a continuous random variable $g(X) = X^2 + X - 2$, where X has the density function

$$f(x) = \begin{cases} 2(x - 1), & 1 < x < 2, \\ 0, & \text{elsewhere.} \end{cases}$$

Find the expected value of the weekly demand for the drink.

Solution: By Theorem 2.6, we write

$$E(X^2 + X - 2) = E(X^2) + E(X) - E(2).$$

From Corollary 2.1, $E(2) = 2$, and by direct integration,

$$E(X) = \int_1^2 2x(x - 1)\, dx = \frac{5}{3} \text{ and } E(X^2) = \int_1^2 2x^2(x - 1)\, dx = \frac{17}{6}.$$

Now

$$E(X^2 + X - 2) = \frac{17}{6} + \frac{5}{3} - 2 = \frac{5}{2},$$

so the average weekly demand for the drink from this chain of efficiency stores is 2500 liters.

Suppose that we have two random variables X and Y with joint probability distribution $f(x, y)$. Two additional properties that will be very useful in succeeding chapters involve the expected values of the sum, difference, and product of these two random variables. First, however, let us give a theorem on the expected value of the sum or difference of functions of the given variables. This, of course, is merely an extension of Theorem 2.6. The proof follows from the use of Definition 2.15 and will be left to the reader.

Theorem 2.7: The expected value of the sum or difference of two or more functions of the random variables X and Y is the sum or difference of the expected values of the functions. That is,

$$E[g(X,Y) \pm h(X,Y)] = E[g(X,Y)] \pm E[h(X,Y)].$$

Corollary 2.3: Setting $g(X,Y) = g(X)$ and $h(X,Y) = h(Y)$, we see that

$$E[g(X) \pm h(Y)] = E[g(X)] \pm E[h(Y)].$$

Corollary 2.4: Setting $g(X,Y) = X$ and $h(X,Y) = Y$, we see that

$$E[X \pm Y] = E[X] \pm E[Y].$$

If X represents the daily production of some item from machine A and Y the daily production of the same kind of item from machine B, then $X + Y$ represents the total number of items produced daily by both machines. Corollary 2.4 states that the average daily production for both machines is equal to the sum of the average daily production of each machine.

Theorem 2.8: Let X and Y be two independent random variables. Then

$$E(XY) = E(X)E(Y).$$

Proof: By Definition 2.15,

$$E(XY) = \int_{-\infty}^{\infty} \int_{-\infty}^{\infty} xy f(x, y) \, dx \, dy.$$

Since X and Y are independent, we may write

$$f(x, y) = g(x)h(y),$$

where $g(x)$ and $h(y)$ are the marginal distributions of X and Y, respectively. Hence,

$$E(XY) = \int_{-\infty}^{\infty} \int_{-\infty}^{\infty} xy g(x)h(y) \, dx \, dy = \int_{-\infty}^{\infty} xg(x) \, dx \int_{-\infty}^{\infty} yh(y) \, dy$$
$$= E(X)E(Y).$$

Corollary 2.5: Let X and Y be two independent random variables. Then $\sigma_{XY} = 0$.

Proof: The proof can be carried out using Theorems 2.4 and 2.8, and Definition 2.14. ∎

Example 2.42: It is known that the ratio of gallium to arsenide does not affect the functioning of gallium-arsenide wafers, which are the main components of some microchips. Let X denote the ratio of gallium to arsenide and Y denote the functional wafers retrieved during a 1-hour period. X and Y are independent random variables with the joint density function

$$f(x,y) = \begin{cases} \frac{x(1+3y^2)}{4}, & 0 < x < 2, \ 0 < y < 1, \\ 0, & \text{elsewhere.} \end{cases}$$

Show that $E(XY) = E(X)E(Y)$, as Theorem 2.8 suggests.

Solution: By definition,

$$E(XY) = \int_0^1 \int_0^2 \frac{x^2 y(1+3y^2)}{4} \ dx \ dy = \frac{5}{6}, \ E(X) = \frac{4}{3}, \ \text{and} \ E(Y) = \frac{5}{8}.$$

Hence,

$$E(X)E(Y) = \left(\frac{4}{3}\right)\left(\frac{5}{8}\right) = \frac{5}{6} = E(XY).$$

We conclude this section by proving one theorem and presenting several corollaries that are useful for calculating variances or standard deviations.

Theorem 2.9: If X and Y are random variables with joint probability distribution $f(x,y)$ and a, b, and c are constants, then

$$\sigma^2_{aX+bY+c} = a^2\sigma^2_X + b^2\sigma^2_Y + 2ab\sigma_{XY}.$$

Proof: By definition, $\sigma^2_{aX+bY+c} = E\{[(aX + bY + c) - \mu_{aX+bY+c}]^2\}$. Now

$$\mu_{aX+bY+c} = E(aX + bY + c) = aE(X) + bE(Y) + c = a\mu_X + b\mu_Y + c,$$

by using Corollary 2.4 followed by Corollary 2.2. Therefore,

$$\begin{aligned} \sigma^2_{aX+bY+c} &= E\{[a(X - \mu_X) + b(Y - \mu_Y)]^2\} \\ &= a^2 E[(X - \mu_X)^2] + b^2 E[(Y - \mu_Y)^2] + 2abE[(X - \mu_X)(Y - \mu_Y)] \\ &= a^2\sigma^2_X + b^2\sigma^2_Y + 2ab\sigma_{XY}. \end{aligned}$$

Using Theorem 2.9, we have the following corollaries.

Corollary 2.6: Setting $b = 0$, we see that

$$\sigma^2_{aX+c} = a^2 \sigma^2_X = a^2 \sigma^2.$$

Corollary 2.7: Setting $a = 1$ and $b = 0$, we see that

$$\sigma^2_{X+c} = \sigma^2_X = \sigma^2.$$

Corollary 2.8: Setting $b = 0$ and $c = 0$, we see that

$$\sigma^2_{aX} = a^2 \sigma^2_X = a^2 \sigma^2.$$

Corollaries 2.6 and 2.7 state that the variance is unchanged if a constant is added to or subtracted from a random variable. The addition or subtraction of a constant simply shifts the values of X to the right or to the left but does not change their variability. However, if a random variable is multiplied or divided by a constant, then Corollaries 2.6 and 2.8 state that the variance is multiplied or divided by the square of the constant.

Corollary 2.9: If X and Y are independent random variables, then

$$\sigma^2_{aX+bY} = a^2 \sigma^2_X + b^2 \sigma^2_Y.$$

The result stated in Corollary 2.9 is obtained from Theorem 2.9 by invoking Corollary 2.5.

Corollary 2.10: If X and Y are independent random variables, then

$$\sigma^2_{aX-bY} = a^2 \sigma^2_X + b^2 \sigma^2_Y.$$

Corollary 2.10 follows when b in Corollary 2.9 is replaced by $-b$. Generalizing to a linear combination of n independent random variables, we have Corollary 2.11.

Corollary 2.11: If $X_1, X_2, \ldots, X_n$ are independent random variables, then

$$\sigma^2_{a_1 X_1 \pm a_2 X_2 \pm \cdots \pm a_n X_n} = a_1^2 \sigma^2_{X_1} + a_2^2 \sigma^2_{X_2} + \cdots + a_n^2 \sigma^2_{X_n}.$$

Example 2.43: If X and Y are random variables with variances $\sigma^2_X = 2$ and $\sigma^2_Y = 4$ and covariance $\sigma_{XY} = -2$, find the variance of the random variable $Z = 3X - 4Y + 8$.

Solution:

$$\sigma^2_Z = \sigma^2_{3X-4Y+8} = \sigma^2_{3X-4Y} \qquad \text{(by Corollary 2.6)}$$
$$= 9\sigma^2_X + 16\sigma^2_Y - 24\sigma_{XY} \qquad \text{(by Theorem 2.9)}$$
$$= (9)(2) + (16)(4) - (24)(-2) = 130.$$

Example 2.44: Let X and Y denote the amounts of two different types of impurities in a batch of a certain chemical product. Suppose that X and Y are independent random variables with variances $\sigma_X^2 = 2$ and $\sigma_Y^2 = 3$. Find the variance of the random variable $Z = 3X - 2Y + 5$.

Solution:

$$\sigma_Z^2 = \sigma_{3X-2Y+5}^2 = \sigma_{3X-2Y}^2 \quad \text{(by Corollary 2.6)}$$
$$= 9\sigma_x^2 + 4\sigma_y^2 \quad \text{(by Corollary 2.10)}$$
$$= (9)(2) + (4)(3) = 30.$$

Exercises

2.91 Suppose that a grocery store purchases 5 cartons of skim milk at the wholesale price of \$1.20 per carton and retails the milk at \$1.65 per carton. After the expiration date, the unsold milk is removed from the shelf and the grocer receives a credit from the distributor equal to three-fourths of the wholesale price. If the probability distribution of the random variable X, the number of cartons that are sold from this lot, is

x	0	1	2	3	4	5
$f(x)$	$\frac{1}{15}$	$\frac{2}{15}$	$\frac{2}{15}$	$\frac{3}{15}$	$\frac{4}{15}$	$\frac{3}{15}$

find the expected profit.

2.92 Repeat Exercise 2.83 on page 88 by applying Theorem 2.5 and Corollary 2.6.

2.93 If a random variable X is defined such that

$$E[(X-1)^2] = 10 \text{ and } E[(X-2)^2] = 6,$$

find μ and σ^2.

2.94 The total time, measured in units of 100 hours, that a teenager runs her hair dryer over a period of one year is a continuous random variable X that has the density function

$$f(x) = \begin{cases} x, & 0 < x < 1, \\ 2 - x, & 1 \le x < 2, \\ 0, & \text{elsewhere.} \end{cases}$$

Use Theorem 2.6 to evaluate the mean of the random variable $Y = 60X^2 + 39X$, where Y is equal to the number of kilowatt hours expended annually.

2.95 Use Theorem 2.7 to evaluate $E(2XY^2 - X^2Y)$ for the joint probability distribution shown in Table 2.1 on page 64.

2.96 If X and Y are independent random variables with variances $\sigma_X^2 = 5$ and $\sigma_Y^2 = 3$, find the variance of the random variable $Z = -2X + 4Y - 3$.

2.97 Repeat Exercise 2.96 if X and Y are not independent and $\sigma_{XY} = 1$.

2.98 Suppose that X and Y are independent random variables with probability densities and

$$g(x) = \begin{cases} \frac{8}{x^3}, & x > 2, \\ 0, & \text{elsewhere,} \end{cases}$$

and

$$h(y) = \begin{cases} 2y, & 0 < y < 1, \\ 0, & \text{elsewhere.} \end{cases}$$

Find the expected value of $Z = XY$.

2.99 Consider Review Exercise 2.117 on page 97. The random variables X and Y represent the number of vehicles that arrive at two separate street corners during a certain 2-minute period in the day. The joint distribution is

$$f(x, y) = \left(\frac{1}{4^{(x+y)}}\right)\left(\frac{9}{16}\right),$$

for $x = 0, 1, 2, \ldots$ and $y = 0, 1, 2, \ldots$.
(a) Give $E(X)$, $E(Y)$, $\text{Var}(X)$, and $\text{Var}(Y)$.
(b) Consider $Z = X + Y$, the sum of the two. Find $E(Z)$ and $\text{Var}(Z)$.

2.100 Consider Review Exercise 2.106 on page 95. There are two service lines. The random variables X and Y are the proportions of time that line 1 and line 2 are in use, respectively. The joint probability density function for (X, Y) is given by

$$f(x, y) = \begin{cases} \frac{3}{2}(x^2 + y^2), & 0 \le x, \ y \le 1, \\ 0, & \text{elsewhere.} \end{cases}$$

(a) Determine whether or not X and Y are independent.

(b) It is of interest to know something about the proportion of $Z = X + Y$, the sum of the two proportions. Find $E(X + Y)$. Also find $E(XY)$.

(c) Find $\text{Var}(X)$, $\text{Var}(Y)$, and $\text{Cov}(X, Y)$.

(d) Find $\text{Var}(X + Y)$.

2.101 The length of time Y, in minutes, required to generate a human reflex to tear gas has the density function

$$f(y) = \begin{cases} \frac{1}{4}e^{-y/4}, & 0 \le y < \infty, \\ 0, & \text{elsewhere.} \end{cases}$$

(a) What is the mean time to reflex?

Review Exercises

2.103 A tobacco company produces blends of tobacco, with each blend containing various proportions of Turkish, domestic, and other tobaccos. The proportions of Turkish and domestic in a blend are random variables with joint density function ($X = $ Turkish and $Y = $ domestic)

$$f(x, y) = \begin{cases} 24xy, & 0 \le x, y \le 1, \ x + y \le 1, \\ 0, & \text{elsewhere.} \end{cases}$$

(a) Find the probability that in a given box the Turkish tobacco accounts for over half the blend.

(b) Find the marginal density function for the proportion of the domestic tobacco.

(c) Find the probability that the proportion of Turkish tobacco is less than 1/8 if it is known that the blend contains 3/4 domestic tobacco.

2.104 An insurance company offers its policyholders a number of different premium payment options. For a randomly selected policyholder, let X be the number of months between successive payments. The cumulative distribution function of X is

$$F(x) = \begin{cases} 0, & \text{if } x < 1, \\ 0.4, & \text{if } 1 \le x < 3, \\ 0.6, & \text{if } 3 \le x < 5, \\ 0.8, & \text{if } 5 \le x < 7, \\ 1.0, & \text{if } x \ge 7. \end{cases}$$

(a) What is the probability mass function of X?

(b) Compute $P(4 < X \le 7)$.

2.105 Two electronic components of a missile system work in harmony for the success of the total system.

(b) Find $E(Y^2)$ and $\text{Var}(Y)$.

2.102 A manufacturing company has developed a machine for cleaning carpet that is fuel-efficient because it delivers carpet cleaner so rapidly. Of interest is a random variable Y, the amount in gallons per minute delivered. It is known that the density function is given by

$$f(y) = \begin{cases} 1, & 7 \le y \le 8, \\ 0, & \text{elsewhere.} \end{cases}$$

(a) Sketch the density function.

(b) Give $E(Y)$, $E(Y^2)$, and $\text{Var}(Y)$.

Let X and Y denote the life in hours of the two components. The joint density of X and Y is

$$f(x, y) = \begin{cases} ye^{-y(1+x)}, & x, y \ge 0, \\ 0, & \text{elsewhere.} \end{cases}$$

(a) Give the marginal density functions for both random variables.

(b) What is the probability that the lives of both components will exceed 2 hours?

2.106 A service facility operates with two service lines. On a randomly selected day, let X be the proportion of time that the first line is in use whereas Y is the proportion of time that the second line is in use. Suppose that the joint probability density function for (X, Y) is

$$f(x, y) = \begin{cases} \frac{3}{2}(x^2 + y^2), & 0 \le x, y \le 1, \\ 0, & \text{elsewhere.} \end{cases}$$

(a) Compute the probability that neither line is busy more than half the time.

(b) Find the probability that the first line is busy more than 75% of the time.

2.107 Let the number of phone calls received by a switchboard during a 5-minute interval be a random variable X with probability function

$$f(x) = \frac{e^{-2}2^x}{x!}, \quad \text{for } x = 0, 1, 2, \dots.$$

(a) Determine the probability that X equals 0, 1, 2, 3, 4, 5, and 6.

(b) Graph the probability mass function for these values of x.

(c) Determine the cumulative distribution function for these values of X.

2.108 An industrial process manufactures items that can be classified as either defective or not defective. The probability that an item is defective is 0.1. An experiment is conducted in which 5 items are drawn randomly from the process. Let the random variable X be the number of defectives in this sample of 5. What is the probability mass function of X?

2.109 The life span in hours of an electrical component is a random variable with cumulative distribution function

$$F(x) = \begin{cases} 1 - e^{-\frac{x}{50}}, & x > 0, \\ 0, & \text{eleswhere.} \end{cases}$$

(a) Determine its probability density function.

(b) Determine the probability that the life span of such a component will exceed 70 hours.

2.110 Pairs of pants are being produced by a particular outlet facility. The pants are checked by a group of 10 workers. The workers inspect pairs of pants taken randomly from the production line. Each inspector is assigned a number from 1 through 10. A buyer selects a pair of pants for purchase. Let the random variable X be the inspector number.

(a) Give a reasonable probability mass function for X.

(b) Plot the cumulative distribution function for X.

2.111 The shelf life of a product is a random variable that is related to consumer acceptance. It turns out that the shelf life Y in days of a certain type of bakery product has a density function

$$f(y) = \begin{cases} \frac{1}{2}e^{-y/2}, & 0 \le y < \infty, \\ 0, & \text{elsewhere.} \end{cases}$$

What fraction of the loaves of this product stocked today would you expect to be sellable 3 days from now?

2.112 Passenger congestion is a service problem in airports. Trains are installed within the airport to reduce the congestion. With the use of the train, the time X in minutes that it takes to travel from the main terminal to a particular concourse has density function

$$f(x) = \begin{cases} \frac{1}{10}, & 0 \le x \le 10, \\ 0, & \text{elsewhere.} \end{cases}$$

(a) Show that the above is a valid probability density function.

(b) Find the probability that the time it takes a passenger to travel from the main terminal to the concourse will not exceed 7 minutes.

2.113 Impurities in a batch of final product of a chemical process often reflect a serious problem. From considerable plant data gathered, it is known that the proportion Y of impurities in a batch has a density function given by

$$f(y) = \begin{cases} 10(1 - y)^9, & 0 \le y \le 1, \\ 0, & \text{elsewhere.} \end{cases}$$

(a) Verify that the above is a valid density function.

(b) A batch is considered not sellable and then not acceptable if the percentage of impurities exceeds 60%. With the current quality of the process, what is the percentage of batches that are not acceptable?

2.114 The time Z in minutes between calls to an electrical supply system has the probability density function

$$f(z) = \begin{cases} \frac{1}{10}e^{-z/10}, & 0 < z < \infty, \\ 0, & \text{elsewhere.} \end{cases}$$

(a) What is the probability that there are no calls within a 20-minute time interval?

(b) What is the probability that the first call comes within 10 minutes of opening?

2.115 A chemical system that results from a chemical reaction has two important components among others in a blend. The joint distribution describing the proportions X_1 and X_2 of these two components is given by

$$f(x_1, x_2) = \begin{cases} 2, & 0 < x_1 < x_2 < 1, \\ 0, & \text{elsewhere.} \end{cases}$$

(a) Give the marginal distribution of X_1.

(b) Give the marginal distribution of X_2.

(c) What is the probability that component proportions produce the results $X_1 < 0.2$ and $X_2 > 0.5$?

(d) Give the conditional distribution $f_{X_1|X_2}(x_1|x_2)$.

2.116 Consider the situation of Review Exercise 2.115. But suppose the joint distribution of the two proportions is given by

$$f(x_1, x_2) = \begin{cases} 6x_2, & 0 < x_2 < x_1 < 1, \\ 0, & \text{elsewhere.} \end{cases}$$

(a) Give the marginal distribution $f_{X_1}(x_1)$ of the proportion X_1 and verify that it is a valid density function.

(b) What is the probability that proportion X_2 is less than 0.5, given that X_1 is 0.7?

2.117 Consider the random variables X and Y that represent the number of vehicles that arrive at two separate street corners during a certain 2-minute period. These street corners are fairly close together so it is important that traffic engineers deal with them jointly if necessary. The joint distribution of X and Y is known to be

$$f(x, y) = \frac{9}{16} \cdot \frac{1}{4^{(x+y)}},$$

for $x = 0, 1, 2, \ldots$ and $y = 0, 1, 2, \ldots$.

(a) Are the two random variables X and Y independent? Explain why or why not.

(b) What is the probability that during the time period in question less than 4 vehicles arrive at the two street corners?

2.118 The behavior of series of components plays a huge role in scientific and engineering reliability problems. The reliability of the entire system is certainly no better than that of the weakest component in the series. In a series system, the components operate independently of each other. In a particular system containing three components, the probabilities of meeting specifications for components 1, 2, and 3, respectively, are 0.95, 0.99, and 0.92. What is the probability that the entire system works?

2.119 Another type of system that is employed in engineering work is a group of parallel components or a parallel system. In this more conservative approach, the probability that the system operates is larger than the probability that any component operates. The system fails only when all components fail. Consider a situation in which there are 4 independent components in a parallel system with probability of operation given by

Component 1: 0.95; Component 2: 0.94;

Component 3: 0.90; Component 4: 0.97.

What is the probability that the system does not fail?

2.120 Consider a system of components in which there are 5 independent components, each of which possesses an operational probability of 0.92. The system does have a redundancy built in such that it does not fail if 3 out of the 5 components are operational. What is the probability that the total system is operational?

2.121 Project: Take 5 class periods to observe the shoe color of individuals in class. Assume the shoe color categories are red, white, black, brown, and other. Complete a frequency table for each color category.

(a) Estimate and interpret the meaning of the probability distribution.

(b) What is the estimated probability that in the next class period a randomly selected student will be wearing a red or a white pair of shoes?

2.122 Referring to the random variables whose joint probability density function is given in Exercise 2.38 on page 73, find the average amount of kerosene left in the tank at the end of the day.

2.123 Assume the length X, in minutes, of a particular type of telephone conversation is a random variable with probability density function

$$f(x) = \begin{cases} \frac{1}{5} e^{-x/5}, & x > 0, \\ 0, & \text{elsewhere.} \end{cases}$$

(a) Determine the mean length $E(X)$ of this type of telephone conversation.

(b) Find the variance and standard deviation of X.

(c) Find $E[(X + 5)^2]$.

2.124 Suppose it is known that the life X of a particular compressor, in hours, has the density function

$$f(x) = \begin{cases} \frac{1}{900} e^{-x/900}, & x > 0, \\ 0, & \text{elsewhere.} \end{cases}$$

(a) Find the mean life of the compressor.

(b) Find $E(X^2)$.

(c) Find the variance and standard deviation of the random variable X.

2.125 Referring to the random variables whose joint density function is given in Exercise 2.32 on page 72,

(a) find μ_X and μ_Y;

(b) find $E[(X + Y)/2]$.

2.126 Show that $\text{Cov}(aX, bY) = ab \, \text{Cov}(X, Y)$.

2.127 Consider Exercise 2.58 on page 79. Can it be said that the ratings given by the two experts are independent? Explain why or why not.

2.128 A company's marketing and accounting departments have determined that if the company markets its newly developed product, the contribution of the product to the firm's profit during the next 6 months will be described by the following:

Profit Contribution	Probability
−$5,000	0.2
$10,000	0.5
$30,000	0.3

What is the company's expected profit?

2.129 In a support system in the U.S. space program, a single crucial component works only 85% of the time. In order to enhance the reliability of the system, it is decided that 3 components will be installed in parallel such that the system fails only if they all fail. Assume the components act independently and that they are equivalent in the sense that all 3 of them have an 85% success rate. Consider the random variable X as the number of components out of 3 that fail.

(a) Write out a probability function for the random variable X.

(b) What is $E(X)$ (i.e., the mean number of components out of 3 that fail)?

(c) What is $\text{Var}(X)$?

(d) What is the probability that the entire system is successful?

(e) What is the probability that the system fails?

(f) If the desire is to have the system be successful with probability 0.99, are three components sufficient? If not, how many are required?

2.130 It is known through data collection and considerable research that the amount of time in seconds that a certain employee of a company is late for work is a random variable X with density function

$$f(x) = \begin{cases} \frac{3}{(4)(50^3)}(50^2 - x^2), & -50 \le x \le 50, \\ 0, & \text{elsewhere.} \end{cases}$$

In other words, he not only is slightly late at times, but also can be early to work.

(a) Find the expected value of the time in seconds that he is late.

(b) Find $E(X^2)$.

(c) What is the standard deviation of the amount of time he is late?

2.131 A delivery truck travels from point A to point B and back using the same route each day. There are four traffic lights on the route. Let X_1 denote the number of red lights the truck encounters going from A to B and X_2 denote the number encountered on the return trip. Data collected over a long period suggest that the joint probability distribution for (X_1, X_2) is given by

x_1	x_2 0	1	2	3	4
0	0.01	0.01	0.03	0.07	0.01
1	0.03	0.05	0.08	0.03	0.02
2	0.03	0.11	0.15	0.01	0.01
3	0.02	0.07	0.10	0.03	0.01
4	0.01	0.06	0.03	0.01	0.01

(a) Give the marginal density of X_1.

(b) Give the marginal density of X_2.

(c) Give the conditional density distribution of X_1 given $X_2 = 3$.

(d) Give $E(X_1)$.

(e) Give $E(X_2)$.

(f) Give $E(X_1 \mid X_2 = 3)$.

(g) Give the standard deviation of X_1.

2.132 A convenience store has two separate locations where customers can be checked out as they leave. These locations each have two cash registers and two employees who check out customers. Let X be the number of cash registers being used at a particular time for location 1 and Y the number being used at the same time for location 2. The joint probability function is given by

x	y 0	1	2
0	0.12	0.04	0.04
1	0.08	0.19	0.05
2	0.06	0.12	0.30

(a) Give the marginal density of both X and Y as well as the probability distribution of X given $Y = 2$.

(b) Give $E(X)$ and $\text{Var}(X)$.

(c) Give $E(X \mid Y = 2)$ and $\text{Var}(X \mid Y = 2)$.

2.133 As we shall illustrate in Chapter 7, statistical methods associated with linear and nonlinear models are very important. In fact, exponential functions are often used in a wide variety of scientific and engineering problems. Consider a model that is fit to a set of data involving measured values k_1 and k_2 and a certain response Y to the measurements. The model postulated is

$$\hat{Y} = e^{b_0 + b_1 k_1 + b_2 k_2},$$

where $\hat{Y}$ denotes the **estimated value of** Y, k_1 and k_2 are fixed values, and $b_0, b_1,$ and b_2 are **estimates** of constants and hence are random variables. Assume that these random variables are independent and use the approximate formula for the variance of a nonlinear function of more than one variable. Give an expression for $\text{Var}(\hat{Y})$. Assume that the means of $b_0, b_1,$ and b_2 are known and are $\beta_0, \beta_1,$ and β_2, and assume that the variances of $b_0, b_1,$ and b_2 are known and are $\sigma_0^2, \sigma_1^2,$ and σ_2^2.

2.134 Consider Review Exercise 2.113 on page 96. It involved Y, the proportion of impurities in a batch, and the density function is given by

$$f(y) = \begin{cases} 10(1-y)^9, & 0 \le y \le 1, \\ 0, & \text{elsewhere.} \end{cases}$$

(a) Find the expected percentage of impurities.

(b) Find the expected value of the proportion of quality material (i.e., find $E(1 - Y)$).

(c) Find the variance of the random variable $Z = 1 - Y$.

2.135 Project: Let $X =$ number of hours each student in the class slept the night before. Create a discrete variable by using the following arbitrary intervals: $X < 3$, $3 \leq X < 6$, $6 \leq X < 9$, and $X \geq 9$.

(a) Estimate the probability distribution for X.

(b) Calculate the estimated mean and variance for X.

2.8 Potential Misconceptions and Hazards; Relationship to Material in Other Chapters

The material in this chapter is extremely fundamental in nature. We focused on general characteristics of a probability distribution and defined important quantities or *parameters* that characterize the general nature of the system. The **mean** of a distribution reflects *central tendency*, and the **variance** or **standard deviation** reflects *variability* in the system. In addition, covariance reflects the tendency for two random variables to "move together" in a system. These important parameters will remain fundamental to all that follows in this text.

The reader should understand that the distribution type is often dictated by the scientific scenario. However, the parameter values often need to be estimated from scientific data. For example, in the case of Review Exercise 2.124, the manufacturer of the compressor may know (material that will be presented in Chapter 3) from experience and knowledge of the type of compressor that the nature of the distribution is as indicated in the exercise. The mean μ would not be known but **estimated** from experimentation on the machine. Though the parameter value of 900 is given as known here, it will not be known in real-life situations without the use of experimental data.

Chapter 3

Some Probability Distributions

3.1 Introduction and Motivation

No matter whether a discrete probability distribution is represented graphically by a histogram, in tabular form, or by means of a formula, the behavior of a random variable is described. Often, the observations generated by different statistical experiments have the same general type of behavior. Consequently, discrete random variables associated with these experiments can be described by essentially the same probability distribution and therefore can be represented by a single formula. In fact, one needs only a handful of important probability distributions to describe many of the discrete random variables encountered in practice.

Such a handful of distributions describe several real-life random phenomena. For instance, in a study involving testing the effectiveness of a new drug, the number of cured patients among all the patients who use the drug approximately follows a binomial distribution (Section 3.2). In an industrial example, when a sample of items selected from a batch of production is tested, the number of defective items in the sample usually can be modeled as a hypergeometric random variable (Section 3.3). In a statistical quality control problem, the experimenter will signal a shift of the process mean when observational data exceed certain limits. The number of samples required to produce a false alarm follows a geometric distribution, which is a special case of the negative binomial distribution (Section 3.4). On the other hand, the number of white cells from a fixed amount of an individual's blood sample is usually random and may be described by a Poisson distribution (Section 3.5). In this chapter, we present these commonly used distributions with various examples.

3.2 Binomial and Multinomial Distributions

An experiment often consists of repeated trials, each with two possible outcomes that may be labeled **success** or **failure**. One obvious application deals with the testing of items as they come off an assembly line, where each trial may indicate a defective or a nondefective item. We may choose to define either outcome as a success. The process is referred to as a **Bernoulli process**. Each trial is called a

Bernoulli trial. Observe that, for example, if one is drawing cards from a deck, the probabilities for repeated trials change if the cards are not replaced. That is, the probability of selecting a heart on the first draw is 1/4, but on the second draw it is a conditional probability having a value of 13/51 or 12/51, depending on whether a heart appeared on the first draw: this, then, would no longer be considered a set of Bernoulli trials.

The Bernoulli Process

Strictly speaking, the Bernoulli process must possess the following properties:

1. The experiment consists of repeated trials.

2. Each trial results in an outcome that may be classified as a success or a failure.

3. The probability of success, denoted by p, remains constant from trial to trial.

4. The repeated trials are independent.

Consider the set of Bernoulli trials where three items are selected at random from a manufacturing process, inspected, and classified as defective or nondefective. A defective item is designated a success. The number of successes is a random variable X assuming integral values from 0 through 3. The eight possible outcomes and the corresponding values of X are

Outcome	NNN	NDN	NND	DNN	NDD	DND	DDN	DDD
x	0	1	1	1	2	2	2	3

Since the items are selected independently and we assume that the process produces 25% defectives, we have

$$P(NDN) = P(N)P(D)P(N) = \left(\frac{3}{4}\right)\left(\frac{1}{4}\right)\left(\frac{3}{4}\right) = \frac{9}{64}.$$

Similar calculations yield the probabilities for the other possible outcomes. The probability distribution of X is therefore

x	0	1	2	3
$f(x)$	$\frac{27}{64}$	$\frac{27}{64}$	$\frac{9}{64}$	$\frac{1}{64}$

Binomial Distribution

The number X of successes in n Bernoulli trials is called a **binomial random variable**. The probability distribution of this discrete random variable is called the **binomial distribution**, and its values will be denoted by $b(x; n, p)$ since they depend on the number of trials and the probability of a success on a given trial. Thus, for the probability distribution of X, the number of defectives is

$$P(X = 2) = f(2) = b\left(2; 3, \frac{1}{4}\right) = \frac{9}{64}.$$

Let us now generalize the above illustration to yield a formula for $b(x; n, p)$. That is, we wish to find a formula that gives the probability of x successes in n trials for a binomial experiment. First, consider the probability of x successes and $n - x$ failures in a specified order. Since the trials are independent, we can multiply all the probabilities corresponding to the different outcomes. Each success occurs with probability p and each failure with probability $q = 1 - p$. Therefore, the probability for the specified order is $p^x q^{n-x}$. We must now determine the total number of sample points in the experiment that have x successes and $n - x$ failures. This number is equal to the number of partitions of n outcomes into two groups with x in one group and $n - x$ in the other and is written $\binom{n}{x}$, as introduced in Section 1.5. Because these partitions are mutually exclusive, we add the probabilities of all the different partitions to obtain the general formula, or simply multiply $p^x q^{n-x}$ by $\binom{n}{x}$.

Binomial Distribution   A Bernoulli trial can result in a success with probability p and a failure with probability $q = 1 - p$. Then the probability distribution of the binomial random variable X, the number of successes in n independent trials, is

$$b(x; n, p) = \binom{n}{x} p^x q^{n-x}, \quad x = 0, 1, 2, \ldots, n.$$

Note that when $n = 3$ and $p = 1/4$, the probability distribution of X, the number of defectives, may be written as

$$b\left(x; 3, \frac{1}{4}\right) = \binom{3}{x} \left(\frac{1}{4}\right)^x \left(\frac{3}{4}\right)^{3-x}, \quad x = 0, 1, 2, 3,$$

rather than in the tabular form on page 102.

Example 3.1:   The probability that a certain kind of component will survive a shock test is 3/4. Find the probability that exactly 2 of the next 4 components tested survive.

Solution: Assuming that the tests are independent and $p = 3/4$ for each of the 4 tests, we obtain

$$b\left(2; 4, \frac{3}{4}\right) = \binom{4}{2} \left(\frac{3}{4}\right)^2 \left(\frac{1}{4}\right)^2 = \left(\frac{4!}{2! \, 2!}\right) \left(\frac{3^2}{4^4}\right) = \frac{27}{128}.$$

Where Does the Name *Binomial* Come From?

The binomial distribution derives its name from the fact that the $n + 1$ terms in the binomial expansion of $(q + p)^n$ correspond to the various values of $b(x; n, p)$ for $x = 0, 1, 2, \ldots, n$. That is,

$$(q + p)^n = \binom{n}{0} q^n + \binom{n}{1} pq^{n-1} + \binom{n}{2} p^2 q^{n-2} + \cdots + \binom{n}{n} p^n$$

$$= b(0; n, p) + b(1; n, p) + b(2; n, p) + \cdots + b(n; n, p).$$

Since $p + q = 1$, we see that

$$\sum_{x=0}^{n} b(x; n, p) = 1,$$

a condition that must hold for any probability distribution.

Frequently, we are interested in problems where it is necessary to find $P(X < r)$ or $P(a \leq X \leq b)$. Binomial sums

$$B(r; n, p) = \sum_{x=0}^{r} b(x; n, p)$$

are given in Table A.1 of the Appendix for $n = 1, 2, \ldots, 20$ for selected values of p from 0.1 to 0.9. We illustrate the use of Table A.1 with the following example.

Example 3.2: The probability that a patient recovers from a rare blood disease is 0.4. If 15 people are known to have contracted this disease, what is the probability that (a) at least 10 survive, (b) from 3 to 8 survive, and (c) exactly 5 survive?

Solution: Let X be the number of people who survive.

(a) $\quad P(X \geq 10) = 1 - P(X < 10) = 1 - \sum_{x=0}^{9} b(x; 15, 0.4) = 1 - 0.9662$

$\qquad\qquad = 0.0338$

(b) $\quad P(3 \leq X \leq 8) = \sum_{x=3}^{8} b(x; 15, 0.4) = \sum_{x=0}^{8} b(x; 15, 0.4) - \sum_{x=0}^{2} b(x; 15, 0.4)$

$\qquad\qquad = 0.9050 - 0.0271 = 0.8779$

(c) $\quad P(X = 5) = b(5; 15, 0.4) = \sum_{x=0}^{5} b(x; 15, 0.4) - \sum_{x=0}^{4} b(x; 15, 0.4)$

$\qquad\qquad = 0.4032 - 0.2173 = 0.1859$

Example 3.3: A large chain retailer purchases a certain kind of electronic device from a manufacturer. The manufacturer indicates that the defective rate of the device is 3%.

(a) The inspector randomly picks 20 items from a shipment. What is the probability that there will be at least one defective item among these 20?

(b) Suppose that the retailer receives 10 shipments in a month and the inspector randomly tests 20 devices per shipment. What is the probability that there will be exactly 3 shipments each containing at least one defective device among the 20 that are selected and tested from the shipment?

Solution: (a) Denote by X the number of defective devices among the 20. Then X follows a $b(x; 20, 0.03)$ distribution. Hence,

$$P(X \geq 1) = 1 - P(X = 0) = 1 - b(0; 20, 0.03)$$
$$= 1 - (0.03)^0 (1 - 0.03)^{20-0} = 0.4562.$$

(b) In this case, each shipment can either contain at least one defective item or not. Hence, testing of each shipment can be viewed as a Bernoulli trial with $p = 0.4562$ from part (a). Assuming independence from shipment to shipment and denoting by Y the number of shipments containing at least one defective item, Y follows another binomial distribution $b(y; 10, 0.4562)$. Therefore,

$$P(Y = 3) = \binom{10}{3} 0.4562^3 (1 - 0.4562)^7 = 0.1602.$$

Areas of Application

From Examples 3.1 through 3.3, it should be clear that the binomial distribution finds applications in many scientific fields. An industrial engineer is keenly interested in the "proportion defective" in an industrial process. Often, quality control measures and sampling schemes for processes are based on the binomial distribution. This distribution applies to any industrial situation where an outcome of a process is dichotomous and the results of the process are independent, with the probability of success being constant from trial to trial. The binomial distribution is also used extensively for medical and military applications. In both fields, a success or failure result is important. For example, "cure" or "no cure" is important in pharmaceutical work, and "hit" or "miss" is often the interpretation of the result of firing a guided missile.

Since the probability distribution of any binomial random variable depends only on the values assumed by the parameters n, p, and q, it would seem reasonable to assume that the mean and variance of a binomial random variable also depend on the values assumed by these parameters. Indeed, this is true, and in the proof of Theorem 3.1 we derive general formulas that can be used to compute the mean and variance of any binomial random variable as functions of n, p, and q.

Theorem 3.1: The mean and variance of the binomial distribution $b(x; n, p)$ are
$$\mu = np \text{ and } \sigma^2 = npq.$$

Proof: Let the outcome on the jth trial be represented by a Bernoulli random variable I_j, which assumes the values 0 and 1 with probabilities q and p, respectively. Therefore, in a binomial experiment the number of successes can be written as the sum of the n independent indicator variables. Hence,

$$X = I_1 + I_2 + \cdots + I_n.$$

The mean of any I_j is $E(I_j) = (0)(q) + (1)(p) = p$. Therefore, using Corollary 2.4 on page 91, the mean of the binomial distribution is

$$\mu = E(X) = E(I_1) + E(I_2) + \cdots + E(I_n) = \underbrace{p + p + \cdots + p}_{n \text{ terms}} = np.$$

The variance of any I_j is $\sigma_{I_j}^2 = E(I_j^2) - p^2 = (0)^2(q) + (1)^2(p) - p^2 = p(1-p) = pq$. Extending Corollary 2.11 to the case of n independent Bernoulli variables gives the variance of the binomial distribution as

$$\sigma_X^2 = \sigma_{I_1}^2 + \sigma_{I_2}^2 + \cdots + \sigma_{I_n}^2 = \underbrace{pq + pq + \cdots + pq}_{n \text{ terms}} = npq.$$

Example 3.4: It is conjectured that an impurity exists in 30% of all drinking wells in a certain rural community. In order to gain some insight into the true extent of the problem, it is determined that some testing is necessary. It is too expensive to test all of the wells in the area, so 10 are randomly selected for testing.

(a) Using the binomial distribution, what is the probability that exactly 3 wells have the impurity, assuming that the conjecture is correct?

(b) What is the probability that more than 3 wells are impure?

Solution: (a) We require

$$b(3; 10, 0.3) = \sum_{x=0}^{3} b(x; 10, 0.3) - \sum_{x=0}^{2} b(x; 10, 0.3) = 0.6496 - 0.3828 = 0.2668.$$

(b) In this case, $P(X > 3) = 1 - 0.6496 = 0.3504$.

There are solutions in which the computation of binomial probabilities may allow us to draw a scientific inference about a population after data are collected. An illustration is given in the next example.

Example 3.5: Consider the situation of Example 3.4. The notion that 30% of the wells are impure is merely a conjecture put forth by the area water board. Suppose 10 wells are randomly selected and 6 are found to contain the impurity. What does this imply about the conjecture? Use a probability statement.

Solution: We must first ask: "If the conjecture is correct, is it likely that we would find 6 or more impure wells?"

$$P(X \geq 6) = \sum_{x=0}^{10} b(x; 10, 0.3) - \sum_{x=0}^{5} b(x; 10, 0.3) = 1 - 0.9527 = 0.0473.$$

As a result, it is very unlikely (4.7% chance) that 6 or more wells would be found impure if only 30% of all are impure. This casts considerable doubt on the conjecture and suggests that the impurity problem is much more severe.

As the reader should realize by now, in many applications there are more than two possible outcomes. To borrow an example from the field of genetics, the color of guinea pigs produced as offspring may be red, black, or white. Often the "defective" or "not defective" dichotomy is truly an oversimplification in engineering situations. Indeed, there are often more than two categories that characterize items or parts coming off an assembly line.

Multinomial Experiments and the Multinomial Distribution

The binomial experiment becomes a **multinomial experiment** if we let each trial have more than two possible outcomes. The classification of a manufactured product as being light, heavy, or acceptable and the recording of accidents at a certain intersection according to the day of the week constitute multinomial experiments. The drawing of a card from a deck *with replacement* is also a multinomial experiment if the 4 suits are the outcomes of interest.

In general, if a given trial can result in any one of k possible outcomes $E_1, E_2, \ldots,$ E_k with probabilities $p_1, p_2, \ldots, p_k$, then the **multinomial distribution** will give the probability that E_1 occurs x_1 times, E_2 occurs x_2 times, ..., and E_k occurs x_k times in n independent trials, where

$$x_1 + x_2 + \cdots + x_k = n.$$

We shall denote this joint probability distribution by

$$f(x_1, x_2, \ldots, x_k; p_1, p_2, \ldots, p_k, n).$$

Clearly, $p_1 + p_2 + \cdots + p_k = 1$, since the result of each trial must be one of the k possible outcomes.

To derive the general formula, we proceed as in the binomial case. Since the trials are independent, any specified order yielding x_1 outcomes for E_1, x_2 for $E_2, \ldots, x_k$ for E_k will occur with probability $p_1^{x_1} p_2^{x_2} \cdots p_k^{x_k}$. The total number of orders yielding similar outcomes for the n trials is equal to the number of partitions of n items into k groups with x_1 in the first group, x_2 in the second group, ..., and x_k in the kth group. This can be done in

$$\binom{n}{x_1, x_2, \ldots, x_k} = \frac{n!}{x_1! \, x_2! \cdots x_k!}$$

ways. Since all the partitions are mutually exclusive and occur with equal probability, we obtain the multinomial distribution by multiplying the probability for a specified order by the total number of partitions.

Multinomial Distribution   If a given trial can result in the k outcomes $E_1, E_2, \ldots, E_k$ with probabilities $p_1, p_2, \ldots, p_k$, then the probability distribution of the random variables $X_1, X_2, \ldots, X_k$, representing the number of occurrences for $E_1, E_2, \ldots, E_k$ in n independent trials, is

$$f(x_1, x_2, \ldots, x_k; p_1, p_2, \ldots, p_k, n) = \binom{n}{x_1, x_2, \ldots, x_k} p_1^{x_1} \, p_2^{x_2} \cdots p_k^{x_k},$$

with

$$\sum_{i=1}^{k} x_i = n \text{ and } \sum_{i=1}^{k} p_i = 1.$$

The multinomial distribution derives its name from the fact that the terms of the multinomial expansion of $(p_1 + p_2 + \cdots + p_k)^n$ correspond to all the possible values of $f(x_1, x_2, \ldots, x_k; p_1, p_2, \ldots, p_k, n)$.

Example 3.6: The complexity of arrivals and departures of planes at an airport is such that computer simulation is often used to model the "ideal" conditions. For a certain airport with three runways, it is known that in the ideal setting the following are the probabilities that the individual runways are accessed by a randomly arriving commercial jet:

$$\begin{array}{ll} \text{Runway 1:} & p_1 = 2/9, \\ \text{Runway 2:} & p_2 = 1/6, \\ \text{Runway 3:} & p_3 = 11/18. \end{array}$$

What is the probability that 6 randomly arriving airplanes are distributed in the following fashion?

Runway 1:   2 airplanes,
Runway 2:   1 airplane,
Runway 3:   3 airplanes

Solution: Using the multinomial distribution, we have

$$f\left(2,1,3;\frac{2}{9},\frac{1}{6},\frac{11}{18},6\right) = \binom{6}{2,1,3}\left(\frac{2}{9}\right)^2\left(\frac{1}{6}\right)^1\left(\frac{11}{18}\right)^3$$

$$= \frac{6!}{2!\,1!\,3!}\cdot\frac{2^2}{9^2}\cdot\frac{1}{6}\cdot\frac{11^3}{18^3} = 0.1127.$$

Exercises

3.1 A random variable X that assumes the values $x_1, x_2, \ldots, x_k$ is called a discrete uniform random variable if its probability mass function is $f(x) = \frac{1}{k}$ for all of $x_1, x_2, \ldots, x_k$ and 0 otherwise. Find the mean and variance of X.

3.2 In a certain city district, the need for money to buy drugs is stated as the reason for 75% of all thefts. Find the probability that among the next 5 theft cases reported in this district,

(a) exactly 2 resulted from the need for money to buy drugs;

(b) at most 3 resulted from the need for money to buy drugs.

3.3 An employee is selected from a staff of 10 to supervise a certain project by means of a tag selected at random from a box containing 10 tags numbered from 1 to 10. Find the formula for the probability distribution of X representing the number on the tag that is drawn. What is the probability that the number drawn is less than 4?

3.4 According to *Chemical Engineering Progress* (November 1990), approximately 30% of all pipework failures in chemical plants are caused by operator error.

(a) What is the probability that out of the next 20 pipework failures at least 10 are due to operator error?

(b) What is the probability that no more than 4 out of 20 such failures are due to operator error?

(c) Suppose that, for a particular plant, out of the random sample of 20 such failures, exactly 5 are due to operator error. Do you feel that the 30% figure stated above applies to this plant? Comment.

3.5 One prominent physician claims that 70% of those with lung cancer are chain smokers. If his assertion is correct,

(a) find the probability that of 10 such patients recently admitted to a hospital, fewer than half are

chain smokers;

(b) find the probability that of 20 such patients recently admitted to a hospital, fewer than half are chain smokers.

3.6 According to a study published by a group of University of Massachusetts sociologists, approximately 60% of the Valium users in the state of Massachusetts first took Valium for psychological problems. Find the probability that among the next 8 users from this state who are interviewed,

(a) exactly 3 began taking Valium for psychological problems;

(b) at least 5 began taking Valium for problems that were not psychological.

3.7 In testing a certain kind of truck tire over rugged terrain, it is found that 25% of the trucks fail to complete the test run without a blowout. Of the next 15 trucks tested, find the probability that

(a) from 3 to 6 have blowouts;

(b) fewer than 4 have blowouts;

(c) more than 5 have blowouts.

3.8 A traffic control engineer reports that 75% of the vehicles passing through a checkpoint are from within the state. What is the probability that fewer than 4 of the next 9 vehicles are from out of state?

3.9 The probability that a patient recovers from a delicate heart operation is 0.9. What is the probability that exactly 5 of the next 7 patients having this operation survive?

3.10 The percentage of wins for the Chicago Bulls basketball team going into the playoffs for the 1996–97 season was 87.7. Round the 87.7 to 90 in order to use Table A.1.

(a) What was the probability that the Bulls would sweep (4-0) the initial best-of-7 playoff series?

(b) What was the probability that the Bulls would win the initial best-of-7 playoff series?

(c) What very important assumption is made in answering parts (a) and (b)?

3.11 It is known that 60% of mice inoculated with a serum are protected from a certain disease. If 5 mice are inoculated, find the probability that

(a) none contracts the disease;

(b) fewer than 2 contract the disease;

(c) more than 3 contract the disease.

3.12 Suppose that airplane engines operate independently and fail with probability equal to 0.4. Assuming that a plane makes a safe flight if at least one-half of its engines run, determine whether a 4-engine plane or a 2-engine plane has the higher probability for a successful flight.

3.13 As a student drives to school, he encounters a traffic signal. This traffic signal stays green for 35 seconds, yellow for 5 seconds, and red for 60 seconds. Assume that the student goes to school each weekday between 8:00 and 8:30 A.M. Let X_1 be the number of times he encounters a green light, X_2 be the number of times he encounters a yellow light, and X_3 be the number of times he encounters a red light. Find the joint distribution of X_1, X_2, and X_3.

3.14 (a) In Exercise 3.7, how many of the 15 trucks would you expect to have blowouts?

(b) What is the variance of the number of blowouts experienced by the 15 trucks? What does that mean?

3.15 According to *USA Today* (March 18, 1997), of 4 million workers in the general workforce, 5.8% tested positive for drugs. Of those testing positive, 22.5% were cocaine users and 54.4% marijuana users.

(a) What is the probability that of 10 workers testing positive, 2 are cocaine users, 5 are marijuana users, and 3 are users of other drugs?

(b) What is the probability that of 10 workers testing positive, all are marijuana users?

(c) What is the probability that of 10 workers testing positive, none is a cocaine user?

3.16 A safety engineer claims that only 40% of all workers wear safety helmets when they eat lunch at the workplace. Assuming that this claim is right, find the probability that 4 of 6 workers randomly chosen will be wearing their helmets while having lunch at the workplace.

3.17 Suppose that for a very large shipment of integrated-circuit chips, the probability of failure for any one chip is 0.10. Assuming that the assumptions underlying the binomial distributions are met, find the probability that at most 3 chips fail in a random sample of 20.

3.18 Assuming that 6 in 10 automobile accidents are due mainly to a speed violation, find the probability that among 8 automobile accidents, 6 will be due mainly to a speed violation

(a) by using the formula for the binomial distribution;

(b) by using Table A.1.

3.19 If the probability that a fluorescent light has a useful life of at least 800 hours is 0.9, find the probabilities that among 20 such lights

(a) exactly 18 will have a useful life of at least 800 hours;

(b) at least 15 will have a useful life of at least 800 hours;

(c) at least 2 will *not* have a useful life of at least 800 hours.

3.20 A manufacturer knows that on average 20% of the electric toasters produced require repairs within 1 year after they are sold. When 20 toasters are randomly selected, find appropriate numbers x and y such that

(a) the probability that at least x of them will require repairs is less than 0.5;

(b) the probability that at least y of them will *not* require repairs is greater than 0.8.

3.3 Hypergeometric Distribution

The simplest way to view the distinction between the binomial distribution of Section 3.2 and the hypergeometric distribution is to note the way the sampling is done. The types of applications for the hypergeometric are very similar to those for the binomial distribution. We are interested in computing probabilities for the number of observations that fall into a particular category. But in the case of the binomial distribution, independence among trials is required. As a result, if that distribution is applied to, say, sampling from a lot of items (deck of cards, batch of production items), the sampling must be done **with replacement** of each item after it is observed. On the other hand, the hypergeometric distribution does not require independence and is based on sampling done **without replacement**.

Applications for the hypergeometric distribution are found in many areas, with heavy use in acceptance sampling, electronic testing, and quality assurance. Obviously, in many of these fields, testing is done at the expense of the item being tested. That is, the item is destroyed and hence cannot be replaced in the sample. Thus, sampling without replacement is necessary. A simple example with playing cards will serve as our first illustration.

If we wish to find the probability of observing 3 red cards in 5 draws from an ordinary deck of 52 playing cards, the binomial distribution of Section 3.2 does not apply unless each card is replaced and the deck reshuffled before the next draw is made. To solve the problem of sampling without replacement, let us restate the problem. If 5 cards are drawn at random, we are interested in the probability of selecting 3 red cards from the 26 available in the deck and 2 black cards from the 26 available in the deck. There are $\binom{26}{3}$ ways of selecting 3 red cards, and for each of these ways we can choose 2 black cards in $\binom{26}{2}$ ways. Therefore, the total number of ways to select 3 red and 2 black cards in 5 draws is the product $\binom{26}{3}\binom{26}{2}$. The total number of ways to select any 5 cards from the 52 that are available is $\binom{52}{5}$. Hence, the probability of selecting 5 cards without replacement of which 3 are red and 2 are black is given by

$$\frac{\binom{26}{3}\binom{26}{2}}{\binom{52}{5}} = \frac{(26!/3!\,23!)(26!/2!\,24!)}{52!/5!\,47!} = 0.3251.$$

In general, we are interested in the probability of selecting x successes from the k items labeled successes and $n - x$ failures from the $N - k$ items labeled failures when a random sample of size n is selected from N items. This is known as a **hypergeometric experiment**, that is, one that possesses the following two properties:

1. A random sample of size n is selected without replacement from N items.

2. Of the N items, k may be classified as successes and $N - k$ are classified as failures.

The number X of successes of a hypergeometric experiment is called a **hypergeometric random variable**. Accordingly, the probability distribution of the hypergeometric variable is called the **hypergeometric distribution**, and its values are denoted by $h(x; N, n, k)$, since they depend on the number of successes k in the set N from which we select n items.

Hypergeometric Distribution in Acceptance Sampling

Like the binomial distribution, the hypergeometric distribution finds applications in acceptance sampling, where lots of materials or parts are sampled in order to determine whether or not the entire lot is accepted.

Example 3.7: A particular part that is used as an injection device is sold in lots of 10. The producer deems a lot acceptable if no more than one defective is in the lot. A sampling plan involves random sampling and testing 3 of the parts out of 10. If none of the 3 is defective, the lot is accepted. Comment on the utility of this plan.

Solution: Let us assume that the lot is truly **unacceptable** (i.e., that 2 out of 10 parts are defective). The probability that the sampling plan finds the lot acceptable is

$$P(X = 0) = \frac{\binom{2}{0}\binom{8}{3}}{\binom{10}{3}} = 0.467.$$

Thus, if the lot is truly unacceptable, with 2 defective parts, this sampling plan will allow acceptance roughly 47% of the time. As a result, this plan should be considered faulty.

Let us now generalize in order to find a formula for $h(x; N, n, k)$. The total number of samples of size n chosen from N items is $\binom{N}{n}$. These samples are assumed to be equally likely. There are $\binom{k}{x}$ ways of selecting x successes from the k that are available, and for each of these ways we can choose the $n - x$ failures in $\binom{N-k}{n-x}$ ways. Thus, the total number of favorable samples among the $\binom{N}{n}$ possible samples is given by $\binom{k}{x}\binom{N-k}{n-x}$. Hence, we have the following definition.

Hypergeometric Distribution   The probability distribution of the hypergeometric random variable X, the number of successes in a random sample of size n selected from N items of which k are labeled **success** and $N - k$ labeled **failure**, is

$$h(x; N, n, k) = \frac{\binom{k}{x}\binom{N-k}{n-x}}{\binom{N}{n}}, \quad \max\{0, n - (N - k)\} \le x \le \min\{n, k\}.$$

The range of x can be determined by the three binomial coefficients in the definition, where x and $n - x$ are no more than k and $N - k$, respectively, and both of them cannot be less than 0. Usually, when both k (the number of successes) and $N - k$ (the number of failures) are larger than the sample size n, the range of a hypergeometric random variable will be $x = 0, 1, \ldots, n$.

Example 3.8: Lots of 40 components each are deemed unacceptable if they contain 3 or more defectives. The procedure for sampling a lot is to select 5 components at random and to reject the lot if a defective is found. What is the probability that exactly 1 defective is found in the sample if there are 3 defectives in the entire lot?

Solution: Using the hypergeometric distribution with $n = 5$, $N = 40$, $k = 3$, and $x = 1$, we find the probability of obtaining 1 defective to be

$$h(1; 40, 5, 3) = \frac{\binom{3}{1}\binom{37}{4}}{\binom{40}{5}} = 0.3011.$$

Once again, this plan is not desirable since it detects a bad lot (3 defectives) only about 30% of the time.

Theorem 3.2: The mean and variance of the hypergeometric distribution $h(x; N, n, k)$ are

$$\mu = \frac{nk}{N} \text{ and } \sigma^2 = \frac{N-n}{N-1} \cdot n \cdot \frac{k}{N}\left(1 - \frac{k}{N}\right).$$

The proof for the mean is shown in Appendix A.12.

Example 3.9: Let us now reinvestigate Example 2.4 on page 51. The purpose of this example was to illustrate the notion of a random variable and the corresponding sample space. In the example, we have a lot of 100 items of which 12 are defective. What is the probability that in a sample of 10, 3 are defective?

Solution: Using the hypergeometric probability function, we have

$$h(3; 100, 10, 12) = \frac{\binom{12}{3}\binom{88}{7}}{\binom{100}{10}} = 0.08.$$

Example 3.10: Find the mean and variance of the random variable of Example 3.8.

Solution: Since Example 3.8 was a hypergeometric experiment with $N = 40$, $n = 5$, and $k = 3$, by Theorem 3.2, we have

$$\mu = \frac{(5)(3)}{40} = \frac{3}{8} = 0.375,$$

and

$$\sigma^2 = \left(\frac{40-5}{39}\right)(5)\left(\frac{3}{40}\right)\left(1 - \frac{3}{40}\right) = 0.3113.$$

Taking the square root of 0.3113, we find that $\sigma = 0.558$.

Relationship to the Binomial Distribution

In this chapter, we discuss several important discrete distributions that have wide applicability. Many of these distributions relate nicely to each other. The beginning student should gain a clear understanding of these relationships. There is an interesting relationship between the hypergeometric and the binomial distribution. As one might expect, if n is small compared to N, the nature of the N items changes very little in each draw. So a binomial distribution can be used to approximate the hypergeometric distribution when n is small compared to N. In fact, as a rule of thumb, the approximation is good when $n/N \leq 0.05$.

Thus, the quantity k/N plays the role of the binomial parameter p. As a result, the binomial distribution may be viewed as a large-population version of the hypergeometric distribution. The mean and variance then come from the formulas

$$\mu = np = \frac{nk}{N} \text{ and } \sigma^2 = npq = n \cdot \frac{k}{N}\left(1 - \frac{k}{N}\right).$$

Comparing these formulas with those of Theorem 3.2, we see that the mean is the same but the variance differs by a correction factor of $(N-n)/(N-1)$, which is negligible when n is small relative to N.

Example 3.11: A manufacturer of automobile tires reports that among a shipment of 5000 sent to a local distributor, 1000 are slightly blemished. If one purchases 10 of these tires at random from the distributor, what is the probability that exactly 3 are blemished?

Solution: Since $N = 5000$ is large relative to the sample size $n = 10$, we shall approximate the desired probability by using the binomial distribution. The probability of obtaining a blemished tire is 0.2. Therefore, the probability of obtaining exactly 3 blemished tires is

$$h(3; 5000, 10, 1000) \approx b(3; 10, 0.2) = 0.8791 - 0.6778 = 0.2013.$$

The exact probability is $h(3; 5000, 10, 1000) = 0.2015$.

Exercises

3.21 To avoid detection at customs, a traveler places 6 narcotic tablets in a bottle containing 9 vitamin tablets that are similar in appearance. If the customs official selects 3 of the tablets at random for analysis, what is the probability that the traveler will be arrested for illegal possession of narcotics?

3.22 From a lot of 10 missiles, 4 are selected at random and fired. If the lot contains 3 defective missiles that will not fire, what is the probability that

(a) all 4 will fire?

(b) at most 2 will not fire?

3.23 A company is interested in evaluating its current inspection procedure for shipments of 50 identical items. The procedure is to take a sample of 5 and pass the shipment if no more than 2 are found to be defective. What proportion of shipments with 20% defectives will be accepted?

3.24 What is the probability that a waitress will refuse to serve alcoholic beverages to only 2 minors if she randomly checks the IDs of 5 among 9 students, 4 of whom are minors?

3.25 Among 150 IRS employees in a large city, only 30 are women. If 10 of the employees are chosen at random to provide free tax assistance for the residents of this city, use the binomial approximation to the hypergeometric distribution to find the probability that at least 3 women are selected.

3.26 A manufacturing company uses an acceptance scheme on items from a production line before they are shipped. The plan is a two-stage one. Boxes of 25 items are readied for shipment, and a sample of 3 items is tested for defectives. If any defectives are found, the entire box is sent back for 100% screening. If no defec-

tives are found, the box is shipped.

(a) What is the probability that a box containing 3 defectives will be shipped?

(b) What is the probability that a box containing only 1 defective will be sent back for screening?

3.27 Suppose that the manufacturing company of Exercise 3.26 decides to change its acceptance scheme. Under the new scheme, an inspector takes 1 item at random, inspects it, and then replaces it in the box; a second inspector does likewise. Finally, a third inspector goes through the same procedure. The box is not shipped if any of the three inspectors find a defective. Answer the questions in Exercise 3.26 for this new plan.

3.28 It is estimated that 4000 of the 10,000 voting residents of a town are against a new sales tax. If 15 eligible voters are selected at random and asked their opinion, what is the probability that at most 7 favor the new tax?

3.29 Biologists doing studies in a particular environment often tag and release animals in order to estimate the size of a population or the prevalence of certain features in the population. Ten animals of a certain population thought to be extinct (or near extinction) are caught, tagged, and released in a certain region. After a period of time, a random sample of 15 of this type of animal is selected in the region. What is the probability that 5 of those selected are tagged if there are 25 animals of this type in the region?

3.30 Find the probability of being dealt a bridge hand of 13 cards containing 5 spades, 2 hearts, 3 diamonds, and 3 clubs.

3.31 A government task force suspects that some

manufacturing companies are in violation of federal pollution regulations with regard to dumping a certain type of product. Twenty firms are under suspicion but not all can be inspected. Suppose that 3 of the firms are in violation.

(a) What is the probability that inspection of 5 firms will find no violations?

(b) What is the probability that the plan above will find two violations?

3.32 A large company has an inspection system for the batches of small compressors purchased from vendors. A batch typically contains 15 compressors. In the inspection system, a random sample of 5 is selected and all are tested. Suppose there are 2 faulty compressors in the batch of 15.

(a) What is the probability that for a given sample there will be 1 faulty compressor?

(b) What is the probability that inspection will discover both faulty compressors?

3.33 Every hour, 10,000 cans of soda are filled by a machine; of these, 300 cans are underfilled. Each hour, a sample of 30 cans is randomly selected and the number of ounces of soda per can is checked. Denote by X the number of cans selected that are underfilled. Find the probability that at least 1 underfilled can will be among those sampled.

3.4   Negative Binomial and Geometric Distributions

Let us consider an experiment where the properties are the same as those listed for a binomial experiment, with the exception that the trials will be repeated until a *fixed* number of successes occur. Therefore, instead of the probability of x successes in n trials, where n is fixed, we are now interested in the probability that the kth success occurs on the xth trial. Experiments of this kind are called **negative binomial experiments**.

As an illustration, consider the use of a drug that is known to be effective in 60% of the cases where it is used. The drug will be considered a success if it is effective in bringing some degree of relief to the patient. We are interested in finding the probability that the fifth patient to experience relief is the seventh patient to receive the drug during a given week. Designating a success by S and a failure by F, a possible order of achieving the desired result is $SFSSSFS$, which occurs with probability

$$(0.6)(0.4)(0.6)(0.6)(0.6)(0.4)(0.6) = (0.6)^5(0.4)^2.$$

We could list all possible orders by rearranging the F's and S's except for the last outcome, which must be the fifth success. The total number of possible orders is equal to the number of partitions of the first 6 trials into two groups with 2 failures assigned to the one group and 4 successes assigned to the other group. This can be done in $\binom{6}{4} = 15$ mutually exclusive ways. Hence, if X represents the outcome on which the fifth success occurs, then

$$P(X = 7) = \binom{6}{4}(0.6)^5(0.4)^2 = 0.1866.$$

What Is the Negative Binomial Random Variable?

The number X of trials required to produce k successes in a negative binomial experiment is called a **negative binomial random variable**, and its probability distribution is called the **negative binomial distribution**. Since its probabilities depend on the number of successes desired and the probability of a success on a given trial, we shall denote them by $b^*(x; k, p)$. To obtain the general formula for $b^*(x; k, p)$, consider the probability of a success on the xth trial preceded by $k - 1$ successes and $x - k$ failures in some specified order. Since the trials are independent, we can multiply all the probabilities corresponding to each desired outcome. Each success occurs with probability p and each failure with probability $q = 1 - p$. Therefore, the probability for the specified order ending in success is

$$p^{k-1}q^{x-k}p = p^k q^{x-k}.$$

The total number of sample points in the experiment ending in a success, after the occurrence of $k - 1$ successes and $x - k$ failures in any order, is equal to the number of partitions of $x - 1$ trials into two groups with $k - 1$ successes corresponding to one group and $x - k$ failures corresponding to the other group. This number is specified by the term $\binom{x-1}{k-1}$, each mutually exclusive and occurring with equal probability $p^k q^{x-k}$. We obtain the general formula by multiplying $p^k q^{x-k}$ by $\binom{x-1}{k-1}$.

Negative Binomial Distribution

If repeated independent trials can result in a success with probability p and a failure with probability $q = 1 - p$, then the probability distribution of the random variable X, the number of the trial on which the kth success occurs, is

$$b^*(x; k, p) = \binom{x-1}{k-1} p^k q^{x-k}, \quad x = k, k+1, k+2, \ldots .$$

Example 3.12: In an NBA (National Basketball Association) championship series, the team that wins four games out of seven is the winner. Suppose that teams A and B face each other in the championship games and that team A has probability 0.55 of winning a game over team B.

(a) What is the probability that team A will win the series in 6 games?

(b) What is the probability that team A will win the series?

(c) If teams A and B were facing each other in a regional playoff series, which is decided by winning 3 out of 5 games, what is the probability that team A would win the series?

Solution: (a) $b^*(6; 4, 0.55) = \binom{5}{3} 0.55^4 (1 - 0.55)^{6-4} = 0.1853$

(b) P(team A wins the championship series) is

$$b^*(4; 4, 0.55) + b^*(5; 4, 0.55) + b^*(6; 4, 0.55) + b^*(7; 4, 0.55)$$
$$= 0.0915 + 0.1647 + 0.1853 + 0.1668 = 0.6083.$$

(c) P(team A wins the playoff) is

$$b^*(3; 3, 0.55) + b^*(4; 3, 0.55) + b^*(5; 3, 0.55)$$
$$= 0.1664 + 0.2246 + 0.2021 = 0.5931.$$

The negative binomial distribution derives its name from the fact that each term in the expansion of $p^k (1 - p)^{-k}$ corresponds to the values of $b^*(x; k, p)$ for $x = k, k+1, k+2, \ldots$. If we consider the special case of the negative binomial distribution where $k = 1$, we have a probability distribution for the number of trials required for a single success. An example would be the tossing of a coin until a head occurs. We might be interested in the probability that the first head occurs on the fourth toss. The negative binomial distribution reduces to the form

$$b^*(x; 1, p) = pq^{x-1}, \quad x = 1, 2, 3, \dots.$$

Since the successive terms constitute a geometric progression, it is customary to refer to this special case as the **geometric distribution** and denote its values by $g(x; p)$.

Geometric
Distribution

If repeated independent trials can result in a success with probability p and a failure with probability $q = 1 - p$, then the probability distribution of the random variable X, the number of the trial on which the first success occurs, is

$$g(x; p) = pq^{x-1}, \quad x = 1, 2, 3, \dots.$$

Example 3.13: For a certain manufacturing process, it is known that, on the average, 1 in every 100 items is defective. What is the probability that the fifth item inspected is the first defective item found?

Solution: Using the geometric distribution with $x = 5$ and $p = 0.01$, we have

$$g(5; 0.01) = (0.01)(0.99)^4 = 0.0096.$$

Example 3.14: At a busy time, a telephone exchange is very near capacity, so callers have difficulty placing their calls. It may be of interest to know the number of attempts necessary in order to make a connection. Suppose that we let $p = 0.05$ be the probability of a connection during a busy time. We are interested in knowing the probability that 5 attempts are necessary for a successful call.

Solution: Using the geometric distribution with $x = 5$ and $p = 0.05$ yields

$$P(X = x) = g(5; 0.05) = (0.05)(0.95)^4 = 0.041.$$

Quite often, in applications dealing with the geometric distribution, the mean and variance are important. For example, in Example 3.14, the *expected* number of calls necessary to make a connection is quite important. The following theorem states without proof the mean and variance of the geometric distribution.

Theorem 3.3: The mean and variance of a random variable following the geometric distribution are

$$\mu = \frac{1}{p} \text{ and } \sigma^2 = \frac{1-p}{p^2}.$$

Applications of Negative Binomial and Geometric Distributions

Areas of application for the negative binomial and geometric distributions become obvious when one focuses on the examples in this section and the exercises devoted to these distributions at the end of Section 3.5. In the case of the geometric distribution, Example 3.14 depicts a situation where engineers or managers are attempting to determine how inefficient a telephone exchange system is during busy times. Clearly, in this case, trials occurring prior to a success represent a cost. If there is a high probability of several attempts being required prior to making a connection, then plans should be made to redesign the system.

Applications of the negative binomial distribution are similar in nature. Suppose attempts are costly in some sense and are *occurring in sequence*. A high probability of needing a "large" number of attempts to experience a fixed number of successes is not beneficial to the scientist or engineer. Consider the scenarios of Review Exercises 3.109 and 3.110. In Review Exercise 3.110, the oil driller defines a certain level of success from sequentially drilling in different locations for oil. If only 6 attempts have been made at the point where the second success is experienced, the profits appear to dominate substantially the investment incurred by the drilling.

3.5 Poisson Distribution and the Poisson Process

Experiments yielding numerical values of a random variable X, the number of outcomes occurring during a given time interval or in a specified region, are called **Poisson experiments**. The given time interval may be of any length, such as a minute, a day, a week, a month, or even a year. For example, a Poisson experiment can generate observations for the random variable X representing the number of telephone calls received per hour by an office, the number of days school is closed due to snow during the winter, or the number of games postponed due to rain during a baseball season. The specified region could be a line segment, an area, a volume, or perhaps a piece of material. In such instances, X might represent the number of field mice per acre, the number of bacteria in a given culture, or the number of typing errors per page. A Poisson experiment is derived from the **Poisson process** and possesses the following properties.

Properties of the Poisson Process

1. The number of outcomes occurring in one time interval or specified region of space is independent of the number that occur in any other disjoint time interval or region. In this sense we say that the Poisson process has no memory.

2. The probability that a single outcome will occur during a very short time interval or in a small region is proportional to the length of the time interval or the size of the region and does not depend on the number of outcomes occurring outside this time interval or region.

3. The probability that more than one outcome will occur in such a short time interval or fall in such a small region is negligible.

The number X of outcomes occurring during a Poisson experiment is called a **Poisson random variable**, and its probability distribution is called the **Poisson distribution**. The mean number of outcomes is computed from $\mu = \lambda t$, where t is the specific "time," "distance," "area," or "volume" of interest. Since the probabilities depend on λ, the rate of occurrence of outcomes, we shall denote them by $p(x; \lambda t)$. The derivation of the formula for $p(x; \lambda t)$, based on the three properties of a Poisson process listed above, is beyond the scope of this book. The following formula is used for computing Poisson probabilities.

Poisson Distribution	The probability distribution of the Poisson random variable X, representing the number of outcomes occurring in a given time interval or specified region denoted by t, is

$$p(x; \lambda t) = \frac{e^{-\lambda t}(\lambda t)^x}{x!}, \quad x = 0, 1, 2, \ldots,$$

where λ is the average number of outcomes per unit time, distance, area, or volume and $e = 2.71828\ldots$.

Table A.2 contains Poisson probability sums,

$$P(r; \lambda t) = \sum_{x=0}^{r} p(x; \lambda t),$$

for selected values of λt ranging from 0.1 to 18.0. We illustrate the use of this table with the following two examples.

Example 3.15: During a laboratory experiment, the average number of radioactive particles passing through a counter in 1 millisecond is 4. What is the probability that 6 particles enter the counter in a given millisecond?

Solution: Using the Poisson distribution with $x = 6$ and $\lambda t = 4$ and referring to Table A.2, we have

$$p(6; 4) = \frac{e^{-4}4^6}{6!} = \sum_{x=0}^{6} p(x; 4) - \sum_{x=0}^{5} p(x; 4) = 0.8893 - 0.7851 = 0.1042.$$

Example 3.16: Ten is the average number of oil tankers arriving each day at a certain port. The facilities at the port can handle at most 15 tankers per day. What is the probability that on a given day tankers have to be turned away?

Solution: Let X be the number of tankers arriving each day. Then, using Table A.2, we have

$$P(X > 15) = 1 - P(X \le 15) = 1 - \sum_{x=0}^{15} p(x; 10) = 1 - 0.9513 = 0.0487.$$

Like the binomial distribution, the Poisson distribution is used for quality control, quality assurance, and acceptance sampling. In addition, certain important continuous distributions used in reliability theory and queuing theory depend on the Poisson process. Some of these distributions are discussed and developed later in the chapter. The proof of the following theorem concerning the Poisson random variable is given in Appendix A.13.

Theorem 3.4:	Both the mean and the variance of the Poisson distribution $p(x; \lambda t)$ are λt.

Nature of the Poisson Probability Function

Like so many discrete and continuous distributions, the form of the Poisson distribution becomes more and more symmetric, even bell-shaped, as the mean grows large. Figure 3.1 illustrates this, showing plots of the probability function for

$\mu = 0.1$, $\mu = 2$, and $\mu = 5$. Note the nearness to symmetry when μ becomes as large as 5. A similar condition exists for the binomial distribution, as will be illustrated later in the text.

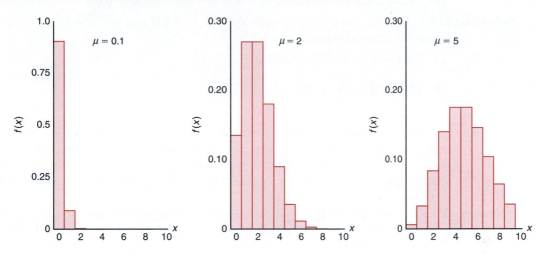

Figure 3.1: Poisson density functions for different means.

Approximation of Binomial Distribution by a Poisson Distribution

It should be evident from the three principles of the Poisson process that the Poisson distribution is related to the binomial distribution. Although the Poisson usually finds applications in space and time problems, as illustrated by Examples 3.15 and 3.16, it can be viewed as a limiting form of the binomial distribution. In the case of the binomial, if n is quite large and p is small, the conditions begin to simulate the *continuous space or time* implications of the Poisson process. The independence among Bernoulli trials in the binomial case is consistent with principle 2 of the Poisson process. Allowing the parameter p to be close to 0 relates to principle 3 of the Poisson process. Indeed, if n is large and p is close to 0, the Poisson distribution can be used, with $\mu = np$, to approximate binomial probabilities. If p is close to 1, we can still use the Poisson distribution to approximate binomial probabilities by interchanging what we have defined to be a success and a failure, thereby changing p to a value close to 0.

Theorem 3.5: Let X be a binomial random variable with probability distribution $b(x; n, p)$. When $n \to \infty$, $p \to 0$, and $np \xrightarrow{n \to \infty} \mu$ remains constant,
$$b(x; n, p) \xrightarrow{n \to \infty} p(x; \mu).$$

Example 3.17: In a certain industrial facility, accidents occur infrequently. It is known that the probability of an accident on any given day is 0.005 and accidents are independent of each other.

(a) What is the probability that in any given period of 400 days there will be an accident on one day?

(b) What is the probability that there are at most three days with an accident?

Solution: Let X be a binomial random variable with $n = 400$ and $p = 0.005$. Thus, $np = 2$. Using the Poisson approximation,

(a) $P(X = 1) = e^{-2}2^1 = 0.271$ and

(b) $P(X \leq 3) = \sum_{x=0}^{3} e^{-2}2^x/x! = 0.857$.

Example 3.18: In a manufacturing process where glass products are made, defects or bubbles occur, occasionally rendering the piece undesirable for marketing. It is known that, on average, 1 in every 1000 of these items produced has one or more bubbles. What is the probability that a random sample of 8000 will yield fewer than 7 items possessing bubbles?

Solution: This is essentially a binomial experiment with $n = 8000$ and $p = 0.001$. Since p is very close to 0 and n is quite large, we shall approximate with the Poisson distribution using

$$\mu = (8000)(0.001) = 8.$$

Hence, if X represents the number of bubbles, we have

$$P(X < 7) = \sum_{x=0}^{6} b(x; 8000, 0.001) \approx p(x; 8) = 0.3134.$$

Exercises

3.34 A scientist inoculates mice, one at a time, with a disease germ until he finds 2 that have contracted the disease. If the probability of contracting the disease is 1/6, what is the probability that 8 mice are required?

3.35 Three people toss a fair coin and the odd one out pays for coffee. If the coins all turn up the same, they are tossed again. Find the probability that fewer than 4 tosses are needed.

3.36 According to a study published by a group of University of Massachusetts sociologists, about two-thirds of the 20 million persons in this country who take Valium are women. Assuming this figure to be a valid estimate, find the probability that on a given day the fifth prescription written by a doctor for Valium is

(a) the first prescribing Valium for a woman;

(b) the third prescribing Valium for a woman.

3.37 An inventory study determines that, on average, demands for a particular item at a warehouse are made 5 times per day. What is the probability that on a given day this item is requested

(a) more than 5 times?

(b) not at all?

3.38 On average, 3 traffic accidents per month occur at a certain intersection. What is the probability that in any given month at this intersection

(a) exactly 5 accidents will occur?

(b) fewer than 3 accidents will occur?

(c) at least 2 accidents will occur?

3.39 The probability that a student pilot passes the written test for a private pilot's license is 0.7. Find the probability that a given student will pass the test

(a) on the third try;

(b) before the fourth try.

3.40 A certain area of the eastern United States is, on average, hit by 6 hurricanes a year. Find the probability that in a given year that area will be hit by

(a) fewer than 4 hurricanes;

(b) anywhere from 6 to 8 hurricanes.

3.41 On average, a textbook author makes two word-processing errors per page on the first draft of her textbook. What is the probability that on the next page she will make

(a) 4 or more errors?

(b) no errors?

3.42 The probability that a student at a local high school fails the screening test for scoliosis (curvature of the spine) is known to be 0.004. Of the next 1875 students at the school who are screened for scoliosis, find the probability that

(a) fewer than 5 fail the test;

(b) 8, 9, or 10 fail the test.

3.43 Suppose that, on average, 1 person in 1000 makes a numerical error in preparing his or her income tax return. If 10,000 returns are selected at random and examined, find the probability that 6, 7, or 8 of them contain an error.

3.44 Find the mean and variance of the random variable X in Exercise 3.43, representing the number of persons among 10,000 who make an error in preparing their income tax returns.

3.45 Find the mean and variance of the random variable X in Exercise 3.40, representing the number of hurricanes per year to hit a certain area of the eastern United States.

3.46 Changes in airport procedures require considerable planning. Arrival rates of aircraft are important factors that must be taken into account. Suppose small aircraft arrive at a certain airport, according to a Poisson process, at the rate of 6 per hour. Thus, the Poisson parameter for arrivals over a period of hours is $\mu = 6t$.

(a) What is the probability that exactly 4 small aircraft arrive during a 1-hour period?

(b) What is the probability that at least 4 arrive during a 1-hour period?

(c) If we define a working day as 12 hours, what is the probability that at least 75 small aircraft arrive during a working day?

3.47 An automobile manufacturer is concerned about a fault in the braking mechanism of a particular model. The fault can, on rare occasions, cause a catastrophe at high speed. The distribution of the number of cars per year that will experience the catastrophe is a Poisson random variable with $\lambda = 5$.

(a) What is the probability that at most 3 cars per year

will experience a catastrophe?

(b) What is the probability that more than 1 car per year will experience a catastrophe?

3.48 Consider Exercise 3.42. What is the mean number of students who fail the test?

3.49 The number of customers arriving per hour at a certain automobile service facility is assumed to follow a Poisson distribution with mean $\lambda = 7$.

(a) Compute the probability that more than 10 customers will arrive in a 2-hour period.

(b) What is the mean number of arrivals during a 2-hour period?

3.50 A company purchases large lots of a certain kind of electronic device. A method is used that rejects a lot if 2 or more defective units are found in a random sample of 100 units.

(a) What is the mean number of defective units found in a sample of 100 units if the lot is 1% defective?

(b) What is the variance?

3.51 The probability that a person will die when he or she contracts a virus infection is 0.001. Of the next 4000 people infected, what is the mean number who will die?

3.52 Potholes on a highway can be a serious problem and are in constant need of repair. With a particular type of terrain and make of concrete, past experience suggests that there are, on the average, 2 potholes per mile after a certain amount of usage. It is assumed that the Poisson process applies to the random variable "number of potholes."

(a) What is the probability that no more than one pothole will appear in a section of 1 mile?

(b) What is the probability that no more than 4 potholes will occur in a given section of 5 miles?

3.53 For a certain type of copper wire, it is known that, on the average, 1.5 flaws occur per millimeter. Assuming that the number of flaws is a Poisson random variable, what is the probability that no flaws occur in a certain portion of wire of length 5 millimeters? What is the mean number of flaws in a portion of length 5 millimeters?

3.54 It is known that 3% of people whose luggage is screened at an airport have questionable objects in their luggage. What is the probability that a string of 15 people pass through screening successfully before an individual is caught with a questionable object? What is the expected number of people to pass through before an individual is stopped?

3.55 Hospital administrators in large cities anguish about traffic in emergency rooms. At a particular hospital in a large city, the staff on hand cannot accommodate the patient traffic if there are more than 10 emergency cases in a given hour. It is assumed that patient arrival follows a Poisson process, and historical data suggest that, on the average, 5 emergencies arrive per hour.

(a) What is the probability that in a given hour the staff cannot accommodate the patient traffic?

(b) What is the probability that more than 20 emergencies arrive during a 3-hour shift?

3.56 Computer technology has produced an environment in which robots operate with the use of microprocessors. The probability that a robot fails during any 6-hour shift is 0.10. What is the probability that a robot will operate through at most 5 shifts before it fails?

3.6 Continuous Uniform Distribution

One of the simplest continuous distributions in all of statistics is the **continuous uniform distribution**. This distribution is characterized by a density function that is "flat," and thus the probability is uniform in a closed interval, say $[A, B]$. Although applications of the continuous uniform distribution are not as abundant as those for other distributions discussed in this chapter, it is appropriate for the novice to begin this introduction to continuous distributions with the uniform distribution.

Uniform
Distribution

The density function of the continuous uniform random variable X on the interval $[A, B]$ is

$$f(x; A, B) = \begin{cases} \frac{1}{B-A}, & A \leq x \leq B, \\ 0, & \text{elsewhere.} \end{cases}$$

The density function forms a rectangle with base $B-A$ and **constant height** $\frac{1}{B-A}$. As a result, the uniform distribution is often called the **rectangular distribution**. Note, however, that the interval may not always be closed: $[A, B]$. It can be (A, B) as well. The density function for a uniform random variable on the interval $[1, 3]$ is shown in Figure 3.2.

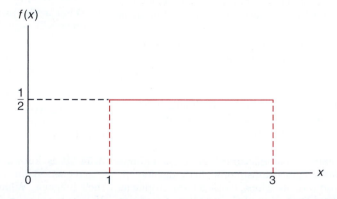

Figure 3.2: The density function for a random variable on the interval $[1, 3]$.

Probabilities are simple to calculate for the uniform distribution because of the simple nature of the density function. However, note that the application of this distribution is based on the assumption that the probability of falling in an interval of fixed length within $[A, B]$ is constant.

Example 3.19: Suppose that a large conference room at a certain company can be reserved for no more than 4 hours. Both long and short conferences occur quite often. In fact, it can be assumed that the length X of a conference has a uniform distribution on the interval $[0, 4]$.

(a) What is the probability density function?

(b) What is the probability that any given conference lasts at least 3 hours?

Solution: (a) The appropriate density function for the uniformly distributed random variable X in this situation is

$$f(x) = \begin{cases} \frac{1}{4}, & 0 \le x \le 4, \\ 0, & \text{elsewhere.} \end{cases}$$

(b) $P[X \ge 3] = \int_3^4 \frac{1}{4}\, dx = \frac{1}{4}.$　　　　　　　　　　　　　　◾

Theorem 3.6: The mean and variance of the uniform distribution are

$$\mu = \frac{A + B}{2} \text{ and } \sigma^2 = \frac{(B - A)^2}{12}.$$

The proofs of the theorems are left to the reader. See Exercise 3.57 on page 135.

3.7   Normal Distribution

The most important continuous probability distribution in the entire field of statistics is the **normal distribution**. Its graph, called the **normal curve**, is the bell-shaped curve of Figure 3.3, which approximately describes many phenomena that occur in nature, industry, and research. For example, physical measurements in areas such as meteorological experiments, rainfall studies, and measurements of manufactured parts are often more than adequately explained with a normal distribution. In addition, errors in scientific measurements are extremely well approximated by a normal distribution. In 1733, Abraham DeMoivre developed the mathematical equation of the normal curve. It provided a basis on which much of the theory of inductive statistics is founded. The normal distribution is often referred to as the **Gaussian distribution**, in honor of Karl Friedrich Gauss (1777–1855), who also derived its equation from a study of errors in repeated measurements of the same quantity.

A continuous random variable X having the bell-shaped distribution of Figure 3.3 is called a **normal random variable**. The mathematical equation for the probability distribution of the normal variable depends on the two parameters μ and σ, its mean and standard deviation, respectively. Hence, we denote the values of the density of X by $n(x; \mu, \sigma)$.

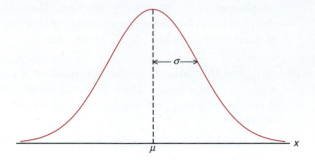

Figure 3.3: The normal curve.

Normal
Distribution

The density of the normal random variable X, with mean μ and variance σ^2, is

$$n(x; \mu, \sigma) = \frac{1}{\sqrt{2\pi}\sigma} e^{-\frac{1}{2\sigma^2}(x-\mu)^2}, \quad -\infty < x < \infty,$$

where $\pi = 3.14159\ldots$ and $e = 2.71828\ldots$.

Once μ and σ are specified, the normal curve is completely determined. For example, if $\mu = 50$ and $\sigma = 5$, then the ordinates $n(x; 50, 5)$ can be computed for various values of x and the curve drawn. In Figure 3.4, we have sketched two normal curves having the same standard deviation but different means. The two curves are identical in form but are centered at different positions along the horizontal axis.

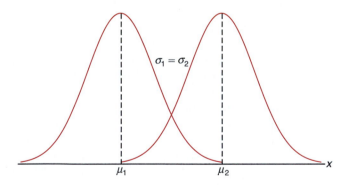

Figure 3.4: Normal curves with $\mu_1 < \mu_2$ and $\sigma_1 = \sigma_2$.

In Figure 3.5, we have sketched two normal curves with the same mean but different standard deviations. This time we see that the two curves are centered at exactly the same position on the horizontal axis, but the curve with the larger standard deviation is lower and spreads out farther. Remember that the area under a probability curve must be equal to 1, and therefore the more variable the set of observations, the lower and wider the corresponding curve will be.

Figure 3.6 shows two normal curves having different means and different stan-

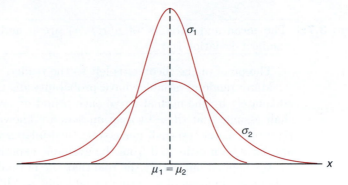

Figure 3.5: Normal curves with $\mu_1 = \mu_2$ and $\sigma_1 < \sigma_2$.

dard deviations. Clearly, they are centered at different positions on the horizontal axis and their shapes reflect the two different values of σ.

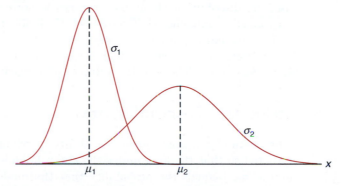

Figure 3.6: Normal curves with $\mu_1 < \mu_2$ and $\sigma_1 < \sigma_2$.

Based on inspection of Figures 3.3 through 3.6 and examination of the first and second derivatives of $n(x; \mu, \sigma)$, we list the following properties of the normal curve:

1. The mode, which is the point on the horizontal axis where the curve is a maximum, occurs at $x = \mu$.

2. The curve is symmetric about a vertical axis through the mean μ.

3. The curve has its points of inflection at $x = \mu \pm \sigma$; it is concave downward if $\mu - \sigma < X < \mu + \sigma$ and is concave upward otherwise.

4. The normal curve approaches the horizontal axis asymptotically as we proceed in either direction away from the mean.

5. The total area under the curve and above the horizontal axis is equal to 1.

Theorem 3.7:	The mean and variance of $n(x; \mu, \sigma)$ are μ and σ^2, respectively. Hence, the standard deviation is σ.

The proof of the theorem is left to the reader.

Many random variables have probability distributions that can be described adequately by the normal curve once μ and σ^2 are specified. In this chapter, we shall assume that these two parameters are known, perhaps from previous investigations. Later, we shall make statistical inferences when μ and σ^2 are unknown and have been estimated from the available experimental data.

We pointed out earlier the role that the normal distribution plays as a reasonable approximation of scientific variables in real-life experiments. There are other applications of the normal distribution that the reader will appreciate as he or she moves on in the book. The normal distribution finds enormous application as a *limiting distribution*. Under certain conditions, the normal distribution provides a good continuous approximation to the binomial and hypergeometric distributions. The case of the approximation to the binomial is covered in Section 3.10. In Chapter 4, the reader will learn about **sampling distributions**. It turns out that the limiting distribution of sample averages is normal. This provides a broad base for statistical inference that proves very valuable to the data analyst interested in estimation and hypothesis testing.

In Section 3.8, examples demonstrate the use of tables of the normal distribution. Section 3.9 follows with examples of applications of the normal distribution.

3.8 Areas under the Normal Curve

The curve of any continuous probability distribution or density function is constructed so that the area under the curve bounded by the two ordinates $x = x_1$ and $x = x_2$ equals the probability that the random variable X assumes a value between $x = x_1$ and $x = x_2$. Thus, for the normal curve in Figure 3.7,

$$P(x_1 < X < x_2) = \int_{x_1}^{x_2} n(x; \mu, \sigma) \, dx = \frac{1}{\sqrt{2\pi}\sigma} \int_{x_1}^{x_2} e^{-\frac{1}{2\sigma^2}(x-\mu)^2} dx$$

is represented by the area of the shaded region.

In Figures 3.4, 3.5, and 3.6 we saw how the normal curve is dependent on the mean and the standard deviation of the distribution under investigation. The area under the curve between any two ordinates must then also depend on the values μ and σ. This is evident in Figure 3.8, where we have shaded regions corresponding to $P(x_1 < X < x_2)$ for two curves with different means and variances. $P(x_1 < X < x_2)$, where X is the random variable describing distribution A, is indicated by the shaded area below the curve of A. If X is the random variable describing distribution B, then $P(x_1 < X < x_2)$ is given by the entire shaded region. Obviously, the two shaded regions are different in size; therefore, the probability associated with each distribution will be different for the two given values of X.

There are many types of statistical software that can be used in calculating areas under the normal curve. The difficulty encountered in solving integrals of normal density functions necessitates the tabulation of normal curve areas for quick

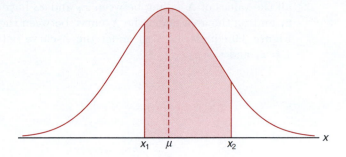

Figure 3.7: $P(x_1 < X < x_2)$ = area of the shaded region.

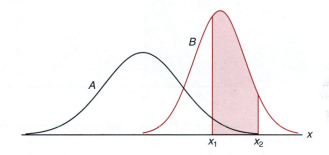

Figure 3.8: $P(x_1 < X < x_2)$ for different normal curves.

reference. However, it would be a hopeless task to attempt to set up separate tables for every conceivable value of μ and σ. Fortunately, we are able to transform all the observations of any normal random variable X into a new set of observations of a normal random variable Z with mean 0 and variance 1. This can be done by means of the transformation

$$Z = \frac{X - \mu}{\sigma}.$$

Whenever X assumes a value x, the corresponding value of Z is given by $z = (x - \mu)/\sigma$. Therefore, if X falls between the values $x = x_1$ and $x = x_2$, the random variable Z will fall between the corresponding values $z_1 = (x_1 - \mu)/\sigma$ and $z_2 = (x_2 - \mu)/\sigma$. Consequently, we may write

$$P(x_1 < X < x_2) = \frac{1}{\sqrt{2\pi}\sigma} \int_{x_1}^{x_2} e^{-\frac{1}{2\sigma^2}(x-\mu)^2} dx = \frac{1}{\sqrt{2\pi}} \int_{z_1}^{z_2} e^{-\frac{1}{2}z^2} dz$$

$$= \int_{z_1}^{z_2} n(z; 0, 1) \, dz = P(z_1 < Z < z_2),$$

where Z is seen to be a normal random variable with mean 0 and variance 1.

Definition 3.1: The distribution of a normal random variable with mean 0 and variance 1 is called a **standard normal distribution**.

The original and transformed distributions are illustrated in Figure 3.9. Since all the values of X falling between x_1 and x_2 have corresponding z values between z_1 and z_2, the area under the X-curve between the ordinates $x = x_1$ and $x = x_2$ in Figure 3.9 equals the area under the Z-curve between the transformed ordinates $z = z_1$ and $z = z_2$.

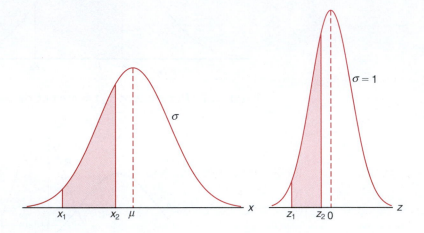

Figure 3.9: The original and transformed normal distributions.

We have now reduced the required number of tables of normal-curve areas to one, that of the standard normal distribution. Table A.3 indicates the area under the standard normal curve corresponding to $P(Z < z)$ for values of z ranging from -3.49 to 3.49. To illustrate the use of this table, let us find the probability that Z is less than 1.74. First, we locate a value of z equal to 1.7 in the left column; then we move across the row to the column under 0.04, where we read 0.9591. Therefore, $P(Z < 1.74) = 0.9591$. To find a z value corresponding to a given probability, the process is reversed. For example, the z value leaving an area of 0.2148 under the curve to the left of z is seen to be -0.79.

Example 3.20: Given a standard normal distribution, find the area under the curve that lies

(a) to the right of $z = 1.84$ and

(b) between $z = -1.97$ and $z = 0.86$.

Solution: See Figure 3.10 for the specific areas.

(a) The area in Figure 3.10(a) to the right of $z = 1.84$ is equal to 1 minus the area in Table A.3 to the left of $z = 1.84$, namely, $1 - 0.9671 = 0.0329$.

(b) The area in Figure 3.10(b) between $z = -1.97$ and $z = 0.86$ is equal to the area to the left of $z = 0.86$ minus the area to the left of $z = -1.97$. From Table A.3 we find the desired area to be $0.8051 - 0.0244 = 0.7807$.

Example 3.21: Given a standard normal distribution, find the value of k such that

(a) $P(Z > k) = 0.3015$ and

(b) $P(k < Z < -0.18) = 0.4197$.

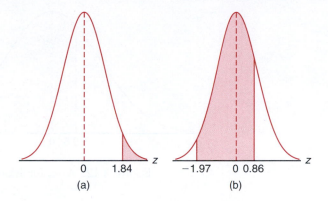

Figure 3.10: Areas for Example 3.20.

Solution: Distributions and the desired areas are shown in Figure 3.11.

(a) In Figure 3.11(a), we see that the k value leaving an area of 0.3015 to the right must then leave an area of 0.6985 to the left. From Table A.3 it follows that $k = 0.52$.

(b) From Table A.3 we note that the total area to the left of -0.18 is equal to 0.4286. In Figure 3.11(b), we see that the area between k and -0.18 is 0.4197, so the area to the left of k must be $0.4286 - 0.4197 = 0.0089$. Hence, from Table A.3, we have $k = -2.37$.

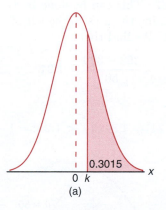
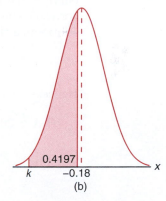

Figure 3.11: Areas for Example 3.21.

Example 3.22: Given a random variable X having a normal distribution with $\mu = 50$ and $\sigma = 10$, find the probability that X assumes a value between 45 and 62.

Solution: The z values corresponding to $x_1 = 45$ and $x_2 = 62$ are

$$z_1 = \frac{45 - 50}{10} = -0.5 \text{ and } z_2 = \frac{62 - 50}{10} = 1.2.$$

Therefore,

$$P(45 < X < 62) = P(-0.5 < Z < 1.2).$$

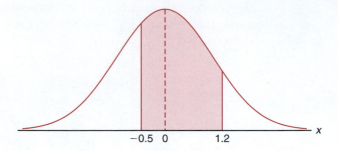

Figure 3.12: Area for Example 3.22.

$P(-0.5 < Z < 1.2)$ is shown by the area of the shaded region in Figure 3.12. This area may be found by subtracting the area to the left of the ordinate $z = -0.5$ from the entire area to the left of $z = 1.2$. Using Table A.3, we have

$$P(45 < X < 62) = P(-0.5 < Z < 1.2) = P(Z < 1.2) - P(Z < -0.5)$$
$$= 0.8849 - 0.3085 = 0.5764.$$

Example 3.23: Given that X has a normal distribution with $\mu = 300$ and $\sigma = 50$, find the probability that X assumes a value greater than 362.

Solution: The normal probability distribution with the desired area shaded is shown in Figure 3.13. To find $P(X > 362)$, we need to evaluate the area under the normal curve to the right of $x = 362$. This can be done by transforming $x = 362$ to the corresponding z value, obtaining the area to the left of z from Table A.3, and then subtracting this area from 1. We find that

$$z = \frac{362 - 300}{50} = 1.24.$$

Hence,

$$P(X > 362) = P(Z > 1.24) = 1 - P(Z < 1.24) = 1 - 0.8925 = 0.1075.$$

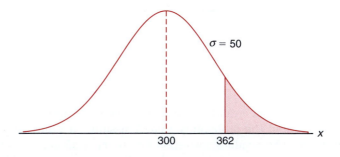

Figure 3.13: Area for Example 3.23.

If the random variable has a normal distribution, the z values corresponding to $x_1 = \mu - 2\sigma$ and $x_2 = \mu + 2\sigma$ are easily computed to be

$$z_1 = \frac{(\mu - 2\sigma) - \mu}{\sigma} = -2 \text{ and } z_2 = \frac{(\mu + 2\sigma) - \mu}{\sigma} = 2.$$

Hence,

$$P(\mu - 2\sigma < X < \mu + 2\sigma) = P(-2 < Z < 2) = P(Z < 2) - P(Z < -2)$$
$$= 0.9772 - 0.0228 = 0.9544.$$

Using the Normal Curve in Reverse

Sometimes, we are required to find the value of z corresponding to a specified probability that falls between values listed in Table A.3 (see Example 3.24). For convenience, we shall always choose the z value corresponding to the tabular probability that comes closest to the specified probability.

The preceding two examples were solved by going first from a value of x to a z value and then computing the desired area. In Example 3.24, we reverse the process and begin with a known area or probability, find the z value, and then determine x by rearranging the formula

$$z = \frac{x - \mu}{\sigma} \quad \text{to give} \quad x = \sigma z + \mu.$$

Example 3.24: Given a normal distribution with $\mu = 40$ and $\sigma = 6$, find the value of x that has

(a) 45% of the area to the left and

(b) 14% of the area to the right.

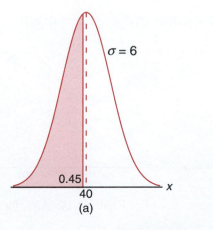
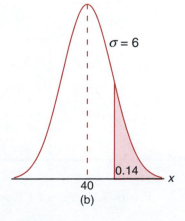

Figure 3.14: Areas for Example 3.24.

Solution: (a) An area of 0.45 to the left of the desired x value is shaded in Figure 3.14(a). We require a z value that leaves an area of 0.45 to the left. From Table A.3 we find $P(Z < -0.13) = 0.45$, so the desired z value is -0.13. Hence,

$$x = (6)(-0.13) + 40 = 39.22.$$

(b) In Figure 3.14(b), we shade an area equal to 0.14 to the right of the desired x value. This time we require a z value that leaves 0.14 of the area to the right and hence an area of 0.86 to the left. Again, from Table A.3, we find $P(Z < 1.08) = 0.86$, so the desired z value is 1.08 and

$$x = (6)(1.08) + 40 = 46.48.$$

3.9 Applications of the Normal Distribution

Some of the many problems for which the normal distribution is applicable are treated in the following examples. The use of the normal curve to approximate binomial probabilities is considered in Section 3.10.

Example 3.25: A certain type of storage battery lasts, on average, 3.0 years with a standard deviation of 0.5 year. Assuming that battery life is normally distributed, find the probability that a given battery will last less than 2.3 years.

Solution: First construct a diagram such as Figure 3.15, showing the given distribution of battery lives and the desired area. To find $P(X < 2.3)$, we need to evaluate the area under the normal curve to the left of 2.3. This is accomplished by finding the area to the left of the corresponding z value. Hence, we find that

$$z = \frac{2.3 - 3}{0.5} = -1.4,$$

and then, using Table A.3, we have

$$P(X < 2.3) = P(Z < -1.4) = 0.0808.$$

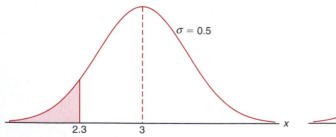 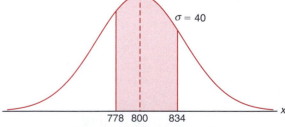

Figure 3.15: Area for Example 3.25.   Figure 3.16: Area for Example 3.26.

Example 3.26: An electrical firm manufactures light bulbs that have a life, before burn-out, that is normally distributed with mean equal to 800 hours and a standard deviation of 40 hours. Find the probability that a bulb burns between 778 and 834 hours.

Solution: The distribution of light bulb life is illustrated in Figure 3.16. The z values corresponding to $x_1 = 778$ and $x_2 = 834$ are

$$z_1 = \frac{778 - 800}{40} = -0.55 \text{ and } z_2 = \frac{834 - 800}{40} = 0.85.$$

Hence,

$$P(778 < X < 834) = P(-0.55 < Z < 0.85) = P(Z < 0.85) - P(Z < -0.55)$$
$$= 0.8023 - 0.2912 = 0.5111.$$

Example 3.27: In an industrial process, the diameter of a ball bearing is an important measurement. The buyer sets specifications for the diameter to be 3.0 ± 0.01 cm. The implication is that no part falling outside these specifications will be accepted. It is known that in the process the diameter of a ball bearing has a normal distribution with mean $\mu = 3.0$ and standard deviation $\sigma = 0.005$. On average, how many manufactured ball bearings will be scrapped?

Solution: The distribution of diameters is illustrated by Figure 3.17. The values corresponding to the specification limits are $x_1 = 2.99$ and $x_2 = 3.01$. The corresponding z values are

$$z_1 = \frac{2.99 - 3.0}{0.005} = -2.0 \text{ and } z_2 = \frac{3.01 - 3.0}{0.005} = +2.0.$$

Hence,

$$P(2.99 < X < 3.01) = P(-2.0 < Z < 2.0).$$

From Table A.3, $P(Z < -2.0) = 0.0228$. Due to symmetry of the normal distribution, we find that

$$P(Z < -2.0) + P(Z > 2.0) = 2(0.0228) = 0.0456.$$

As a result, it is anticipated that, on average, 4.56% of manufactured ball bearings will be scrapped.

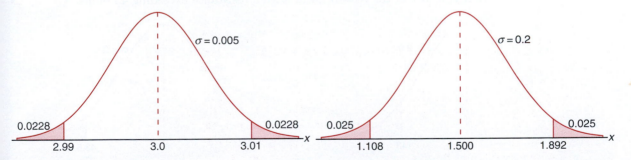

Figure 3.17: Area for Example 3.27.   Figure 3.18: Specifications for Example 3.28.

Example 3.28: Gauges are used to reject all components for which a certain dimension is not within the specification $1.50 \pm d$. It is known that this measurement is normally distributed with mean 1.50 and standard deviation 0.2. Determine the value d such that the specifications "cover" 95% of the measurements.

Solution: From Table A.3 we know that

$$P(-1.96 < Z < 1.96) = 0.95.$$

Therefore,

$$1.96 = \frac{(1.50 + d) - 1.50}{0.2},$$

from which we obtain

$$d = (0.2)(1.96) = 0.392.$$

An illustration of the specifications is shown in Figure 3.18.

Example 3.29: A certain machine makes electrical resistors having a mean resistance of 40 ohms and a standard deviation of 2 ohms. Assuming that the resistance follows a normal distribution and can be measured to any degree of accuracy, what percentage of resistors will have a resistance exceeding 43 ohms?

Solution: A percentage is found by multiplying the relative frequency by 100%. Since the relative frequency for an interval is equal to the probability of a value falling in the interval, we must find the area to the right of $x = 43$ in Figure 3.19. This can be done by transforming $x = 43$ to the corresponding z value, obtaining the area to the left of z from Table A.3, and then subtracting this area from 1. We find

$$z = \frac{43 - 40}{2} = 1.5.$$

Therefore,

$$P(X > 43) = P(Z > 1.5) = 1 - P(Z < 1.5) = 1 - 0.9332 = 0.0668.$$

Hence, 6.68% of the resistors will have a resistance exceeding 43 ohms.

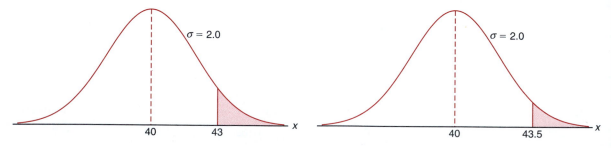

Figure 3.19: Area for Example 3.29.   Figure 3.20: Area for Example 3.30.

Example 3.30: Find the percentage of resistances exceeding 43 ohms for Example 3.29 if resistance is measured to the nearest ohm.

Solution: This problem differs from that in Example 3.29 in that we now assign a measurement of 43 ohms to all resistors whose resistances are greater than 42.5 and less than 43.5. We are actually approximating a discrete distribution by means of a continuous normal distribution. The required area is the region shaded to the right of 43.5 in Figure 3.20. We now find that

$$z = \frac{43.5 - 40}{2} = 1.75.$$

Hence,

$$P(X > 43.5) = P(Z > 1.75) = 1 - P(Z < 1.75) = 1 - 0.9599 = 0.0401.$$

Therefore, 4.01% of the resistances exceed 43 ohms when measured to the nearest ohm. The difference $6.68\% - 4.01\% = 2.67\%$ between this answer and that of Example 3.29 represents all those resistance values greater than 43 and less than 43.5 that are now being recorded as 43 ohms. ◢

Exercises

3.57 Given a continuous uniform distribution, show that

(a) $\mu = \frac{A+B}{2}$ and

(b) $\sigma^2 = \frac{(B-A)^2}{12}$.

3.58 Suppose X follows a continuous uniform distribution from 1 to 5. Determine the conditional probability $P(X > 2.5 \mid X \leq 4)$.

3.59 The daily amount of coffee, in liters, dispensed by a machine located in an airport lobby is a random variable X having a continuous uniform distribution with $A = 7$ and $B = 10$. Find the probability that on a given day the amount of coffee dispensed by this machine will be

(a) at most 8.8 liters;

(b) more than 7.4 liters but less than 9.5 liters;

(c) at least 8.5 liters.

3.60 Find the value of z if the area under a standard normal curve

(a) to the right of z is 0.3622;

(b) to the left of z is 0.1131;

(c) between 0 and z, with $z > 0$, is 0.4838;

(d) between $-z$ and z, with $z > 0$, is 0.9500.

3.61 Given a standard normal distribution, find the area under the curve that lies

(a) to the left of $z = -1.39$;

(b) to the right of $z = 1.96$;

(c) between $z = -2.16$ and $z = -0.65$;

(d) to the left of $z = 1.43$;

(e) to the right of $z = -0.89$;

(f) between $z = -0.48$ and $z = 1.74$.

3.62 Given a standard normal distribution, find the value of k such that

(a) $P(Z > k) = 0.2946$;

(b) $P(Z < k) = 0.0427$;

(c) $P(-0.93 < Z < k) = 0.7235$.

3.63 Given the normally distributed variable X with mean 18 and standard deviation 2.5, find

(a) $P(X < 15)$;

(b) the value of k such that $P(X < k) = 0.2236$;

(c) the value of k such that $P(X > k) = 0.1814$;

(d) $P(17 < X < 21)$.

3.64 Given a normal distribution with $\mu = 30$ and $\sigma = 6$, find

(a) the normal curve area to the right of $x = 17$;

(b) the normal curve area to the left of $x = 22$;

(c) the normal curve area between $x = 32$ and $x = 41$;

(d) the value of x that has 80% of the normal curve area to the left;

(e) the two values of x that contain the middle 75% of the normal curve area.

3.65 A soft-drink machine is regulated so that it discharges an average of 200 milliliters per cup. If the amount of drink is normally distributed with a standard deviation equal to 15 milliliters,

(a) what fraction of the cups will contain more than 224 milliliters?

(b) what is the probability that a cup contains between 191 and 209 milliliters?

(c) how many cups will probably overflow if 230-milliliter cups are used for the next 1000 drinks?

(d) below what value do we get the smallest 25% of the drinks?

3.66 The loaves of rye bread distributed to local stores by a certain bakery have an average length of 30 centimeters and a standard deviation of 2 centimeters. Assuming that the lengths are normally distributed, what percentage of the loaves are

(a) longer than 31.7 centimeters?

(b) between 29.3 and 33.5 centimeters in length?

(c) shorter than 25.5 centimeters?

3.67 A research scientist reports that mice will live an average of 40 months when their diets are sharply restricted and then enriched with vitamins and proteins. Assuming that the lifetimes of such mice are normally distributed with a standard deviation of 6.3 months, find the probability that a given mouse will live

(a) more than 32 months;

(b) less than 28 months;

(c) between 37 and 49 months.

3.68 The finished inside diameter of a piston ring is normally distributed with a mean of 10 centimeters and a standard deviation of 0.03 centimeter.

(a) What proportion of rings will have inside diameters exceeding 10.075 centimeters?

(b) What is the probability that a piston ring will have an inside diameter between 9.97 and 10.03 centimeters?

(c) Below what value of inside diameter will 15% of the piston rings fall?

3.69 A lawyer commutes daily from his suburban home to his midtown office. The average time for a one-way trip is 24 minutes, with a standard deviation of 3.8 minutes. Assume the trip times to be normally distributed.

(a) What is the probability that a trip will take at least 1/2 hour?

(b) If the office opens at 9:00 A.M. and the lawyer leaves his house at 8:45 A.M. daily, what percentage of the time is he late for work?

(c) If he leaves the house at 8:35 A.M. and coffee is served at the office from 8:50 A.M. until 9:00 A.M., what is the probability that he misses coffee?

(d) Find the length of time above which we find the slowest 15% of the trips.

(e) Find the probability that 2 of the next 3 trips will take at least 1/2 hour.

3.70 In the November 1990 issue of *Chemical Engineering Progress*, a study discussed the percent purity of oxygen from a certain supplier. Assume that the mean was 99.61 with a standard deviation of 0.08. Assume that the distribution of percent purity was approximately normal.

(a) What percentage of the purity values would you expect to be between 99.5 and 99.7?

(b) What purity value would you expect to exceed exactly 5% of the population?

3.71 The average life of a certain type of small motor is 10 years with a standard deviation of 2 years. The manufacturer replaces free all motors that fail while under guarantee. If she is willing to replace only 3% of the motors that fail, how long a guarantee should be offered? Assume that the lifetime of a motor follows a normal distribution.

3.72 The heights of 1000 students are normally distributed with a mean of 174.5 centimeters and a standard deviation of 6.9 centimeters. Assuming that the heights are recorded to the nearest half-centimeter, how many of these students would you expect to have heights

(a) less than 160.0 centimeters?

(b) between 171.5 and 182.0 centimeters inclusive?

(c) equal to 175.0 centimeters?

(d) greater than or equal to 188.0 centimeters?

3.73 The tensile strength of a certain metal component is normally distributed with a mean of 10,000 kilograms per square centimeter and a standard deviation of 100 kilograms per square centimeter. Measurements are recorded to the nearest 50 kilograms per square centimeter.

(a) What proportion of these components exceed 10,150 kilograms per square centimeter in tensile strength?

(b) If specifications require that all components have tensile strength between 9800 and 10,200 kilograms per square centimeter inclusive, what proportion of pieces would we expect to scrap?

3.74 The weights of a large number of miniature poodles are approximately normally distributed with a mean of 8 kilograms and a standard deviation of 0.9 kilogram. If measurements are recorded to the nearest tenth of a kilogram, find the fraction of these poodles with weights

(a) over 9.5 kilograms;

(b) of at most 8.6 kilograms;

(c) between 7.3 and 9.1 kilograms inclusive.

3.75 The IQs of 600 applicants to a certain college are approximately normally distributed with a mean of 115 and a standard deviation of 12. If the college requires an IQ of at least 95, how many of these students will be rejected on the basis of IQ, regardless of their other qualifications? Note that IQs are recorded to the nearest integers.

3.76 If a set of observations is normally distributed, what percent of these differ from the mean by

(a) more than 1.3σ?

(b) less than 0.52σ?

3.10　Normal Approximation to the Binomial

Probabilities associated with binomial experiments are readily obtainable from the formula $b(x; n, p)$ of the binomial distribution or from Table A.1 when n is small. In addition, binomial probabilities are readily available in many computer software packages. However, it is instructive to learn the relationship between the binomial and the normal distribution. In Section 3.5, we illustrated how the Poisson distribution can be used to approximate binomial probabilities when n is quite large and p is very close to 0 or 1. Both the binomial and the Poisson distributions are discrete. The first application of a continuous probability distribution to approximate probabilities over a discrete sample space was demonstrated in Example 3.30, where the normal curve was used. The normal distribution is often a good approximation to a discrete distribution when the latter takes on a symmetric bell shape. From a theoretical point of view, some distributions converge to the normal as their parameters approach certain limits. The normal distribution is a convenient approximating distribution because the cumulative distribution function is so easily tabled. The binomial distribution is nicely approximated by the normal in practical problems when one works with the cumulative distribution function. We now state a theorem that allows us to use areas under the normal curve to approximate binomial properties when n is sufficiently large.

Theorem 3.8: If X is a binomial random variable with mean $\mu = np$ and variance $\sigma^2 = npq$, then the limiting form of the distribution of

$$Z = \frac{X - np}{\sqrt{npq}},$$

as $n \to \infty$, is the standard normal distribution $n(z; 0, 1)$.

It turns out that the normal distribution with $\mu = np$ and $\sigma^2 = np(1 - p)$ not only provides a very accurate approximation to the binomial distribution when n is large and p is not extremely close to 0 or 1 but also provides a fairly good approximation even when n is small and p is reasonably close to $1/2$.

To illustrate the normal approximation to the binomial distribution, we first draw the histogram for $b(x; 15, 0.4)$ and then superimpose the particular normal curve having the same mean and variance as the binomial variable X. Hence, we draw a normal curve with

$$\mu = np = (15)(0.4) = 6 \text{ and } \sigma^2 = npq = (15)(0.4)(0.6) = 3.6.$$

The histogram of $b(x; 15, 0.4)$ and the corresponding superimposed normal curve, which is completely determined by its mean and variance, are illustrated in Figure 3.21.

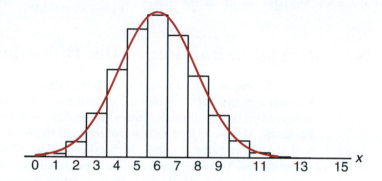

Figure 3.21: Normal approximation of $b(x; 15, 0.4)$.

The exact probability that the binomial random variable X assumes a given value x is equal to the area of the bar whose base is centered at x. For example, the exact probability that X assumes the value 4 is equal to the area of the rectangle with base centered at $x = 4$. Using Table A.1, we find this area to be

$$P(X = 4) = b(4; 15, 0.4) = 0.1268,$$

which is approximately equal to the area of the shaded region under the normal curve between the two ordinates $x_1 = 3.5$ and $x_2 = 4.5$ in Figure 3.22. Converting to z values, we have

$$z_1 = \frac{3.5 - 6}{1.897} = -1.32 \quad \text{and} \quad z_2 = \frac{4.5 - 6}{1.897} = -0.79.$$

If X is a binomial random variable and Z a standard normal variable, then

$$P(X = 4) = b(4; 15, 0.4) \approx P(-1.32 < Z < -0.79)$$
$$= P(Z < -0.79) - P(Z < -1.32) = 0.2148 - 0.0934 = 0.1214.$$

This agrees very closely with the exact value of 0.1268.

The normal approximation is most useful in calculating binomial sums for large values of n. Referring to Figure 3.22, we might be interested in the probability that X assumes a value from 7 to 9 inclusive. The exact probability is given by

$$P(7 \leq X \leq 9) = \sum_{x=0}^{9} b(x; 15, 0.4) - \sum_{x=0}^{6} b(x; 15, 0.4)$$
$$= 0.9662 - 0.6098 = 0.3564,$$

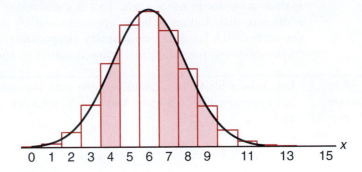

Figure 3.22: Normal approximation of $b(x; 15, 0.4)$ and $\sum_{x=7}^{9} b(x; 15, 0.4)$.

which is equal to the sum of the areas of the rectangles with bases centered at $x = 7$, 8, and 9. For the normal approximation, we find the area of the shaded region under the curve between the ordinates $x_1 = 6.5$ and $x_2 = 9.5$ in Figure 3.22. The corresponding z values are

$$z_1 = \frac{6.5 - 6}{1.897} = 0.26 \quad \text{and} \quad z_2 = \frac{9.5 - 6}{1.897} = 1.85.$$

Now,

$$P(7 \leq X \leq 9) \approx P(0.26 < Z < 1.85) = P(Z < 1.85) - P(Z < 0.26)$$
$$= 0.9678 - 0.6026 = 0.3652.$$

Once again, the normal curve approximation provides a value that agrees very closely with the exact value of 0.3564. The degree of accuracy, which depends on how well the curve fits the histogram, will increase as n increases. This is particularly true when p is not very close to $1/2$ and the histogram is no longer symmetric. Figures 3.23 and 3.24 show the histograms for $b(x; 6, 0.2)$ and $b(x; 15, 0.2)$, respectively. It is evident that a normal curve would fit the histogram considerably better when $n = 15$ than when $n = 6$.

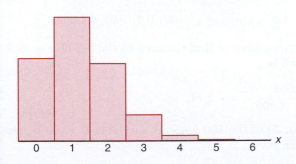

Figure 3.23: Histogram for $b(x; 6, 0.2)$.

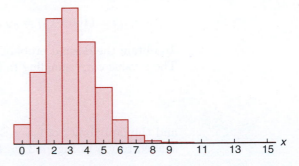

Figure 3.24: Histogram for $b(x; 15, 0.2)$.

In our illustrations of the normal approximation to the binomial, it became apparent that if we seek the area under the normal curve to the left of, say, x, it is more accurate to use $x + 0.5$. This is a correction to accommodate the fact that a discrete distribution is being approximated by a continuous distribution. The correction $+0.5$ is called a **continuity correction**. The foregoing discussion leads to the following formal normal approximation to the binomial.

Normal Approximation to the Binomial Distribution

Let X be a binomial random variable with parameters n and p. For large n, X has approximately a normal distribution with $\mu = np$ and $\sigma^2 = npq = np(1-p)$ and

$$P(X \le x) = \sum_{k=0}^{x} b(k; n, p)$$

$$\approx \text{ area under normal curve to the left of } x + 0.5$$

$$= P\left(Z \le \frac{x + 0.5 - np}{\sqrt{npq}}\right),$$

and the approximation will be good if np and $n(1-p)$ are greater than or equal to 5.

As we indicated earlier, the quality of the approximation is quite good for large n. If p is close to $1/2$, a moderate or small sample size will be sufficient for a reasonable approximation. We offer Table 3.1 as an indication of the quality of the approximation. Both the normal approximation and the true binomial cumulative probabilities are given. Notice that at $p = 0.05$ and $p = 0.10$, the approximation is fairly crude for $n = 10$. However, even for $n = 10$, note the improvement for $p = 0.50$. On the other hand, when p is fixed at $p = 0.05$, note the improvement of the approximation as we go from $n = 20$ to $n = 100$.

Example 3.31: The probability that a patient recovers from a rare blood disease is 0.4. If 100 people are known to have contracted this disease, what is the probability that fewer than 30 survive?

Solution: Let the binomial variable X represent the number of patients who survive. Since $n = 100$, we should obtain fairly accurate results using the normal-curve approximation with

$$\mu = np = (100)(0.4) = 40 \text{ and } \sigma = \sqrt{npq} = \sqrt{(100)(0.4)(0.6)} = 4.899.$$

To obtain the desired probability, we have to find the area to the left of $x = 29.5$. The z value corresponding to 29.5 is

$$z = \frac{29.5 - 40}{4.899} = -2.14,$$

and the probability of fewer than 30 of the 100 patients surviving is given by the shaded region in Figure 3.25. Hence,

$$P(X < 30) \approx P(Z < -2.14) = 0.0162.$$

Example 3.32: A multiple-choice quiz has 200 questions, each with 4 possible answers of which only 1 is correct. What is the probability that sheer guesswork yields from 25 to

Table 3.1: Normal Approximation and True Cumulative Binomial Probabilities

	$p = 0.05$, $n = 10$		$p = 0.10$, $n = 10$		$p = 0.50$, $n = 10$	
r	Binomial	Normal	Binomial	Normal	Binomial	Normal
0	0.5987	0.5000	0.3487	0.2981	0.0010	0.0022
1	0.9139	0.9265	0.7361	0.7019	0.0107	0.0136
2	0.9885	0.9981	0.9298	0.9429	0.0547	0.0571
3	0.9990	1.0000	0.9872	0.9959	0.1719	0.1711
4	1.0000	1.0000	0.9984	0.9999	0.3770	0.3745
5			1.0000	1.0000	0.6230	0.6255
6					0.8281	0.8289
7					0.9453	0.9429
8					0.9893	0.9864
9					0.9990	0.9978
10					1.0000	0.9997

	$p = 0.05$					
	$n = 20$		$n = 50$		$n = 100$	
r	Binomial	Normal	Binomial	Normal	Binomial	Normal
0	0.3585	0.3015	0.0769	0.0968	0.0059	0.0197
1	0.7358	0.6985	0.2794	0.2578	0.0371	0.0537
2	0.9245	0.9382	0.5405	0.5000	0.1183	0.1251
3	0.9841	0.9948	0.7604	0.7422	0.2578	0.2451
4	0.9974	0.9998	0.8964	0.9032	0.4360	0.4090
5	0.9997	1.0000	0.9622	0.9744	0.6160	0.5910
6	1.0000	1.0000	0.9882	0.9953	0.7660	0.7549
7			0.9968	0.9994	0.8720	0.8749
8			0.9992	0.9999	0.9369	0.9463
9			0.9998	1.0000	0.9718	0.9803
10			1.0000	1.0000	0.9885	0.9941

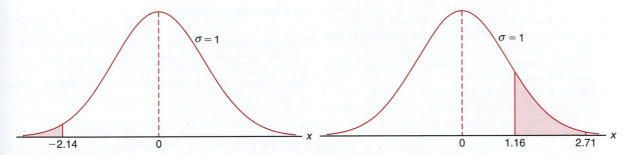

Figure 3.25: Area for Example 3.31.　　　　Figure 3.26: Area for Example 3.32.

30 correct answers for the 80 of the 200 problems about which the student has no knowledge?

Solution: The probability of guessing a correct answer for each of the 80 questions is $p = 1/4$.

If X represents the number of correct answers resulting from guesswork, then

$$P(25 \leq X \leq 30) = \sum_{x=25}^{30} b(x; 80, 1/4).$$

Using the normal curve approximation with

$$\mu = np = (80)\left(\frac{1}{4}\right) = 20$$

and

$$\sigma = \sqrt{npq} = \sqrt{(80)(1/4)(3/4)} = 3.873,$$

we need the area between $x_1 = 24.5$ and $x_2 = 30.5$. The corresponding z values are

$$z_1 = \frac{24.5 - 20}{3.873} = 1.16 \text{ and } z_2 = \frac{30.5 - 20}{3.873} = 2.71.$$

The probability of correctly guessing from 25 to 30 questions is given by the shaded region in Figure 3.26. From Table A.3 we find that

$$P(25 \leq X \leq 30) = \sum_{x=25}^{30} b(x; 80, 0.25) \approx P(1.16 < Z < 2.71)$$

$$= P(Z < 2.71) - P(Z < 1.16) = 0.9966 - 0.8770 = 0.1196.$$

Exercises

3.77 A process for manufacturing an electronic component yields items of which 1% are defective. A quality control plan is to select 100 items from the process, and if none are defective, the process continues. Use the normal approximation to the binomial to find

(a) the probability that the process continues given the sampling plan described;

(b) the probability that the process continues even if the process has gone bad (i.e., if the frequency of defective components has shifted to 5.0% defective).

3.78 A process yields 10% defective items. If 100 items are randomly selected from the process, what is the probability that the number of defectives

(a) exceeds 13?

(b) is less than 8?

3.79 The probability that a patient recovers from a delicate heart operation is 0.9. Of the next 100 patients having this operation, what is the probability that

(a) between 84 and 95 inclusive survive?

(b) fewer than 86 survive?

3.80 Researchers at George Washington University and the National Institutes of Health claim that approximately 75% of people believe "tranquilizers work very well to make a person more calm and relaxed." Of the next 80 people interviewed, what is the probability that

(a) at least 50 are of this opinion?

(b) at most 56 are of this opinion?

3.81 A company produces component parts for an engine. Parts specifications suggest that 95% of items meet specifications. The parts are shipped to customers in lots of 100.

(a) What is the probability that more than 2 items in a given lot will be defective?

(b) What is the probability that more than 10 items in a lot will be defective?

3.82 A pharmaceutical company knows that approximately 5% of its birth-control pills have an ingredient that is below the minimum strength, thus rendering the pill ineffective. What is the probability that fewer than 10 in a sample of 200 pills will be ineffective?

3.83 Statistics released by the National Highway Traffic Safety Administration and the National Safety Council show that on an average weekend night, 1 out of every 10 drivers on the road is drunk. If 400 drivers are randomly checked next Saturday night, what is the probability that the number of drunk drivers will be

(a) less than 32?

(b) more than 49?

(c) at least 35 but less than 47?

3.84 A drug manufacturer claims that a certain drug cures a blood disease, on the average, 80% of the time. To check the claim, government testers use the drug on a sample of 100 individuals and decide to accept the claim if 75 or more are cured.

(a) What is the probability that the claim will be rejected when the cure probability is, in fact, 0.8?

(b) What is the probability that the claim will be accepted by the government when the cure probability is as low as 0.7?

3.85 The serum cholesterol level X in 14-year-old boys has approximately a normal distribution with mean 170 and standard deviation 30.

(a) Find the probability that the serum cholesterol level of a randomly chosen 14-year-old boy exceeds 230.

(b) In a middle school there are 300 14-year-old boys. Find the probability that at least 8 boys have a serum cholesterol level that exceeds 230.

3.86 A common practice of airline companies is to sell more tickets for a particular flight than there are seats on the plane, because customers who buy tickets do not always show up for the flight. Suppose that the percentage of no-shows at flight time is 2%. For a particular flight with 197 seats, a total of 200 tickets were sold. What is the probability that the airline overbooked this flight?

3.87 A telemarketing company has a special letter-opening machine that opens and removes the contents of an envelope. If the envelope is fed improperly into the machine, the contents of the envelope may not be removed or may be damaged. In this case, the machine is said to have "failed."

(a) If the machine has a probability of failure of 0.01, what is the probability of more than 1 failure occurring in a batch of 20 envelopes?

(b) If the probability of failure of the machine is 0.01 and a batch of 500 envelopes is to be opened, what is the probability that more than 8 failures will occur?

3.11 Gamma and Exponential Distributions

Although the normal distribution can be used to solve many problems in engineering and science, there are still numerous situations that require different types of density functions. Two such density functions, the **gamma** and **exponential distributions**, are discussed in this section.

It turns out that the exponential distribution is a special case of the gamma distribution. Both find a large number of applications. The exponential and gamma distributions play an important role in both queuing theory and reliability problems. Time between arrivals at service facilities and time to failure of component parts and electrical systems often are nicely modeled by the exponential distribution. The relationship between the gamma and the exponential allows the gamma to be used in similar types of problems. More details and illustrations will be supplied later in the section.

The gamma distribution derives its name from the well-known **gamma function**, studied in many areas of mathematics. Before we proceed to the gamma distribution, let us review this function and some of its important properties.

Definition 3.2: The **gamma function** is defined by

$$\Gamma(\alpha) = \int_0^\infty x^{\alpha-1}e^{-x}\,dx, \qquad \text{for } \alpha > 0.$$

The following are a few simple properties of the gamma function.

(a) $\Gamma(n) = (n-1)(n-2)\cdots(1)\Gamma(1)$, for a positive integer n.

To see the proof, integrating by parts with $u = x^{\alpha-1}$ and $dv = e^{-x}\,dx$, we obtain

$$\Gamma(\alpha) = -e^{-x}\,x^{\alpha-1}\Big|_0^\infty + \int_0^\infty e^{-x}(\alpha-1)x^{\alpha-2}\,dx = (\alpha-1)\int_0^\infty x^{\alpha-2}e^{-x}\,dx,$$

for $\alpha > 1$, which yields the recursion formula

$$\Gamma(\alpha) = (\alpha-1)\Gamma(\alpha-1).$$

The result follows after repeated application of the recursion formula. Using this result, we can easily show the following two properties.

(b) $\Gamma(n) = (n-1)!$ for a positive integer n.

(c) $\Gamma(1) = 1$.

Furthermore, we have the following property of $\Gamma(\alpha)$, which is left for the reader to verify.

(d) $\Gamma(1/2) = \sqrt{\pi}$.

The following is the definition of the **gamma distribution**.

Gamma Distribution The continuous random variable X has a **gamma distribution**, with parameters α and β, if its density function is given by

$$f(x;\alpha,\beta) = \begin{cases} \frac{1}{\beta^\alpha\Gamma(\alpha)}x^{\alpha-1}e^{-x/\beta}, & x > 0, \\ 0, & \text{elsewhere,} \end{cases}$$

where $\alpha > 0$ and $\beta > 0$.

Graphs of several gamma distributions are shown in Figure 3.27 for certain specified values of the parameters α and β. The special gamma distribution for which $\alpha = 1$ is called the **exponential distribution**.

Exponential Distribution The continuous random variable X has an **exponential distribution**, with parameter β, if its density function is given by

$$f(x;\beta) = \begin{cases} \frac{1}{\beta}e^{-x/\beta}, & x > 0, \\ 0, & \text{elsewhere,} \end{cases}$$

where $\beta > 0$.

The following theorem and corollary give the mean and variance of the gamma and exponential distributions.

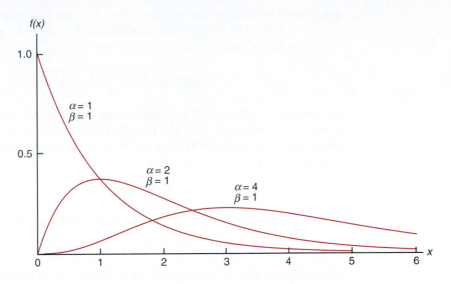

Figure 3.27: Gamma distributions.

Theorem 3.9: The mean and variance of the gamma distribution are

$$\mu = \alpha\beta \quad \text{and} \quad \sigma^2 = \alpha\beta^2.$$

The proof of this theorem is found in Appendix A.14.

Corollary 3.1: The mean and variance of the exponential distribution are

$$\mu = \beta \quad \text{and} \quad \sigma^2 = \beta^2.$$

Relationship to the Poisson Process

We shall pursue applications of the exponential distribution and then return to the gamma distribution. The most important applications of the exponential distribution are situations where the Poisson process applies (see Section 3.5). The reader should recall that the Poisson process allows for the use of the discrete distribution called the Poisson distribution. Recall that the Poisson distribution is used to compute the probability of specific numbers of "events" during a particular *period of time or span of space*. In many applications, the time period or span of space is the random variable. For example, an industrial engineer may be interested in modeling the time T between arrivals at a congested intersection during rush hour in a large city. An arrival represents the Poisson event.

The relationship between the exponential distribution (often called the negative exponential) and the Poisson process is quite simple. In Section 3.5, the Poisson distribution was developed as a single-parameter distribution with parameter λ, where λ may be interpreted as the mean number of events *per unit "time."* Consider now the *random variable* described by the time required for the first event to occur. Using the Poisson distribution, we find that the probability of no events occurring in the span up to time t is given by

$$p(0; \lambda t) = \frac{e^{-\lambda t}(\lambda t)^0}{0!} = e^{-\lambda t}.$$

We can now make use of the above and let X be the time to the first Poisson event. The probability that the length of time until the first event will exceed x is the same as the probability that no Poisson events will occur in x. The latter, of course, is given by $e^{-\lambda x}$. As a result,

$$P(X > x) = e^{-\lambda x}.$$

Thus, the cumulative distribution function for X is given by

$$P(0 \leq X \leq x) = 1 - e^{-\lambda x}.$$

Now, in order that we may recognize the presence of the exponential distribution, we differentiate the cumulative distribution function above to obtain the density function

$$f(x) = \lambda e^{-\lambda x},$$

which is the density function of the exponential distribution with $\lambda = 1/\beta$.

Applications of the Exponential and Gamma Distributions

In the foregoing, we provided the foundation for the application of the exponential distribution in "time to arrival" or time to Poisson event problems. We will illustrate some applications here and then proceed to discuss the role of the gamma distribution in these modeling applications. Notice that the mean of the exponential distribution is the parameter β, the reciprocal of the parameter in the Poisson distribution. The reader should recall that it is often said that the Poisson distribution has no memory, implying that occurrences in successive time periods are independent. The important parameter β is the mean time between events. In reliability theory, where equipment failure often conforms to this Poisson process, β is called **mean time between failures**. Many equipment breakdowns do follow the Poisson process, and thus the exponential distribution does apply. Other applications include survival times in biomedical experiments and computer response time.

In the following example, we show a simple application of the exponential distribution to a problem in reliability. The binomial distribution also plays a role in the solution.

Example 3.33: Suppose that a system contains a certain type of component whose time, in years, to failure is given by T. The random variable T is modeled nicely by the exponential distribution with mean time to failure $\beta = 5$. If 5 of these components are installed in different systems, what is the probability that at least 2 are still functioning at the end of 8 years?

Solution: The probability that a given component is still functioning after 8 years is given by

$$P(T > 8) = \frac{1}{5} \int_8^\infty e^{-t/5} \, dt = e^{-8/5} \approx 0.2.$$

Let X represent the number of components functioning after 8 years. Then using the binomial distribution, we have

$$P(X \geq 2) = \sum_{x=2}^{5} b(x; 5, 0.2) = 1 - \sum_{x=0}^{1} b(x; 5, 0.2) = 1 - 0.7373 = 0.2627.$$

There are exercises in Chapter 2 where the reader has already encountered the exponential distribution. Some of the other applications will be found in the exercises and review exercises at the end of this chapter.

The Memoryless Property and Its Effect on the Exponential Distribution

The types of applications of the exponential distribution in reliability and component or machine lifetime problems are influenced by the **memoryless** (or lack-of-memory) **property** of the exponential distribution. For example, in the case of, say, an electronic component where lifetime has an exponential distribution, the probability that the component lasts, say, t hours, that is, $P(X \geq t)$, is the same as the conditional probability

$$P(X \geq t_0 + t \mid X \geq t_0).$$

So if the component "makes it" to t_0 hours, the probability of lasting an additional t hours is the same as the probability of lasting t hours. There is no "punishment" through wear that may have ensued for lasting the first t_0 hours. Thus, the exponential distribution is more appropriate when the memoryless property is justified. But if the failure of the component is a result of gradual or slow wear (as in mechanical wear), then the exponential does not apply and the gamma distribution may be more appropriate.

The importance of the gamma distribution lies in the fact that it defines a family of which other distributions are special cases. But the gamma itself has important applications in waiting time and reliability theory. Whereas the exponential distribution describes the time until the occurrence of a Poisson event (or the time between Poisson events), the time (or space) occurring until a *specified number of Poisson events occur* is a random variable whose density function is described by the gamma distribution. This specific number of events is the parameter α in the gamma density function. Thus, it becomes easy to understand that when $\alpha = 1$, the special case of the exponential distribution occurs. The gamma density can be developed from its relationship to the Poisson process in much the same manner as we developed the exponential density. The details are left to the reader. The following is a numerical example of the use of the gamma distribution in a waiting-time application.

Example 3.34: Suppose that telephone calls arriving at a particular switchboard follow a Poisson process with an average of 5 calls coming per minute. What is the probability that up to a minute will elapse by the time 2 calls have come in to the switchboard?

Solution: The Poisson process applies, with time until 2 Poisson events following a gamma distribution with $\beta = 1/5$ and $\alpha = 2$. Denote by X the time in minutes that transpires before 2 calls come. The required probability is given by

$$P(X \le 1) = \int_0^1 \frac{1}{\beta^2} x e^{-x/\beta} \, dx = 25 \int_0^1 x e^{-5x} \, dx = 1 - e^{-5}(1 + 5) = 0.96.$$

While the origin of the gamma distribution deals in time (or space) until the occurrence of α Poisson events, there are many instances where a gamma distribution works very well even though there is no clear Poisson structure. This is particularly true for **survival time** problems in both engineering and biomedical applications.

Example 3.35: In a biomedical study with rats, a dose-response investigation is used to determine the effect of the dose of a toxicant on their survival time. The toxicant is one that is frequently discharged into the atmosphere from jet fuel. For a certain dose of the toxicant, the study determines that the survival time, in weeks, has a gamma distribution with $\alpha = 5$ and $\beta = 10$. What is the probability that a rat survives no longer than 60 weeks?

Solution: Let the random variable X be the survival time (time to death). The required probability is

$$P(X \le 60) = \frac{1}{\beta^5} \int_0^{60} \frac{x^{\alpha-1} e^{-x/\beta}}{\Gamma(5)} \, dx.$$

The integral above can be solved through the use of the **incomplete gamma function**, which becomes the cumulative distribution function for the gamma distribution. This function is written as

$$F(x; \alpha) = \int_0^x \frac{y^{\alpha-1} e^{-y}}{\Gamma(\alpha)} \, dy.$$

If we let $y = x/\beta$, so $x = \beta y$, we have

$$P(X \le 60) = \int_0^6 \frac{y^4 e^{-y}}{\Gamma(5)} \, dy,$$

which is denoted as $F(6; 5)$ in the table of the incomplete gamma function in Appendix A.11. Note that this allows a quick computation of probabilities for the gamma distribution. Indeed, for this problem, the probability that the rat survives no longer than 60 days is given by

$$P(X \le 60) = F(6; 5) = 0.715.$$

Example 3.36: It is known, from previous data, that the length of time in months between customer complaints about a certain product is a gamma distribution with $\alpha = 2$ and $\beta = 4$. Changes were made to tighten quality control requirements. Following these changes, 20 months passed before the first complaint. Does it appear as if the quality control tightening was effective?

Solution: Let X be the time to the first complaint, which, under conditions prior to the changes, followed a gamma distribution with $\alpha = 2$ and $\beta = 4$. The question centers around how rare $X \geq 20$ is, given that α and β remain at values 2 and 4, respectively. In other words, under the prior conditions is a "time to complaint" as large as 20 months reasonable? Thus, following the solution to Example 3.35,

$$P(X \geq 20) = 1 - \frac{1}{\beta^\alpha} \int_0^{20} \frac{x^{\alpha-1} e^{-x/\beta}}{\Gamma(\alpha)} \, dx.$$

Again, using $y = x/\beta$, we have

$$P(X \geq 20) = 1 - \int_0^5 \frac{y e^{-y}}{\Gamma(2)} \, dy = 1 - F(5; 2) = 1 - 0.96 = 0.04,$$

where $F(5; 2) = 0.96$ is found from Table A.11.

As a result, we could conclude that the conditions of the gamma distribution with $\alpha = 2$ and $\beta = 4$ are not supported by the data that an observed time to complaint is as large as 20 months. Thus, it is reasonable to conclude that the quality control work was effective.

Example 3.37: Consider Exercise 2.23 on page 61. Based on extensive testing, it is determined that the time Y in years before a major repair is required for a certain washing machine is characterized by the density function

$$f(y) = \begin{cases} \frac{1}{4} e^{-y/4}, & y \geq 0, \\ 0, & \text{elsewhere.} \end{cases}$$

Note that Y is an exponential random variable with $\mu = 4$ years. The machine is considered a bargain if it is unlikely to require a major repair before the sixth year. What is the probability $P(Y > 6)$? What is the probability that a major repair is required in the first year?

Solution: Consider the cumulative distribution function $F(y)$ for the exponential distribution,

$$F(y) = \frac{1}{\beta} \int_0^y e^{-t/\beta} \, dt = 1 - e^{-y/\beta}.$$

Then

$$P(Y > 6) = 1 - F(6) = e^{-3/2} = 0.2231.$$

Thus, the probability that the washing machine will require major repair after year six is 0.223. Of course, it will require repair before year six with probability 0.777. Thus, one might conclude the machine is not really a bargain. The probability that a major repair is necessary in the first year is

$$P(Y < 1) = 1 - e^{-1/4} = 1 - 0.779 = 0.221.$$

3.12 Chi-Squared Distribution

Another very important special case of the gamma distribution is obtained by letting $\alpha = v/2$ and $\beta = 2$, where v is a positive integer. The result is called **the chi-squared distribution**. The distribution has a single parameter, v, called the **degrees of freedom**.

Chi-Squared
Distribution

The continuous random variable X has a **chi-squared distribution**, with v **degrees of freedom**, if its density function is given by

$$f(x;v) = \begin{cases} \frac{1}{2^{v/2}\Gamma(v/2)}x^{v/2-1}e^{-x/2}, & x > 0, \\ 0, & \text{elsewhere,} \end{cases}$$

where v is a positive integer.

The chi-squared distribution plays a vital role in statistical inference. It has considerable applications in both methodology and theory. While we do not discuss applications in detail in this chapter, it is important to understand that Chapters 4 and 5 contain important applications. The chi-squared distribution is an important component of statistical hypothesis testing and estimation.

Topics dealing with sampling distributions, analysis of variance, and nonparametric statistics involve extensive use of the chi-squared distribution.

Theorem 3.10: The mean and variance of the chi-squared distribution are

$$\mu = v \text{ and } \sigma^2 = 2v.$$

Exercises

3.88 In a certain city, the daily consumption of water (in millions of liters) follows approximately a gamma distribution with $\alpha = 2$ and $\beta = 3$. If the daily capacity of that city is 9 million liters of water, what is the probability that on any given day the water supply is inadequate?

3.89 If a random variable X has the gamma distribution with $\alpha = 2$ and $\beta = 1$, find $P(1.8 < X < 2.4)$.

3.90 Suppose that the time, in hours, required to repair a heat pump is a random variable X having a gamma distribution with parameters $\alpha = 2$ and $\beta = 1/2$. What is the probability that on the next service call

(a) at most 1 hour will be required to repair the heat pump?

(b) at least 2 hours will be required to repair the heat pump?

3.91 Find the mean and variance of the daily water consumption in Exercise 3.88.

3.92 In a certain city, the daily consumption of electric power, in millions of kilowatt hours, is a random variable X having a gamma distribution with mean $\mu = 6$ and variance $\sigma^2 = 12$.

(a) Find the values of α and β.

(b) Find the probability that on any given day the daily power consumption will exceed 12 million kilowatt hours.

3.93 The length of time for one individual to be served at a cafeteria is a random variable having an exponential distribution with a mean of 4 minutes. What is the probability that a person will be served in less than 3 minutes on at least 4 of the next 6 days?

3.94 The life, in years, of a certain type of electrical switch has an exponential distribution with an average life $\beta = 2$. If 100 of these switches are installed in different systems, what is the probability that at most 30 fail during the first year?

3.95 In a biomedical research study, it was determined that the survival time, in weeks, of an animal subjected to a certain exposure of gamma radiation has a gamma distribution with $\alpha = 5$ and $\beta = 10$.

(a) What is the mean survival time of a randomly selected animal of the type used in the experiment?

(b) What is the standard deviation of survival time?

(c) What is the probability that an animal survives more than 30 weeks?

3.96 The lifetime, in weeks, of a certain type of transistor is known to follow a gamma distribution with mean 10 weeks and standard deviation $\sqrt{50}$ weeks.

(a) What is the probability that a transistor of this type will last at most 50 weeks?

(b) What is the probability that a transistor of this type will not survive the first 10 weeks?

3.97 Computer response time is an important application of the gamma and exponential distributions. Suppose that a study of a certain computer system reveals that the response time, in seconds, has an exponential distribution with a mean of 3 seconds.

(a) What is the probability that response time exceeds 5 seconds?

(b) What is the probability that response time exceeds 10 seconds?

3.98 The number of automobiles that arrive at a certain intersection per minute has a Poisson distribution with a mean of 5. Interest centers around the time that elapses before 10 automobiles appear at the intersection.

(a) What is the probability that more than 10 automobiles appear at the intersection during any given minute of time?

(b) What is the probability that more than 2 minutes elapse before 10 cars arrive?

3.99 Consider the information in Exercise 3.98.

(a) What is the probability that more than 1 minute elapses between arrivals?

(b) What is the mean number of minutes that elapse between arrivals?

Review Exercises

3.100 During a manufacturing process, 15 units are randomly selected each day from the production line to check the percent defective. From historical information it is known that the probability of a defective unit is 0.05. Any time 2 or more defectives are found in the sample of 15, the process is stopped. This procedure is used to provide a signal in case the probability of a defective has increased.

(a) What is the probability that on any given day the production process will be stopped? (Assume 5% defective.)

(b) Suppose that the probability of a defective has increased to 0.07. What is the probability that on any given day the production process will not be stopped?

3.101 An automatic welding machine is being considered for use in a production process. It will be considered for purchase if it is successful on 99% of its welds. Otherwise, it will not be considered efficient. A test is to be conducted with a prototype that is to perform 100 welds. The machine will be accepted if it misses no more than 3 welds.

(a) What is the probability that a good machine will be rejected?

(b) What is the probability that an inefficient machine with 95% welding success will be accepted?

3.102 Service calls come to a maintenance center according to a Poisson process, and on average, 2.7 calls are received per minute. Find the probability that

(a) no more than 4 calls come in any minute;

(b) fewer than 2 calls come in any minute;

(c) more than 10 calls come in a 5-minute period.

3.103 An electronics firm claims that the proportion of defective units from a certain process is 5%. A buyer has a standard procedure of inspecting 15 units selected randomly from a large lot. On a particular occasion, the buyer found 5 items defective.

(a) What is the probability of this occurrence, given that the claim of 5% defective is correct?

(b) What would be your reaction if you were the buyer?

3.104 An electronic switching device occasionally malfunctions, but the device is considered satisfactory if it makes, on average, no more than 0.20 error per hour. A particular 5-hour period is chosen for testing the device. If no more than 1 error occurs during the time period, the device will be considered satisfactory.

(a) What is the probability that a satisfactory device will be considered unsatisfactory on the basis of the test? Assume a Poisson process.

(b) What is the probability that a device will be accepted as satisfactory when, in fact, the mean number of errors is 0.25? Again, assume a Poisson process.

3.105 A company generally purchases large lots of a certain kind of electronic device. A method is used that rejects a lot if 2 or more defective units are found in a random sample of 100 units.

(a) What is the probability of rejecting a lot that is 1% defective?

(b) What is the probability of accepting a lot that is 5% defective?

3.106 Imperfections in computer circuit boards and computer chips lend themselves to statistical treatment. For a particular type of board, the probability of a diode failure is 0.03 and the board contains 200 diodes.

(a) What is the mean number of failures among the diodes?

(b) What is the variance?

(c) The board will work if there are no defective diodes. What is the probability that a board will work?

3.107 The potential buyer of a particular engine requires (among other things) that the engine start successfully 10 consecutive times. Suppose the probability of a successful start is 0.990. Let us assume that the outcomes of attempted starts are independent.

(a) What is the probability that the engine is accepted after only 10 starts?

(b) What is the probability that 12 attempted starts are made during the acceptance process?

3.108 The acceptance scheme for purchasing lots containing a large number of batteries is to test no more than 75 randomly selected batteries and to reject a lot if a single battery fails. Suppose the probability of a failure is 0.001.

(a) What is the probability that a lot is accepted?

(b) What is the probability that a lot is rejected on the 20th test?

(c) What is the probability that it is rejected in 10 or fewer trials?

3.109 An oil drilling company ventures into various locations, and its success or failure is independent from one location to another. Suppose the probability of a success at any specific location is 0.25.

(a) What is the probability that the driller drills at 10 locations and has 1 success?

(b) The driller will go bankrupt if it drills 10 times before the first success occurs. What are the driller's prospects for bankruptcy?

3.110 Consider the information in Review Exercise 3.109. The drilling engineer feels that the driller will "hit it big" if the second success occurs on or before the sixth attempt. What is the probability that the driller will hit it big?

3.111 It is known by researchers that 1 in 100 people carries a gene that leads to the inheritance of a certain chronic disease. In a random sample of 1000 individ-

uals, what is the probability that fewer than 7 individuals carry the gene? Use a Poisson approximation. Again, using the approximation, what is the approximate mean number of people out of 1000 carrying the gene?

3.112 A production process produces electronic component parts. It is presumed that the probability of a defective part is 0.01. During a test of this presumption, 500 parts are sampled randomly and 15 defectives are observed.

(a) What is your response to the presumption that the process is 1% defective? Be sure that a computed probability accompanies your comment.

(b) Under the presumption of a 1% defective process, what is the probability that only 3 parts will be found defective?

(c) Do parts (a) and (b) again using the Poisson approximation.

3.113 A production process outputs items in lots of 50. Sampling plans exist in which lots are pulled aside periodically and exposed to a certain type of inspection. It is usually assumed that the proportion defective is very small. It is important to the company that lots containing defectives be a rare event. The current inspection plan is to periodically sample randomly 10 out of the 50 items in a lot and, if none are defective, to perform no intervention.

(a) Suppose in a lot chosen at random, 2 out of 50 are defective. What is the probability that at least 1 in the sample of 10 from the lot is defective?

(b) From your answer to part (a), comment on the quality of this sampling plan.

(c) What is the mean number of defects found out of 10 items sampled?

3.114 Consider the situation of Review Exercise 3.113. It has been determined that the sampling plan should be extensive enough that there is a high probability, say 0.9, that if as many as 2 defectives exist in the lot of 50 being sampled, at least 1 will be found in the sampling. With these restrictions, how many of the 50 items should be sampled?

3.115 National security requires that defense technology be able to detect incoming projectiles or missiles. To make the defense system successful, multiple radar screens are required. Suppose that three independent screens are to be operated and the probability that any one screen will detect an incoming missile is 0.8. Obviously, if no screens detect an incoming projectile, the system is untrustworthy and must be improved.

(a) What is the probability that an incoming missile will not be detected by any of the three screens?

(b) What is the probability that the missile will be detected by only one screen?

(c) What is the probability that it will be detected by at least two out of three screens?

3.116 Suppose it is important that the overall missile defense system be as near perfect as possible.

(a) Assuming the quality of the screens is as indicated in Review Exercise 3.115, how many are needed to ensure that the probability that a missile gets through undetected is 0.0001?

(b) Suppose it is decided to stay with only 3 screens and attempt to improve the screen detection ability. What must the individual screen effectiveness (i.e., probability of detection) be in order to achieve the effectiveness required in part (a)?

3.117 Go back to Review Exercise 3.113(a). Recompute the probability using the binomial distribution. Comment.

3.118 There are two vacancies in a certain university statistics department. Five individuals apply. Two have expertise in linear models, and one has expertise in applied probability. The search committee is instructed to choose the two applicants randomly.

(a) What is the probability that the two chosen are those with expertise in linear models?

(b) What is the probability that of the two chosen, one has expertise in linear models and one has expertise in applied probability?

3.119 The manufacturer of a tricycle for children has received complaints about defective brakes in the product. According to the design of the product and considerable preliminary testing, it had been determined that the probability of the kind of defect in the complaint was 1 in 10,000 (i.e., 0.0001). After a thorough investigation of the complaints, it was determined that during a certain period of time, 200 products were randomly chosen from production and 5 had defective brakes.

(a) Comment on the "1 in 10,000" claim by the manufacturer. Use a probabilistic argument. Use the binomial distribution for your calculations.

(b) Repeat part (a) using the Poisson approximation.

3.120 Group Project: Divide the class into two groups of approximately equal size. The students in group 1 will each toss a coin 10 times (n_1) and count the number of heads obtained. The students in group 2 will each toss a coin 40 times (n_2) and again count the number of heads. The students in each group should individually compute the proportion of heads observed, which is an estimate of p, the probability of observing a head. Thus, there will be a set of values of p_1 (from group 1) and a set of values p_2 (from group 2). All of the values of p_1 and p_2 are estimates of 0.5, which is the true value of the probability of observing a head for a fair coin.

(a) Which set of values is consistently closer to 0.5, the values of p_1 or p_2? Consider the proof of Theorem 3.1 on page 105 with regard to the estimates of the parameter $p = 0.5$. The values of p_1 were obtained with $n = n_1 = 10$, and the values of p_2 were obtained with $n = n_2 = 40$. Using the notation of the proof, the estimates are given by

$$p_1 = \frac{x_1}{n_1} = \frac{I_1 + \cdots + I_{n_1}}{n_1},$$

where $I_1, \ldots, I_{n_1}$ are 0s and 1s and $n_1 = 10$, and

$$p_2 = \frac{x_2}{n_2} = \frac{I_1 + \cdots + I_{n_2}}{n_2},$$

where $I_1, \ldots, I_{n_2}$, again, are 0s and 1s and $n_2 = 40$.

(b) Referring again to Theorem 3.1, show that

$$E(p_1) = E(p_2) = p = 0.5.$$

(c) Show that $\sigma_{p_1}^2 = \frac{\sigma_{X_1}^2}{n_1}$ is 4 times the value of $\sigma_{p_2}^2 = \frac{\sigma_{X_2}^2}{n_2}$. Then explain further why the values of p_2 from group 2 are more consistently closer to the true value, $p = 0.5$, than the values of p_1 from group 1.

3.121 According to a study published by a group of sociologists at the University of Massachusetts, approximately 49% of the Valium users in the state of Massachusetts are white-collar workers. What is the probability that between 482 and 510, inclusive, of the next 1000 randomly selected Valium users from this state are white-collar workers?

3.122 The exponential distribution is frequently applied to the waiting times between successes in a Poisson process. If the number of calls received per hour by a telephone answering service is a Poisson random variable with parameter $\lambda = 6$, we know that the time, in hours, between successive calls has an exponential distribution with parameter $\beta = 1/6$. What is the probability of waiting more than 15 minutes between any two successive calls?

3.123 A manufacturer of a certain type of large machine wishes to buy rivets from one of two manufacturers. It is important that the breaking strength of each rivet exceed 10,000 psi. Two manufacturers (A and B) offer this type of rivet and both have rivets whose breaking strength is normally distributed. The mean breaking strengths for manufacturers A and B are 14,000 psi and 13,000 psi, respectively. The standard deviations are 2000 psi and 1000 psi, respectively. Which manufacturer will produce, on the average, the fewest number of defective rivets?

3.124 A certain type of device has an advertised failure rate of 0.01 per hour. The failure rate is constant and the exponential distribution applies.

(a) What is the mean time to failure?

(b) What is the probability that 200 hours will pass before a failure is observed?

3.125 In a chemical processing plant, it is important that the yield of a certain type of batch product stay above 80%. If it stays below 80% for an extended period of time, the company loses money. Occasional defective batches are of little concern. But if several batches per day are defective, the plant shuts down and adjustments are made. It is known that the yield is normally distributed with standard deviation 4%.

(a) What is the probability of a "false alarm" (yield below 80%) when the mean yield is 85%?

(b) What is the probability that a batch will have a yield that exceeds 80% when in fact the mean yield is 79%?

3.126 For an electrical component with a failure rate of once every 5 hours, it is important to consider the time that it takes for 2 components to fail.

(a) Assuming that the gamma distribution applies, what is the mean time that it takes for 2 components to fail?

(b) What is the probability that 12 hours will elapse before 2 components fail?

3.127 The elongation of a steel bar under a particular load has been established to be normally distributed with a mean of 0.05 inch and $\sigma = 0.01$ inch. Find the probability that the elongation is

(a) above 0.1 inch;

(b) below 0.04 inch;

(c) between 0.025 and 0.065 inch.

3.128 A controlled satellite is known to have an error (distance from target) that is normally distributed with mean zero and standard deviation 4 feet. The manufacturer of the satellite defines a success as a launch in which the satellite comes within 10 feet of the target. Compute the probability that the satellite fails.

3.129 A technician plans to test a certain type of resin developed in the laboratory to determine the nature of the time required before bonding takes place.

It is known that the mean time to bonding is 3 hours and the standard deviation is 0.5 hour. It will be considered an undesirable product if the bonding time is either less than 1 hour or more than 4 hours. Comment on the utility of the resin. How often would its performance be considered undesirable? Assume that time to bonding is normally distributed.

3.130 The average rate of water usage (thousands of gallons per hour) by a certain community is known to involve the lognormal distribution with parameters $\mu = 5$ and $\sigma = 2$. It is important for planning purposes to get a sense of periods of high usage. What is the probability that, for any given hour, 50,000 gallons of water are used?

3.131 For Review Exercise 3.130, what is the mean of the average water usage per hour in thousands of gallons?

3.132 In Exercise 3.96 on page 151, the lifetime of a transistor is assumed to have a gamma distribution with mean 10 weeks and standard deviation $\sqrt{50}$ weeks. Suppose that the gamma distribution assumption is incorrect. Assume that the distribution is normal.

(a) What is the probability that a transistor will last at most 50 weeks?

(b) What is the probability that a transistor will not survive for the first 10 weeks?

(c) Comment on the difference between your results here and those found in Exercise 3.96 on page 151.

3.133 Consider now Review Exercise 2.114 on page 96. The density function of the time Z in minutes between calls to an electrical supply store is given by

$$f(z) = \begin{cases} \frac{1}{10}e^{-z/10}, & 0 < z < \infty, \\ 0, & \text{elsewhere.} \end{cases}$$

(a) What is the mean time between calls?

(b) What is the variance in the time between calls?

(c) What is the probability that the time between calls exceeds the mean?

3.134 Consider Review Exercise 3.133. Given the assumption of the exponential distribution, what is the mean number of calls per hour? What is the variance in the number of calls per hour?

3.135 In a human factor experimental project, it has been determined that the reaction time of a pilot to a visual stimulus is normally distributed with a mean of 1/2 second and standard deviation of 2/5 second.

(a) What is the probability that a reaction from the pilot takes more than 0.3 second?

(b) What reaction time is exceeded 95% of the time?

3.136 The length of time between breakdowns of an essential piece of equipment is important in the decision as to whether to use auxiliary equipment. An engineer thinks that the best model for time between breakdowns of a generator is the exponential distribution with a mean of 15 days.

(a) If the generator has just broken down, what is the probability that it will break down in the next 21 days?

(b) What is the probability that the generator will operate for 30 days without a breakdown?

3.137 The length of time, in seconds, that a computer user takes to read his or her e-mail is distributed as a lognormal random variable with $\mu = 1.8$ and $\sigma^2 = 4.0$.

(a) What is the probability that a user reads e-mail for more than 20 seconds? More than a minute?

(b) What is the probability that a user reads e-mail for a length of time that is equal to the mean of the underlying lognormal distribution?

3.138 Group Project: Have groups of students observe the number of people who enter a specific coffee shop or fast-food restaurant over the course of an hour, beginning at the same time every day, for two weeks. The hour should be a time of peak traffic at the shop or restaurant. The data collected will be the number of customers who enter the shop in each half hour of time. Thus, two data points will be collected each day. Let us assume that the random variable X, the number of people entering each half hour, follows a Poisson distribution. The students should calculate the sample mean and variance of X using the 28 data points collected.

(a) What evidence indicates that the Poisson distribution assumption may or may not be correct?

(b) Given that X is Poisson, what is the distribution of T, the time between arrivals into the shop during a half hour period? Give a numerical estimate of the parameter of that distribution.

(c) Give an estimate of the probability that the time between two arrivals is less than 15 minutes.

(d) What is the estimated probability that the time between two arrivals is more than 10 minutes?

(e) What is the estimated probability that 20 minutes after the start of data collection not one customer has appeared?

3.13 Potential Misconceptions and Hazards; Relationship to Material in Other Chapters

The discrete distributions discussed in this chapter occur with great frequency in engineering and the biological and physical sciences. The exercises and examples certainly suggest this. Industrial sampling plans and many engineering judgments are based on the binomial and Poisson distributions as well as on the hypergeometric distribution. While the geometric and negative binomial distributions are used to a somewhat lesser extent, they also find applications. In particular, a negative binomial random variable can be viewed as a mixture of Poisson and gamma random variables, which are also discussed in this chapter.

Despite the rich heritage that these distributions find in real life, they can be misused unless the scientific practitioner is prudent and cautious. Of course, any probability calculation for the distributions discussed in this chapter is made under the assumption that the parameter value is known. Real-world applications often result in a parameter value that may "move around" due to factors that are difficult to control in the process or because of interventions in the process that have not been taken into account.

For the continuous distribution case, one of the biggest misuses of statistics is the assumption of an underlying normal distribution in carrying out a type of statistical inference when indeed the distribution is not normal. The reader will be exposed to tests of hypotheses in Chapters 6 through 9 in which the normality assumption is made. In addition, however, the reader will be reminded that there are **tests of goodness of fit** discussed in Chapter 6 that allow for checks on data to determine if the normality assumption is reasonable.

Similar warnings should be conveyed regarding assumptions that are often made concerning other distributions, apart from the normal. This chapter has presented examples in which one is required to calculate probabilities to failure of a certain item or the probability that one observes a complaint during a certain time period. Assumptions are made concerning a certain distribution type as well as values of parameters of the distributions. Note that parameter values (for example, the value of β for the exponential distribution) were given in the example problems. However, in real-life problems, parameter values must be estimates from real-life experience or data.

Chapter 4

Sampling Distributions and Data Descriptions

4.1 Random Sampling

The outcome of a statistical experiment may be recorded either as a numerical value or as a descriptive representation. When a pair of dice is tossed and the total is the outcome of interest, we record a numerical value. However, if the students of a certain school are given blood tests and the type of blood is of interest, then a descriptive representation might be more useful. A person's blood can be classified in 8 ways: AB, A, B, or O, each with a plus or minus sign, depending on the presence or absence of the Rh antigen.

In this chapter, we focus on sampling from distributions or populations and study such important quantities as the sample mean and sample variance, which will be of vital importance in future chapters. In addition, we attempt to give the reader an introduction to the role that the sample mean and variance will play in statistical inference in later chapters. The use of modern high-speed computers allows the scientist or engineer to greatly enhance his or her use of formal statistical inference with graphical techniques. Much of the time, formal inference appears quite dry and perhaps even abstract to the practitioner or to the manager who wishes to let statistical analysis be a guide to decision-making.

Populations and Samples

We begin this section by discussing the notions of *populations* and *samples*. Both are mentioned in a broad fashion in Chapter 1. However, much more needs to be presented about them here, particularly in the context of the concept of random variables. The totality of observations with which we are concerned, whether their number be finite or infinite, constitutes what we call a **population**. There was a time when the word *population* referred to observations obtained from statistical studies about people. Today, statisticians use the term to refer to observations relevant to anything of interest, whether it be groups of people, animals, or all possible outcomes from some complicated biological or engineering system.

Definition 4.1: | A **population** consists of the totality of the observations with which we are concerned.

The number of observations in the population is defined to be the size of the population. If there are 600 students in the school whom we classified according to blood type, we say that we have a population of size 600. The numbers on the cards in a deck, the heights of residents in a certain city, and the lengths of fish in a particular lake are examples of populations with finite size. In each case, the total number of observations is a finite number. The observations obtained by measuring the atmospheric pressure every day, from the past on into the future, or all measurements of the depth of a lake, from any conceivable position, are examples of populations whose sizes are infinite. Some finite populations are so large that in theory we assume them to be infinite. For example, this is true in the case of the population of lifetimes of a certain type of storage battery being manufactured for mass distribution throughout the country.

Each observation in a population is a value of a random variable X having some probability distribution $f(x)$. If one is inspecting items coming off an assembly line for defects, then each observation in the population might be a value 0 or 1 of the Bernoulli random variable X with probability distribution

$$b(x; 1, p) = p^x q^{1-x}, \qquad x = 0, 1$$

where 0 indicates a nondefective item and 1 indicates a defective item. Of course, it is assumed that p, the probability of any item being defective, remains constant from trial to trial. In the blood-type experiment, the random variable X represents the type of blood and is assumed to take on values from 1 to 8. Each student is given one of the values of the discrete random variable. The lives of the storage batteries are values assumed by a continuous random variable having perhaps a normal distribution. When we refer hereafter to a "binomial population," a "normal population," or, in general, the "population $f(x)$," we shall mean a population whose observations are values of a random variable having a binomial distribution, a normal distribution, or the probability distribution $f(x)$. Hence, the mean and variance of a random variable or probability distribution are also referred to as the mean and variance of the corresponding population.

In the field of statistical inference, statisticians are interested in arriving at conclusions concerning a population when it is impossible or impractical to observe the entire set of observations that make up the population. For example, in attempting to determine the average length of life of a certain brand of light bulb, it would be impossible to test all such bulbs if we are to have any left to sell. Exorbitant costs can also be a prohibitive factor in studying an entire population. Therefore, we must depend on a subset of observations from the population to help us make inferences concerning that population. This brings us to consider the notion of sampling.

Definition 4.2: | A **sample** is a subset of a population.

If our inferences from the sample to the population are to be valid, we must obtain samples that are representative of the population. All too often we are tempted to choose a sample by selecting the most convenient members of the population. Such a procedure may lead to erroneous inferences concerning the population. Any sampling procedure that produces inferences that consistently overestimate or consistently underestimate some characteristic of the population is said to be **biased**. To eliminate any possibility of bias in the sampling procedure, it is desirable to choose a **random sample** in the sense that the observations are made independently and at random.

In selecting a random sample of size n from a population $f(x)$, let us define the random variable X_i, $i = 1, 2, \ldots, n$, to represent the ith measurement or sample value that we observe. The random variables $X_1, X_2, \ldots, X_n$ will then constitute a random sample from the population $f(x)$ with numerical values $x_1, x_2, \ldots, x_n$ if the measurements are obtained by repeating the experiment n independent times under essentially the same conditions. Because of the identical conditions under which the elements of the sample are selected, it is reasonable to assume that the n random variables $X_1, X_2, \ldots, X_n$ are independent and that each has the same probability distribution $f(x)$. That is, the probability distributions of $X_1, X_2, \ldots, X_n$ are, respectively, $f(x_1), f(x_2), \ldots, f(x_n)$, and their joint probability distribution is $f(x_1, x_2, \ldots, x_n) = f(x_1)f(x_2) \cdots f(x_n)$. The concept of a random sample is described formally by the following definition.

Definition 4.3: Let $X_1, X_2, \ldots, X_n$ be n independent random variables, each having the same probability distribution $f(x)$. Define $X_1, X_2, \ldots, X_n$ to be a **random sample** of size n from the population $f(x)$ and write its joint probability distribution as

$$f(x_1, x_2, \ldots, x_n) = f(x_1)f(x_2) \cdots f(x_n).$$

If one makes a random selection of $n = 8$ storage batteries from a manufacturing process that has maintained the same specification throughout and records the length of life for each battery, with the first measurement x_1 being a value of X_1, the second measurement x_2 a value of X_2, and so forth, then $x_1, x_2, \ldots, x_8$ are the values of the random sample $X_1, X_2, \ldots, X_8$. If we assume the population of battery lives to be normal, the possible values of any X_i, $i = 1, 2, \ldots, 8$, will be precisely the same as those in the original population, and hence X_i has the same normal distribution as X.

4.2 Some Important Statistics

Our main purpose in selecting random samples is to elicit information about the unknown population parameters. Suppose, for example, that we wish to arrive at a conclusion concerning the proportion of coffee-drinkers in the United States who prefer a certain brand of coffee. It would be impossible to question every coffee-drinking American in order to compute the value of the parameter p representing the population proportion. Instead, a large random sample is selected and the proportion $\hat{p}$ of people in this sample favoring the brand of coffee in question is calculated. The value $\hat{p}$ is now used to make an inference concerning the true proportion p.

Now, $\hat{p}$ is a function of the observed values in the random sample; since many random samples are possible from the same population, we would expect $\hat{p}$ to vary somewhat from sample to sample. That is, $\hat{p}$ is a value of a random variable that we represent by P. Such a random variable is called a **statistic**.

Definition 4.4: Any function of the random variables constituting a random sample is called a **statistic**.

Location Measures of a Sample: The Sample Mean, Median, and Mode

In Chapter 2 we introduced the two parameters μ and σ^2, which are the mean and the variability of a probability distribution. These are constant population parameters and are in no way affected or influenced by the observations of a random sample. We shall, however, define some important statistics that describe corresponding measures of a random sample. The most commonly used statistics for measuring the center of a set of data are the **mean**, **median**, and **mode**. Suppose X_1 assumes the value x_1, X_2 assumes the value x_2, and so forth. Let

$$\bar{X} = \frac{1}{n} \sum_{i=1}^{n} X_i$$

represent the sample mean of the n random variables. The three measurements of location are defined as follows.

(a) The sample mean, denoted by $\bar{x}$, is

$$\bar{x} = \frac{1}{n} \sum_{i=1}^{n} x_i.$$

The term *sample mean* is applied to both the statistic $\bar{X}$ and its computed value $\bar{x}$.

(b) Given that the observations in a sample are $x_1, x_2, \ldots, x_n$, arranged in increasing order of magnitude, with x_1 being the smallest and x_n the largest, then the sample median, denoted by $\tilde{x}$, is

$$\tilde{x} = \begin{cases} x_{(n+1)/2}, & \text{if } n \text{ is odd,} \\ \frac{1}{2}(x_{n/2} + x_{n/2+1}), & \text{if } n \text{ is even.} \end{cases}$$

The sample median is also a location measure that shows the middle value of the sample.

(c) The sample mode is the value of the sample that occurs most often. If the highest frequency is shared by multiple values, then there are multiple modes.

Example 4.1: Suppose a data set consists of the following observations:

$$0.32 \ 0.53 \ 0.28 \ 0.37 \ 0.47 \ 0.43 \ 0.36 \ 0.42 \ 0.38 \ 0.43.$$

The sample mode is 0.43, since this value occurs more than any other value.

As we suggested in Chapter 2, a measure of location or central tendency in a sample does not by itself give a clear indication of the nature of the sample. Thus, a measure of variability in the sample must also be considered.

Variability Measures of a Sample: The Sample Variance, Standard Deviation, and Range

The variability in a sample displays how the observations spread out from the average. The reader is referred to Chapter 2 for more discussion. It is possible to have two sets of observations with the same mean or median that differ considerably in the variability of their measurements about the average.

Consider the following measurements, in liters, for two samples of orange juice bottled by companies A and B:

Sample A	0.97	1.00	0.94	1.03	1.06
Sample B	1.06	1.01	0.88	0.91	1.14

Both samples have the same mean, 1.00 liter. It is obvious that company A bottles orange juice with a more uniform content than company B. We say that the **variability**, or the **dispersion**, of the observations from the average is less for sample A than for sample B. Therefore, in buying orange juice, we can feel more confident that the bottle we select will be close to the advertised average if we buy from company A.

At this point, we shall introduce several measures of sampling variability. Let $X_1, \ldots, X_n$ represent n random variables.

(a) Sample variance:

$$S^2 = \frac{1}{n-1} \sum_{i=1}^{n} (X_i - \bar{X})^2. \tag{4.2.1}$$

The computed value of S^2 for a given sample is denoted by s^2. Note that S^2 is essentially defined to be the average of the squares of the deviations of the observations from their mean. The reason for using $n-1$ as a divisor rather than the more obvious choice n will become apparent in Chapter 5.

Example 4.2: A comparison of coffee prices at 4 randomly selected grocery stores in San Diego showed increases from the previous month of 12, 15, 17, and 20 cents for a 1-pound bag. Find the variance of this random sample of price increases.

Solution: Calculating the sample mean, we get

$$\bar{x} = \frac{12 + 15 + 17 + 20}{4} = 16 \text{ cents.}$$

Therefore,

$$s^2 = \frac{1}{3} \sum_{i=1}^{4} (x_i - 16)^2 = \frac{(12-16)^2 + (15-16)^2 + (17-16)^2 + (20-16)^2}{3}$$

$$= \frac{(-4)^2 + (-1)^2 + (1)^2 + (4)^2}{3} = \frac{34}{3}.$$

Whereas the expression for the sample variance best illustrates that S^2 is a measure of variability, an alternative expression does have some merit and thus the reader should be aware of it. The following theorem contains this expression.

Theorem 4.1: If S^2 is the variance of a random sample of size n, we may write

$$S^2 = \frac{1}{n(n-1)}\left[n\sum_{i=1}^{n}X_i^2 - \left(\sum_{i=1}^{n}X_i\right)^2\right].$$

Proof: By definition,

$$S^2 = \frac{1}{n-1}\sum_{i=1}^{n}(X_i - \bar{X})^2 = \frac{1}{n-1}\sum_{i=1}^{n}(X_i^2 - 2\bar{X}X_i + \bar{X}^2)$$

$$= \frac{1}{n-1}\left[\sum_{i=1}^{n}X_i^2 - 2\bar{X}\sum_{i=1}^{n}X_i + n\bar{X}^2\right].$$

The **sample standard deviation** and the **sample range** are defined below.

(b) Sample standard deviation:

$$S = \sqrt{S^2},$$

where S^2 is the sample variance.

(c) Sample range:

$$R = X_{\max} - X_{\min},$$

where $X_{\max}$ denotes the largest of the X_i values and $X_{\min}$ the smallest.

Example 4.3: Find the variance of the data 3, 4, 5, 6, 6, and 7, representing the number of trout caught by a random sample of 6 fishermen on June 19, 1996, at Lake Muskoka.

Solution: We find that $\sum_{i=1}^{6}x_i^2 = 171$, $\sum_{i=1}^{6}x_i = 31$, and $n = 6$. Hence,

$$s^2 = \frac{1}{(6)(5)}[(6)(171) - (31)^2] = \frac{13}{6}.$$

Thus, the sample standard deviation $s = \sqrt{13/6} = 1.47$ and the sample range is $7 - 3 = 4$.

Exercises

4.1 Define suitable populations from which the following samples are selected:

(a) Persons in 200 homes in the city of Richmond are called on the phone and asked to name the candidate they favor for election to the school board.

(b) A coin is tossed 100 times and 34 tails are recorded.

(c) Two hundred pairs of a new type of tennis shoe were tested on the professional tour and, on aver-

age, lasted 4 months.

(d) On five different occasions it took a lawyer 21, 26, 24, 22, and 21 minutes to drive from her suburban home to her midtown office.

4.2 The lengths of time, in minutes, that 10 patients waited in a doctor's office before receiving treatment were recorded as follows: 5, 11, 9, 5, 10, 15, 6, 10, 5, and 10. Treating the data as a random sample, find

(a) the mean;

(b) the median;

(c) the mode.

4.3 A random sample of employees from a local manufacturing plant pledged the following donations, in dollars, to the United Fund: 100, 40, 75, 15, 20, 100, 75, 50, 30, 10, 55, 75, 25, 50, 90, 80, 15, 25, 45, and 100. Calculate

(a) the mean;

(b) the mode.

4.4 According to ecology writer Jacqueline Killeen, phosphates contained in household detergents pass right through our sewer systems, causing lakes to turn into swamps that eventually dry up into deserts. The following data show the amount of phosphates per load of laundry, in grams, for a random sample of various types of detergents used according to the prescribed directions:

Laundry Detergent	Phosphates per Load (grams)
A&P Blue Sail	48
Dash	47
Concentrated All	42
Cold Water All	42
Breeze	41
Oxydol	34
Ajax	31
Sears	30
Fab	29
Cold Power	29
Bold	29
Rinso	26

For the given phosphate data, find

(a) the mean;

(b) the median;

(c) the mode.

4.5 Considering the data in Exercise 4.2, find

(a) the range;

(b) the standard deviation.

4.6 The numbers of tickets issued for traffic violations by 8 state troopers during the Memorial Day weekend are 5, 4, 7, 7, 6, 3, 8, and 6.

(a) If these values represent the number of tickets is-

sued by a random sample of 8 state troopers from Montgomery County in Virginia, define a suitable population.

(b) If the values represent the number of tickets issued by a random sample of 8 state troopers from South Carolina, define a suitable population.

4.7 For the data of Exercise 4.4, calculate the variance using the formula

(a) of form (4.2.1);

(b) in Theorem 4.1.

4.8 The tar contents of 8 brands of cigarettes selected at random from the latest list released by the Federal Trade Commission are as follows: 7.3, 8.6, 10.4, 16.1, 12.2, 15.1, 14.5, and 9.3 milligrams. Calculate

(a) the mean;

(b) the variance.

4.9 The grade-point averages of 20 college seniors selected at random from a graduating class are as follows:

$$3.2 \quad 1.9 \quad 2.7 \quad 2.4 \quad 2.8$$
$$2.9 \quad 3.8 \quad 3.0 \quad 2.5 \quad 3.3$$
$$1.8 \quad 2.5 \quad 3.7 \quad 2.8 \quad 2.0$$
$$3.2 \quad 2.3 \quad 2.1 \quad 2.5 \quad 1.9$$

Calculate the standard deviation.

4.10 (a) Show that the sample variance is unchanged if a constant c is added to or subtracted from each value in the sample.

(b) Show that the sample variance becomes c^2 times its original value if each observation in the sample is multiplied by c.

4.11 Verify that the variance of the sample 4, 9, 3, 6, 4, and 7 is 5.1, and using this fact, along with the results of Exercise 4.10, find

(a) the variance of the sample 12, 27, 9, 18, 12, and 21;

(b) the variance of the sample 9, 14, 8, 11, 9, and 12.

4.12 In the 2004-05 football season, University of Southern California had the following score differences for the 13 games it played:

$$11 \; 49 \; 32 \; 3 \; 6 \; 38 \; 38 \; 30 \; 8 \; 40 \; 31 \; 5 \; 36$$

Find

(a) the mean score difference;

(b) the median score difference.

4.3   Sampling Distributions

The field of statistical inference is basically concerned with generalizations and predictions. For example, we might claim, based on the opinions of several people interviewed on the street, that in a forthcoming election 60% of the eligible voters in the city of Detroit favor a certain candidate. In this case, we are dealing with a random sample of opinions from a very large finite population. As a second illustration we might state that the average cost to build a residence in Charleston, South Carolina, is between $330,000 and $335,000, based on the estimates of 3 contractors selected at random from the 30 now building in this city. The population being sampled here is again finite but very small. Finally, let us consider a soft-drink machine designed to dispense, on average, 240 milliliters per drink. A company analyst who computes the mean of 40 drinks obtains $\bar{x} = 236$ milliliters and, on the basis of this value, decides that the machine is still dispensing drinks with an average content of $\mu = 240$ milliliters. The 40 drinks represent a sample from the infinite population of possible drinks that will be dispensed by this machine.

Inference about the Population from Sample Information

In each of the examples above, we computed a statistic from a sample selected from the population, and from this statistic we made various statements concerning the values of population parameters that may or may not be true. The company official made the decision that the soft-drink machine dispenses drinks with an average content of 240 milliliters, even though the sample mean was 236 milliliters, because he knows from sampling theory that, if $\mu = 240$ milliliters, such a sample value could easily occur. In fact, if he ran similar tests, say every hour, he would expect the values of the statistic $\bar{x}$ to fluctuate above and below $\mu = 240$ milliliters. Only when the value of $\bar{x}$ is substantially different from 240 milliliters will the company analyst initiate action to adjust the machine.

Since a statistic is a random variable that depends only on the observed sample, it must have a probability distribution.

Definition 4.5: | The probability distribution of a statistic is called a **sampling distribution**.

The sampling distribution of a statistic depends on the distribution of the population, the size of the samples, and the method of choosing the samples. In the remainder of this chapter we study several of the important sampling distributions of frequently used statistics. Applications of these sampling distributions to problems of statistical inference are considered throughout most of the remaining chapters. The probability distribution of $\bar{X}$ is called the **sampling distribution of the mean**.

What Is the Sampling Distribution of $\bar{X}$?

We should view the sampling distributions of $\bar{X}$ and S^2 as the mechanisms from which we will be able to make inferences on the parameters μ and σ^2. The sampling distribution of $\bar{X}$ with sample size n is the distribution that results when an **experiment is conducted over and over** (always with sample size n) **and the many values of $\bar{X}$ result**. This sampling distribution, then, describes the variability of sample averages around the population mean μ. In the case of the soft-drink machine, knowledge of the sampling distribution of $\bar{X}$ arms the analyst with the knowledge of a "typical" discrepancy between an observed $\bar{x}$ value and true μ. The same principle applies in the case of the distribution of S^2. The sampling distribution produces information about the variability of s^2 values around σ^2 in repeated experiments.

4.4 Sampling Distribution of Means and the Central Limit Theorem

The first important sampling distribution to be considered is that of the mean $\bar{X}$. Suppose that a random sample of n observations is taken from a normal population with mean μ and variance σ^2. Each observation X_i, $i = 1, 2, \ldots, n$, of the random sample will then have the same normal distribution as the population being sampled. It can be shown that

$$\bar{X} = \frac{1}{n}(X_1 + X_2 + \cdots + X_n)$$

has a normal distribution with mean

$$\mu_{\bar{X}} = \frac{1}{n}(\underbrace{\mu + \mu + \cdots + \mu}_{n \text{ terms}}) = \mu \text{ and variance } \sigma_{\bar{X}}^2 = \frac{1}{n^2}(\underbrace{\sigma^2 + \sigma^2 + \cdots + \sigma^2}_{n \text{ terms}}) = \frac{\sigma^2}{n}.$$

If we are sampling from a population with unknown distribution, either finite or infinite, the sampling distribution of $\bar{X}$ will still be approximately normal with mean μ and variance σ^2/n, provided that the sample size is large. This amazing result is an immediate consequence of the following theorem, called the Central Limit Theorem. Figure 4.1 shows just how amazing this theorem is.

The Central Limit Theorem

Theorem 4.2: **Central Limit Theorem:** If $\bar{X}$ is the mean of a random sample of size n taken from a population with mean μ and finite variance σ^2, then the limiting form of the distribution of

$$Z = \frac{\bar{X} - \mu}{\sigma/\sqrt{n}},$$

as $n \to \infty$, is the standard normal distribution $n(z; 0, 1)$.

The normal approximation for $\bar{X}$ will generally be good if $n \geq 30$, provided the population distribution is not terribly skewed. If $n < 30$, the approximation is good only if the population is not too different from a normal distribution and, as stated above, if the population is known to be normal, the sampling distribution of $\bar{X}$ will follow a normal distribution exactly, no matter how small the size of the samples.

The sample size $n = 30$ is a guideline to use for the Central Limit Theorem. However, as the statement of the theorem implies, the presumption of normality on the distribution of $\bar{X}$ becomes more accurate as n grows larger. In fact, Figure 4.1 illustrates how the theorem works. It shows how the distribution of $\bar{X}$ becomes closer to normal as n grows larger, beginning with the clearly nonsymmetric distribution of an individual observation ($n = 1$), i.e., the population. It also illustrates that the mean of $\bar{X}$ remains μ for any sample size and the variance of $\bar{X}$ gets smaller as n increases.

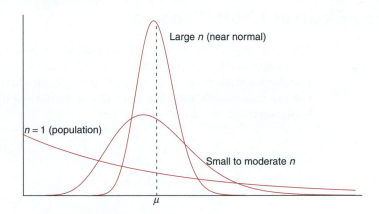

Figure 4.1: Illustration of the Central Limit Theorem (distribution of $\bar{X}$ for $n = 1$, small to moderate n, and large n).

The following example shows the use of the Central Limit Theorem to find the probability of $\bar{X}$ values.

Example 4.4: An electrical firm manufactures light bulbs that have a length of life that is approximately normally distributed, with mean equal to 800 hours and a standard deviation of 40 hours. Find the probability that a random sample of 16 bulbs will have an average life of less than 775 hours.

Solution: The sampling distribution of $\bar{X}$ will be approximately normal, with $\mu_{\bar{X}} = 800$ and $\sigma_{\bar{X}} = 40/\sqrt{16} = 10$, where the sample standard deviation is used to estimate the population standard deviation. The desired probability is given by the area of the shaded region in Figure 4.2.

Corresponding to $\bar{x} = 775$, we find that

$$z = \frac{\bar{X} - \mu}{s/\sqrt{n}} = \frac{775 - 800}{10} = -2.5,$$

and therefore

$$P(\bar{X} < 775) = P(Z < -2.5) = 0.0062,$$

where the probability of $Z < -2.5$ is taken from Table A.3 in the Appendix.

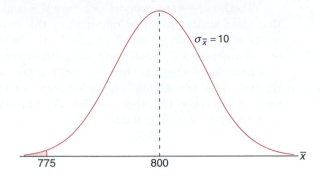

Figure 4.2: Area for Example 4.4.

Inferences on the Population Mean

One very important application of the Central Limit Theorem is the determination of reasonable values of the population mean μ. Topics such as hypothesis testing, estimation, quality control, and many others make use of the Central Limit Theorem. The following example illustrates the use of the Central Limit Theorem with regard to its relationship with μ, the mean of the population, although the formal application to the foregoing topics is relegated to future chapters.

In the following case study, an inference is drawn that makes use of the sampling distribution of $\bar{X}$. In this simple illustration, μ and σ are both known. The Central Limit Theorem and the general notion of sampling distributions are often used to produce evidence about some important aspect of a distribution such as a parameter of the distribution. In the case of the Central Limit Theorem, the parameter of interest is the mean μ. The inference made concerning μ may take one of many forms. Often there is a desire on the part of the analyst that the data (in the form of $\bar{x}$) support (or not) some predetermined conjecture concerning the value of μ. The use of what we know about the sampling distribution can contribute to answering this type of question. In this case study, the concept of hypothesis testing leads to a formal objective that we will highlight in future chapters.

Case Study 4.1: **Automobile Parts**: An important manufacturing process produces cylindrical component parts for the automotive industry. It is important that the process produce parts having a mean diameter of 5.0 millimeters. The engineer involved conjectures that the population mean is 5.0 millimeters. An experiment is conducted in which 100 parts produced by the process are selected randomly and the diameter measured on each. It is known that the population standard deviation is $\sigma = 0.1$ millimeter. The experiment indicates a sample average diameter of $\bar{x} = 5.027$ millimeters. Does this sample information appear to support or refute the engineer's conjecture?

Solution: This example reflects the kind of problem often posed and solved with hypothesis testing machinery introduced in future chapters. We will not use the formality associated with hypothesis testing here, but we will illustrate the principles and logic used.

Whether the data support or refute the conjecture depends on the probability that data similar to those obtained in this experiment ($\bar{x} = 5.027$) can readily occur when in fact $\mu = 5.0$ (Figure 4.3). In other words, how likely is it that one can obtain $\bar{x} \geq 5.027$ with $n = 100$ if the population mean is $\mu = 5.0$? If this probability suggests that $\bar{x} = 5.027$ is not unreasonable, the conjecture is not refuted. If the probability is quite low, one can certainly argue that the data do not support the conjecture that $\mu = 5.0$. The probability that we choose to compute is given by $P(|\bar{X} - 5| \geq 0.027)$.

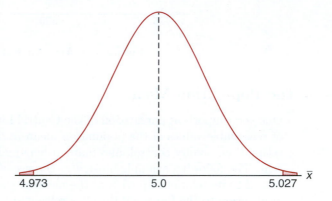

Figure 4.3: Area for Case Study 4.1.

In other words, if the mean μ is 5, what is the chance that $\bar{X}$ will deviate by as much as 0.027 millimeter?

$$P(|\bar{X} - 5| \geq 0.027) = P(\bar{X} - 5 \geq 0.027) + P(\bar{X} - 5 \leq -0.027)$$

$$= 2P\left(\frac{\bar{X} - 5}{0.1/\sqrt{100}} \geq 2.7\right).$$

Here we are simply standardizing $\bar{X}$ according to the Central Limit Theorem. If the conjecture $\mu = 5.0$ is true, $\frac{\bar{X}-5}{0.1/\sqrt{100}}$ should follow $N(0, 1)$. Thus,

$$2P\left(\frac{\bar{X} - 5}{0.1/\sqrt{100}} \geq 2.7\right) = 2P(Z \geq 2.7) = 2(0.0035) = 0.007.$$

Therefore, one would expect by chance that an $\bar{x}$ would be 0.027 millimeter from the mean in only 7 in 1000 experiments. As a result, this experiment with $\bar{x} = 5.027$ certainly does not give supporting evidence to the conjecture that $\mu = 5.0$. In fact, it strongly refutes the conjecture!

Example 4.5: Traveling between two campuses of a university in a city via shuttle bus takes, on average, 28 minutes with a standard deviation of 5 minutes. In a given week, a bus transported passengers 40 times. What is the probability that the average transport time, i.e., the average for 40 trips, was more than 30 minutes? Assume the mean time is measured to the nearest minute.

Solution: In this case, $\mu = 28$ and $\sigma = 5$. We need to calculate the probability $P(\bar{X} > 30)$ with $n = 40$. Since the time is measured on a continuous scale to the nearest minute, an $\bar{x}$ greater than 30 is equivalent to $\bar{x} \geq 30.5$. Hence,

$$P(\bar{X} > 30) = P\left(\frac{\bar{X} - 28}{5/\sqrt{40}} \geq \frac{30.5 - 28}{5/\sqrt{40}}\right) = P(Z \geq 3.16) = 0.0008.$$

There is only a slight chance that the average time of one bus trip will exceed 30 minutes. An illustrative graph is shown in Figure 4.4.

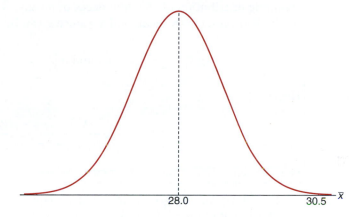

Figure 4.4: Area for Example 4.5.

Sampling Distribution of the Difference between Two Means

The illustration in Case Study 4.1 deals with notions of statistical inference on a single mean μ. The engineer was interested in supporting a conjecture regarding a single population mean. A far more important application involves two populations. A scientist or engineer may be interested in a comparative experiment in which two manufacturing methods, 1 and 2, are to be compared. The basis for that comparison is $\mu_1 - \mu_2$, the difference in the population means.

Suppose that we have two populations, the first with mean μ_1 and variance σ_1^2, and the second with mean μ_2 and variance σ_2^2. Let the statistic $\bar{X}_1$ represent the mean of a random sample of size n_1 selected from the first population, and the statistic $\bar{X}_2$ represent the mean of a random sample of size n_2 selected from the second population, independent of the sample from the first population. What can we say about the sampling distribution of the difference $\bar{X}_1 - \bar{X}_2$ for repeated samples of size n_1 and n_2? According to Theorem 4.2, the variables $\bar{X}_1$ and $\bar{X}_2$ are both approximately normally distributed with means μ_1 and μ_2 and variances σ_1^2/n_1 and σ_2^2/n_2, respectively. This approximation improves as n_1 and n_2 increase. By choosing independent samples from the two populations we ensure that the variables $\bar{X}_1$ and $\bar{X}_2$ will be independent, and we can conclude, based on the material in Chapter 2, that $\bar{X}_1 - \bar{X}_2$ is approximately normally distributed with mean

$$\mu_{\bar{X}_1 - \bar{X}_2} = \mu_{\bar{X}_1} - \mu_{\bar{X}_2} = \mu_1 - \mu_2$$

and variance

$$\sigma^2_{\bar{X}_1 - \bar{X}_2} = \sigma^2_{\bar{X}_1} + \sigma^2_{\bar{X}_2} = \frac{\sigma^2_1}{n_1} + \frac{\sigma^2_2}{n_2}.$$

The Central Limit Theorem can be easily extended to the two-sample, two-population case.

Theorem 4.3: If independent samples of size n_1 and n_2 are drawn at random from two populations with means μ_1 and μ_2 and variances σ^2_1 and σ^2_2, respectively, then the sampling distribution of the differences of means, $\bar{X}_1 - \bar{X}_2$, is approximately normally distributed with mean and variance given by

$$\mu_{\bar{X}_1 - \bar{X}_2} = \mu_1 - \mu_2 \text{ and } \sigma^2_{\bar{X}_1 - \bar{X}_2} = \frac{\sigma^2_1}{n_1} + \frac{\sigma^2_2}{n_2}.$$

Hence,

$$Z = \frac{(\bar{X}_1 - \bar{X}_2) - (\mu_1 - \mu_2)}{\sqrt{(\sigma^2_1/n_1) + (\sigma^2_2/n_2)}}$$

is approximately a standard normal variable.

If both n_1 and n_2 are greater than or equal to 30, the normal approximation for the distribution of $\bar{X}_1 - \bar{X}_2$ is very good when the underlying distributions are not too far away from normal. However, even when n_1 and n_2 are less than 30, the normal approximation is reasonably good except when the populations are decidedly nonnormal. Of course, if both populations are normal, then $\bar{X}_1 - \bar{X}_2$ has a normal distribution no matter what the sizes of n_1 and n_2 are.

The utility of the sampling distribution of the difference between two sample averages is very similar to that described in Case Study 4.1 on page 167 for the case of a single mean. Case Study 4.2 that follows focuses on the use of the difference between two sample means to support (or not) the conjecture that two population means are the same.

Case Study 4.2: **Paint Drying Time**: Two independent experiments are run in which two different types of paint are compared. Eighteen specimens are painted using type A, and the drying time, in hours, is recorded for each. The same is done with type B. The population standard deviations are both known to be 1.0 hour.

Assuming that the mean drying time is equal for the two types of paint, find $P(\bar{X}_A - \bar{X}_B > 1.0)$, where $\bar{X}_A$ and $\bar{X}_B$ are average drying times for samples of size $n_A = n_B = 18$.

Solution: From the sampling distribution of $\bar{X}_A - \bar{X}_B$, we know that the distribution is approximately normal with mean

$$\mu_{\bar{X}_A - \bar{X}_B} = \mu_A - \mu_B = 0$$

and variance

$$\sigma^2_{\bar{X}_A - \bar{X}_B} = \frac{\sigma^2_A}{n_A} + \frac{\sigma^2_B}{n_B} = \frac{1}{18} + \frac{1}{18} = \frac{1}{9}.$$

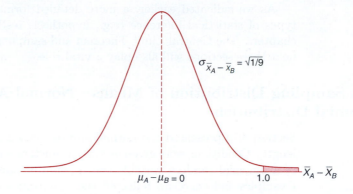

Figure 4.5: Area for Case Study 4.2.

The desired probability is given by the shaded region in Figure 4.5. Corresponding to the value $\bar{X}_A - \bar{X}_B = 1.0$, we have

$$z = \frac{1 - (\mu_A - \mu_B)}{\sqrt{1/9}} = \frac{1 - 0}{\sqrt{1/9}} = 3.0;$$

so

$$P(Z > 3.0) = 1 - P(Z < 3.0) = 1 - 0.9987 = 0.0013.$$

What Do We Learn from Case Study 4.2?

The machinery in the calculation is based on the presumption that $\mu_A = \mu_B$. Suppose, however, that the experiment is actually conducted for the purpose of drawing an inference regarding the equality of μ_A and μ_B, the two population mean drying times. If the two averages differ by as much as 1 hour (or more), this clearly is evidence that would lead one to conclude that the population mean drying time is not equal for the two types of paint, because the probability of such a result is very low if the means are indeed equal. On the other hand, suppose that the difference in the two sample averages is as small as, say, 15 minutes. If $\mu_A = \mu_B$,

$$P[(\bar{X}_A - \bar{X}_B) > 0.25 \text{ hour}] = P\left(\frac{\bar{X}_A - \bar{X}_B - 0}{\sqrt{1/9}} > \frac{3}{4}\right)$$

$$= P\left(Z > \frac{3}{4}\right) = 1 - P(Z < 0.75)$$

$$= 1 - 0.7734 = 0.2266.$$

Since this probability is not low, one would conclude that a difference in sample means of 15 minutes can happen by chance (i.e., it happens frequently even though $\mu_A = \mu_B$). As a result, that type of difference in average drying times certainly *is not a clear signal* that $\mu_A \neq \mu_B$.

As we indicated earlier, a more detailed formalism regarding this and other types of statistical inference (e.g., hypothesis testing) will be supplied in future chapters. The Central Limit Theorem and sampling distributions discussed in the next three sections will also play a vital role.

More on Sampling Distribution of Means—Normal Approximation to the Binomial Distribution

Section 3.10 presented the normal approximation to the binomial distribution at length. Conditions were given on the parameters n and p for which the distribution of a binomial random variable can be approximated by the normal distribution. Examples and exercises reflected the importance of the concept of the "normal approximation." It turns out that the Central Limit Theorem sheds even more light on how and why this approximation works. We certainly know that a binomial random variable is the number X of successes in n independent trials, where the outcome of each trial is binary. We also illustrated in Chapter 1 that the proportion computed in such an experiment is an average of a set of 0s and 1s. Indeed, while the proportion X/n is an average, X is the sum of this set of 0s and 1s, and both X and X/n are approximately normal if n is sufficiently large. Of course, from what we learned in Chapter 3, we know that there are conditions on n and p that affect the quality of the approximation, namely $np \geq 5$ and $nq \geq 5$.

Exercises

4.13 If all possible samples of size 16 are drawn from a normal population with mean equal to 50 and standard deviation equal to 5, what is the probability that a sample mean $\bar{X}$ will fall in the interval from $\mu_{\bar{X}} - 1.9\sigma_{\bar{X}}$ to $\mu_{\bar{X}} - 0.4\sigma_{\bar{X}}$? Assume that the sample means can be measured to any degree of accuracy.

4.14 A certain type of thread is manufactured with a mean tensile strength of 78.3 kilograms and a standard deviation of 5.6 kilograms. How is the variance of the sample mean changed when the sample size is

(a) increased from 64 to 196?

(b) decreased from 784 to 49?

4.15 A soft-drink machine is regulated so that the amount of drink dispensed averages 240 milliliters with a standard deviation of 15 milliliters. Periodically, the machine is checked by taking a sample of 40 drinks and computing the average content. If the mean of the 40 drinks is a value within the interval $\mu_{\bar{X}} \pm 2\sigma_{\bar{X}}$, the machine is thought to be operating satisfactorily; otherwise, adjustments are made. In Section 4.3, the company analyst found the mean of 40 drinks to be $\bar{x} = 236$ milliliters and concluded that the machine needed no adjustment. Was this a reasonable decision?

4.16 The heights of 1000 students are approximately normally distributed with a mean of 174.5 centimeters and a standard deviation of 6.9 centimeters. Suppose 200 random samples of size 25 are drawn from this population and the means recorded to the nearest tenth of a centimeter. Determine

(a) the mean and standard deviation of the sampling distribution of $\bar{X}$;

(b) the number of sample means that fall between 172.5 and 175.8 centimeters inclusive;

(c) the number of sample means falling below 172.0 centimeters.

4.17 The random variable X, representing the number of cherries in a cherry puff, has the following probability distribution:

x	4	5	6	7
$P(X = x)$	0.2	0.4	0.3	0.1

(a) Find the mean μ and the variance σ^2 of X.

(b) Find the mean $\mu_{\bar{X}}$ and the variance $\sigma^2_{\bar{X}}$ of the mean $\bar{X}$ for random samples of 36 cherry puffs.

(c) Find the probability that the average number of cherries in 36 cherry puffs will be less than 5.5.

4.18 If a certain machine makes electrical resistors

having a mean resistance of 40 ohms and a standard deviation of 2 ohms, what is the probability that a random sample of 36 of these resistors will have a combined resistance of more than 1458 ohms?

4.19 The average life of a bread-making machine is 7 years, with a standard deviation of 1 year. Assuming that the lives of these machines follow approximately a normal distribution, find

(a) the probability that the mean life of a random sample of 9 such machines falls between 6.4 and 7.2 years;

(b) the value of x to the right of which 15% of the means computed from random samples of size 9 would fall.

4.20 The amount of time that a drive-through bank teller spends on a customer is a random variable with a mean $\mu = 3.2$ minutes and a standard deviation $\sigma = 1.6$ minutes. If a random sample of 64 customers is observed, find the probability that their mean time at the teller's window is

(a) at most 2.7 minutes;

(b) more than 3.5 minutes;

(c) at least 3.2 minutes but less than 3.4 minutes.

4.21 In a chemical process, the amount of a certain type of impurity in the output is difficult to control and is thus a random variable. Speculation is that the population mean amount of the impurity is 0.20 gram per gram of output. It is known that the standard deviation is 0.1 gram per gram. An experiment is conducted to gain more insight regarding the speculation that $\mu = 0.2$. The process is run on a lab scale 50 times and the sample average $\bar{x}$ turns out to be 0.23 gram per gram. Comment on the speculation that the mean amount of impurity is 0.20 gram per gram. Make use of the Central Limit Theorem in your work.

4.22 The distribution of heights of a certain breed of terrier has a mean of 72 centimeters and a standard deviation of 10 centimeters, whereas the distribution of heights of a certain breed of poodle has a mean of 28 centimeters with a standard deviation of 5 centimeters. Assuming that the sample means can be measured to any degree of accuracy, find the probability that the sample mean for a random sample of heights of 64 terriers exceeds the sample mean for a random sample of heights of 100 poodles by at most 44.2 centimeters.

4.23 Consider Case Study 4.2 on page 170. Suppose 18 specimens were used for each type of paint in an experiment and $\bar{x}_A - \bar{x}_B$, the actual difference in mean drying time, turned out to be 1.0.

(a) Does this seem to be a reasonable result if the two population mean drying times truly are equal?

Make use of the result in the solution to Case Study 4.2.

(b) If someone did the experiment 10,000 times under the condition that $\mu_A = \mu_B$, in how many of those 10,000 experiments would there be a difference $\bar{x}_A - \bar{x}_B$ that was as large as (or larger than) 1.0?

4.24 Two different box-filling machines are used to fill cereal boxes on an assembly line. The critical measurement influenced by these machines is the weight of the product in the boxes. Engineers are quite certain that the variance of the weight of product is $\sigma^2 = 1$ ounce. Experiments are conducted using both machines with sample sizes of 36 each. The sample averages for machines A and B are $\bar{x}_A = 4.5$ ounces and $\bar{x}_B = 4.7$ ounces. Engineers are surprised that the two sample averages for the filling machines are so different.

(a) Use the Central Limit Theorem to determine

$$P(\bar{X}_B - \bar{X}_A \geq 0.2)$$

under the condition that $\mu_A = \mu_B$.

(b) Do the aforementioned experiments seem to, in any way, strongly support a conjecture that the population means for the two machines are different? Explain using your answer in (a).

4.25 The chemical benzene is highly toxic to humans. However, it is used in the manufacture of many medicine dyes, leather, and coverings. Government regulations dictate that for any production process involving benzene, the water in the output of the process must not exceed 7950 parts per million (ppm) of benzene. For a particular process of concern, the water sample was collected by a manufacturer 25 times randomly and the sample average $\bar{x}$ was 7960 ppm. It is known from historical data that the standard deviation σ is 100 ppm.

(a) What is the probability that the sample average in this experiment would exceed the government limit if the population mean is equal to the limit? Use the Central Limit Theorem.

(b) Is an observed $\bar{x} = 7960$ in this experiment firm evidence that the population mean for the process exceeds the government limit? Answer your question by computing

$$P(\bar{X} \geq 7960 \mid \mu = 7950).$$

Assume that the distribution of benzene concentration is normal.

4.26 Two alloys A and B are being used to manufacture a certain steel product. An experiment needs to be designed to compare the two in terms of maximum load capacity in tons (the maximum weight that can be tolerated without breaking). It is known that the two standard deviations in load capacity are equal at 5 tons each. An experiment is conducted in which 30 specimens of each alloy (A and B) are tested and the results recorded as follows:

$$\bar{x}_A = 49.5, \quad \bar{x}_B = 45.5; \qquad \bar{x}_A - \bar{x}_B = 4.$$

The manufacturers of alloy A are convinced that this evidence shows conclusively that $\mu_A > \mu_B$ and strongly supports the claim that their alloy is superior. Manufacturers of alloy B claim that the experiment could easily have given $\bar{x}_A - \bar{x}_B = 4$ *even if* the two population means are equal. In other words, "the results are inconclusive!"

(a) Make an argument that manufacturers of alloy B are wrong. Do it by computing

$$P(\bar{X}_A - \bar{X}_B > 4 \mid \mu_A = \mu_B).$$

(b) Do you think these data strongly support alloy A?

4.27 Consider the situation described in Example 4.4 on page 166. Do these results prompt you to question the premise that $\mu = 800$ hours? Give a probabilistic result that indicates how *rare* an event $\bar{X} \leq 775$ is when $\mu = 800$. On the other hand, how rare would it be if μ truly were, say, 760 hours?

4.5   Sampling Distribution of S^2

In the preceding section we learned about the sampling distribution of $\bar{X}$. The Central Limit Theorem allowed us to make use of the fact that

$$\frac{\bar{X} - \mu}{\sigma/\sqrt{n}}$$

tends toward $N(0, 1)$ as the sample size grows large. *Sampling distributions of important statistics* allow us to learn information about parameters. Usually, the parameters are the counterpart to the statistics in question. For example, if an engineer is interested in the population mean resistance of a certain type of resistor, the sampling distribution of $\bar{X}$ will be exploited once the sample information is gathered. On the other hand, if the variability in resistance is to be studied, clearly the sampling distribution of S^2 will be used in learning about the parametric counterpart, the population variance σ^2.

If a random sample of size n is drawn from a normal population with mean μ and variance σ^2, and the sample variance is computed, we obtain a value of the statistic S^2.

Theorem 4.4: If S^2 is the variance of a random sample of size n taken from a normal population having the variance σ^2, then the statistic

$$\chi^2 = \frac{(n-1)S^2}{\sigma^2} = \sum_{i=1}^{n} \frac{(X_i - \bar{X})^2}{\sigma^2}$$

has a chi-squared distribution with $v = n - 1$ degrees of freedom.

The values of the random variable χ^2 are calculated from each sample by the formula

$$\chi^2 = \frac{(n-1)s^2}{\sigma^2}.$$

The probability that a random sample produces a χ^2 value greater than some specified value is equal to the area under the curve to the right of this value. It is customary to let χ_α^2 represent the χ^2 value above which we find an area of α. This is illustrated by the shaded region in Figure 4.6.

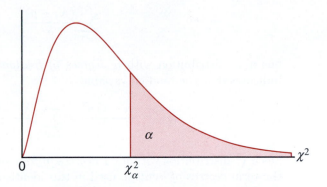

Figure 4.6: The chi-squared distribution.

Table A.5 gives values of χ_α^2 for various values of α and v. The areas, α, are the column headings; the degrees of freedom, v, are given in the left column; and the table entries are the χ^2 values. Hence, the χ^2 value with 7 degrees of freedom, leaving an area of 0.05 to the right, is $\chi_{0.05}^2 = 14.067$. Owing to lack of symmetry, we must also use the tables to find $\chi_{0.95}^2 = 2.167$ for $v = 7$.

Exactly 95% of a chi-squared distribution lies between $\chi_{0.975}^2$ and $\chi_{0.025}^2$. A χ^2 value falling to the right of $\chi_{0.025}^2$, i.e., greater than $\chi_{0.025}^2$, is not likely to occur unless our assumed value of σ^2 is too small. Similarly, a χ^2 value falling to the left of $\chi_{0.975}^2$ is unlikely unless our assumed value of σ^2 is too large. In other words, it is possible to have a χ^2 value to the left of $\chi_{0.975}^2$ or to the right of $\chi_{0.025}^2$ when σ^2 is correct, but if this should occur, it is more probable that the assumed value of σ^2 is in error.

Example 4.6: A manufacturer of car batteries guarantees that the batteries will last, on average, 3 years with a standard deviation of 1 year. If five of these batteries have lifetimes of 1.9, 2.4, 3.0, 3.5, and 4.2 years, should the manufacturer still be convinced that the batteries have a standard deviation of 1 year? Assume that the battery lifetime follows a normal distribution.

Solution: We first find the sample variance using Theorem 4.1,

$$s^2 = \frac{(5)(48.26) - (15)^2}{(5)(4)} = 0.815.$$

Then

$$\chi^2 = \frac{(4)(0.815)}{1} = 3.26$$

is a value from a chi-squared distribution with 4 degrees of freedom. Since 95% of the χ^2 values with 4 degrees of freedom fall between 0.484 and 11.143, the computed value with $\sigma^2 = 1$ is reasonable, and therefore the manufacturer has no reason to suspect that the standard deviation is other than 1 year.

Degrees of Freedom as a Measure of Sample Information

The statistic

$$\sum_{i=1}^{n} \frac{(X_i - \mu)^2}{\sigma^2}$$

has a χ^2-distribution with n *degrees of freedom*. Note also Theorem 4.4, which indicates that the random variable

$$\frac{(n-1)S^2}{\sigma^2} = \sum_{i=1}^{n} \frac{(X_i - \bar{X})^2}{\sigma^2}$$

has a χ^2-distribution with $n-1$ *degrees of freedom*. The reader may also recall that the term *degrees of freedom*, used in this identical context, is discussed in Chapter 3.

Although the proof of Theorem 4.4 was not given, the reader can view Theorem 4.4 as indicating that when μ is not known and one considers the distribution of

$$\sum_{i=1}^{n} \frac{(X_i - \bar{X})^2}{\sigma^2},$$

there is **1 less degree of freedom**, or a degree of freedom is lost in the estimation of μ (i.e., when μ is replaced by $\bar{x}$). In other words, there are n degrees of freedom, or independent *pieces of information*, in the random sample from the normal distribution. When the data (the values in the sample) are used to compute the mean, there is 1 less degree of freedom in the information used to estimate σ^2.

4.6   *t*-Distribution

In Section 4.4, we discussed the utility of the Central Limit Theorem. Its applications revolve around inferences on a population mean or the difference between two population means. Use of the Central Limit Theorem and the normal distribution is certainly helpful in this context. However, it was assumed that the population standard deviation is known. This assumption may not be unreasonable in situations where the engineer is quite familiar with the system or process. However, in many experimental scenarios, assuming knowledge of σ is certainly no more reasonable than assuming knowledge of the population mean μ. Often, in fact, an estimate of σ must be supplied by the same sample information that produced the sample average $\bar{x}$. As a result, a natural statistic to consider to deal with inferences on μ is

$$T = \frac{\bar{X} - \mu}{S/\sqrt{n}},$$

since S is the sample analog to σ. If the sample size is small, the values of S^2 fluctuate considerably from sample to sample (see Exercise 4.34 on page 191) and the distribution of T deviates appreciably from that of a standard normal distribution.

If the sample size is large enough, say $n \geq 30$, the distribution of T does not differ considerably from the standard normal. However, for $n < 30$, it is useful to deal with the exact distribution of T. In developing the sampling distribution of T, we shall assume that our random sample was selected from a normal population. We can then write

$$T = \frac{(\bar{X} - \mu)/(\sigma/\sqrt{n})}{\sqrt{S^2/\sigma^2}} = \frac{Z}{\sqrt{V/(n-1)}},$$

where

$$Z = \frac{\bar{X} - \mu}{\sigma/\sqrt{n}}$$

has the standard normal distribution and

$$V = \frac{(n-1)S^2}{\sigma^2}$$

has a chi-squared distribution with $v = n-1$ degrees of freedom. In sampling from normal populations, we can show that $\bar{X}$ and S^2 are independent, and consequently so are Z and V. The following theorem gives the definition of a random variable T as a function of Z (standard normal) and χ^2. For completeness, the density function of the t-distribution is given.

Theorem 4.5: Let Z be a standard normal random variable and V a chi-squared random variable with v degrees of freedom. If Z and V are independent, then the distribution of the random variable T, where

$$T = \frac{Z}{\sqrt{V/v}},$$

is given by the density function

$$h(t) = \frac{\Gamma[(v+1)/2]}{\Gamma(v/2)\sqrt{\pi v}} \left(1 + \frac{t^2}{v}\right)^{-(v+1)/2}, \quad -\infty < t < \infty.$$

This is known as the **Student t-distribution** with v degrees of freedom.

From the foregoing and the theorem above we have the following corollary.

Corollary 4.1: Let $X_1, X_2, \ldots, X_n$ be independent random variables that are all normal with mean μ and standard deviation σ. Let

$$\bar{X} = \frac{1}{n}\sum_{i=1}^{n} X_i \quad \text{and} \quad S^2 = \frac{1}{n-1}\sum_{i=1}^{n}(X_i - \bar{X})^2.$$

Then the random variable $T = \frac{\bar{X}-\mu}{S/\sqrt{n}}$ has a Student t-distribution with $v = n-1$ degrees of freedom. This will often be referred to as merely the t-distribution.

What Does the *t*-Distribution Look Like?

The distribution of T is similar to the distribution of Z in that they both are symmetric about a mean of zero. Both distributions are bell shaped, but the *t*-distribution is more variable, owing to the fact that the T-values depend on the fluctuations of two quantities, $\bar{X}$ and S^2, whereas the Z-values depend only on the changes in $\bar{X}$ from sample to sample. The distribution of T differs from that of Z in that the variance of T depends on the sample size n and is always greater than 1. Only when the sample size $n \to \infty$ will the two distributions become the same. In Figure 4.7, we show the relationship between a standard normal distribution ($v = \infty$) and *t*-distributions with 2 and 5 degrees of freedom. The percentage points of the *t*-distribution are given in Table A.4.

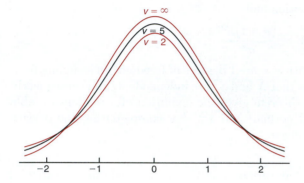

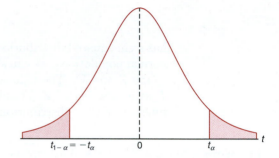

Figure 4.7: The *t*-distribution curves for $v = 2, 5$, and ∞.

Figure 4.8: Symmetry property (about 0) of the *t*-distribution.

It is customary to let t_α represent the *t*-value above which we find an area equal to α. Hence, the *t*-value with 10 degrees of freedom leaving an area of 0.025 to the right is $t = 2.228$. Since the *t*-distribution is symmetric about a mean of zero, we have $t_{1-\alpha} = -t_\alpha$; that is, the *t*-value leaving an area of $1 - \alpha$ to the right and therefore an area of α to the left is equal to the negative *t*-value that leaves an area of α in the right tail of the distribution (see Figure 4.8). That is, $t_{0.95} = -t_{0.05}$, $t_{0.99} = -t_{0.01}$, and so forth.

Example 4.7: The *t*-value with $v = 14$ degrees of freedom that leaves an area of 0.025 to the left, and therefore an area of 0.975 to the right, is

$$t_{0.975} = -t_{0.025} = -2.145.$$

Example 4.8: Find $P(-t_{0.025} < T < t_{0.05})$.

Solution: Since $t_{0.05}$ leaves an area of 0.05 to the right, and $-t_{0.025}$ leaves an area of 0.025 to the left, we find a total area of

$$1 - 0.05 - 0.025 = 0.925$$

between $-t_{0.025}$ and $t_{0.05}$. Hence

$$P(-t_{0.025} < T < t_{0.05}) = 0.925.$$

Example 4.9: Find k such that $P(k < T < -1.761) = 0.045$ for a random sample of size 15 selected from a normal distribution and $T = \frac{\bar{X}-\mu}{s/\sqrt{n}}$.

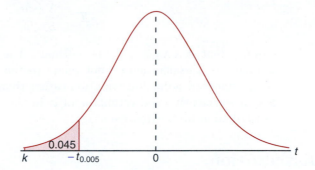

Figure 4.9: The t-values for Example 4.9.

Solution: From Table A.4 we note that 1.761 corresponds to $t_{0.05}$ when $v = 14$. Therefore, $-t_{0.05} = -1.761$. Since k in the original probability statement is to the left of $-t_{0.05} = -1.761$, let $k = -t_\alpha$. Then, from Figure 4.9, we have

$$0.045 = 0.05 - \alpha, \text{ or } \alpha = 0.005.$$

Hence, from Table A.4 with $v = 14$,
$$k = -t_{0.005} = -2.977 \text{ and } P(-2.977 < T < -1.761) = 0.045.$$

Example 4.10: A chemical engineer claims that the population mean yield of a certain batch process is 500 grams per liter of raw material. To check this claim he samples 25 batches each month. If the computed t-value falls between $-t_{0.05}$ and $t_{0.05}$, he is satisfied with this claim. What conclusion should he draw from a sample that has a mean $\bar{x} = 518$ grams per liter and a sample standard deviation $s = 40$ grams? Assume the distribution of yields to be approximately normal.

Solution: From Table A.4 we find that $t_{0.05} = 1.711$ for 24 degrees of freedom. Therefore, the engineer can be satisfied with his claim if a sample of 25 batches yields a t-value between -1.711 and 1.711. If $\mu = 500$, then

$$t = \frac{518 - 500}{40/\sqrt{25}} = 2.25,$$

a value well above 1.711. The probability of obtaining a t-value, with $v = 24$, equal to or greater than 2.25 is approximately 0.02. If $\mu > 500$, the value of t computed from the sample is more reasonable. Hence, the engineer is likely to conclude that the process produces a better product than he thought.

What Is the *t*-Distribution Used For?

The t-distribution is used extensively in problems that deal with inference about the population mean (as illustrated in Example 4.10) or in problems that involve comparative samples (i.e., in cases where one is trying to determine if means from two samples are significantly different). The use of the distribution will be extended in Chapters 5, 6, and 7. The reader should note that use of the t-distribution for the statistic

$$T = \frac{\bar{X} - \mu}{S/\sqrt{n}}$$

requires that $X_1, X_2, \ldots, X_n$ be normal. The use of the t-distribution and the sample size consideration do not relate to the Central Limit Theorem. The use of the standard normal distribution rather than T for $n \geq 30$ merely implies that S is a sufficiently good estimator of σ in this case. In chapters that follow the t-distribution finds extensive usage.

4.7   *F*-Distribution

We have motivated the t-distribution in part by its application to problems in which there is comparative sampling (i.e., a comparison between two sample means). For example, some of our examples in future chapters will take a more formal approach: a chemical engineer collects data on two catalysts, a biologist collects data on two growth media, or a chemist gathers data on two methods of coating material to inhibit corrosion. While it is of interest to let sample information shed light on two population means, it is often the case that a comparison of variability is equally important, if not more so. The F-distribution finds enormous application in comparing sample variances. Applications of the F-distribution are found in problems involving two or more samples.

The statistic F is defined to be the ratio of two independent chi-squared random variables, each divided by its number of degrees of freedom. Hence, we can write

$$F = \frac{U/v_1}{V/v_2},$$

where U and V are independent random variables having chi-squared distributions with v_1 and v_2 degrees of freedom, respectively. We shall now state the sampling distribution of F.

Theorem 4.6: Let U and V be two independent random variables having chi-squared distributions with v_1 and v_2 degrees of freedom, respectively. Then the distribution of the random variable $F = \frac{U/v_1}{V/v_2}$ is given by the density function

$$h(f) = \begin{cases} \frac{\Gamma[(v_1+v_2)/2](v_1/v_2)^{v_1/2}}{\Gamma(v_1/2)\Gamma(v_2/2)} \frac{f^{(v_1/2)-1}}{(1+v_1 f/v_2)^{(v_1+v_2)/2}}, & f > 0, \\ 0, & f \leq 0. \end{cases}$$

This is known as the **F-distribution** with v_1 and v_2 degrees of freedom (d.f.).

We will make considerable use of the random variable F in future chapters. However, the density function will not be used and is given only for completeness;

instead, we will take F values from Table A.6 in the Appendix. The curve of the F-distribution depends not only on the two parameters v_1 and v_2 but also on the order in which we state them. Once these two values are given, we can identify the curve. Typical F-distributions are shown in Figure 4.10.

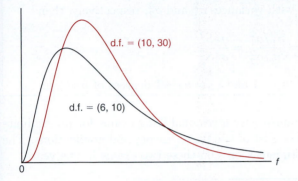

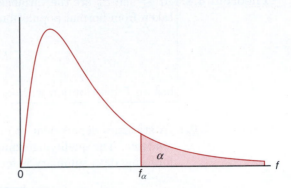

Figure 4.10: Typical F-distributions.

Figure 4.11: Illustration of the f_α for the F-distribution.

Let f_α be the f-value above which we find an area equal to α. This is illustrated by the shaded region in Figure 4.11. Table A.6 gives values of f_α only for $\alpha = 0.05$ and $\alpha = 0.01$ for various combinations of the degrees of freedom v_1 and v_2. Hence, the f-value with 6 and 10 degrees of freedom, leaving an area of 0.05 to the right, is $f_{0.05} = 3.22$. By means of the following theorem, Table A.6 can also be used to find values of $f_{0.95}$ and $f_{0.99}$. The proof is left for the reader.

Theorem 4.7: | Writing $f_\alpha(v_1, v_2)$ for f_α with v_1 and v_2 degrees of freedom, we obtain

$$f_{1-\alpha}(v_1, v_2) = \frac{1}{f_\alpha(v_2, v_1)}.$$

Thus, the f-value with 6 and 10 degrees of freedom, leaving an area of 0.95 to the right, is

$$f_{0.95}(6, 10) = \frac{1}{f_{0.05}(10, 6)} = \frac{1}{4.06} = 0.246.$$

The F-Distribution with Two Sample Variances

Suppose that random samples of size n_1 and n_2 are selected from two normal populations with variances σ_1^2 and σ_2^2, respectively. From Theorem 4.4, we know that

$$\chi_1^2 = \frac{(n_1 - 1)S_1^2}{\sigma_1^2} \text{ and } \chi_2^2 = \frac{(n_2 - 1)S_2^2}{\sigma_2^2}$$

are random variables having chi-squared distributions with $v_1 = n_1 - 1$ and $v_2 = n_2 - 1$ degrees of freedom. Furthermore, since the samples are selected at random, we are dealing with independent random variables. Then, using Theorem 4.6 with $\chi_1^2 = U$ and $\chi_2^2 = V$, we obtain the following result.

Theorem 4.8: If S_1^2 and S_2^2 are the variances of independent random samples of size n_1 and n_2 taken from normal populations with variances σ_1^2 and σ_2^2, respectively, then

$$F = \frac{S_1^2/\sigma_1^2}{S_2^2/\sigma_2^2} = \frac{\sigma_2^2 S_1^2}{\sigma_1^2 S_2^2}$$

has an F-distribution with $v_1 = n_1 - 1$ and $v_2 = n_2 - 1$ degrees of freedom.

Example 4.11: A statistics department at one university gives qualifying exams for its graduate students. The qualifying exams consist of two parts: theory and applications. The following data summary is for student scores on these two exams in one year.

Exam	n	$\bar{x}$	s
Theory	16	76.2	13.6
Application	16	88.3	7.5

Assuming normality on both samples, is it reasonable to believe that variances for the scores of both subjects are equal?

Solution: Suppose both variances, denoted respectively by σ_1^2 and σ_2^2, are the same. Using Theorem 4.8, we conclude that

$$F = \frac{S_1^2/\sigma_1^2}{S_2^2/\sigma_2^2} = \frac{S_1^2}{S_2^2}$$

follows an F-distribution with degrees of freedom 15 and 15. Based on the data, we compute

$$f = \frac{13.6^2}{7.5^2} = 3.288.$$

From Table A.6, we find out that $f_{0.05;15,15} = 2.33$, which means that the probability of S_1^2/S_2^2 being larger than 2.33 is only 5%. Since our observed f-value is 3.288, which is larger than 2.33, it would not be reasonable to assume that the variances for the scores of both subjects are equal.

What Is the F-Distribution Used For?

We answered this question, in part, at the beginning of this section. The F-distribution is used in two-sample situations to draw inferences about the population variances. This involves the application of Theorem 4.8. However, the F-distribution can also be applied to many other types of problems involving sample variances. In fact, the F-distribution is called the *variance ratio distribution*. As an illustration, consider Case Study 4.2, in which two paints, A and B, were compared with regard to mean drying time. The normal distribution applies nicely (assuming that σ_A and σ_B are known). However, suppose that there are three types of paints to compare, say A, B, and C. We wish to determine if the population means are equivalent. Suppose that important summary information from the experiment is as follows:

Paint	Sample Mean	Sample Variance	Sample Size
A	$\bar{X}_A = 4.5$	$s_A^2 = 0.20$	10
B	$\bar{X}_B = 5.5$	$s_B^2 = 0.14$	10
C	$\bar{X}_C = 6.5$	$s_C^2 = 0.11$	10

The problem centers around whether or not the sample averages $(\bar{x}_A, \bar{x}_B, \bar{x}_C)$ are far enough apart. The implication of "far enough apart" is very important. It would seem reasonable that if the variability between sample averages is larger than what one would expect by chance, the data do not support the conclusion that $\mu_A = \mu_B = \mu_C$. Whether these sample averages could have occurred by chance depends on the *variability within samples*, as quantified by s_A^2, s_B^2, and s_C^2.

An analysis that involves all of the data would attempt to determine if the variability between the sample averages *and* the variability within the samples could have occurred jointly *if in fact the populations have a common mean*. Notice that the key to this analysis centers around the two following sources of variability:

(1) Variability within samples (between observations in distinct samples)

(2) Variability between samples (between sample averages)

Clearly, if the variability in (1) is considerably larger than that in (2), there will be considerable overlap in the sample data, a signal that the data could all have come from a common distribution.

4.8 Graphical Presentation

Often the end result of a statistical analysis is the estimation of parameters of a **postulated model**. This is natural for scientists and engineers since they often deal in modeling. A statistical model is not deterministic but, rather, must entail some probabilistic aspects. A model form is often the foundation of **assumptions** that are made by the analyst. For example, the scientist in Example 1.2 may wish to draw some level of distinction between the nitrogen and no-nitrogen populations through the sample information. The analysis may require a certain model for the data, for example, that the two samples come from **normal** or **Gaussian distributions**. See Chapter 3 for a discussion of the normal distribution.

Obviously, the user of statistical methods cannot generate sufficient information or experimental data to characterize the population totally. But sets of data are often used to learn about certain properties of the population. Scientists and engineers are accustomed to dealing with data sets. The importance of characterizing or *summarizing* the nature of collections of data should be obvious. Often a summary of a collection of data via a graphical display can provide insight regarding the system from which the data were taken.

In this section, the role of sampling and the display of data for enhancement of **statistical inference** is explored in detail. We introduce some simple but often effective displays that complement the study of statistical populations.

Scatter Plot

At times the model postulated may take on a somewhat complicated form. Consider, for example, a textile manufacturer who designs an experiment where cloth specimens that contain various percentages of cotton are produced. Consider the data in Table 4.1.

Table 4.1: Tensile Strength

Cotton Percentage	Tensile Strength
15	7, 7, 9, 8, 10
20	19, 20, 21, 20, 22
25	21, 21, 17, 19, 20
30	8, 7, 8, 9, 10

Five cloth specimens are manufactured for each of the four cotton percentages. In this case, both the model for the experiment and the type of analysis used should take into account the goal of the experiment and important input from the textile scientist. Some simple graphics can shed important light on the clear distinction between the samples. See Figure 4.12; the sample means and variability are depicted nicely in the scatter plot.

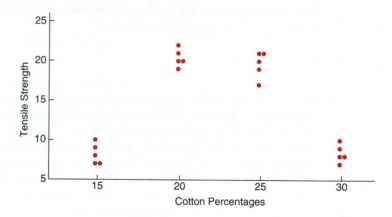

Figure 4.12: Scatter plot of tensile strength and cotton percentages.

One possible goal of this experiment is simply to determine which cotton percentages are truly distinct from the others. In other words, as in the case of the nitrogen/no-nitrogen data, for which cotton percentages are there clear distinctions between the populations or, more specifically, between the population means? In this case, perhaps a reasonable model is that each sample comes from a normal distribution. Here the goal is very much like that of the nitrogen/no-nitrogen data except that more samples are involved. The formalism of the analysis involves notions of hypothesis testing discussed in Chapter 6. Incidentally, this formality is perhaps not necessary in light of the diagnostic plot. But does this describe the real goal of the experiment and hence the proper approach to data analysis?

It is likely that the scientist anticipates the existence of a *maximum population mean tensile strength* in the range of cotton concentration in the experiment. Here the analysis of the data should revolve around a different type of model, one that postulates a type of structure relating the population mean tensile strength to the cotton concentration. In other words, a model may be written

$$\mu_{t,c} = \beta_0 + \beta_1 C + \beta_2 C^2,$$

where $\mu_{t,c}$ is the population mean tensile strength, which varies with the amount of cotton in the product C. The implication of this model is that for a fixed cotton level, there is a population of tensile strength measurements and the population mean is $\mu_{t,c}$. This type of model, called a **regression model**, is discussed in Chapter 7. The functional form is chosen by the scientist. At times the data analysis may suggest that the model be changed. Then the data analyst "entertains" a model that may be altered after some analysis is done. The use of an empirical model is accompanied by **estimation theory**, where β_0, β_1, and β_2 are estimated by the data. Further, statistical inference can then be used to determine model adequacy.

Two points become evident from the two data illustrations here: (1) The type of model used to describe the data often depends on the goal of the experiment; and (2) the structure of the model should take advantage of nonstatistical scientific input. A selection of a model represents a **fundamental assumption** upon which the resulting statistical inference is based.

It will become apparent throughout the book how important graphics can be. Often, plots can illustrate information that allows the results of the formal statistical inference to be better communicated to the scientist or engineer. At times, plots or **exploratory data analysis** can teach the analyst something not retrieved from the formal analysis. Almost any formal analysis requires assumptions that evolve from the model of the data. Graphics can nicely highlight **violation of assumptions** that would otherwise go unnoticed. Throughout the book, graphics are used extensively to supplement formal data analysis. The following sections reveal some graphical tools that are useful in exploratory or descriptive data analysis.

Stem-and-Leaf Plot

Statistical data, generated in large masses, can be very useful for studying the behavior of the distribution if presented in a combined tabular and graphic display called a **stem-and-leaf plot**.

To illustrate the construction of a stem-and-leaf plot, consider the data of Table 4.2, which specifies the "life" of 40 similar car batteries recorded to the nearest tenth of a year. The batteries are guaranteed to last 3 years. First, split each observation into two parts consisting of a stem and a leaf such that the stem represents the digit preceding the decimal and the leaf corresponds to the decimal part of the number. In other words, for the number 3.7, the digit 3 is designated the stem and the digit 7 is the leaf. The four stems 1, 2, 3, and 4 for our data are listed vertically on the left side in Table 4.3; the leaves are recorded on the right side opposite the appropriate stem value. Thus, the leaf 6 of the number 1.6 is recorded opposite the stem 1; the leaf 5 of the number 2.5 is recorded opposite the stem 2; and so forth. The number of leaves recorded opposite each stem is summarized under the frequency column.

Table 4.2: Car Battery Life

2.2	4.1	3.5	4.5	3.2	3.7	3.0	2.6
3.4	1.6	3.1	3.3	3.8	3.1	4.7	3.7
2.5	4.3	3.4	3.6	2.9	3.3	3.9	3.1
3.3	3.1	3.7	4.4	3.2	4.1	1.9	3.4
4.7	3.8	3.2	2.6	3.9	3.0	4.2	3.5

Table 4.3: Stem-and-Leaf Plot of Battery Life

Stem	Leaf	Frequency
1	69	2
2	25669	5
3	0011112223334445567778899	25
4	11234577	8

The stem-and-leaf plot of Table 4.3 contains only four stems and consequently does not provide an adequate picture of the distribution. To remedy this problem, we need to increase the number of stems in our plot. One simple way to accomplish this is to write each stem value twice and then record the leaves 0, 1, 2, 3, and 4 opposite the appropriate stem value where it appears for the first time, and the leaves 5, 6, 7, 8, and 9 opposite this same stem value where it appears for the second time. This modified double-stem-and-leaf plot is illustrated in Table 4.4, where the stems corresponding to leaves 0 through 4 have been coded by the symbol $\star$ and the stems corresponding to leaves 5 through 9 by the symbol $\cdot$.

Table 4.4: Double-Stem-and-Leaf Plot of Battery Life

Stem	Leaf	Frequency
1·	69	2
2$\star$	2	1
2·	5669	4
3$\star$	001111222333444	15
3·	5567778899	10
4$\star$	11234	5
4·	577	3

In any given problem, we must decide on the appropriate stem values. This decision is made somewhat arbitrarily, although we are guided by the size of our sample. Usually, we choose between 5 and 20 stems. The smaller the number of data available, the smaller is our choice for the number of stems. For example, if the data consist of numbers from 1 to 21 representing the number of people in a cafeteria line on 40 randomly selected workdays and we choose a double-stem-and-leaf plot, the stems will be 0$\star$, 0·, 1$\star$, 1·, and 2$\star$ so that the smallest observation

1 has stem 0⋆ and leaf 1, the number 18 has stem 1· and leaf 8, and the largest observation 21 has stem 2⋆ and leaf 1. On the other hand, if the data consist of numbers from \$18,800 to \$19,600 representing the best possible deals on 100 new automobiles from a certain dealership and we choose a single-stem-and-leaf plot, the stems will be 188, 189, 190, . . . , 196 and the leaves will now each contain two digits. A car that sold for \$19,385 would have a stem value of 193 and the two-digit leaf 85. Multiple-digit leaves belonging to the same stem are usually separated by commas in the stem-and-leaf plot. Decimal points in the data are generally ignored when all the digits to the right of the decimal represent the leaf. Such was the case in Tables 4.3 and 4.4. However, if the data consist of numbers ranging from 21.8 to 74.9, we might choose the digits 2, 3, 4, 5, 6, and 7 as our stems so that a number such as 48.3 would have a stem value of 4 and a leaf of 8.3.

The stem-and-leaf plot represents an effective way to summarize data. Another way is through the use of the **frequency distribution**, where the data, grouped into different classes or intervals, can be constructed by counting the leaves belonging to each stem and noting that each stem defines a class interval. In Table 4.3, the stem 1 with 2 leaves defines the interval 1.0–1.9 containing 2 observations; the stem 2 with 5 leaves defines the interval 2.0–2.9 containing 5 observations; the stem 3 with 25 leaves defines the interval 3.0–3.9 with 25 observations; and the stem 4 with 8 leaves defines the interval 4.0–4.9 containing 8 observations. For the double-stem-and-leaf plot of Table 4.4, the stems define the seven class intervals 1.5–1.9, 2.0–2.4, 2.5–2.9, 3.0–3.4, 3.5–3.9, 4.0–4.4, and 4.5–4.9 with frequencies 2, 1, 4, 15, 10, 5, and 3, respectively.

Histogram

Dividing each class frequency by the total number of observations, we obtain the proportion of the set of observations in each of the classes. A table listing relative frequencies is called a **relative frequency distribution**. The relative frequency distribution for the data of Table 4.2, showing the midpoint of each class interval, is given in Table 4.5.

Table 4.5: Relative Frequency Distribution of Battery Life

Class Interval	Class Midpoint	Frequency, f	Relative Frequency
1.5–1.9	1.7	2	0.050
2.0–2.4	2.2	1	0.025
2.5–2.9	2.7	4	0.100
3.0–3.4	3.2	15	0.375
3.5–3.9	3.7	10	0.250
4.0–4.4	4.2	5	0.125
4.5–4.9	4.7	3	0.075

The information provided by a relative frequency distribution in tabular form is easier to grasp if presented graphically. Using the midpoint of each interval and the corresponding relative frequency, we construct a **relative frequency histogram** (Figure 4.13).

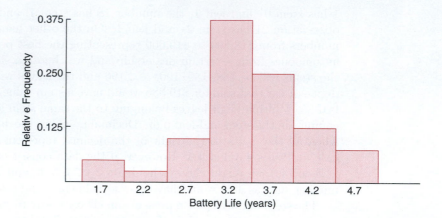

Figure 4.13: Relative frequency histogram.

Many continuous frequency distributions can be represented graphically by the characteristic bell-shaped curve of Figure 4.14. Graphical tools such as what we see in Figures 4.13 and 4.14 aid in the characterization of the nature of the population.

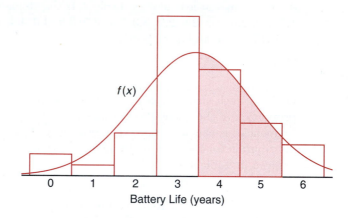

Figure 4.14: Estimating frequency distribution.

A distribution is said to be **symmetric** if it can be folded along a vertical axis so that the two sides coincide. A distribution that lacks symmetry with respect to a vertical axis is said to be **skewed**. The distribution illustrated in Figure 4.15(a) is said to be skewed to the right since it has a long right tail and a much shorter left tail. In Figure 4.15(b) we see that the distribution is symmetric, while in Figure 4.15(c) it is skewed to the left.

If we rotate a stem-and-leaf plot counterclockwise through an angle of 90°, we observe that the resulting columns of leaves form a picture that is similar to a histogram. Consequently, if our primary purpose in looking at the data is to determine the general shape or form of the distribution, it will seldom be necessary to construct a relative frequency histogram.

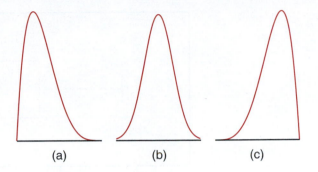

Figure 4.15: Skewness of data.

Box-and-Whisker Plot or Box Plot

Another display that is helpful for reflecting properties of a sample is the **box-and-whisker plot**. This plot encloses the *interquartile range* of the data in a box that has the median displayed within. The interquartile range has as its extremes the 75th percentile (upper quartile) and the 25th percentile (lower quartile). In addition to the box, "whiskers" extend, showing extreme observations in the sample. For reasonably large samples, the display shows center of location, variability, and the degree of asymmetry.

In addition, a variation called a **box plot** can provide the viewer with information regarding which observations may be **outliers**. Outliers are observations that are considered to be unusually far from the bulk of the data. There are many statistical tests that are designed to detect outliers. Technically, one may view an outlier as being an observation that represents a "rare event" (there is a small probability of obtaining a value that far from the bulk of the data). The concept of outliers resurfaces in Chapter 7 in the context of regression analysis.

The visual information in the box-and-whisker plot or box plot is not intended to be a formal test for outliers. Rather, it is viewed as a diagnostic tool. While the determination of which observations are outliers varies with the type of software that is used, one common procedure is to use a **multiple of the interquartile range**. For example, if its distance from the box exceeds 1.5 times the interquartile range (in either direction), an observation may be labeled an outlier.

Example 4.12: Nicotine content was measured in a random sample of 40 cigarettes. The data are displayed in Table 4.6.

Table 4.6: Nicotine Data for Example 4.12

1.09	1.92	2.31	1.79	2.28	1.74	1.47	1.97
0.85	1.24	1.58	2.03	1.70	2.17	2.55	2.11
1.86	1.90	1.68	1.51	1.64	0.72	1.69	1.85
1.82	1.79	2.46	1.88	2.08	1.67	1.37	1.93
1.40	1.64	2.09	1.75	1.63	2.37	1.75	1.69

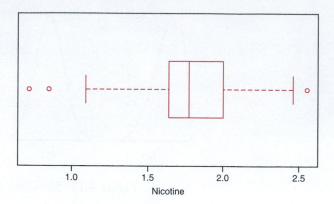

Figure 4.16: Box-and-whisker plot for Example 4.12.

```
The decimal point is 1 digit(s) to the left of the |
  7 | 2
  8 | 5
  9 |
 10 | 9
 11 |
 12 | 4
 13 | 7
 14 | 07
 15 | 18
 16 | 3447899
 17 | 045599
 18 | 2568
 19 | 0237
 20 | 389
 21 | 17
 22 | 8
 23 | 17
 24 | 6
 25 | 5
```

Figure 4.17: Stem-and-leaf plot for the nicotine data.

Figure 4.16 shows the box-and-whisker plot of the data, depicting the observations 0.72 and 0.85 as mild outliers in the lower tail, whereas the observation 2.55 is a mild outlier in the upper tail. In this example, the interquartile range is 0.365, and 1.5 times the interquartile range is 0.5475. Figure 4.17 provides a stem-and-leaf plot of the data.

Exercises

4.28 For a chi-squared distribution, find
(a) $\chi^2_{0.005}$ when $v = 5$;
(b) $\chi^2_{0.05}$ when $v = 19$;

(c) $\chi^2_{0.01}$ when $v = 12$.

4.29 For a chi-squared distribution, find

(a) $\chi^2_{0.025}$ when $v = 15$;

(b) $\chi^2_{0.01}$ when $v = 7$;

(c) $\chi^2_{0.05}$ when $v = 24$.

4.30 For a chi-squared distribution, find χ^2_α such that

(a) $P(X^2 > \chi^2_\alpha) = 0.01$ when $v = 21$;

(b) $P(X^2 < \chi^2_\alpha) = 0.95$ when $v = 6$;

(c) $P(\chi^2_\alpha < X^2 < 23.209) = 0.015$ when $v = 10$.

4.31 For a chi-squared distribution, find χ^2_α such that

(a) $P(X^2 > \chi^2_\alpha) = 0.99$ when $v = 4$;

(b) $P(X^2 > \chi^2_\alpha) = 0.025$ when $v = 19$;

(c) $P(37.652 < X^2 < \chi^2_\alpha) = 0.045$ when $v = 25$.

4.32 The scores on a placement test given to college freshmen for the past five years are approximately normally distributed with a mean $\mu = 74$ and a variance $\sigma^2 = 8$. Would you still consider $\sigma^2 = 8$ to be a valid value of the variance if a random sample of 20 students who take the placement test this year obtain a value of $s^2 = 20$?

4.33 Assume the sample variances to be continuous measurements. Find the probability that a random sample of 25 observations, from a normal population with variance $\sigma^2 = 6$, will have a sample variance S^2

(a) greater than 9.1;

(b) between 3.462 and 10.745.

4.34 Show that the variance of S^2 for random samples of size n from a normal population decreases as n becomes large. [*Hint:* First find the variance of $(n-1)S^2/\sigma^2$.]

4.35 (a) Find $P(T < 2.365)$ when $v = 7$.

(b) Find $P(T > 1.318)$ when $v = 24$.

(c) Find $P(-1.356 < T < 2.179)$ when $v = 12$.

(d) Find $P(T > -2.567)$ when $v = 17$.

4.36 (a) Find $t_{0.025}$ when $v = 14$.

(b) Find $-t_{0.10}$ when $v = 10$.

(c) Find $t_{0.995}$ when $v = 7$.

4.37 Given a random sample of size 24 from a normal distribution, find k such that

(a) $P(-2.069 < T < k) = 0.965$;

(b) $P(k < T < 2.807) = 0.095$;

(c) $P(-k < T < k) = 0.90$.

4.38 (a) Find $P(-t_{0.005} < T < t_{0.01})$ for $v = 20$.

(b) Find $P(T > -t_{0.025})$.

4.39 A manufacturing firm claims that the batteries used in its electronic games will last an average of 30 hours. To maintain this average, 16 batteries are tested each month. If the computed t-value falls between $-t_{0.025}$ and $t_{0.025}$, the firm is satisfied with its claim. What conclusion should the firm draw from a sample that has a mean of $\bar{x} = 27.5$ hours and a standard deviation of $s = 5$ hours? Assume the distribution of battery lives to be approximately normal.

4.40 A maker of a certain brand of low-fat cereal bars claims that the average saturated fat content is 0.5 gram. In a random sample of 8 cereal bars of this brand, the saturated fat content was 0.6, 0.7, 0.7, 0.3, 0.4, 0.5, 0.4, and 0.2. Would you agree with the claim? Assume a normal distribution.

4.41 For an F-distribution, find

(a) $f_{0.05}$ with $v_1 = 7$ and $v_2 = 15$;

(b) $f_{0.05}$ with $v_1 = 15$ and $v_2 = 7$:

(c) $f_{0.01}$ with $v_1 = 24$ and $v_2 = 19$;

(d) $f_{0.95}$ with $v_1 = 19$ and $v_2 = 24$;

(e) $f_{0.99}$ with $v_1 = 28$ and $v_2 = 12$.

4.42 Construct a box-and-whisker plot of these data, which represent the lifetimes, in hours, of fifty 40-watt, 110-volt internally frosted incandescent lamps taken from forced life tests:

919	1196	785	1126	936	918
1156	920	948	1067	1092	1162
1170	929	950	905	972	1035
1045	855	1195	1195	1340	1122
938	970	1237	956	1102	1157
978	832	1009	1157	1151	1009
765	958	902	1022	1333	811
1217	1085	896	958	1311	1037
702	923				

4.43 Consider the following measurements of the heat-producing capacity of the coal produced by two mines (in millions of calories per ton):

  Mine 1: 8260 8130 8350 8070 8340

  Mine 2: 7950 7890 7900 8140 7920 7840

Can it be concluded that the two population variances are equal?

4.44 Pull-strength tests on 10 soldered leads for a semiconductor device yield the following results, in pounds of force required to rupture the bond:

19.8	12.7	13.2	16.9	10.6
18.8	11.1	14.3	17.0	12.5

Another set of 8 leads was tested after encapsulation to determine whether the pull strength had been increased by encapsulation of the device, with the following results:

    24.9 22.8 23.6 22.1 20.4 21.6 21.8 22.5

Review Exercises

4.45 The following data represent the length of life, in seconds, of 50 fruit flies subject to a new spray in a controlled laboratory experiment:

17	20	10	9	23	13	12	19	18	24
12	14	6	9	13	6	7	10	13	7
16	18	8	13	3	32	9	7	10	11
13	7	18	7	10	4	27	19	16	8
7	10	5	14	15	10	9	6	7	15

Construct a box-and-whisker plot and comment on the nature of the sample. Compute the sample mean and sample standard deviation.

4.46 In testing for carbon monoxide in a certain brand of cigarette, the data, in milligrams per cigarette, were coded by subtracting 12 from each observation. Use the results of Exercise 4.10 on page 163 to find the sample standard deviation for the carbon monoxide content of a random sample of 15 cigarettes of this brand if the coded measurements are 3.8, -0.9, 5.4, 4.5, 5.2, 5.6, 2.7, -0.1, -0.3, -1.7, 5.7, 3.3, 4.4, -0.5, and 1.9.

4.47 If S_1^2 and S_2^2 represent the variances of independent random samples of size $n_1 = 8$ and $n_2 = 12$, taken from normal populations with equal variances, find $P(S_1^2/S_2^2 < 4.89)$.

4.48 If the number of hurricanes that hit a certain area of the eastern United States per year is a random variable having a Poisson distribution with $\mu = 6$, find the probability that this area will be hit by

(a) exactly 15 hurricanes in 2 years;

(b) at most 9 hurricanes in 2 years.

4.49 The following data represent the length of life in years, measured to the nearest tenth, of 30 similar fuel pumps:

2.0	3.0	0.3	3.3	1.3	0.4
0.2	6.0	5.5	6.5	0.2	2.3
1.5	4.0	5.9	1.8	4.7	0.7
4.5	0.3	1.5	0.5	2.5	5.0
1.0	6.0	5.6	6.0	1.2	0.2

Construct a box-and-whisker plot. Comment on the outliers in the data.

4.50 If S_1^2 and S_2^2 represent the variances of independent random samples of size $n_1 = 25$ and $n_2 = 31$, taken from normal populations with variances $\sigma_1^2 = 10$

and $\sigma_2^2 = 15$, respectively, find

$$P(S_1^2/S_2^2 > 1.26).$$

4.51 Consider Example 4.12 on page 189. Comment on any outliers.

4.52 The breaking strength X of a certain rivet used in a machine engine has a mean 5000 psi and standard deviation 400 psi. A random sample of 36 rivets is taken. Consider the distribution of $\bar{X}$, the sample mean breaking strength.

(a) What is the probability that the sample mean falls between 4800 psi and 5200 psi?

(b) What sample n would be necessary in order to have

$$P(4900 < \bar{X} < 5100) = 0.99?$$

4.53 A taxi company tests a random sample of 10 steel-belted radial tires of a certain brand and records the following tread wear: 48,000, 53,000, 45,000, 61,000, 59,000, 56,000, 63,000, 49,000, 53,000, and 54,000 kilometers. The marketing claim for the tires is that, on the average, the tires last for 53,000 kilometers of use. In your answer, compute

$$t = \frac{\bar{x} - 53,000}{s/\sqrt{10}}$$

and determine from Table A.4 (with 9 degrees of freedom) whether the computed t-value is reasonable or appears to be a rare event.

4.54 Two distinct solid fuel propellants, type A and type B, are being considered for a space program activity. Burning rates of the propellant are crucial. Random samples of 20 specimens of the two propellants are taken with sample means 20.5 cm/sec for propellant A and 24.50 cm/sec for propellant B. It is generally assumed that the variability in burning rate is roughly the same for the two propellants and is given by a population standard deviation of 5 cm/sec. Assume that the burning rates for the propellants are approximately normal and hence make use of the Central Limit Theorem. Nothing is known about the two population mean burning rates, and it is hoped that this experiment might shed some light on them.

(a) If, indeed, $\mu_A = \mu_B$, what is $P(\bar{X}_B - \bar{X}_A \geq 4.0)$?

(b) Use your answer in (a) to shed some light on the proposition that $\mu_A = \mu_B$.

4.55 The concentration of an active ingredient in the output of a chemical reaction is strongly influenced by the catalyst that is used in the reaction. It is believed that when catalyst A is used, the population mean concentration exceeds 65%. The standard deviation is known to be $\sigma = 5\%$. A sample of outputs from 30 independent experiments gives the average concentration of $\bar{x}_A = 64.5\%$.

(a) Does this sample information with an average concentration of $\bar{x}_A = 64.5\%$ provide disturbing information that perhaps μ_A is not 65%, but less than 65%? Support your answer with a probability statement.

(b) Suppose a similar experiment is done with the use of another catalyst, catalyst B. The standard deviation σ is still assumed to be 5% and $\bar{x}_B$ turns out to be 70%. Comment on whether or not the sample information on catalyst B strongly suggests that μ_B is truly greater than μ_A. Support your answer by computing

$$P(\bar{X}_B - \bar{X}_A \geq 5.5 \mid \mu_B = \mu_A).$$

(c) Under the condition that $\mu_A = \mu_B = 65\%$, give the approximate distribution of the following quantities (with mean and variance of each). Make use of the Central Limit Theorem.
   i) $\bar{X}_B$;
   ii) $\bar{X}_A - \bar{X}_B$;
   iii) $\frac{\bar{X}_A - \bar{X}_B}{\sigma \sqrt{2/30}}$.

4.56 From the information in Review Exercise 4.55, compute (assuming $\mu_B = 65\%$) $P(\bar{X}_B \geq 70)$.

4.57 Given a normal random variable X with mean 20 and variance 9, and a random sample of size n taken from the distribution, what sample size n is necessary in order that

$$P(19.9 \leq \bar{X} \leq 20.1) = 0.95?$$

4.58 In Chapter 5, the concept of **parameter estimation** will be discussed at length. Suppose X is a random variable with mean μ and variance $\sigma^2 = 1.0$. Suppose also that a random sample of size n is to be taken and $\bar{x}$ is to be used as an *estimate* of μ. When the data are taken and the sample mean is measured, we wish it to be within 0.05 unit of the true mean with

probability 0.99. That is, we want there to be a good chance that the computed $\bar{x}$ from the sample is "very close" to the population mean (wherever it is!), so we wish

$$P(|\bar{X} - \mu| > 0.05) = 0.99.$$

What sample size is required?

4.59 Suppose a filling machine is used to fill cartons with a liquid product. The specification that is strictly enforced for the filling machine is 9 ± 1.5 oz. If any carton is produced with weight outside these bounds, it is considered by the supplier to be defective. It is hoped that at least 99% of cartons will meet these specifications. With the conditions $\mu = 9$ and $\sigma = 1$, what proportion of cartons from the process are defective? If changes are made to reduce variability, what must σ be reduced to in order to meet specifications with probability 0.99? Assume a normal distribution for the weight.

4.60 Consider the situation in Review Exercise 4.59. Suppose a considerable effort is conducted to "tighten" the variability in the system. Following the effort, a random sample of size 40 is taken from the new assembly line and the sample variance is $s^2 = 0.188$ ounces^2. Do we have strong numerical evidence that σ^2 has been reduced below 1.0? Consider the probability

$$P(S^2 \leq 0.188 \mid \sigma^2 = 1.0),$$

and give your conclusion.

4.61 Group Project: The class should be divided into groups of four people. The four students in each group should go to the college gym or a local fitness center. The students should ask each person who comes through the door his or her height in inches. Each group will then divide the height data by gender and work together to answer the following questions.

(a) Construct a histogram for each gender of the data. Based on these plots, do the data appear to follow a normal distribution?

(b) Use the estimated sample variance as the true variance for each gender. Assume that the population mean height for male students is actually 3 inches more than that of female students. What is the probability that the average height of the male students will be 4 inches more than that of the female students in your sample?

(c) What factors could render these results misleading?

4.9   Potential Misconceptions and Hazards; Relationship to Material in Other Chapters

The Central Limit Theorem is one of the most powerful tools in all of statistics, and even though this chapter is relatively short, it contains a wealth of fundamental information about tools that will be used throughout the balance of the text.

The notion of a sampling distribution is one of the most important fundamental concepts in all of statistics, and the student at this point in his or her training should gain a clear understanding of it before proceeding beyond this chapter. All chapters that follow will make considerable use of sampling distributions. Suppose one wants to use the statistic $\bar{X}$ to draw inferences about the population mean μ. This will be done by using the observed value $\bar{x}$ from a single sample of size n. Then any inference made must be accomplished by taking into account not just the single value but rather the theoretical structure, or **distribution of all $\bar{x}$ values that could be observed from samples of size n**. Thus, the concept of a *sampling distribution* comes to the surface. This distribution is the basis for the Central Limit Theorem. The t, χ^2, and F-distributions are also used in the context of sampling distributions. For example, the t-distribution, pictured in Figure 4.7, represents the structure that occurs if all of the values of $\frac{\bar{x}-\mu}{s/\sqrt{n}}$ are formed, where $\bar{x}$ and s are taken from samples of size n from a $n(x; \mu, \sigma)$ distribution. Similar remarks can be made about χ^2 and F, and the reader should not forget that the sample information forming the statistics for all of these distributions is the normal. So it can be said that **where there is a t, F, or χ^2, the source was a sample from a normal distribution**.

The three distributions described above may appear to have been introduced in a rather self-contained fashion with no indication of what they are about. However, they will appear in practical problem-solving throughout the balance of the text.

Now, there are three things that one must bear in mind, lest confusion set in regarding these fundamental sampling distributions:

(i)   One cannot use the Central Limit Theorem unless σ is known. When σ is not known, it should be replaced by s, the sample standard deviation, in order to use the Central Limit Theorem.

(ii) The T statistic is **not** a result of the Central Limit Theorem and $x_1, x_2, \ldots, x_n$ must come from a $n(x; \mu, \sigma)$ distribution in order for $\frac{\bar{x}-\mu}{s/\sqrt{n}}$ to be a t-distribution; s is, of course, merely an estimate of σ.

(iii) While the notion of **degrees of freedom** is new at this point, the concept should be very intuitive, since it is reasonable that the nature of the distribution of S and also t should depend on the amount of information in the sample $x_1, x_2, \ldots, x_n$.

Chapter 5

One- and Two-Sample Estimation Problems

5.1 Introduction

In previous chapters, we emphasized sampling properties of the sample mean and variance. We also emphasized displays of data in various forms. The purpose of these presentations is to build a foundation that allows us to draw conclusions about the population parameters from experimental data. For example, the Central Limit Theorem provides information about the distribution of the sample mean $\bar{X}$. The distribution involves the population mean μ. Thus, any conclusions concerning μ drawn from an observed sample average must depend on knowledge of this sampling distribution. Similar comments apply to S^2 and σ^2. Clearly, any conclusions we draw about the variance of a normal distribution will likely involve the sampling distribution of S^2.

In this chapter, we begin by formally outlining the purpose of statistical inference. We follow this by discussing the problem of **estimation of population parameters**. We confine our formal developments of specific estimation procedures to problems involving one and two samples.

5.2 Statistical Inference

In Chapter 1, we discussed the general philosophy of formal statistical inference. **Statistical inference** consists of those methods by which one makes inferences or generalizations about a population. The trend today is to distinguish between the **classical method** of estimating a population parameter, whereby inferences are based strictly on information obtained from a random sample selected from the population, and the **Bayesian method**, which utilizes prior subjective knowledge about the probability distribution of the unknown parameters in conjunction with the information provided by the sample data. Throughout most of this chapter, we shall use classical methods to estimate unknown population parameters such as the mean, the proportion, and the variance by computing statistics from random samples and applying the theory of sampling distributions, much of which was covered in Chapter 4.

Statistical inference may be divided into two major areas: **estimation** and **tests of hypotheses**. We treat these two areas separately, dealing with theory and applications of estimation in this chapter and hypothesis testing in Chapter 6. To distinguish clearly between the two areas, consider the following examples. A candidate for public office may wish to estimate the true proportion of voters favoring him by obtaining opinions from a random sample of 100 eligible voters. The fraction of voters in the sample favoring the candidate could be used as an estimate of the true proportion in the population of voters. A knowledge of the sampling distribution of a proportion enables one to establish the degree of accuracy of such an estimate. This problem falls in the area of estimation.

Now consider the case in which one is interested in finding out whether brand A floor wax is more scuff-resistant than brand B floor wax. He or she might hypothesize that brand A is better than brand B and, after proper testing, accept or reject this hypothesis. In this example, we do not attempt to estimate a parameter, but instead we try to arrive at a correct decision about a prestated hypothesis. Once again we are dependent on sampling theory and the use of data to provide us with some measure of accuracy for our decision.

5.3   Classical Methods of Estimation

A **point estimate** of some population parameter θ is a single value $\hat{\theta}$ of a statistic $\hat{\Theta}$. For example, the value $\bar{x}$ of the statistic $\bar{X}$, computed from a sample of size n, is a point estimate of the population parameter μ. Similarly, $\hat{p} = x/n$ is a point estimate of the true proportion p for a binomial experiment.

An estimator is not expected to estimate the population parameter without error. We do not expect $\bar{X}$ to estimate μ exactly, but we certainly hope that it is not far off. For a particular sample, it is possible to obtain a closer estimate of μ by using the sample median $\tilde{X}$ as an estimator. Consider, for instance, a sample consisting of the values 2, 5, and 11 from a population whose mean is 4 but is supposedly unknown. We would estimate μ to be $\bar{x} = 6$, using the sample mean as our estimate, or $\tilde{x} = 5$, using the sample median as our estimate. In this case, the estimator $\tilde{X}$ produces an estimate closer to the true parameter than does the estimator $\bar{X}$. On the other hand, if our random sample contains the values 2, 6, and 7, then $\bar{x} = 5$ and $\tilde{x} = 6$, so $\bar{X}$ is the better estimator. Not knowing the true value of μ, we must decide in advance whether to use $\bar{X}$ or $\tilde{X}$ as our estimator.

Unbiased Estimator

What are the desirable properties of a "good" decision function that would influence us to choose one estimator rather than another? Let $\hat{\Theta}$ be an estimator whose value $\hat{\theta}$ is a point estimate of some unknown population parameter θ. Certainly, we would like the sampling distribution of $\hat{\Theta}$ to have a mean equal to the parameter estimated. An estimator possessing this property is said to be **unbiased**.

Definition 5.1: A statistic $\hat{\Theta}$ is said to be an **unbiased estimator** of the parameter θ if

$$\mu_{\hat{\Theta}} = E(\hat{\Theta}) = \theta.$$

Example 5.1: If a sample $X_1, \ldots, X_n$ has an unknown population mean μ, then the sample mean $\bar{X}$ is an unbiased estimator for μ.

Solution: This result can be easily shown as

$$E(\bar{X}) = E\left(\frac{1}{n}\sum_{i=1}^{n} X_i\right) = \frac{1}{n}\sum_{i=1}^{n} E(X_i) = \frac{1}{n}\sum_{i=1}^{n} \mu = \mu.$$

This means that the sample mean is always unbiased to the population mean. ∎

Example 5.2: Show that S^2 is an unbiased estimator of the parameter σ^2.

Solution: One can show that (see Exercise 5.10)

$$\sum_{i=1}^{n}(X_i - \bar{X})^2 = \sum_{i=1}^{n}(X_i - \mu)^2 - n(\bar{X} - \mu)^2.$$

Now

$$E(S^2) = E\left[\frac{1}{n-1}\sum_{i=1}^{n}(X_i - \bar{X})^2\right]$$

$$= \frac{1}{n-1}\left[\sum_{i=1}^{n} E(X_i - \mu)^2 - nE(\bar{X} - \mu)^2\right] = \frac{1}{n-1}\left(\sum_{i=1}^{n}\sigma_{X_i}^2 - n\sigma_{\bar{X}}^2\right).$$

However,

$$\sigma_{X_i}^2 = \sigma^2, \text{ for } i = 1, 2, \ldots, n, \text{ and } \sigma_{\bar{X}}^2 = \frac{\sigma^2}{n}.$$

Therefore,

$$E(S^2) = \frac{1}{n-1}\left(n\sigma^2 - n\frac{\sigma^2}{n}\right) = \sigma^2.$$

This result shows that the sample variance is always unbiased to the population variance σ^2. ∎

Although S^2 is an unbiased estimator of σ^2, S, on the other hand, is usually a biased estimator of σ, with the bias becoming insignificant for large samples. This example illustrates **why we divide by $n - 1$** rather than n when the variance is estimated.

Variance of a Point Estimator

If $\hat{\Theta}_1$ and $\hat{\Theta}_2$ are two unbiased estimators of the same population parameter θ, we want to choose the estimator whose sampling distribution has the smaller variance. Hence, if $\sigma_{\hat{\theta}_1}^2 < \sigma_{\hat{\theta}_2}^2$, we say that $\hat{\Theta}_1$ is a **more efficient estimator** of θ than $\hat{\Theta}_2$.

Definition 5.2: | If we consider all possible unbiased estimators of some parameter θ, the one with the smallest variance is called the **most efficient estimator** of θ.

Figure 5.1 illustrates the sampling distributions of three different estimators, $\hat{\Theta}_1$, $\hat{\Theta}_2$, and $\hat{\Theta}_3$, all estimating θ. It is clear that only $\hat{\Theta}_1$ and $\hat{\Theta}_2$ are unbiased, since their distributions are centered at θ. The estimator $\hat{\Theta}_1$ has a smaller variance than $\hat{\Theta}_2$ and is therefore more efficient. Hence, our choice for an estimator of θ, among the three considered, would be $\hat{\Theta}_1$.

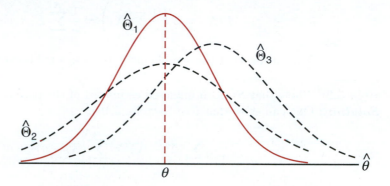

Figure 5.1: Sampling distributions of different estimators of θ.

For normal populations, one can show that both $\bar{X}$ and $\tilde{X}$ are unbiased estimators of the population mean μ, but the variance of $\bar{X}$ is smaller than the variance of $\tilde{X}$. Thus, both estimates $\bar{x}$ and $\tilde{x}$ will, on average, equal the population mean μ, but $\bar{x}$ is likely to be closer to μ for a given sample, and thus $\bar{X}$ is more efficient than $\tilde{X}$.

Interval Estimation

Even the most efficient unbiased estimator is unlikely to estimate the population parameter exactly. It is true that estimation accuracy increases with large samples, but there is still no reason we should expect a **point estimate** from a given sample to be exactly equal to the population parameter it is supposed to estimate. There are many situations in which it is preferable to determine an interval within which we would expect to find the value of the parameter. Such an interval is called an **interval estimate**.

An interval estimate of a population parameter θ is an interval of the form $\hat{\theta}_L < \theta < \hat{\theta}_U$, where $\hat{\theta}_L$ and $\hat{\theta}_U$ depend on the value of the statistic $\hat{\Theta}$ for a particular sample and also on the sampling distribution of $\hat{\Theta}$. For example, a random sample of SAT verbal scores for students in the entering freshman class might produce an interval from 530 to 550, within which we expect to find the true average of all SAT verbal scores for the freshman class. The values of the endpoints, 530 and 550, will depend on the computed sample mean $\bar{x}$ and the sampling distribution of $\bar{X}$. As the sample size increases, we know that $\sigma_{\bar{X}}^2 = \sigma^2/n$ decreases, and consequently our estimate is likely to be closer to the parameter μ, resulting in a shorter interval. Thus, the interval estimate indicates, by its length, the accuracy of the point estimate. An engineer will gain some insight into the population proportion defective by taking a sample and computing the *sample proportion defective*. But an interval estimate might be more informative.

Interpretation of Interval Estimates

Since different samples will generally yield different values of $\hat{\Theta}$ and, therefore, different values for $\hat{\theta}_L$ and $\hat{\theta}_U$, these endpoints of the interval are values of corresponding random variables $\hat{\Theta}_L$ and $\hat{\Theta}_U$. From the sampling distribution of $\hat{\Theta}$ we shall be able to determine $\hat{\Theta}_L$ and $\hat{\Theta}_U$ such that $P(\hat{\Theta}_L < \theta < \hat{\Theta}_U)$ is equal to any positive fractional value we care to specify. If, for instance, we find $\hat{\Theta}_L$ and $\hat{\Theta}_U$ such that

$$P(\hat{\Theta}_L < \theta < \hat{\Theta}_U) = 1 - \alpha,$$

for $0 < \alpha < 1$, then we have a probability of $1-\alpha$ of selecting a random sample that will produce an interval containing θ. The interval $\hat{\theta}_L < \theta < \hat{\theta}_U$, computed from the selected sample, is called a $100(1 - \alpha)\%$ **confidence interval**, the fraction $1 - \alpha$ is called the **confidence coefficient** or the **degree of confidence**, and the endpoints, $\hat{\theta}_L$ and $\hat{\theta}_U$, are called the lower and upper **confidence limits**. Thus, when $\alpha = 0.05$, we have a 95% confidence interval, and when $\alpha = 0.01$, we obtain a wider 99% confidence interval. The wider the confidence interval is, the more confident we can be that the interval contains the unknown parameter. Of course, it is better to be 95% confident that the average life of a certain television transistor is between 6 and 7 years than to be 99% confident that it is between 3 and 10 years. Ideally, we prefer a short interval with a high degree of confidence. Sometimes, restrictions on the size of our sample prevent us from achieving short intervals without sacrificing some degree of confidence.

In the sections that follow, we pursue the notions of point and interval estimation, with each section presenting a different special case. The reader should notice that while point and interval estimation represent different approaches to gaining information regarding a parameter, they are related in the sense that confidence interval estimators are based on point estimators. In the following section, for example, we will see that $\bar{X}$ is a very reasonable point estimator of μ. As a result, the important confidence interval estimator of μ depends on knowledge of the sampling distribution of $\bar{X}$.

5.4 Single Sample: Estimating the Mean

The sampling distribution of $\bar{X}$ is centered at μ, and in most applications the variance is smaller than that of any other unbiased estimators of μ. Thus, the sample mean $\bar{x}$ will be used as a point estimate for the population mean μ. Recall that $\sigma_{\bar{X}}^2 = \sigma^2/n$, so a large sample will yield a value of $\bar{X}$ that comes from a sampling distribution with a small variance. Hence, $\bar{x}$ is likely to be a very accurate estimate of μ when n is large.

Let us now consider the interval estimate of μ. If our sample is selected from a normal population or, failing this, if n is sufficiently large, we can establish a confidence interval for μ by considering the sampling distribution of $\bar{X}$.

According to the Central Limit Theorem, we can expect the sampling distribution of $\bar{X}$ to be approximately normal with mean $\mu_{\bar{X}} = \mu$ and standard deviation $\sigma_{\bar{X}} = \sigma/\sqrt{n}$. Writing $z_{\alpha/2}$ for the z-value above which we find an area of $\alpha/2$ under the normal curve, we can see from Figure 5.2 that

$$P(-z_{\alpha/2} < Z < z_{\alpha/2}) = 1 - \alpha,$$

where

$$Z = \frac{\bar{X} - \mu}{\sigma/\sqrt{n}}.$$

Hence,

$$P\left(-z_{\alpha/2} < \frac{\bar{X} - \mu}{\sigma/\sqrt{n}} < z_{\alpha/2}\right) = 1 - \alpha.$$

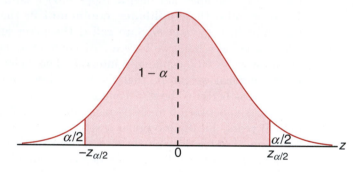

Figure 5.2: $P(-z_{\alpha/2} < Z < z_{\alpha/2}) = 1 - \alpha.$

Multiplying each term in the inequality by $\sigma/\sqrt{n}$ and then subtracting $\bar{X}$ from each term and multiplying by -1 (reversing the sense of the inequalities), we obtain

$$P\left(\bar{X} - z_{\alpha/2}\frac{\sigma}{\sqrt{n}} < \mu < \bar{X} + z_{\alpha/2}\frac{\sigma}{\sqrt{n}}\right) = 1 - \alpha.$$

A random sample of size n is selected from a population whose variance σ^2 is known, and the mean $\bar{x}$ is computed to give the $100(1 - \alpha)\%$ confidence interval below. It is important to emphasize that we have invoked the Central Limit Theorem above. As a result, it is important to note the conditions for applications that follow.

Confidence Interval on μ, σ^2 Known

If $\bar{x}$ is the mean of a random sample of size n from a population with known variance σ^2, a $100(1 - \alpha)\%$ confidence interval for μ is given by

$$\bar{x} - z_{\alpha/2}\frac{\sigma}{\sqrt{n}} < \mu < \bar{x} + z_{\alpha/2}\frac{\sigma}{\sqrt{n}},$$

where $z_{\alpha/2}$ is the z-value leaving an area of $\alpha/2$ to the right.

For small samples selected from nonnormal populations, we cannot expect our degree of confidence to be accurate. However, for samples of size $n \geq 30$, with the shape of the distributions not too skewed, sampling theory guarantees good results.

Clearly, the values of the random variables $\hat{\Theta}_L$ and $\hat{\Theta}_U$, defined in Section 5.3, are the confidence limits

$$\hat{\theta}_L = \bar{x} - z_{\alpha/2}\frac{\sigma}{\sqrt{n}} \quad \text{and} \quad \hat{\theta}_U = \bar{x} + z_{\alpha/2}\frac{\sigma}{\sqrt{n}}.$$

Different samples will yield different values of $\bar{x}$ and therefore produce different interval estimates of the parameter μ, as shown in Figure 5.3. The dot at the center of each interval indicates the position of the point estimate $\bar{x}$ for that random sample. Note that all of these intervals are of the same width, since their widths depend only on the choice of $z_{\alpha/2}$ once $\bar{x}$ is determined. The larger the value we choose for $z_{\alpha/2}$, the wider we make all the intervals and the more confident we can be that the particular sample selected will produce an interval that contains the unknown parameter μ. In general, for a selection of $z_{\alpha/2}$, $100(1 - \alpha)\%$ of the intervals will cover μ.

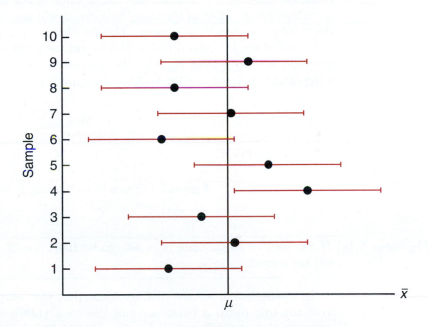

Figure 5.3: Interval estimates of μ for different samples.

Example 5.3: The average zinc concentration recovered from a sample of measurements taken in 36 different locations in a river is found to be 2.6 grams per milliliter. Find the 95% and 99% confidence intervals for the mean zinc concentration in the river. Assume that the population standard deviation is 0.3 gram per milliliter.

Solution: The point estimate of μ is $\bar{x} = 2.6$. The z-value leaving an area of 0.025 to the right, and therefore an area of 0.975 to the left, is $z_{0.025} = 1.96$ (Table A.3). Hence,

the 95% confidence interval is

$$2.6 - (1.96)\left(\frac{0.3}{\sqrt{36}}\right) < \mu < 2.6 + (1.96)\left(\frac{0.3}{\sqrt{36}}\right),$$

which reduces to $2.50 < \mu < 2.70$. To find a 99% confidence interval, we find the z-value leaving an area of 0.005 to the right and 0.995 to the left. From Table A.3 again, $z_{0.005} = 2.575$, and the 99% confidence interval is

$$2.6 - (2.575)\left(\frac{0.3}{\sqrt{36}}\right) < \mu < 2.6 + (2.575)\left(\frac{0.3}{\sqrt{36}}\right),$$

or simply

$$2.47 < \mu < 2.73.$$

We now see that a longer interval is required to estimate μ with a higher degree of confidence.

The $100(1-\alpha)\%$ confidence interval provides an estimate of the accuracy of our point estimate. If μ is actually the center value of the interval, then $\bar{x}$ estimates μ without error. Most of the time, however, $\bar{x}$ will not be exactly equal to μ and the point estimate will be in error. The size of this error will be the absolute value of the difference between μ and $\bar{x}$, and we can be $100(1 - \alpha)\%$ confident that this difference will not exceed $z_{\alpha/2}\frac{\sigma}{\sqrt{n}}$. We can readily see this if we draw a diagram of a hypothetical confidence interval, as in Figure 5.4.

Figure 5.4: Error in estimating μ by $\bar{x}$.

Theorem 5.1: If $\bar{x}$ is used as an estimate of μ, we can be $100(1 - \alpha)\%$ confident that the error will not exceed $z_{\alpha/2}\frac{\sigma}{\sqrt{n}}$.

In Example 5.3, we are 95% confident that the sample mean $\bar{x} = 2.6$ differs from the true mean μ by an amount less than $(1.96)(0.3)/\sqrt{36} = 0.1$ and 99% confident that the difference is less than $(2.575)(0.3)/\sqrt{36} = 0.13$.

Frequently, we wish to know how large a sample is necessary to ensure that the error in estimating μ will be less than a specified amount e. By Theorem 5.1, we must choose n such that $z_{\alpha/2}\frac{\sigma}{\sqrt{n}} = e$. Solving this equation gives the following formula for n.

Theorem 5.2: If $\bar{x}$ is used as an estimate of μ, we can be $100(1-\alpha)\%$ confident that the error will not exceed a specified amount e when the sample size is

$$n = \left(\frac{z_{\alpha/2}\sigma}{e}\right)^2.$$

When solving for the sample size, n, we round all fractional values up to the next whole number. By adhering to this principle, we can be sure that our degree of confidence never falls below $100(1-\alpha)\%$.

Strictly speaking, the formula in Theorem 5.2 is applicable only if we know the variance of the population from which we select our sample. Lacking this information, we could take a preliminary sample of size $n \geq 30$ to provide an estimate of σ. Then, using s as an approximation for σ in Theorem 5.2, we could determine approximately how many observations are needed to provide the desired degree of accuracy.

Example 5.4: How large a sample is required if we want to be 95% confident that our estimate of μ in Example 5.3 is off by less than 0.05?

Solution: The population standard deviation is $\sigma = 0.3$. Then, by Theorem 5.2,

$$n = \left[\frac{(1.96)(0.3)}{0.05}\right]^2 = 138.3.$$

Therefore, we can be 95% confident that a random sample of size 139 will provide an estimate $\bar{x}$ differing from μ by an amount less than 0.05. ⌐

One-Sided Confidence Bounds

The confidence intervals and resulting confidence bounds discussed thus far are *two-sided* (i.e., both upper and lower bounds are given). However, there are many applications in which only one bound is sought. For example, if the measurement of interest is tensile strength, the engineer receives better information from a lower bound only. This bound communicates the worst-case scenario. On the other hand, if the measurement is something for which a relatively large value of μ is not profitable or desirable, then an upper confidence bound is of interest. An example would be a case in which inferences need to be made concerning the mean mercury concentration in a river. An upper bound is very informative in this case.

One-sided confidence bounds are developed in the same fashion as two-sided intervals. However, the source is a one-sided probability statement that makes use of the Central Limit Theorem:

$$P\left(\frac{\bar{X} - \mu}{\sigma/\sqrt{n}} < z_\alpha\right) = 1 - \alpha.$$

One can then manipulate the probability statement much as before and obtain

$$P(\mu > \bar{X} - z_\alpha \sigma/\sqrt{n}) = 1 - \alpha.$$

Similar manipulation of $P\left(\frac{\bar{X}-\mu}{\sigma/\sqrt{n}} > -z_\alpha\right) = 1 - \alpha$ gives

$$P(\mu < \bar{X} + z_\alpha \sigma/\sqrt{n}) = 1 - \alpha.$$

As a result, the upper and lower one-sided bounds follow.

One-Sided Confidence Bounds on μ, σ^2 Known	If $\bar{X}$ is the mean of a random sample of size n from a population with variance σ^2, the one-sided $100(1 - \alpha)\%$ confidence bounds for μ are given by

upper one-sided bound: $\bar{x} + z_\alpha \sigma/\sqrt{n}$;

lower one-sided bound: $\bar{x} - z_\alpha \sigma/\sqrt{n}$.

Example 5.5: In a psychological testing experiment, 25 subjects are selected randomly and their reaction time, in seconds, to a particular stimulus is measured. Past experience suggests that the variance in reaction times to these types of stimuli is 4 sec^2 and that the distribution of reaction times is approximately normal. The average time for the subjects is 6.2 seconds. Give an upper 95% bound for the mean reaction time.

Solution: The upper 95% bound is given by

$$\bar{x} + z_\alpha \sigma/\sqrt{n} = 6.2 + (1.645)\sqrt{4/25} = 6.2 + 0.658$$
$$= 6.858 \text{ seconds.}$$

Hence, we are 95% confident that the mean reaction time is less than 6.858 seconds.

The Case of σ Unknown

Frequently, we must attempt to estimate the mean of a population when the variance is unknown. The reader should recall learning in Chapter 4 that if we have a random sample from a *normal distribution*, then the random variable

$$T = \frac{\bar{X} - \mu}{S/\sqrt{n}}$$

has a Student t-distribution with $n - 1$ degrees of freedom. Here S is the sample standard deviation. In this situation, with σ unknown, T can be used to construct a confidence interval on μ. The procedure is the same as that with σ known except that σ is replaced by S and the standard normal distribution is replaced by the t-distribution. Referring to Figure 5.5, we can assert that

$$P(-t_{\alpha/2} < T < t_{\alpha/2}) = 1 - \alpha,$$

where $t_{\alpha/2}$ is the t-value with $n-1$ degrees of freedom, above which we find an area of $\alpha/2$. Because of symmetry, an equal area of $\alpha/2$ will fall to the left of $-t_{\alpha/2}$. Substituting for T, we write

$$P\left(-t_{\alpha/2} < \frac{\bar{X} - \mu}{S/\sqrt{n}} < t_{\alpha/2}\right) = 1 - \alpha.$$

Multiplying each term in the inequality by $S/\sqrt{n}$, and then subtracting $\bar{X}$ from each term and multiplying by -1, we obtain

$$P\left(\bar{X} - t_{\alpha/2}\frac{S}{\sqrt{n}} < \mu < \bar{X} + t_{\alpha/2}\frac{S}{\sqrt{n}}\right) = 1 - \alpha.$$

For a particular random sample of size n, the mean $\bar{x}$ and standard deviation s are computed and the following $100(1-\alpha)\%$ confidence interval for μ is obtained.

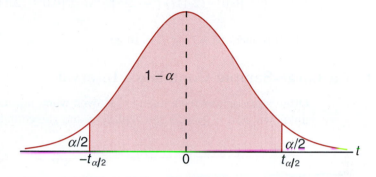

Figure 5.5: $P(-t_{\alpha/2} < T < t_{\alpha/2}) = 1 - \alpha.$

Confidence Interval on μ, σ^2 Unknown   If $\bar{x}$ and s are the mean and standard deviation of a random sample from a normal population with unknown variance σ^2, a $100(1-\alpha)\%$ confidence interval for μ is

$$\bar{x} - t_{\alpha/2}\frac{s}{\sqrt{n}} < \mu < \bar{x} + t_{\alpha/2}\frac{s}{\sqrt{n}},$$

where $t_{\alpha/2}$ is the t-value with $v = n - 1$ degrees of freedom, leaving an area of $\alpha/2$ to the right.

We have made a distinction between the cases of σ known and σ unknown in computing confidence interval estimates. We should emphasize that for σ known we exploited the Central Limit Theorem, whereas for σ unknown we made use of the sampling distribution of the random variable T. However, the use of the t-distribution is based on the premise that the sampling is from a normal distribution. As long as the distribution is approximately bell shaped, confidence intervals can be computed when σ^2 is unknown by using the t-distribution and we may expect very good results.

Computed one-sided confidence bounds for μ with σ unknown are as the reader would expect, namely

$$\bar{x} + t_{\alpha}\frac{s}{\sqrt{n}} \quad \text{and} \quad \bar{x} - t_{\alpha}\frac{s}{\sqrt{n}}.$$

They are the upper and lower $100(1-\alpha)\%$ bounds, respectively. Here t_α is the t-value having an area of α to the right.

Example 5.6: The contents of seven similar containers of sulfuric acid are 9.8, 10.2, 10.4, 9.8, 10.0, 10.2, and 9.6 liters. Find a 95% confidence interval for the mean contents of all such containers, assuming an approximately normal distribution.

Solution: The sample mean and standard deviation for the given data are

$$\bar{x} = 10.0 \quad \text{and} \quad s = 0.283.$$

Using Table A.4, we find $t_{0.025} = 2.447$ for $v = 6$ degrees of freedom. Hence, the 95% confidence interval for μ is

$$10.0 - (2.447)\left(\frac{0.283}{\sqrt{7}}\right) < \mu < 10.0 + (2.447)\left(\frac{0.283}{\sqrt{7}}\right),$$

which reduces to $9.74 < \mu < 10.26$.

Concept of a Large-Sample Confidence Interval

Often statisticians recommend that even when normality cannot be assumed, σ is unknown, and $n \geq 30$, s can replace σ and the confidence interval

$$\bar{x} \pm z_{\alpha/2}\frac{s}{\sqrt{n}}$$

may be used. This is often referred to as a *large-sample confidence interval*. The justification lies only in the presumption that with a sample as large as 30 and the population distribution not too skewed, s will be very close to the true σ and thus the Central Limit Theorem prevails. It should be emphasized that this is only an approximation and the quality of the result becomes better as the sample size grows larger.

Example 5.7: Scholastic Aptitude Test (SAT) mathematics scores of a random sample of 500 high school seniors in the state of Texas are collected, and the sample mean and standard deviation are found to be 501 and 112, respectively. Find a 99% confidence interval on the mean SAT mathematics score for seniors in the state of Texas.

Solution: Since the sample size is large, it is reasonable to use the normal approximation. Using Table A.3, we find $z_{0.005} = 2.575$. Hence, a 99% confidence interval for μ is

$$501 \pm (2.575)\left(\frac{112}{\sqrt{500}}\right) = 501 \pm 12.9,$$

which yields $488.1 < \mu < 513.9$.

5.5 Standard Error of a Point Estimate

We have made a rather sharp distinction between the goal of a point estimate and that of a confidence interval estimate. The former supplies a single number extracted from a set of experimental data, and the latter provides an interval that is reasonable for the parameter, *given the experimental data*; that is, $100(1 - \alpha)\%$ of such computed intervals "cover" the parameter.

These two approaches to estimation are related to each other. The common thread is the sampling distribution of the point estimator. Consider, for example, the estimator $\bar{X}$ of μ with σ known. We indicated earlier that a measure of the quality of an unbiased estimator is its variance. The variance of $\bar{X}$ is

$$\sigma_{\bar{X}}^2 = \frac{\sigma^2}{n}.$$

Thus, the standard deviation of $\bar{X}$, or *standard error* of $\bar{X}$, is $\sigma/\sqrt{n}$. Simply put, the standard error of an estimator is its standard deviation. For $\bar{X}$, the computed confidence limit

$$\bar{x} \pm z_{\alpha/2}\frac{\sigma}{\sqrt{n}} \text{ is written as } \bar{x} \pm z_{\alpha/2} \text{ s.e.}(\bar{x}),$$

where "s.e." is the "standard error." The important point is that the width of the confidence interval on μ is dependent on the quality of the point estimator through its standard error. In the case where σ is unknown and sampling is from a normal distribution, s replaces σ and the *estimated standard error* $s/\sqrt{n}$ is involved. Thus, the confidence limits on μ are as follows.

Confidence Limits on μ, σ^2 Unknown	$$\bar{x} \pm t_{\alpha/2}\frac{s}{\sqrt{n}} = \bar{x} \pm t_{\alpha/2} \text{ s.e.}(\bar{x})$$

Again, the confidence interval is *no better* (in terms of width) *than the quality of the point estimate*, in this case through its estimated standard error. Computer packages often refer to estimated standard errors simply as "standard errors."

As we move to more complex confidence intervals, there is a prevailing notion that widths of confidence intervals become shorter as the quality of the corresponding point estimate becomes better, although it is not always quite as simple as we have illustrated here. It can be argued that a confidence interval is merely an augmentation of the point estimate to take into account the precision of the point estimate.

5.6 Prediction Intervals

The point and interval estimations of the mean in Sections 5.4 and 5.5 provide good information about the unknown parameter μ of a normal distribution or a nonnormal distribution from which a large sample is drawn. Sometimes, other than the population mean, the experimenter may also be interested in predicting the possible **value of a future observation**. For instance, in quality control, the experimenter may need to use the observed data to predict a new observation. A process that produces a metal part may be evaluated on the basis of whether the part meets specifications on tensile strength. On certain occasions, a customer may be interested in purchasing a **single part**. In this case, a confidence interval on the mean tensile strength does not capture the required information. The customer requires a statement regarding the uncertainty of a **single observation**. This type of requirement is nicely fulfilled by the construction of a **prediction interval**.

It is quite simple to obtain a prediction interval for the situations we have considered so far. Assume that the random sample comes from a normal population with unknown mean μ and known variance σ^2. A natural point estimator of a new observation is $\bar{X}$. It is known, from Section 4.4, that the variance of $\bar{X}$ is σ^2/n. However, to predict a new observation, not only do we need to account for the variation due to estimating the mean, but also we should account for the **variation of a future observation**. From the assumption, we know that the variance of the random error in a new observation is σ^2. The development of a prediction interval is best illustrated by beginning with a normal random variable $x_0 - \bar{x}$, where x_0 is the new observation and $\bar{x}$ comes from the sample. Since x_0 and $\bar{x}$ are independent, we know that

$$z = \frac{x_0 - \bar{x}}{\sqrt{\sigma^2 + \sigma^2/n}} = \frac{x_0 - \bar{x}}{\sigma\sqrt{1 + 1/n}}$$

is $n(z; 0, 1)$. As a result, if we use the probability statement

$$P(-z_{\alpha/2} < Z < z_{\alpha/2}) = 1 - \alpha$$

with the z-statistic above and place x_0 in the center of the probability statement, we have the following event occurring with probability $1 - \alpha$:

$$\bar{x} - z_{\alpha/2}\sigma\sqrt{1 + 1/n} < x_0 < \bar{x} + z_{\alpha/2}\sigma\sqrt{1 + 1/n}.$$

Computation of the prediction interval is formalized as follows.

Prediction Interval of a Future Observation, σ^2 Known

For a normal distribution of measurements with unknown mean μ and known variance σ^2, a $100(1 - \alpha)\%$ **prediction interval** of a future observation x_0 is

$$\bar{x} - z_{\alpha/2}\sigma\sqrt{1 + 1/n} < x_0 < \bar{x} + z_{\alpha/2}\sigma\sqrt{1 + 1/n},$$

where $z_{\alpha/2}$ is the z-value leaving an area of $\alpha/2$ to the right.

Example 5.8: Due to the decrease in interest rates, the First Citizens Bank received a lot of mortgage applications. A recent sample of 50 mortgage loans resulted in an average loan amount of $257,300. Assume a population standard deviation of $25,000. For the next customer who fills out a mortgage application, find a 95% prediction interval for the loan amount.

Solution: The point prediction of the next customer's loan amount is $\bar{x} = \$257,300$. The z-value here is $z_{0.025} = 1.96$. Hence, a 95% prediction interval for the future loan amount is

$$257,300 - (1.96)(25,000)\sqrt{1 + 1/50} < x_0 < 257,300 + (1.96)(25,000)\sqrt{1 + 1/50},$$

which gives the interval ($207,812.43, $306,787.57). ∎

The prediction interval provides a good estimate of the location of a future observation, which is quite different from the estimate of the sample mean value. It should be noted that the variation of this prediction is the sum of the variation due to an estimation of the mean and the variation of a single observation. However, as in the past, we first consider the case with known variance. It is also important to deal with the prediction interval of a future observation in the situation where the variance is unknown. Indeed a Student t-distribution may be used in this case, as described in the following result. The normal distribution is merely replaced by the t-distribution.

Prediction Interval of a Future Observation, σ^2 Unknown	For a normal distribution of measurements with unknown mean μ and unknown variance σ^2, a $100(1-\alpha)\%$ **prediction interval** of a future observation x_0 is $$\bar{x} - t_{\alpha/2}s\sqrt{1+1/n} < x_0 < \bar{x} + t_{\alpha/2}s\sqrt{1+1/n},$$ where $t_{\alpha/2}$ is the t-value with $v = n-1$ degrees of freedom, leaving an area of $\alpha/2$ to the right.

One-sided prediction intervals can also be constructed. Upper prediction bounds apply in cases where focus must be placed on future large observations. Concern over future small observations calls for the use of lower prediction bounds. The upper bound is given by

$$\bar{x} + t_\alpha s\sqrt{1+1/n}$$

and the lower bound by

$$\bar{x} - t_\alpha s\sqrt{1+1/n}.$$

Example 5.9: A meat inspector has randomly selected 30 packs of 95% lean beef. The sample resulted in a mean of 96.2% with a sample standard deviation of 0.8%. Find a 99% prediction interval for the leanness of a new pack. Assume normality.

Solution: For $v = 29$ degrees of freedom, $t_{0.005} = 2.756$. Hence, a 99% prediction interval for a new observation x_0 is

$$96.2 - (2.756)(0.8)\sqrt{1 + \frac{1}{30}} < x_0 < 96.2 + (2.756)(0.8)\sqrt{1 + \frac{1}{30}},$$

which reduces to $(93.96, 98.44)$.

Use of Prediction Limits for Outlier Detection

To this point in the text very little attention has been paid to the concept of **outliers**, or aberrant observations. The majority of scientific investigators are keenly sensitive to the existence of outlying observations or so-called faulty or "bad" data. It is certainly of interest here since there is an important relationship between outlier detection and prediction intervals.

It is convenient for our purposes to view an outlying observation as one that comes from a population with a mean that is different from the mean that governs the rest of the sample of size n being studied. The prediction interval produces a bound that "covers" a future single observation with probability $1 - \alpha$ if it comes from the population from which the sample was drawn. As a result, a methodology for outlier detection involves the rule that **an observation is an outlier if it falls outside the prediction interval computed without including the questionable observation in the sample.** As a result, for the prediction interval of Example 5.9, if a new pack of beef is measured and its leanness is outside the interval $(93.96, 98.44)$, that observation can be viewed as an outlier.

5.7   Tolerance Limits

As discussed in Section 5.6, the scientist or engineer may be less interested in estimating parameters than in gaining a notion about where an individual *observation* or measurement might fall. Such situations call for the use of prediction intervals. However, there is yet a third type of interval that is of interest in many applications. Once again, suppose that interest centers around the manufacturing of a component part and specifications exist on a dimension of that part. In addition, there is little concern about the mean of the dimension. But unlike in the scenario in Section 5.6, one may be less interested in a single observation and more interested in where the majority of the population falls. If process specifications are important, the manager of the process is concerned about long-range performance, **not the next observation**. One must attempt to determine bounds that, in some probabilistic sense, "cover" values in the population (i.e., the measured values of the dimension).

One method of establishing the desired bounds is to determine a confidence interval on a *fixed proportion* of the measurements. This is best motivated by visualizing a situation in which we are doing random sampling from a normal distribution with known mean μ and variance σ^2. Clearly, a bound that covers the middle 95% of the population of observations is

$$\mu \pm 1.96\sigma.$$

This is called a **tolerance interval**, and indeed its coverage of 95% of measured observations is exact. However, in practice, μ and σ are seldom known; thus, the user must apply

$$\bar{x} \pm ks.$$

Now, of course, the interval is a random variable, and hence the *coverage* of a proportion of the population by the interval is not exact. As a result, a $100(1-\gamma)\%$ confidence interval must be used since $\bar{x} \pm ks$ cannot be expected to cover any specified proportion all the time. As a result, we have the following definition.

Tolerance Limits	For a normal distribution of measurements with unknown mean μ and unknown standard deviation σ, **tolerance limits** are given by $\bar{x} \pm ks$, where k is determined such that one can assert with $100(1 - \gamma)\%$ confidence that the given limits contain at least the proportion $1 - \alpha$ of the measurements.

Table A.7 gives values of k for $1 - \alpha = 0.90, 0.95, 0.99$; $\gamma = 0.05, 0.01$; and selected values of n from 2 to 300.

Example 5.10:   Consider Example 5.9. With the information given, find a tolerance interval that gives two-sided 95% bounds on 90% of the distribution of packages of 95% lean beef. Assume the data came from an approximately normal distribution.

Solution: Recall from Example 5.9 that $n = 30$, the sample mean is 96.2%, and the sample standard deviation is 0.8%. From Table A.7, $k = 2.14$. Using

$$\bar{x} \pm ks = 96.2 \pm (2.14)(0.8),$$

we find that the lower and upper bounds are 94.5 and 97.9.

We are 95% confident that the above range covers the central 90% of the distribution of 95% lean beef packages. ▟

Distinction among Confidence Intervals, Prediction Intervals, and Tolerance Intervals

It is important to reemphasize the difference among the three types of intervals discussed and illustrated in the preceding sections. The computations are straightforward, but interpretation can be confusing. In real-life applications, these intervals are not interchangeable because their interpretations are quite distinct.

In the case of confidence intervals, one is attentive only to the **population mean**. For example, Exercise 5.11 on page 213 deals with an engineering process that produces shearing pins. A specification will be set on Rockwell hardness, below which a customer will not accept any pins. Here, a population parameter must take a backseat. It is important that the engineer know where the *majority of the values of Rockwell hardness are going to be*. Thus, tolerance limits should be used. Surely, when tolerance limits on any process output are tighter than process specifications, that is good news for the process manager.

It is true that the tolerance limit interpretation is somewhat related to the confidence interval. The $100(1-\alpha)\%$ tolerance interval on, say, the proportion 0.95 can be viewed as a confidence interval **on the middle 95%** of the corresponding normal distribution. One-sided tolerance limits are also relevant. In the case of the Rockwell hardness problem, it is desirable to have a lower bound of the form $\bar{x} - ks$ such that there is 99% confidence that at least 99% of Rockwell hardness values will exceed the computed value.

Prediction intervals are applicable when it is important to determine a bound on a **single value**. The mean is not the issue here, nor is the location of the majority of the population. Rather, the location of a single new observation is required.

Case Study 5.1: **Machine Quality**: A machine produces metal pieces that are cylindrical in shape. A sample of these pieces is taken, and the diameters are found to be 1.01, 0.97, 1.03, 1.04, 0.99, 0.98, 0.99, 1.01, and 1.03 centimeters. Use these data to calculate three interval types and draw interpretations that illustrate the distinction between them in the context of the system. For all computations, assume an approximately normal distribution. The sample mean and standard deviation for the given data are $\bar{x} = 1.0056$ and $s = 0.0246$.

(a) Find a 99% confidence interval on the mean diameter.

(b) Compute a 99% prediction interval on a measured diameter of a single metal piece taken from the machine.

(c) Find the 99% tolerance limits that will contain 95% of the metal pieces produced by this machine.

Solution: (a) The 99% confidence interval for the mean diameter is given by

$$\bar{x} \pm t_{0.005}s/\sqrt{n} = 1.0056 \pm (3.355)(0.0246/3) = 1.0056 \pm 0.0275.$$

Thus, the 99% confidence bounds are 0.9781 and 1.0331.

(b) The 99% prediction interval for a future observation is given by

$$\bar{x} \pm t_{0.005} s \sqrt{1 + 1/n} = 1.0056 \pm (3.355)(0.0246)\sqrt{1 + 1/9},$$

with the bounds being 0.9186 and 1.0926.

(c) From Table A.7, for $n = 9$, $1 - \gamma = 0.99$, and $1 - \alpha = 0.95$, we find $k = 4.550$ for two-sided limits. Hence, the 99% tolerance limits are given by

$$\bar{x} + ks = 1.0056 \pm (4.550)(0.0246),$$

with the bounds being 0.8937 and 1.1175. We are 99% confident that the tolerance interval from 0.8937 to 1.1175 will contain the central 95% of the distribution of diameters produced.

This case study illustrates that the three types of limits can give appreciably different results even though they are all 99% bounds. In the case of the confidence interval on the mean, 99% of such intervals cover the population mean diameter. Thus, we say that we are 99% confident that the mean diameter produced by the process is between 0.9781 and 1.0331 centimeters. Emphasis is placed on the mean, with less concern about a single reading or the general nature of the distribution of diameters in the population. In the case of the prediction limits, the bounds 0.9186 and 1.0926 are based on the distribution of a single "new" metal piece taken from the process, and again 99% of such limits will cover the diameter of a new measured piece. On the other hand, the tolerance limits, as suggested in the previous section, give the engineer a sense of where the "majority," say the central 95%, of the diameters of measured pieces in the population reside. The 99% tolerance limits, 0.8937 and 1.1175, are numerically quite different from the other two bounds. If these bounds appear alarmingly wide to the engineer, it reflects negatively on process quality. On the other hand, if the bounds represent a desirable result, the engineer may conclude that a majority (95% in this case) of the diameters are in a desirable range. Again, a confidence interval interpretation may be used: namely, 99% of such calculated bounds will cover the middle 95% of the population of diameters.

Exercises

5.1 A UCLA researcher claims that the life span of mice can be extended by as much as 25% when the calories in their diet are reduced by approximately 40% from the time they are weaned. The restricted diet is enriched to normal levels by vitamins and protein. Assuming that it is known from previous studies that $\sigma = 5.8$ months, how many mice should be included in our sample if we wish to be 99% confident that the mean life span of the sample will be within 2 months of the population mean for all mice subjected to this reduced diet?

5.2 An electrical firm manufactures light bulbs that have a length of life that is approximately normally distributed with a standard deviation of 40 hours. If a sample of 30 bulbs has an average life of 780 hours, find a 96% confidence interval for the population mean of all bulbs produced by this firm.

5.3 Many cardiac patients wear an implanted pacemaker to control their heartbeat. A plastic connector module mounts on the top of the pacemaker. Assuming a standard deviation of 0.0015 inch and an approximately normal distribution, find a 95% confidence interval for the mean of the depths of all connector modules made by a certain manufacturing company. A random sample of 75 modules has an average depth of 0.310 inch.

5.4 The heights of a random sample of 50 college students showed a mean of 174.5 centimeters and a standard deviation of 6.9 centimeters.

(a) Construct a 98% confidence interval for the mean height of all college students.

(b) What can we assert with 98% confidence about the possible size of our error if we estimate the mean height of all college students to be 174.5 centimeters?

5.5 A random sample of 100 automobile owners in the state of Virginia shows that an automobile is driven on average 23,500 kilometers per year with a standard deviation of 3900 kilometers. Assume the distribution of measurements to be approximately normal.

(a) Construct a 99% confidence interval for the average number of kilometers an automobile is driven annually in Virginia.

(b) What can we assert with 99% confidence about the possible size of our error if we estimate the average number of kilometers driven by car owners in Virginia to be 23,500 kilometers per year?

5.6 How large a sample is needed in Exercise 5.2 if we wish to be 96% confident that our sample mean will be within 10 hours of the true mean?

5.7 How large a sample is needed in Exercise 5.3 if we wish to be 95% confident that our sample mean will be within 0.0005 inch of the true mean?

5.8 An efficiency expert wishes to determine the average time that it takes to drill three holes in a certain metal clamp. How large a sample will she need to be 95% confident that her sample mean will be within 15 seconds of the true mean? Assume that it is known from previous studies that $\sigma = 40$ seconds.

5.9 Regular consumption of presweetened cereals contributes to tooth decay, heart disease, and other degenerative diseases, according to studies conducted by Dr. W. H. Bowen of the National Institute of Health and Dr. J. Yudben, Professor of Nutrition and Dietetics at the University of London. In a random sample consisting of 20 similar single servings of Alpha-Bits, the average sugar content was 11.3 grams with a standard deviation of 2.45 grams. Assuming that the sugar contents are normally distributed, construct a 95% confidence interval for the mean sugar content for single servings of Alpha-Bits.

5.10 For a random sample $X_1, \ldots, X_n$, show that

$$\sum_{i=1}^{n}(X_i - \mu)^2 = \sum_{i=1}^{n}(X_i - \bar{X})^2 + n(\bar{X} - \mu)^2.$$

5.11 A random sample of 12 shearing pins is taken in a study of the Rockwell hardness of the pin head. Measurements on the Rockwell hardness are made for each of the 12, yielding an average value of 48.50 with a sample standard deviation of 1.5. Assuming the measurements to be normally distributed, construct a 90% confidence interval for the mean Rockwell hardness.

5.12 The following measurements were recorded for the drying time, in hours, of a certain brand of latex paint:

3.4	2.5	4.8	2.9	3.6
2.8	3.3	5.6	3.7	2.8
4.4	4.0	5.2	3.0	4.8

Assuming that the measurements represent a random sample from a normal population, find a 95% prediction interval for the drying time for the next trial of the paint.

5.13 Referring to Exercise 5.5, construct a 99% prediction interval for the kilometers traveled annually by an automobile owner in Virginia.

5.14 Consider Exercise 5.9. Compute a 95% prediction interval for the sugar content of the next single serving of Alpha-Bits.

5.15 A random sample of 25 tablets of buffered aspirin contains, on average, 325.05 mg of aspirin per tablet, with a standard deviation of 0.5 mg. Find the 95% tolerance limits that will contain 90% of the tablet contents for this brand of buffered aspirin. Assume that the aspirin content is normally distributed.

5.16 Referring to Exercise 5.11, construct a 95% tolerance interval containing 90% of the measurements.

5.17 In a study conducted by the Zoology department at Virginia Tech, fifteen samples of water were collected from a certain station in the James River in order to gain some insight regarding the amount of orthophosphorus in the river. The concentration of the chemical is measured in milligrams per liter. Let us suppose that the mean at the station is not as important as the upper extreme of the distribution of the concentration of the chemical at the station. Concern centers around whether the concentration at the extreme is too large. Readings for the fifteen water samples gave a sample mean of 3.84 milligrams per liter and a sample standard deviation of 3.07 milligrams per liter. Assume that the readings are a random sample from a normal distribution. Calculate a prediction interval (upper 95% prediction limit) and a tolerance limit (95% upper tolerance limit that exceeds 95% of the population of values). Interpret both; that is, tell what each communicates about the upper extreme of the distribution of orthophosphorus at the sampling station.

5.18 Consider the situation of Case Study 5.1 on page 211. Estimation of the mean diameter, while important, is not nearly as important as trying to pin down the location of the majority of the distribution of diameters. Find the 95% tolerance limits that contain 95% of the diameters.

5.19 Consider the situation of Case Study 5.1 with a larger sample of metal pieces. The diameters are as follows: 1.01, 0.97, 1.03, 1.04, 0.99, 0.98, 1.01, 1.03, 0.99, 1.00, 1.00, 0.99, 0.98, 1.01, 1.02, 0.99 centimeters. Once again the normality assumption may be made. Do the following and compare your results to those of the case study. Discuss how they are different and why.

(a) Compute a 99% confidence interval on the mean diameter.

(b) Compute a 99% prediction interval on the next diameter to be measured.

(c) Compute a 99% tolerance interval for coverage of the central 95% of the distribution of diameters.

5.20 A type of thread is being studied for its tensile strength properties. Fifty pieces were tested under similar conditions, and the results showed an average tensile strength of 78.3 kilograms and a standard deviation of 5.6 kilograms. Assuming a normal distribution of tensile strengths, give a lower 95% prediction limit on a single observed tensile strength value. In addition, give a lower 95% tolerance limit that is exceeded by 99% of the tensile strength values.

5.21 Refer to Exercise 5.20. Why are the quantities requested in the exercise likely to be more important to the manufacturer of the thread than, say, a confidence interval on the mean tensile strength?

5.22 Refer to Exercise 5.20 again. Suppose that specifications by a buyer of the thread are that the tensile strength of the material must be at least 62 kilograms. The manufacturer is satisfied if at most 5% of the manufactured pieces have tensile strength less than 62 kilograms. Is there cause for concern? Use a one-sided 99% tolerance limit that is exceeded by 95% of the tensile strength values.

5.23 Consider the drying time measurements in Exercise 5.12. Suppose the 15 observations in the data set are supplemented by a 16th value of 6.9 hours. In the context of the original 15 observations, is the 16th value an outlier? Show work.

5.24 Consider the data in Exercise 5.11. Suppose the manufacturer of the shearing pins insists that the Rockwell hardness of the product be less than or equal to 44.0 only 5% of the time. What is your reaction? Use a tolerance limit calculation as the basis for your judgment.

5.8 Two Samples: Estimating the Difference between Two Means

If we have two populations with means μ_1 and μ_2 and variances σ_1^2 and σ_2^2, respectively, a point estimator of the difference between μ_1 and μ_2 is given by the statistic $\bar{X}_1 - \bar{X}_2$. Therefore, to obtain a point estimate of $\mu_1 - \mu_2$, we shall select two independent random samples, one from each population, of sizes n_1 and n_2, and compute $\bar{x}_1 - \bar{x}_2$, the difference of the sample means. Clearly, we must consider the sampling distribution of $\bar{X}_1 - \bar{X}_2$.

According to Theorem 4.3, we can expect the sampling distribution of $\bar{X}_1 - \bar{X}_2$ to be approximately normal with mean $\mu_{\bar{X}_1 - \bar{X}_2} = \mu_1 - \mu_2$ and standard deviation $\sigma_{\bar{X}_1 - \bar{X}_2} = \sqrt{\sigma_1^2/n_1 + \sigma_2^2/n_2}$. Therefore, we can assert with a probability of $1 - \alpha$ that the standard normal variable

$$Z = \frac{(\bar{X}_1 - \bar{X}_2) - (\mu_1 - \mu_2)}{\sqrt{\sigma_1^2/n_1 + \sigma_2^2/n_2}}$$

will fall between $-z_{\alpha/2}$ and $z_{\alpha/2}$. Referring once again to Figure 5.2 on page 200, we write

$$P(-z_{\alpha/2} < Z < z_{\alpha/2}) = 1 - \alpha.$$

Substituting for Z, we state equivalently that

$$P\left(-z_{\alpha/2} < \frac{(\bar{X}_1 - \bar{X}_2) - (\mu_1 - \mu_2)}{\sqrt{\sigma_1^2/n_1 + \sigma_2^2/n_2}} < z_{\alpha/2}\right) = 1 - \alpha,$$

which leads to the following $100(1 - \alpha)\%$ confidence interval for $\mu_1 - \mu_2$.

Confidence Interval for $\mu_1 - \mu_2$, σ_1^2 and σ_2^2 Known	If $\bar{x}_1$ and $\bar{x}_2$ are means of independent random samples of sizes n_1 and n_2 from populations with known variances σ_1^2 and σ_2^2, respectively, a $100(1 - \alpha)\%$ confidence interval for $\mu_1 - \mu_2$ is given by $$(\bar{x}_1 - \bar{x}_2) - z_{\alpha/2}\sqrt{\frac{\sigma_1^2}{n_1} + \frac{\sigma_2^2}{n_2}} < \mu_1 - \mu_2 < (\bar{x}_1 - \bar{x}_2) + z_{\alpha/2}\sqrt{\frac{\sigma_1^2}{n_1} + \frac{\sigma_2^2}{n_2}},$$ where $z_{\alpha/2}$ is the z-value leaving an area of $\alpha/2$ to the right.

The degree of confidence is exact when samples are selected from normal populations. For nonnormal populations, the Central Limit Theorem allows for a good approximation for reasonable size samples.

Variances Unknown but Equal

Consider the case where σ_1^2 and σ_2^2 are unknown. If $\sigma_1^2 = \sigma_2^2 = \sigma^2$, we obtain a standard normal variable of the form

$$Z = \frac{(\bar{X}_1 - \bar{X}_2) - (\mu_1 - \mu_2)}{\sqrt{\sigma^2[(1/n_1) + (1/n_2)]}}.$$

According to Theorem 4.4, the two random variables

$$\frac{(n_1 - 1)S_1^2}{\sigma^2} \quad \text{and} \quad \frac{(n_2 - 1)S_2^2}{\sigma^2}$$

have chi-squared distributions with $n_1 - 1$ and $n_2 - 1$ degrees of freedom, respectively. Furthermore, they are independent chi-squared variables, since the random samples were selected independently. Consequently, their sum

$$V = \frac{(n_1 - 1)S_1^2}{\sigma^2} + \frac{(n_2 - 1)S_2^2}{\sigma^2} = \frac{(n_1 - 1)S_1^2 + (n_2 - 1)S_2^2}{\sigma^2}$$

has a chi-squared distribution with $v = n_1 + n_2 - 2$ degrees of freedom.

Since the preceding expressions for Z and V can be shown to be independent, it follows from Theorem 4.5 that the statistic

$$T = \frac{(\bar{X}_1 - \bar{X}_2) - (\mu_1 - \mu_2)}{\sqrt{\sigma^2[(1/n_1) + (1/n_2)]}} \bigg/ \sqrt{\frac{(n_1 - 1)S_1^2 + (n_2 - 1)S_2^2}{\sigma^2(n_1 + n_2 - 2)}}$$

has the t-distribution with $v = n_1 + n_2 - 2$ degrees of freedom.

A point estimate of the unknown common variance σ^2 can be obtained by pooling the sample variances. Denoting the pooled estimator by S_p^2, we have the following.

Pooled Estimate of Variance	$$S_p^2 = \frac{(n_1 - 1)S_1^2 + (n_2 - 1)S_2^2}{n_1 + n_2 - 2}$$

Substituting S_p^2 in the T statistic, we obtain the less cumbersome form

$$T = \frac{(\bar{X}_1 - \bar{X}_2) - (\mu_1 - \mu_2)}{S_p\sqrt{(1/n_1) + (1/n_2)}}.$$

Using the T statistic, we have

$$P(-t_{\alpha/2} < T < t_{\alpha/2}) = 1 - \alpha,$$

where $t_{\alpha/2}$ is the t-value with $n_1 + n_2 - 2$ degrees of freedom, above which we find an area of $\alpha/2$. Substituting for T in the inequality, we write

$$P\left[-t_{\alpha/2} < \frac{(\bar{X}_1 - \bar{X}_2) - (\mu_1 - \mu_2)}{S_p\sqrt{(1/n_1) + (1/n_2)}} < t_{\alpha/2}\right] = 1 - \alpha.$$

After the usual mathematical manipulations, the difference of the sample means $\bar{x}_1 - \bar{x}_2$ and the pooled variance are computed and then the following $100(1 - \alpha)\%$ confidence interval for $\mu_1 - \mu_2$ is obtained. The value of s_p^2 is easily seen to be a weighted average of the two sample variances s_1^2 and s_2^2, where the weights are the degrees of freedom.

Confidence Interval for $\mu_1 - \mu_2$, $\sigma_1^2 = \sigma_2^2$ but Both Unknown	If $\bar{x}_1$ and $\bar{x}_2$ are the means of independent random samples of sizes n_1 and n_2, respectively, from approximately normal populations with unknown but equal variances, a $100(1 - \alpha)\%$ confidence interval for $\mu_1 - \mu_2$ is given by $$(\bar{x}_1 - \bar{x}_2) - t_{\alpha/2}s_p\sqrt{\frac{1}{n_1} + \frac{1}{n_2}} < \mu_1 - \mu_2 < (\bar{x}_1 - \bar{x}_2) + t_{\alpha/2}s_p\sqrt{\frac{1}{n_1} + \frac{1}{n_2}},$$ where s_p is the pooled estimate of the population standard deviation and $t_{\alpha/2}$ is the t-value with $v = n_1 + n_2 - 2$ degrees of freedom, leaving an area of $\alpha/2$ to the right.

Example 5.11: The article "Macroinvertebrate Community Structure as an Indicator of Acid Mine Pollution," published in the *Journal of Environmental Pollution*, reports on an investigation undertaken in Cane Creek, Alabama, to determine the relationship between selected physiochemical parameters and different measures of macroinvertebrate community structure. One facet of the investigation was an evaluation of the effectiveness of a numerical species diversity index to indicate aquatic degradation due to acid mine drainage. Conceptually, a high index of macroinvertebrate species diversity should indicate an unstressed aquatic system, while a low diversity index should indicate a stressed aquatic system.

Two independent sampling stations were chosen for this study, one located downstream from the acid mine discharge point and the other located upstream. For 12 monthly samples collected at the downstream station, the species diversity index had a mean value $\bar{x}_1 = 3.11$ and a standard deviation $s_1 = 0.771$, while 10 monthly samples collected at the upstream station had a mean index value $\bar{x}_2 = 2.04$ and a standard deviation $s_2 = 0.448$. Find a 90% confidence interval for the difference between the population means for the two locations, assuming that the populations are approximately normally distributed with equal variances.

Solution: Let μ_1 and μ_2 represent the population means, respectively, for the species diversity indices at the downstream and upstream stations. We wish to find a 90% confidence interval for $\mu_1 - \mu_2$. Our point estimate of $\mu_1 - \mu_2$ is

$$\bar{x}_1 - \bar{x}_2 = 3.11 - 2.04 = 1.07.$$

The pooled estimate, s_p^2, of the common variance, σ^2, is

$$s_p^2 = \frac{(n_1 - 1)s_1^2 + (n_2 - 1)s_2^2}{n_1 + n_2 - 2} = \frac{(11)(0.771^2) + (9)(0.448^2)}{12 + 10 - 2} = 0.417.$$

Taking the square root, we obtain $s_p = 0.646$. Using $\alpha = 0.1$, we find in Table A.4 that $t_{0.05} = 1.725$ for $v = n_1 + n_2 - 2 = 20$ degrees of freedom. Therefore, the 90% confidence interval for $\mu_1 - \mu_2$ is

$$1.07 - (1.725)(0.646)\sqrt{\frac{1}{12} + \frac{1}{10}} < \mu_1 - \mu_2 < 1.07 + (1.725)(0.646)\sqrt{\frac{1}{12} + \frac{1}{10}},$$

which simplifies to $0.593 < \mu_1 - \mu_2 < 1.547$.   ⌐

Unknown and Unequal Variances

Let us now consider the problem of finding an interval estimate of $\mu_1 - \mu_2$ when the unknown population variances are not likely to be equal. The statistic most often used in this case is

$$T' = \frac{(\bar{X}_1 - \bar{X}_2) - (\mu_1 - \mu_2)}{\sqrt{(S_1^2/n_1) + (S_2^2/n_2)}},$$

which has approximately a t-distribution with v degrees of freedom, where

$$v = \frac{(s_1^2/n_1 + s_2^2/n_2)^2}{[(s_1^2/n_1)^2/(n_1 - 1)] + [(s_2^2/n_2)^2/(n_2 - 1)]}.$$

Since v is seldom an integer, we *round it down* to the nearest whole number. The above estimate of the degrees of freedom is called the Satterthwaite approximation (Satterthwaite, 1946, in the Bibliography).

Using the statistic T', we write

$$P(-t_{\alpha/2} < T' < t_{\alpha/2}) \approx 1 - \alpha,$$

where $t_{\alpha/2}$ is the value of the t-distribution with v degrees of freedom, above which we find an area of $\alpha/2$. Substituting for T' in the inequality and following the same steps as before, we state the final result.

Confidence
Interval for
$\mu_1 - \mu_2$, $\sigma_1^2 \neq \sigma_2^2$
and Both
Unknown

If $\bar{x}_1$ and s_1^2 and $\bar{x}_2$ and s_2^2 are the means and variances of independent random samples of sizes n_1 and n_2, respectively, from approximately normal populations with unknown and unequal variances, an approximate $100(1 - \alpha)\%$ confidence interval for $\mu_1 - \mu_2$ is given by

$$(\bar{x}_1 - \bar{x}_2) - t_{\alpha/2}\sqrt{\frac{s_1^2}{n_1} + \frac{s_2^2}{n_2}} < \mu_1 - \mu_2 < (\bar{x}_1 - \bar{x}_2) + t_{\alpha/2}\sqrt{\frac{s_1^2}{n_1} + \frac{s_2^2}{n_2}},$$

where $t_{\alpha/2}$ is the t-value with

$$v = \frac{(s_1^2/n_1 + s_2^2/n_2)^2}{[(s_1^2/n_1)^2/(n_1 - 1)] + [(s_2^2/n_2)^2/(n_2 - 1)]}$$

degrees of freedom, leaving an area of $\alpha/2$ to the right.

Note that the expression for v above involves random variables, and thus v is an *estimate* of the degrees of freedom. In applications, this estimate will not result in a whole number, and thus the analyst must round down to the nearest integer to achieve the desired confidence.

Before we illustrate the above confidence interval with an example, we should point out that all the confidence intervals on $\mu_1 - \mu_2$ are of the same general form as those on a single mean; namely, they can be written as

$$\text{point estimate} \pm t_{\alpha/2}\, \widehat{\text{s.e.}}(\text{point estimate})$$

or

$$\text{point estimate} \pm z_{\alpha/2}\, \text{s.e.}(\text{point estimate}).$$

For example, in the case where $\sigma_1 = \sigma_2 = \sigma$, the estimated standard error of $\bar{x}_1 - \bar{x}_2$ is $s_p\sqrt{1/n_1 + 1/n_2}$. For the case where $\sigma_1^2 \neq \sigma_2^2$,

$$\widehat{\text{s.e.}}(\bar{x}_1 - \bar{x}_2) = \sqrt{\frac{s_1^2}{n_1} + \frac{s_2^2}{n_2}}.$$

Example 5.12: A study was conducted by the Department of Zoology at Virginia Tech to estimate the difference in the amounts of the chemical orthophosphorus measured at two different stations on the James River. Orthophosphorus was measured in milligrams per liter. Fifteen samples were collected from station 1, and 12 samples were obtained from station 2. The 15 samples from station 1 had an average orthophosphorus content of 3.84 milligrams per liter and a standard deviation of 3.07 milligrams per liter, while the 12 samples from station 2 had an average content of 1.49 milligrams per liter and a standard deviation of 0.80 milligram per liter. Find a 95% confidence interval for the difference in the true average orthophosphorus contents at these two stations, assuming that the observations came from normal populations with different variances.

Solution: For station 1, we have $\bar{x}_1 = 3.84$, $s_1 = 3.07$, and $n_1 = 15$. For station 2, $\bar{x}_2 = 1.49$, $s_2 = 0.80$, and $n_2 = 12$. We wish to find a 95% confidence interval for $\mu_1 - \mu_2$.

Since the population variances are assumed to be unequal, we can only find an approximate 95% confidence interval based on the t-distribution with v degrees of freedom, where

$$v = \frac{(3.07^2/15 + 0.80^2/12)^2}{[(3.07^2/15)^2/14] + [(0.80^2/12)^2/11]} = 16.3 \approx 16.$$

Our point estimate of $\mu_1 - \mu_2$ is

$$\bar{x}_1 - \bar{x}_2 = 3.84 - 1.49 = 2.35.$$

Using $\alpha = 0.05$, we find in Table A.4 that $t_{0.025} = 2.120$ for $v = 16$ degrees of freedom. Therefore, the 95% confidence interval for $\mu_1 - \mu_2$ is

$$2.35 - 2.120\sqrt{\frac{3.07^2}{15} + \frac{0.80^2}{12}} < \mu_1 - \mu_2 < 2.35 + 2.120\sqrt{\frac{3.07^2}{15} + \frac{0.80^2}{12}},$$

which simplifies to $0.60 < \mu_1 - \mu_2 < 4.10$. Hence, we are 95% confident that the interval from 0.60 to 4.10 milligrams per liter contains the difference of the true average orthophosphorus contents for these two locations. ⌐

When two population variances are unknown, the assumption of equal variances or unequal variances may be precarious.

5.9 Paired Observations

At this point, we shall consider estimation procedures for the difference of two means when the samples are not independent and the variances of the two populations are not necessarily equal. The situation considered here deals with a very special experimental condition, namely that of *paired observations*. Unlike in the situations described earlier, the conditions of the two populations are not assigned randomly to experimental units. Rather, each homogeneous experimental unit receives both population conditions; as a result, each experimental unit has a pair of observations, one for each population. For example, if we run a test on a new diet using 15 individuals, the weights before and after going on the diet form the information for our two samples. The two populations are "before" and "after," and the experimental unit is the individual. Obviously, the observations in a pair have something in common. To determine if the diet is effective, we consider the differences $d_1, d_2, \ldots, d_n$ in the paired observations. These differences are the values of a random sample $D_1, D_2, \ldots, D_n$ from a population of differences that we shall assume to be normally distributed with mean $\mu_D = \mu_1 - \mu_2$ and variance σ_D^2. We estimate σ_D^2 by s_d^2, the variance of the differences that constitute our sample. The point estimator of μ_D is given by $\bar{D}$.

Tradeoff between Reducing Variance and Losing Degrees of Freedom

Comparing the confidence intervals obtained with and without pairing makes apparent that there is a tradeoff involved. Although pairing should indeed reduce variance and hence reduce the standard error of the point estimate, the degrees of freedom are reduced by reducing the problem to a one-sample problem. As a result, the $t_{\alpha/2}$ point attached to the standard error is adjusted accordingly. Thus, pairing may be counterproductive. This would certainly be the case if one experienced only a modest reduction in variance (through σ_D^2) by pairing.

Another illustration of pairing involves choosing n pairs of subjects, with each pair having a similar characteristic such as IQ, age, or breed, and then selecting one member of each pair at random to yield a value of X_1, leaving the other member to provide the value of X_2. In this case, X_1 and X_2 might represent the grades obtained by two individuals of equal IQ when one of the individuals is assigned at random to a class using the conventional lecture approach while the other individual is assigned to a class using programmed materials.

A $100(1 - \alpha)\%$ confidence interval for μ_D can be established by writing

$$P(-t_{\alpha/2} < T < t_{\alpha/2}) = 1 - \alpha,$$

where $T = \frac{\bar{D} - \mu_D}{S_d/\sqrt{n}}$ and $t_{\alpha/2}$, as before, is a value of the t-distribution with $n - 1$ degrees of freedom.

It is now a routine procedure to replace T by its definition in the inequality above and carry out the mathematical steps that lead to the following $100(1 - \alpha)\%$ confidence interval for $\mu_1 - \mu_2 = \mu_D$.

Confidence Interval for $\mu_D = \mu_1 - \mu_2$ for Paired Observations

If $\bar{d}$ and s_d are the mean and standard deviation, respectively, of the normally distributed differences of n random pairs of measurements, a $100(1 - \alpha)\%$ confidence interval for $\mu_D = \mu_1 - \mu_2$ is

$$\bar{d} - t_{\alpha/2}\frac{s_d}{\sqrt{n}} < \mu_D < \bar{d} + t_{\alpha/2}\frac{s_d}{\sqrt{n}},$$

where $t_{\alpha/2}$ is the t-value with $v = n - 1$ degrees of freedom, leaving an area of $\alpha/2$ to the right.

Example 5.13: A study was undertaken at Virginia Tech to determine if fire can be used as a viable management tool to increase the amount of forage available to deer during the critical months in late winter and early spring. Calcium is a required element for plants and animals. The amount taken up and stored in plants is closely correlated to the amount present in the soil. It was hypothesized that a fire may change the calcium levels present in the soil and thus affect the amount available to deer. A large tract of land in the Fishburn Forest was selected for a prescribed burn. Soil samples were taken from 12 plots of equal area just prior to the burn and analyzed for calcium. Postburn calcium levels were analyzed from the same plots. These values, in kilograms per plot, are presented in Table 5.1.

Construct a 95% confidence interval for the mean difference in calcium levels in the soil prior to and after the prescribed burn. Assume the distribution of differences in calcium levels to be approximately normal.

Solution: In this problem, we wish to find a 95% confidence interval for the mean difference in calcium levels between the preburn and postburn. Since the observations are paired, we define $\mu_{\text{Preburn}} - \mu_{\text{Postburn}} = \mu_D$. The sample size is $n = 12$, the point estimate of μ_D is $\bar{d} = 40.58$, and the sample standard deviation, s_d, of the differences

Table 5.1: Data for Example 5.13

Plot	Calcium Level (kg/plot) Preburn	Postburn	Plot	Calcium Level (kg/plot) Preburn	Postburn
1	50	9	7	77	32
2	50	18	8	54	9
3	82	45	9	23	18
4	64	18	10	45	9
5	82	18	11	36	9
6	73	9	12	54	9

is

$$s_d = \sqrt{\frac{1}{n-1}\sum_{i=1}^{n}(d_i - \bar{d})^2} = 15.791.$$

Using $\alpha = 0.05$, we find in Table A.4 that $t_{0.025} = 2.201$ for $v = n-1 = 11$ degrees of freedom. Therefore, the 95% confidence interval is

$$40.58 - (2.201)\left(\frac{15.791}{\sqrt{12}}\right) < \mu_D < 40.58 + (2.201)\left(\frac{15.791}{\sqrt{12}}\right),$$

or simply $30.55 < \mu_D < 50.61$, from which we can conclude that there is significant difference in calcium levels between the soil prior and post to the prescribed burns. Hence there is significant reduction in the level of calcium after the burn.

Exercises

5.25 A study was conducted to determine if a certain treatment has any effect on the amount of metal removed in a pickling operation. A random sample of 100 pieces was immersed in a bath for 24 hours without the treatment, yielding an average of 12.2 millimeters of metal removed and a sample standard deviation of 1.1 millimeters. A second sample of 200 pieces was exposed to the treatment, followed by the 24-hour immersion in the bath, resulting in an average removal of 9.1 millimeters of metal with a sample standard deviation of 0.9 millimeter. Compute a 98% confidence interval estimate for the difference between the population means. Does the treatment appear to reduce the mean amount of metal removed?

5.26 Two kinds of thread are being compared for strength. Fifty pieces of each type of thread are tested under similar conditions. Brand A has an average tensile strength of 78.3 kilograms with a standard deviation of 5.6 kilograms, while brand B has an average tensile strength of 87.2 kilograms with a standard deviation of 6.3 kilograms. Construct a 95% confidence interval for the difference of the population means.

5.27 Two catalysts in a batch chemical process are being compared for their effect on the output of the process reaction. A sample of 12 batches was prepared using catalyst 1, and a sample of 10 batches was prepared using catalyst 2. The 12 batches for which catalyst 1 was used in the reaction gave an average yield of 85 with a sample standard deviation of 4, and the 10 batches for which catalyst 2 was used gave an average yield of 81 and a sample standard deviation of 5. Find a 90% confidence interval for the difference between the population means, assuming that the populations are approximately normally distributed with equal variances.

5.28 In a study conducted at Virginia Tech on the development of ectomycorrhizal, a symbiotic relationship between the roots of trees and a fungus, in which minerals are transferred from the fungus to the trees and sugars from the trees to the fungus, 20 northern red oak seedlings exposed to the fungus *Pisolithus tinctorus* were grown in a greenhouse. All seedlings were planted in the same type of soil and received the same amount of sunshine and water. Half received no nitrogen at planting time, to serve as a control, and the other half received 368 ppm of nitrogen in the form $NaNO_3$. The stem weights, in grams, at the end of 140 days were recorded as follows:

No Nitrogen	Nitrogen
0.32	0.26
0.53	0.43
0.28	0.47
0.37	0.49
0.47	0.52
0.43	0.75
0.36	0.79
0.42	0.86
0.38	0.62
0.43	0.46

Construct a 95% confidence interval for the difference in the mean stem weight between seedlings that receive no nitrogen and those that receive 368 ppm of nitrogen. Assume the populations to be normally distributed with equal variances.

5.29 The following data represent the length of time, in days, to recovery for patients randomly treated with one of two medications to clear up severe bladder infections:

Medication 1	Medication 2
$n_1 = 14$	$n_2 = 16$
$\bar{x}_1 = 17$	$\bar{x}_2 = 19$
$s_1^2 = 1.5$	$s_2^2 = 1.8$

Find a 99% confidence interval for the difference $\mu_2 - \mu_1$ in the mean recovery times for the two medications, assuming normal populations with equal variances.

5.30 An experiment reported in *Popular Science* compared fuel economies for two types of similarly equipped diesel mini-trucks. Let us suppose that 12 Volkswagen and 10 Toyota trucks were tested in 90-kilometer-per-hour steady-paced trials. If the 12 Volkswagen trucks averaged 16 kilometers per liter with a standard deviation of 1.0 kilometer per liter and the 10 Toyota trucks averaged 11 kilometers per liter with a standard deviation of 0.8 kilometer per liter, construct a 90% confidence interval for the difference between the average kilometers per liter for these two mini-trucks. Assume that the distances per liter for the truck models are approximately normally distributed with equal variances.

5.31 A taxi company is trying to decide whether to purchase brand A or brand B tires for its fleet of taxis. To estimate the difference in the two brands, an experiment is conducted using 12 of each brand. The tires are run until they wear out. The results are

Brand A: $\bar{x}_1 = 36{,}300$ kilometers,
             $s_1 = 5000$ kilometers.
Brand B: $\bar{x}_2 = 38{,}100$ kilometers,
             $s_2 = 6100$ kilometers.

Compute a 95% confidence interval for $\mu_A - \mu_B$ assuming the populations to be approximately normally distributed. You may not assume that the variances are equal.

5.32 Referring to Exercise 5.31, find a 99% confidence interval for $\mu_1 - \mu_2$ if tires of the two brands are assigned at random to the left and right rear wheels of 8 taxis and the following distances, in kilometers, are recorded:

Taxi	Brand A	Brand B
1	34,400	36,700
2	45,500	46,800
3	36,700	37,700
4	32,000	31,100
5	48,400	47,800
6	32,800	36,400
7	38,100	38,900
8	30,100	31,500

Assume that the differences of the distances are approximately normally distributed.

5.33 The federal government awarded grants to the agricultural departments of 9 universities to test the yield capabilities of two new varieties of wheat. Each variety was planted on a plot of equal area at each university, and the yields, in kilograms per plot, were recorded as follows:

	University								
Variety	1	2	3	4	5	6	7	8	9
1	38	23	35	41	44	29	37	31	38
2	45	25	31	38	50	33	36	40	43

Find a 95% confidence interval for the mean difference between the yields of the two varieties, assuming the differences of yields to be approximately normally distributed. Explain why pairing is necessary in this problem.

5.34 The following data represent the running times of films produced by two motion-picture companies.

Company	Time (minutes)					
I	103	94	110	87	98	
II	97	82	123	92	175	88	118

Compute a 90% confidence interval for the difference between the average running times of films produced by the two companies. Assume that the running-time differences are approximately normally distributed with unequal variances.

5.35 *Fortune* magazine (March 1997) reported the total returns to investors for the 10 years prior to 1996 and also for 1996 for 431 companies. The total returns for 9 of the companies and the S&P 500 are listed below. Find a 95% confidence interval for the mean change in percent return to investors.

Company	Total Return to Investors 1986–96	1996
Coca-Cola	29.8%	43.3%
Mirage Resorts	27.9%	25.4%
Merck	22.1%	24.0%
Microsoft	44.5%	88.3%
Johnson & Johnson	22.2%	18.1%
Intel	43.8%	131.2%
Pfizer	21.7%	34.0%
Procter & Gamble	21.9%	32.1%
Berkshire Hathaway	28.3%	6.2%
S&P 500	11.8%	20.3%

5.36 An automotive company is considering two types of batteries for its automobile. Sample information on battery life is collected for 20 batteries of type A and 20 batteries of type B. The summary statistics are $\bar{x}_A = 32.91$, $\bar{x}_B = 30.47$, $s_A = 1.57$, and $s_B = 1.74$. Assume the data on each battery are normally distributed and assume $\sigma_A = \sigma_B$.

(a) Find a 95% confidence interval on $\mu_A - \mu_B$.

(b) Draw a conclusion from (a) that provides insight into whether A or B should be adopted.

5.37 Two different brands of latex paint are being considered for use. Fifteen specimens of each type of paint were selected, and the drying times, in hours, were as follows:

Paint A					Paint B				
3.5	2.7	3.9	4.2	3.6	4.7	3.9	4.5	5.5	4.0
2.7	3.3	5.2	4.2	2.9	5.3	4.3	6.0	5.2	3.7
4.4	5.2	4.0	4.1	3.4	5.5	6.2	5.1	5.4	4.8

Assume the drying time is normally distributed with $\sigma_A = \sigma_B$. Find a 95% confidence interval on $\mu_B - \mu_A$, where μ_A and μ_B are the mean drying times.

5.38 Two levels (low and high) of insulin doses are given to two groups of diabetic rats to check the insulin-binding capacity, yielding the following data:

Low dose:	$n_1 = 8$	$\bar{x}_1 = 1.98$	$s_1 = 0.51$
High dose:	$n_2 = 13$	$\bar{x}_2 = 1.30$	$s_2 = 0.35$

Assume that the variances are equal. Give a 95% confidence interval for the difference in the true average insulin-binding capacity between the two samples.

5.10 Single Sample: Estimating a Proportion

A point estimator of the proportion p in a binomial experiment is given by the statistic $\widehat{P} = X/n$, where X represents the number of successes in n trials. Therefore, the sample proportion $\hat{p} = x/n$ will be used as the point estimate of the parameter p.

If the unknown proportion p is not expected to be too close to 0 or 1, we can establish a confidence interval for p by considering the sampling distribution of $\widehat{P}$. Designating a failure in each binomial trial by the value 0 and a success by the value 1, the number of successes, x, can be interpreted as the sum of n values consisting only of 0 and 1s, and $\hat{p}$ is just the sample mean of these n values. Hence, by the Central Limit Theorem, for n sufficiently large, $\widehat{P}$ is approximately normally distributed with mean

$$\mu_{\widehat{P}} = E(\widehat{P}) = E\left(\frac{X}{n}\right) = \frac{np}{n} = p$$

and variance

$$\sigma_{\widehat{P}}^2 = \sigma_{X/n}^2 = \frac{\sigma_X^2}{n^2} = \frac{npq}{n^2} = \frac{pq}{n}.$$

Therefore, we can assert that

$$P(-z_{\alpha/2} < Z < z_{\alpha/2}) = 1 - \alpha, \text{ with } Z = \frac{\widehat{P} - p}{\sqrt{pq/n}},$$

and $z_{\alpha/2}$ is the value above which we find an area of $\alpha/2$ under the standard normal curve.

For a random sample of size n, the sample proportion $\hat{p} = x/n$ is computed, and the following approximate $100(1-\alpha)\%$ confidence intervals for p can be obtained.

Large-Sample Confidence Intervals for p   If $\hat{p}$ is the proportion of successes in a random sample of size n and $\hat{q} = 1 - \hat{p}$, an approximate $100(1-\alpha)\%$ confidence interval for the binomial parameter p is given by

$$\hat{p} - z_{\alpha/2}\sqrt{\frac{\hat{p}\hat{q}}{n}} < p < \hat{p} + z_{\alpha/2}\sqrt{\frac{\hat{p}\hat{q}}{n}},$$

where $z_{\alpha/2}$ is the z-value leaving an area of $\alpha/2$ to the right.

When n is small and the unknown proportion p is believed to be close to 0 or to 1, the confidence-interval procedure established here is unreliable and, therefore, should not be used. To be on the safe side, one should require both $n\hat{p}$ and $n\hat{q}$ to be greater than or equal to 5. The methods for finding a confidence interval for the binomial parameter p are also applicable when the binomial distribution is being used to approximate the hypergeometric distribution, that is, when n is small relative to N, as illustrated by Example 5.14.

Example 5.14: In a random sample of $n = 500$ families owning television sets in the city of Hamilton, Canada, it is found that $x = 340$ subscribe to HBO. Find a 95% confidence interval for the actual proportion of families with television sets in this city that subscribe to HBO.

Solution: The point estimate of p is $\hat{p} = 340/500 = 0.68$. Using Table A.3, we find that $z_{0.025} = 1.96$. Therefore, the 95% confidence interval for p is

$$0.68 - 1.96\sqrt{\frac{(0.68)(0.32)}{500}} < p < 0.68 + 1.96\sqrt{\frac{(0.68)(0.32)}{500}},$$

which simplifies to $0.6391 < p < 0.7209$.   ⌐

Theorem 5.3: If $\hat{p}$ is used as an estimate of p, we can be $100(1-\alpha)\%$ confident that the error will not exceed $z_{\alpha/2}\sqrt{\hat{p}\hat{q}/n}$.

In Example 5.14, we are 95% confident that the sample proportion $\hat{p} = 0.68$ differs from the true proportion p by an amount not exceeding 0.04.

Choice of Sample Size

Let us now determine how large a sample is necessary to ensure that the error in estimating p will be less than a specified amount e. By Theorem 5.3, we must choose n such that $z_{\alpha/2}\sqrt{\hat{p}\hat{q}/n} = e$.

Theorem 5.4: If $\hat{p}$ is used as an estimate of p, we can be $100(1 - \alpha)\%$ confident that the error will be less than a specified amount e when the sample size is approximately

$$n = \frac{z_{\alpha/2}^2 \hat{p}\hat{q}}{e^2}.$$

Theorem 5.4 is somewhat misleading in that we must use $\hat{p}$ to determine the sample size n, but $\hat{p}$ is computed from the sample. If a crude estimate of p can be made without taking a sample, this value can be used to determine n. Lacking such an estimate, we could take a preliminary sample of size $n \geq 30$ to provide an estimate of p. Using Theorem 5.4, we could determine approximately how many observations are needed to provide the desired degree of accuracy. Note that fractional values of n are rounded up to the next whole number.

Example 5.15: How large a sample is required if we want to be 95% confident that our estimate of p in Example 5.14 is within 0.02 of the true value?

Solution: Let us treat the 500 families as a preliminary sample, providing an estimate $\hat{p} = 0.68$. Then, by Theorem 5.4,

$$n = \frac{(1.96)^2 (0.68)(0.32)}{(0.02)^2} = 2089.8 \approx 2090.$$

Therefore, if we base our estimate of p on a random sample of size 2090, we can be 95% confident that our sample proportion will not differ from the true proportion by more than 0.02.

Occasionally, it will be impractical to obtain an estimate of p to be used for determining the sample size for a specified degree of confidence. If this happens, an upper bound for n is established by noting that $\hat{p}\hat{q} = \hat{p}(1 - \hat{p})$, which must be at most $1/4$, since $\hat{p}$ must lie between 0 and 1. This fact may be verified by completing the square. Hence

$$\hat{p}(1 - \hat{p}) = -(\hat{p}^2 - \hat{p}) = \frac{1}{4} - \left(\hat{p}^2 - \hat{p} + \frac{1}{4} \right) = \frac{1}{4} - \left(\hat{p} - \frac{1}{2} \right)^2,$$

which is always less than $1/4$ except when $\hat{p} = 1/2$, and then $\hat{p}\hat{q} = 1/4$. Therefore, if we substitute $\hat{p} = 1/2$ into the formula for n in Theorem 5.4 when, in fact, p actually differs from $1/2$, n will turn out to be larger than necessary for the specified degree of confidence; as a result, our degree of confidence will increase.

Theorem 5.5: If $\hat{p}$ is used as an estimate of p, we can be **at least** $100(1 - \alpha)\%$ confident that the error will not exceed a specified amount e when the sample size is

$$n = \frac{z_{\alpha/2}^2}{4e^2}.$$

Example 5.16: How large a sample is required if we want to be at least 95% confident that our estimate of p in Example 5.14 is within 0.02 of the true value?

Solution: Unlike in Example 5.15, we shall now assume that no preliminary sample has been taken to provide an estimate of p. Consequently, we can be at least 95% confident that our sample proportion will not differ from the true proportion by more than 0.02 if we choose a sample of size

$$n = \frac{(1.96)^2}{(4)(0.02)^2} = 2401.$$

Comparing the results of Examples 5.15 and 5.16, we see that information concerning p, provided by a preliminary sample or from experience, enables us to choose a smaller sample while maintaining our required degree of accuracy.

5.11   Two Samples: Estimating the Difference between Two Proportions

Consider the problem where we wish to estimate the difference between two binomial parameters p_1 and p_2. For example, p_1 might be the proportion of smokers with lung cancer and p_2 the proportion of nonsmokers with lung cancer, and the problem is to estimate the difference between these two proportions. First, we select independent random samples of sizes n_1 and n_2 from the two binomial populations with means $n_1 p_1$ and $n_2 p_2$ and variances $n_1 p_1 q_1$ and $n_2 p_2 q_2$, respectively; then we determine the numbers x_1 and x_2 of people in each sample with lung cancer and form the proportions $\hat{p}_1 = x_1/n$ and $\hat{p}_2 = x_2/n$. A point estimator of the difference between the two proportions, $p_1 - p_2$, is given by the statistic $\hat{P}_1 - \hat{P}_2$. Therefore, the difference of the sample proportions, $\hat{p}_1 - \hat{p}_2$, will be used as the point estimate of $p_1 - p_2$.

We use concepts identical to those covered in Section 5.8 in the discussion of the confidence interval for the difference $\mu_1 - \mu_2$. Note that the Z statistic for $\mu_1 - \mu_2$ has an equivalent in the case of $p_1 - p_2$, namely,

$$Z = \frac{(\hat{P}_1 - \hat{P}_2) - (p_1 - p_2)}{\sqrt{p_1 q_1/n_1 + p_2 q_2/n_2}}.$$

As a result, we have the Large Sample Confidence Interval for $p_1 - p_2$.

Large-Sample Confidence Interval for $p_1 - p_2$	If $\hat{p}_1$ and $\hat{p}_2$ are the proportions of successes in random samples of sizes n_1 and n_2, respectively, $\hat{q}_1 = 1 - \hat{p}_1$, and $\hat{q}_2 = 1 - \hat{p}_2$, an approximate $100(1 - \alpha)\%$ confidence interval for the difference of two binomial parameters, $p_1 - p_2$, is given by $$(\hat{p}_1 - \hat{p}_2) - z_{\alpha/2}\sqrt{\frac{\hat{p}_1\hat{q}_1}{n_1} + \frac{\hat{p}_2\hat{q}_2}{n_2}} < p_1 - p_2 < (\hat{p}_1 - \hat{p}_2) + z_{\alpha/2}\sqrt{\frac{\hat{p}_1\hat{q}_1}{n_1} + \frac{\hat{p}_2\hat{q}_2}{n_2}},$$ where $z_{\alpha/2}$ is the z-value leaving an area of $\alpha/2$ to the right.

Example 5.17: A certain change in a process for manufacturing component parts is being considered. Samples are taken from both the existing and the new process so as to determine if the new process results in an improvement. If 75 of 1500 items

from the existing process are found to be defective and 80 of 2000 items from the new process are found to be defective, find a 90% confidence interval for the true difference in the proportion of defectives between the existing and the new process.

Solution: Let p_1 and p_2 be the true proportions of defectives for the existing and new processes, respectively. Hence, $\hat{p}_1 = 75/1500 = 0.05$ and $\hat{p}_2 = 80/2000 = 0.04$, and the point estimate of $p_1 - p_2$ is

$$\hat{p}_1 - \hat{p}_2 = 0.05 - 0.04 = 0.01.$$

Using Table A.3, we find $z_{0.05} = 1.645$. Therefore, substituting into the formula, with

$$1.645\sqrt{\frac{(0.05)(0.95)}{1500} + \frac{(0.04)(0.96)}{2000}} = 0.0117,$$

we find the 90% confidence interval to be $-0.0017 < p_1 - p_2 < 0.0217$. Since the interval contains the value 0, there is no reason to believe that the new process produces a significant decrease in the proportion of defectives over the existing method.

Up to this point, all confidence intervals presented were of the form

point estimate $\pm$ K s.e.(point estimate),

where K is a constant (either t or normal percent point). This form is valid when the parameter is a mean, a difference between means, a proportion, or a difference between proportions, due to the symmetry of the t- and Z-distributions.

Exercises

5.39 In a random sample of 1000 homes in a certain city, it is found that 228 are heated by oil. Find 99% confidence intervals for the proportion of homes in this city that are heated by oil.

5.40 Compute 95% confidence intervals for the proportion of defective items in a process when it is found that a sample of size 100 yields 8 defectives.

5.41 (a) A random sample of 200 voters in a town is selected, and 114 are found to support an annexation suit. Find the 96% confidence interval for the proportion of the voting population favoring the suit.

(b) What can we assert with 96% confidence about the possible size of our error if we estimate the proportion of voters favoring the annexation suit to be 0.57?

5.42 A manufacturer of MP3 players conducts a set of comprehensive tests on the electrical functions of its product. All MP3 players must pass all tests prior to

being sold. Of a random sample of 500 MP3 players, 15 failed one or more tests. Find a 90% confidence interval for the proportion of MP3 players from the population that pass all tests.

5.43 A new rocket-launching system is being considered for deployment of small, short-range rockets. The existing system has $p = 0.8$ as the probability of a successful launch. A sample of 40 experimental launches is made with the new system, and 34 are successful.

(a) Construct a 95% confidence interval for p.

(b) Would you conclude that the new system is better?

5.44 A geneticist is interested in the proportion of African males who have a certain minor blood disorder. In a random sample of 100 African males, 24 are found to be afflicted.

(a) Compute a 99% confidence interval for the proportion of African males who have this blood disorder.

(b) What can we assert with 99% confidence about the possible size of our error if we estimate the proportion of African males with this blood disorder to be 0.24?

5.45 How large a sample is needed if we wish to be 96% confident that our sample proportion in Exercise 5.41 will be within 0.02 of the true fraction of the voting population?

5.46 How large a sample is needed if we wish to be 99% confident that our sample proportion in Exercise 5.39 will be within 0.05 of the true proportion of homes in the city that are heated by oil?

5.47 How large a sample is needed in Exercise 5.40 if we wish to be 98% confident that our sample proportion will be within 0.05 of the true proportion defective?

5.48 How large a sample is needed to estimate the percentage of citizens in a certain town who favor having their water fluoridated if one wishes to be at least 99% confident that the estimate is within 1% of the true percentage?

5.49 A study is to be made to estimate the proportion of residents of a certain city and its suburbs who favor the construction of a nuclear power plant near the city. How large a sample is needed if one wishes to be at least 95% confident that the estimate is within 0.04 of the true proportion of residents who favor the construction of the nuclear power plant?

5.50 Ten engineering schools in the United States were surveyed. The sample contained 250 electrical en-

gineers, 80 being women, and 175 chemical engineers, 40 being women. Compute a 90% confidence interval for the difference between the proportions of women in these two fields of engineering. Is there a significant difference between the two proportions?

5.51 A certain geneticist is interested in the proportion of males and females in the population who have a minor blood disorder. In a random sample of 1000 males, 250 are found to be afflicted, whereas 275 of 1000 females tested appear to have the disorder. Compute a 95% confidence interval for the difference between the proportions of males and females who have the blood disorder.

5.52 In the study *Germination and Emergence of Broccoli*, conducted by the Department of Horticulture at Virginia Tech, a researcher found that at 5°C, 10 broccoli seeds out of 20 germinated, while at 15°C, 15 out of 20 germinated. Compute a 95% confidence interval for the difference between the proportions of germination at the two different temperatures and decide if there is a significant difference.

5.53 A clinical trial was conducted to determine if a certain type of inoculation has an effect on the incidence of a certain disease. A sample of 1000 rats was kept in a controlled environment for a period of 1 year, and 500 of the rats were given the inoculation. In the group not inoculated, there were 120 incidences of the disease, while 98 of the rats in the inoculated group contracted it. If p_1 is the probability of incidence of the disease in uninoculated rats and p_2 the probability of incidence in inoculated rats, compute a 90% confidence interval for $p_1 - p_2$.

5.12   Single Sample: Estimating the Variance

If a sample of size n is drawn from a normal population with variance σ^2 and the sample variance s^2 is computed, we obtain a value of the statistic S^2. This computed sample variance is used as a point estimate of σ^2. Hence, the statistic S^2 is called an estimator of σ^2.

An interval estimate of σ^2 can be established by using the statistic

$$X^2 = \frac{(n-1)S^2}{\sigma^2}.$$

According to Theorem 4.4, the statistic X^2 has a chi-squared distribution with $n-1$ degrees of freedom when samples are chosen from a normal population. We may write (see Figure 5.6)

$$P(\chi^2_{1-\alpha/2} < X^2 < \chi^2_{\alpha/2}) = 1 - \alpha,$$

where $\chi^2_{1-\alpha/2}$ and $\chi^2_{\alpha/2}$ are values of the chi-squared distribution with $n-1$ degrees of freedom, leaving areas of $1-\alpha/2$ and $\alpha/2$, respectively, to the right. Substituting for X^2, we write

$$P\left[\chi^2_{1-\alpha/2} < \frac{(n-1)S^2}{\sigma^2} < \chi^2_{\alpha/2}\right] = 1 - \alpha.$$

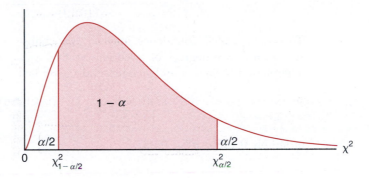

Figure 5.6: $P(\chi^2_{1-\alpha/2} < X^2 < \chi^2_{\alpha/2}) = 1 - \alpha.$

Dividing each term in the inequality by $(n-1)S^2$ and then inverting each term (thereby changing the sense of the inequalities), we obtain

$$P\left[\frac{(n-1)S^2}{\chi^2_{\alpha/2}} < \sigma^2 < \frac{(n-1)S^2}{\chi^2_{1-\alpha/2}}\right] = 1 - \alpha.$$

For a random sample of size n from a normal population, the sample variance s^2 is computed, and the following $100(1-\alpha)\%$ confidence interval for σ^2 is obtained.

Confidence Interval for σ^2   If s^2 is the variance of a random sample of size n from a normal population, a $100(1-\alpha)\%$ confidence interval for σ^2 is

$$\frac{(n-1)s^2}{\chi^2_{\alpha/2}} < \sigma^2 < \frac{(n-1)s^2}{\chi^2_{1-\alpha/2}},$$

where $\chi^2_{\alpha/2}$ and $\chi^2_{1-\alpha/2}$ are χ^2-values with $v = n-1$ degrees of freedom, leaving areas of $\alpha/2$ and $1-\alpha/2$, respectively, to the right.

An approximate $100(1-\alpha)\%$ confidence interval for σ is obtained by taking the square root of each endpoint of the interval for σ^2.

Example 5.18:   The following are the weights, in decagrams, of 10 packages of grass seed distributed by a certain company: 46.4, 46.1, 45.8, 47.0, 46.1, 45.9, 45.8, 46.9, 45.2, and 46.0. Find a 95% confidence interval for the variance of the weights of all such packages of grass seed distributed by this company, assuming a normal population.

Solution: First we find

$$s^2 = \frac{n \sum_{i=1}^{n} x_i^2 - \left(\sum_{i=1}^{n} x_i\right)^2}{n(n-1)}$$

$$= \frac{(10)(21{,}273.12) - (461.2)^2}{(10)(9)} = 0.286.$$

To obtain a 95% confidence interval, we choose $\alpha = 0.05$. Then, using Table A.5 with $v = 9$ degrees of freedom, we find $\chi^2_{0.025} = 19.023$ and $\chi^2_{0.975} = 2.700$. Therefore, the 95% confidence interval for σ^2 is

$$\frac{(9)(0.286)}{19.023} < \sigma^2 < \frac{(9)(0.286)}{2.700},$$

or simply $0.135 < \sigma^2 < 0.953$.

Exercises

5.54 A random sample of 20 students yielded a mean of $\bar{x} = 72$ and a variance of $s^2 = 16$ for scores on a college placement test in mathematics. Assuming the scores to be normally distributed, construct a 98% confidence interval for σ^2.

5.55 A manufacturer of car batteries claims that the batteries will last, on average, 3 years with a variance of 1 year. If 5 of these batteries have lifetimes of 1.9, 2.4, 3.0, 3.5, and 4.2 years, construct a 95% confidence interval for σ^2 and decide if the manufacturer's claim

that $\sigma^2 = 1$ is valid. Assume the population of battery lives to be approximately normally distributed.

5.56 Construct a 99% confidence interval for σ^2 in Case Study 5.1 on page 211.

5.57 Construct a 95% confidence interval for σ^2 in Exercise 5.9 on page 213.

5.58 Construct a 90% confidence interval for σ in Exercise 5.11 on page 213.

Review Exercises

5.59 According to the *Roanoke Times*, in a particular year McDonald's sold 42.1% of the market share of hamburgers. A random sample of 75 burgers sold resulted in 28 of them being from McDonald's. Use material in Section 5.10 to determine if this information supports the claim in the *Roanoke Times*.

5.60 It is claimed that a new diet will reduce a person's weight by 4.5 kilograms on average in a period of 2 weeks. The weights of 7 women who followed this diet were recorded before and after the 2-week period. Test the claim about the diet by computing a 95% confidence interval for the mean difference in weights. Assume the differences of weights to be approximately normally distributed.

Woman	Weight Before	Weight After
1	58.5	60.0
2	60.3	54.9
3	61.7	58.1
4	69.0	62.1
5	64.0	58.5
6	62.6	59.9
7	56.7	54.4

5.61 A health spa claims that a new exercise program will reduce a person's waist size by 2 centimeters on average over a 5-day period. The waist sizes, in centimeters, of 6 men who participated in this exercise program are recorded before and after the 5-day period in the following table:

Man	Waist Size Before	Waist Size After
1	90.4	91.7
2	95.5	93.9
3	98.7	97.4
4	115.9	112.8
5	104.0	101.3
6	85.6	84.0

By computing a 95% confidence interval for the mean reduction in waist size, determine whether the health spa's claim is valid. Assume the distribution of differences in waist sizes before and after the program to be approximately normal.

5.62 The Department of Civil Engineering at Virginia Tech compared a modified (M-5 hr) assay technique for recovering fecal coliforms in stormwater runoff from an urban area to a most probable number (MPN) technique. A total of 12 runoff samples were collected and analyzed by the two techniques. Fecal coliform counts per 100 milliliters are recorded in the following table.

Sample	MPN Count	M-5 hr Count
1	2300	2010
2	1200	930
3	450	400
4	210	436
5	270	4100
6	450	2090
7	154	219
8	179	169
9	192	194
10	230	174
11	340	274
12	194	183

Construct a 90% confidence interval for the difference in the mean fecal coliform counts between the M-5 hr and the MPN techniques. Assume that the count differences are approximately normally distributed.

5.63 An experiment was conducted to determine whether surface finish has an effect on the endurance limit of steel. There is a theory that polishing increases the average endurance limit (for reverse bending). From a practical point of view, polishing should not have any effect on the standard deviation of the endurance limit, which is known from numerous endurance limit experiments to be 4000 psi. An experiment was performed on 0.4% carbon steel using both unpolished and polished smooth-turned specimens. The data are given below. Find a 95% confidence interval for the difference between the population means for the two methods, assuming that the populations are approximately normally distributed.

Endurance Limit (psi)	
Polished 0.4% Carbon	Unpolished 0.4% Carbon
85,500	82,600
91,900	82,400
89,400	81,700
84,000	79,500
89,900	79,400
78,700	69,800
87,500	79,900
83,100	83,400

5.64 An anthropologist is interested in the proportion of individuals in two Indian tribes with double occipital hair whorls. Suppose that independent samples are taken from each of the two tribes, and it is found that 24 of 100 Indians from tribe A and 36 of 120 Indians from tribe B possess this characteristic. Construct a 95% confidence interval for the difference $p_B - p_A$ between the proportions of these two tribes with occipital hair whorls.

5.65 A manufacturer of electric irons produces these items in two plants. Both plants have the same suppliers of small parts. A saving can be made by purchasing thermostats for plant B from a local supplier. A single lot was purchased from the local supplier, and a test was conducted to see whether or not these new thermostats were as accurate as the old. The thermostats were tested on tile irons on the 550°F setting, and the actual temperatures were read to the nearest 0.1°F with a thermocouple. The data are as follows:

New Supplier (°F)					
530.3	559.3	549.4	544.0	551.7	566.3
549.9	556.9	536.7	558.8	538.8	543.3
559.1	555.0	538.6	551.1	565.4	554.9
550.0	554.9	554.7	536.1	569.1	

Old Supplier (°F)					
559.7	534.7	554.8	545.0	544.6	538.0
550.7	563.1	551.1	553.8	538.8	564.6
554.5	553.0	538.4	548.3	552.9	535.1
555.0	544.8	558.4	548.7	560.3	

Find 95% confidence intervals for σ_1^2/σ_2^2 and for σ_1/σ_2, where σ_1^2 and σ_2^2 are the population variances of the thermostat readings for the new and old suppliers, respectively.

5.66 It is argued that the resistance of wire A is greater than the resistance of wire B. An experiment on the wires shows the results (in ohms) given here. Assuming equal variances, what conclusions do you draw? Justify your answer.

Wire A	Wire B
0.140	0.135
0.138	0.140
0.143	0.136
0.142	0.142
0.144	0.138
0.137	0.140

5.67 A survey was done with the hope of comparing salaries of chemical plant managers employed in two areas of the country, the northern and west central regions. An independent random sample of 300 plant managers was selected from each of the two regions. These managers were asked their annual salaries. The results are as follows

North	West Central
$\bar{x}_1 = \$102,300$	$\bar{x}_2 = \$98,500$
$s_1 = \$5700$	$s_2 = \$3800$

(a) Construct a 99% confidence interval for $\mu_1 - \mu_2$, the difference in the mean salaries.

(b) What assumption did you make in (a) about the distribution of annual salaries for the two regions? Is the assumption of normality necessary? Why or why not?

(c) What assumption did you make about the two variances? Is the assumption of equality of variances reasonable? Explain!

5.68 Consider Review Exercise 5.67. Let us assume that the data have not been collected yet and that previous statistics suggest that $\sigma_1 = \sigma_2 = \$4000$. Are the sample sizes in Review Exercise 5.67 sufficient to produce a 95% confidence interval on $\mu_1 - \mu_2$ having a width of only $1000? Show all work.

5.69 A labor union is becoming defensive about gross absenteeism by its members. The union leaders had always claimed that, in a typical month, 95% of its members were absent less than 10 hours. The union decided to check this by monitoring a random sample of 300 of its members. The number of hours absent was recorded for each of the 300 members. The results were $\bar{x} = 6.5$ hours and $s = 2.5$ hours. Use the data to respond to this claim, using a one-sided tolerance limit and choosing the confidence level to be 99%. Be sure to interpret what you learn from the tolerance limit calculation.

5.70 A random sample of 30 firms dealing in wireless products was selected to determine the proportion of such firms that have implemented new software to improve productivity. It turned out that 8 of the 30 had implemented such software. Find a 95% confidence interval on p, the true proportion of such firms that have implemented new software.

5.71 Refer to Review Exercise 5.70. Suppose there is concern about whether the point estimate $\hat{p} = 8/30$ is accurate enough because the confidence interval around p is not sufficiently narrow. Using $\hat{p}$ as the estimate of p, how many companies would need to be sampled in order to have a 95% confidence interval with a width of only 0.05?

5.72 A manufacturer turns out a product item that is labeled either "defective" or "not defective." In order to estimate the proportion defective, a random sample of 100 items is taken from production, and 10 are found to be defective. Following implementation of a quality improvement program, the experiment is conducted again. A new sample of 100 is taken, and this time only 6 are found to be defective.

(a) Give a 95% confidence interval on $p_1 - p_2$, where p_1 is the population proportion defective before improvement and p_2 is the proportion defective after improvement.

(b) Is there information in the confidence interval found in (a) that would suggest that $p_1 > p_2$? Explain.

5.73 A machine is used to fill boxes with product in an assembly line operation. Much concern centers around the variability in the number of ounces of product in a box. The standard deviation in weight of product is known to be 0.3 ounce. An improvement is implemented, after which a random sample of 20 boxes is selected and the sample variance is found to be 0.045 ounce^2. Find a 95% confidence interval on the variance in the weight of the product. Does it appear from the range of the confidence interval that the improvement of the process enhanced quality as far as variability is concerned? Assume normality on the distribution of weights of product.

5.74 A consumer group is interested in comparing operating costs for two different types of automobile engines. The group is able to find 15 owners whose cars have engine type A and 15 whose cars have engine type B. All 30 owners bought their cars at roughly the same time, and all have kept good records for a certain 12-month period. In addition, these owners drove roughly the same number of miles. The cost statistics are $\bar{y}_A = \$87.00/1000$ miles, $\bar{y}_B = \$75.00/1000$ miles, $s_A = \$5.99$, and $s_B = \$4.85$. Compute a 95% confidence interval to estimate $\mu_A - \mu_B$, the difference in the mean operating costs. Assume normality and equal variances.

5.75 A group of human factor researchers are concerned about reaction to a stimulus by airplane pilots in a certain cockpit arrangement. An experiment was conducted in a simulation laboratory, and 15 pilots were used with average reaction time of 3.2 seconds with a sample standard deviation of 0.6 second. It is of interest to characterize the extreme (i.e., worst-case scenario). To that end, do the following:

(a) Give a particular important one-sided 99% confidence bound on the mean reaction time. What assumption, if any, must you make on the distribution of reaction times?

(b) Give a 99% one-sided prediction interval and give an interpretation of what it means. Must you make an assumption about the distribution of reaction times to compute this bound?

(c) Compute a one-sided tolerance bound with 99% confidence that involves 95% of reaction times. Again, give an interpretation and assumptions about the distribution, if any. (Note: The one-sided tolerance limit values are also included in Table A.7.)

5.76 A certain supplier manufactures a type of rubber mat that is sold to automotive companies. The material used to produce the mats must have certain hardness characteristics. Defective mats are occasionally discovered and rejected. The supplier claims that the proportion defective is 0.05. A challenge was made by one of the clients who purchased the mats, so an experiment was conducted in which 400 mats were tested and 17 were found defective.

(a) Compute a 95% two-sided confidence interval on the proportion defective.

(b) Compute an appropriate 95% one-sided confidence interval on the proportion defective.

(c) Interpret both intervals from (a) and (b) and comment on the claim made by the supplier.

5.13 Potential Misconceptions and Hazards; Relationship to Material in Other Chapters

The concept of a *large-sample confidence interval* on a population parameter is often confusing to the beginning student. It is based on the notion that even when σ is unknown and one is not convinced that the distribution being sampled is normal, a confidence interval on μ can be computed from

$$\bar{x} \pm z_{\alpha/2}\frac{s}{\sqrt{n}}.$$

In practice, this formula is often used when the sample is too small. The genesis of this large sample interval is, of course, the Central Limit Theorem (CLT), under which normality is not necessary in practice. Here the CLT requires a known σ, of which s is only an estimate. Thus, n must be at least as large as 30 and the underlying distribution must be close to symmetric, in which case the interval remains an approximation.

There are instances in which the appropriateness of the practical application of material in this chapter depends very much on the specific context. One very important illustration is the use of the t-distribution for the confidence interval on μ when σ is unknown. Strictly speaking, the use of the t-distribution requires that the distribution sampled from be normal. However, it is well known that any application of the t-distribution is reasonably insensitive (i.e., **robust**) to the normality assumption. This represents one of those fortunate situations which occur often in the field of statistics in which a basic assumption does not hold and yet "everything turns out all right!"

It is our experience that one of the most serious misuses of statistics in practice evolves from confusion about distinctions in the interpretation of the types of statistical intervals. Thus, the subsection in this chapter where differences among the three types of intervals are discussed is important. It is very likely that in practice the **confidence interval is heavily overused**. That is, it is used when there is really no interest in the mean; rather, the question is "Where is the next observation going to fall?" or often, more importantly, "Where is the large bulk of the distribution?" These are crucial questions that are not answered by computing an interval on the mean.

The interpretation of a confidence interval is often misunderstood. It is tempting to conclude that the parameter falls inside the interval with probability 0.95. A confidence interval merely suggests that if the experiment is conducted and data are observed again and again, about 95% of such intervals will contain the true parameter. Any beginning student of practical statistics should be very clear on the difference among these statistical intervals.

Another potential serious misuse of statistics centers around the use of the χ^2-distribution for a confidence interval on a single variance. Again, normality of the distribution from which the sample is drawn is assumed. Unlike the use of the t-distribution, the use of the χ^2 test for this application is **not robust to the normality assumption** (i.e., the sampling distribution of $\frac{(n-1)S^2}{\sigma^2}$ deviates far from χ^2 if the underlying distribution is not normal).

Chapter 6

One- and Two-Sample Tests of Hypotheses

6.1 Statistical Hypotheses: General Concepts

Often, the problem confronting the scientist or engineer is not so much the estimation of a population parameter, as discussed in Chapter 5, but rather the formation of a data-based decision procedure that can produce a conclusion about some scientific system. For example, a medical researcher may decide on the basis of experimental evidence whether coffee drinking increases the risk of cancer in humans; an engineer might have to decide on the basis of sample data whether there is a difference between the accuracy of two kinds of gauges; or a sociologist might wish to collect appropriate data to enable him or her to decide whether a person's blood type and eye color are independent variables. In each of these cases, the scientist or engineer *postulates* or *conjectures* something about a system. In addition, each must make use of experimental data and make a decision based on the data. In each case, the conjecture can be put in the form of a statistical hypothesis. Procedures that lead to the acceptance or rejection of statistical hypotheses such as these comprise a major area of statistical inference. First, let us define precisely what we mean by a **statistical hypothesis**.

Definition 6.1: | A **statistical hypothesis** is an assertion or conjecture concerning one or more populations.

The truth or falsity of a statistical hypothesis is never known with absolute certainty unless we examine the entire population. This, of course, would be impractical in most situations. Instead, we take a random sample from the population of interest and use the data contained in this sample to provide evidence that either supports or does not support the hypothesis. Evidence from the sample that is inconsistent with the stated hypothesis leads to a rejection of the hypothesis.

The Role of Probability in Hypothesis Testing

It should be made clear to the reader that the decision procedure must include an awareness of the *probability of a wrong conclusion.* For example, suppose that the hypothesis postulated by the engineer is that the fraction defective p in a certain process is 0.10. The experiment is to observe a random sample of the product in question. Suppose that 100 items are tested and 12 items are found defective. It is reasonable to conclude that this evidence does not refute the condition that the binomial parameter $p = 0.10$, and thus it may lead one not to reject the hypothesis. However, it also does not refute $p = 0.12$ or perhaps even $p = 0.15$. As a result, the reader must be accustomed to understanding that **rejection of a hypothesis implies that the sample evidence refutes it**. Put another way, **rejection means that there is a small probability of obtaining the sample information observed when, in fact, the hypothesis is true**. For example, for our proportion-defective hypothesis, a sample of 100 revealing 20 defective items is certainly evidence for rejection. Why? If, indeed, $p = 0.10$, the probability of obtaining 20 or more defectives is approximately 0.002. With the resulting small risk of a wrong conclusion, it would seem safe to **reject the hypothesis** that $p = 0.10$. In other words, rejection of a hypothesis tends to all but "rule out" the hypothesis. On the other hand, it is very important to emphasize that acceptance or, rather, failure to reject does not rule out other possibilities. As a result, the *firm conclusion is established by the data analyst when a hypothesis is rejected.*

The Null and Alternative Hypotheses

The structure of hypothesis testing will be formulated with the use of the term **null hypothesis**, which refers to any hypothesis we wish to test and is denoted by H_0. The rejection of H_0 leads to the acceptance of an **alternative hypothesis**, denoted by H_1. An understanding of the different roles played by the null hypothesis (H_0) and the alternative hypothesis (H_1) is crucial to one's understanding of the rudiments of hypothesis testing. The alternative hypothesis H_1 usually represents the *question to be answered or the theory to be tested,* and thus its specification is crucial. The null hypothesis H_0 *nullifies or opposes* H_1 and is often the logical complement to H_1. As the reader gains more understanding of hypothesis testing, he or she should note that the analyst arrives at one of the two following conclusions:

reject H_0 in favor of H_1 because of sufficient evidence in the data or

fail to reject H_0 because of insufficient evidence in the data.

Note that the *conclusions do not involve a formal and literal "accept H_0."* The statement of H_0 often represents the "status quo" in opposition to the new idea, conjecture, and so on, stated in H_1, while failure to reject H_0 represents the proper conclusion. In our binomial example, the practical issue may be a concern that the historical defective probability of 0.10 no longer is true. Indeed, the conjecture may be that p exceeds 0.10. We may then state

$$H_0: \ p = 0.10,$$
$$H_1: \ p > 0.10.$$

Now 12 defective items out of 100 does not refute $p = 0.10$, so the conclusion is "fail to reject H_0." However, if the data produce 20 out of 100 defective items, then the conclusion is "reject H_0" in favor of H_1: $p > 0.10$.

Though the applications of hypothesis testing are quite abundant in scientific and engineering work, perhaps the best illustration for a novice lies in the predicament encountered in a jury trial. The null and alternative hypotheses are

$$H_0: \text{defendant is not guilty,}$$

$$H_1: \text{defendant is guilty.}$$

The indictment comes because of suspicion of guilt. The hypothesis H_0 (the status quo) stands in opposition to H_1 and is maintained unless H_1 is supported by evidence "beyond a reasonable doubt." However, "failure to reject H_0" in this case does not imply innocence, but merely that the evidence was insufficient to convict. So the jury does not necessarily *accept H_0* but *fails to reject H_0*.

6.2 Testing a Statistical Hypothesis

To illustrate the concepts used in testing a statistical hypothesis about a population, we present the following example. A certain type of cold vaccine is known to be only 25% effective after a period of 2 years. To determine if a new and somewhat more expensive vaccine is superior in providing protection against the same virus for a longer period of time, suppose that 20 people are chosen at random and inoculated. (In an actual study of this type, the participants receiving the new vaccine might number several thousand. The number 20 is being used here only to demonstrate the basic steps in carrying out a statistical test.) If more than 8 of those receiving the new vaccine surpass the 2-year period without contracting the virus, the new vaccine will be considered superior to the one presently in use. The requirement that the number exceed 8 is somewhat arbitrary but appears reasonable in that it represents a modest gain over the 5 people who could be expected to receive protection if the 20 people had been inoculated with the vaccine already in use. We are essentially testing the null hypothesis that the new vaccine is equally effective after a period of 2 years as the one now commonly used. The alternative hypothesis is that the new vaccine is in fact superior. This is equivalent to testing the hypothesis that the binomial parameter for the probability of a success on a given trial is $p = 1/4$ against the alternative that $p > 1/4$. This is usually written as follows:

$$H_0: \ p = 0.25,$$

$$H_1: \ p > 0.25.$$

The Test Statistic

The **test statistic** on which we base our decision is X, the number of individuals in our test group who receive protection from the new vaccine for a period of at least 2 years. The possible values of X, from 0 to 20, are divided into two groups: those numbers less than or equal to 8 and those greater than 8. All possible scores greater than 8 constitute the **critical region**. The last number that we observe in passing into the critical region is called the **critical value**. In our illustration, the critical value is the number 8. Therefore, if $x > 8$, we reject H_0 in favor of the alternative hypothesis H_1. If $x \leq 8$, we fail to reject H_0. This decision criterion is illustrated in Figure 6.1.

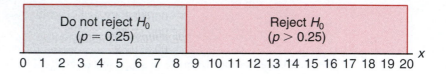

Figure 6.1: Decision criterion for testing $p = 0.25$ versus $p > 0.25$.

The Probability of a Type I Error

The decision procedure just described could lead to either of two wrong conclusions. For instance, the new vaccine may be no better than the one now in use (H_0 true) and yet, in this particular randomly selected group of individuals, more than 8 surpass the 2-year period without contracting the virus. We would be committing an error by rejecting H_0 in favor of H_1 when, in fact, H_0 is true. Such an error is called a **type I error**.

Definition 6.2: Rejection of the null hypothesis when it is true is called a **type I error**.

A second kind of error is committed if 8 or fewer of the group surpass the 2-year period successfully and we are unable to conclude that the vaccine is better when it actually is better (H_1 true). Thus, in this case, we fail to reject H_0 when in fact H_0 is false. This is called a **type II error**.

Definition 6.3: Nonrejection of the null hypothesis when it is false is called a **type II error**.

In testing any statistical hypothesis, there are four possible situations that determine whether our decision is correct or in error. These four situations are summarized in Table 6.1.

Table 6.1: Possible Situations for Testing a Statistical Hypothesis

	H_0 **is true**	H_0 **is false**
Do not reject H_0	Correct decision	Type II error
Reject H_0	Type I error	Correct decision

The probability of committing a type I error, also called the **level of significance**, is denoted by the Greek letter α. In our illustration, a type I error will occur when more than 8 individuals inoculated with the new vaccine surpass the 2-year period without contracting the virus and researchers conclude that the new vaccine is better when it is actually equivalent to the one in use. Hence, if X is the number of individuals who remain free of the virus for at least 2 years,

$$\alpha = P(\text{type I error}) = P\left(X > 8 \text{ when } p = \frac{1}{4}\right) = \sum_{x=9}^{20} b\left(x; 20, \frac{1}{4}\right)$$

$$= 1 - \sum_{x=0}^{8} b\left(x; 20, \frac{1}{4}\right) = 1 - 0.9591 = 0.0409.$$

We say that the null hypothesis, $p = 1/4$, is being tested at the $\alpha = 0.0409$ level of significance. Sometimes the level of significance is called the **size of the test**. A critical region of size 0.0409 is very small, and therefore it is unlikely that a type I error will be committed. Consequently, it would be most unusual for more than 8 individuals to remain immune to a virus for a 2-year period using a new vaccine that is essentially equivalent to the one now on the market.

The Probability of a Type II Error

The probability of committing a type II error, denoted by β, is impossible to compute unless we have a specific alternative hypothesis. If we test the null hypothesis that $p = 1/4$ against the alternative hypothesis that $p = 1/2$, then we are able to compute the probability of not rejecting H_0 when it is false. We simply find the probability of obtaining 8 or fewer in the group that surpass the 2-year period when $p = 1/2$. In this case,

$$\beta = P(\text{type II error}) = P\left(X \le 8 \text{ when } p = \frac{1}{2}\right)$$

$$= \sum_{x=0}^{8} b\left(x; 20, \frac{1}{2}\right) = 0.2517.$$

This is a rather high probability, indicating a test procedure in which it is quite likely that we shall reject the new vaccine when, in fact, it is superior to what is now in use. Ideally, we like to use a test procedure for which the type I and type II error probabilities are both small.

It is possible that the director of the testing program is willing to make a type II error if the more expensive vaccine is not significantly superior. In fact, the only time he wishes to guard against the type II error is when the true value of p is at least 0.7. If $p = 0.7$, this test procedure gives

$$\beta = P(\text{type II error}) = P(X \le 8 \text{ when } p = 0.7)$$

$$= \sum_{x=0}^{8} b(x; 20, 0.7) = 0.0051.$$

With such a small probability of committing a type II error, it is extremely unlikely that the new vaccine would be rejected when it was 70% effective after a period of 2 years. As the alternative hypothesis approaches unity, the value of β diminishes to zero.

The Role of α, β, and Sample Size

Let us assume that the director of the testing program is unwilling to commit a type II error when the alternative hypothesis $p = 1/2$ is true, even though we have found the probability of such an error to be $\beta = 0.2517$. It is always possible to reduce β by increasing the size of the critical region. For example, consider what happens to the values of α and β when we change our critical value to 7 so that all scores greater than 7 fall in the critical region and those less than or equal to 7 fall in the nonrejection region. Now, in testing $p = 1/4$ against the alternative hypothesis that $p = 1/2$, we find that

$$\alpha = \sum_{x=8}^{20} b\left(x; 20, \frac{1}{4}\right) = 1 - \sum_{x=0}^{7} b\left(x; 20, \frac{1}{4}\right) = 1 - 0.8982 = 0.1018$$

and

$$\beta = \sum_{x=0}^{7} b\left(x; 20, \frac{1}{2}\right) = 0.1316.$$

By adopting a new decision procedure, we have reduced the probability of committing a type II error at the expense of increasing the probability of committing a type I error. For a fixed sample size, a decrease in the probability of one error will usually result in an increase in the probability of the other error. Fortunately, **the probability of committing both types of error can be reduced by increasing the sample size.** Consider the same problem using a random sample of 100 individuals. If more than 36 of the group surpass the 2-year period, we reject the null hypothesis that $p = 1/4$ and accept the alternative hypothesis that $p > 1/4$. The critical value is now 36. All possible scores above 36 constitute the critical region, and all possible scores less than or equal to 36 fall in the acceptance region.

To determine the probability of committing a type I error, we shall use the normal curve approximation with

$$\mu = np = (100)\left(\frac{1}{4}\right) = 25 \quad \text{and} \quad \sigma = \sqrt{npq} = \sqrt{(100)(1/4)(3/4)} = 4.33.$$

Referring to Figure 6.2, we need the area under the normal curve to the right of $x = 36.5$. The corresponding z-value is

$$z = \frac{36.5 - 25}{4.33} = 2.66.$$

From Table A.3 we find that

$$\alpha = P(\text{type I error}) = P\left(X > 36 \text{ when } p = \frac{1}{4}\right) \approx P(Z > 2.66)$$

$$= 1 - P(Z < 2.66) = 1 - 0.9961 = 0.0039.$$

If H_0 is false and the true value of H_1 is $p = 1/2$, we can determine the probability of a type II error using the normal curve approximation with

$$\mu = np = (100)(1/2) = 50 \quad \text{and} \quad \sigma = \sqrt{npq} = \sqrt{(100)(1/2)(1/2)} = 5.$$

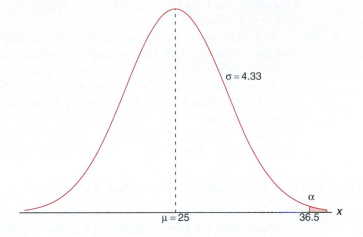

Figure 6.2: Probability of a type I error.

The probability of a value falling in the nonrejection region when H_0 is false and H_1 is true, with $p = 1/2$, is given by the area of the shaded region to the left of $x = 36.5$ in Figure 6.3. The two normal distributions in the figure show the null hypothesis on the left and the alternative hypothesis on the right. The z-value corresponding to $x = 36.5$ is

$$z = \frac{36.5 - 50}{5} = -2.7.$$

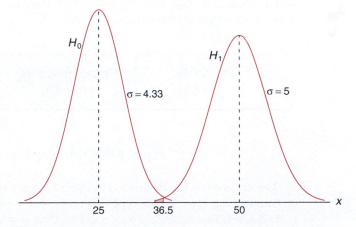

Figure 6.3: Probability of a type II error.

Therefore,

$$\beta = P(\text{type II error}) = P\left(X \le 36 \text{ when } p = \frac{1}{2}\right) \approx P(Z < -2.7) = 0.0035.$$

Obviously, the type I and type II errors will rarely occur if the experiment consists of 100 individuals.

The illustration above underscores the strategy of the scientist in hypothesis testing. After the null and alternative hypotheses are stated, it is important to consider the sensitivity of the test procedure. By this we mean that there should be a determination, for a fixed α, of a reasonable value for the probability of wrongly accepting H_0 (i.e., the value of β) when the true situation represents some *important deviation from H_0*. A value for the sample size can usually be determined for which there is a reasonable balance between the values of α and β computed in this fashion. The vaccine problem provides an illustration.

Illustration with a Continuous Random Variable

The concepts discussed here for a discrete population can be applied equally well to continuous random variables. Consider the null hypothesis that the average weight of male students in a certain college is 68 kilograms against the alternative hypothesis that it is unequal to 68. That is, we wish to test

$$H_0:\ \mu = 68,$$
$$H_1:\ \mu \neq 68.$$

The alternative hypothesis allows for the possibility that $\mu < 68$ or $\mu > 68$.

A sample mean that falls close to the hypothesized value of 68 would be considered evidence in favor of H_0. On the other hand, a sample mean that is considerably less than or more than 68 would be evidence inconsistent with H_0 and therefore favoring H_1. The sample mean is the test statistic in this case. A critical region for the test statistic might arbitrarily be chosen to be the two intervals $\bar{x} < 67$ and $\bar{x} > 69$. The nonrejection region will then be the interval $67 \leq \bar{x} \leq 69$. This decision criterion is illustrated in Figure 6.4.

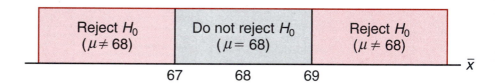

Figure 6.4: Critical region (in blue).

Let us now use the decision criterion of Figure 6.4 to calculate the probabilities of committing type I and type II errors when testing the null hypothesis that $\mu = 68$ kilograms against the alternative that $\mu \neq 68$ kilograms.

Assume the standard deviation of the population of weights to be $\sigma = 3.6$. For large samples, we may substitute s for σ if no other estimate of σ is available. Our decision statistic, based on a random sample of size $n = 36$, will be $\bar{X}$, the most efficient estimator of μ. From the Central Limit Theorem, we know that the sampling distribution of $\bar{X}$ is approximately normal with standard deviation $\sigma_{\bar{X}} = \sigma/\sqrt{n} = 3.6/6 = 0.6$.

The probability of committing a type I error, or the level of significance of our test, is equal to the sum of the areas that have been shaded in each tail of the distribution in Figure 6.5. Therefore,

$$\alpha = P(\bar{X} < 67 \text{ when } \mu = 68) + P(\bar{X} > 69 \text{ when } \mu = 68).$$

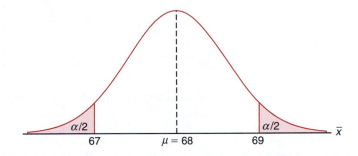

Figure 6.5: Critical region for testing $\mu = 68$ versus $\mu \neq 68$.

The z-values corresponding to $\bar{x}_1 = 67$ and $\bar{x}_2 = 69$ when H_0 is true are

$$z_1 = \frac{67 - 68}{0.6} = -1.67 \quad \text{and} \quad z_2 = \frac{69 - 68}{0.6} = 1.67.$$

Therefore,

$$\alpha = P(Z < -1.67) + P(Z > 1.67) = 2P(Z < -1.67) = 0.0950.$$

Thus, 9.5% of all samples of size 36 would lead us to reject $\mu = 68$ kilograms when, in fact, it is true. To reduce α, we have a choice of increasing the sample size or widening the fail-to-reject region. Suppose that we increase the sample size to $n = 64$. Then $\sigma_{\bar{X}} = 3.6/8 = 0.45$. Now

$$z_1 = \frac{67 - 68}{0.45} = -2.22 \quad \text{and} \quad z_2 = \frac{69 - 68}{0.45} = 2.22.$$

Hence,

$$\alpha = P(Z < -2.22) + P(Z > 2.22) = 2P(Z < -2.22) = 0.0264.$$

The reduction in α is not sufficient by itself to guarantee a good testing procedure. We must also evaluate β for various alternative hypotheses. If it is important to reject H_0 when the true mean is some value $\mu \geq 70$ or $\mu \leq 66$, then the probability of committing a type II error should be computed and examined for the alternatives $\mu = 66$ and $\mu = 70$. Because of symmetry, it is only necessary to consider the probability of not rejecting the null hypothesis that $\mu = 68$ when the alternative $\mu = 70$ is true. A type II error will result when the sample mean $\bar{x}$ falls between 67 and 69 when H_1 is true. Therefore, referring to Figure 6.6, we find that

$$\beta = P(67 \leq \bar{X} \leq 69 \text{ when } \mu = 70).$$

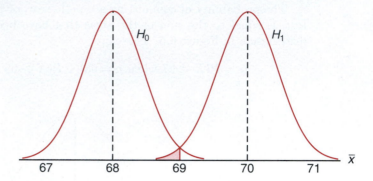

Figure 6.6: Probability of type II error for testing $\mu = 68$ versus $\mu = 70$.

The z-values corresponding to $\bar{x}_1 = 67$ and $\bar{x}_2 = 69$ when H_1 is true are

$$z_1 = \frac{67 - 70}{0.45} = -6.67 \quad \text{and} \quad z_2 = \frac{69 - 70}{0.45} = -2.22.$$

Therefore,

$$\beta = P(-6.67 < Z < -2.22) = P(Z < -2.22) - P(Z < -6.67)$$
$$= 0.0132 - 0.0000 = 0.0132.$$

If the true value of μ is the alternative $\mu = 66$, the value of β will again be 0.0132. For all possible values of $\mu < 66$ or $\mu > 70$, the value of β will be even smaller when $n = 64$, and consequently there would be little chance of not rejecting H_0 when it is false.

The probability of committing a type II error increases rapidly when the true value of μ approaches, but is not equal to, the hypothesized value. Of course, this is usually the situation where we do not mind making a type II error. For example, if the alternative hypothesis $\mu = 68.5$ is true, we do not mind committing a type II error by concluding that the true answer is $\mu = 68$. The probability of making such an error will be high when $n = 64$. Referring to Figure 6.7, we have

$$\beta = P(67 \leq \bar{X} \leq 69 \text{ when } \mu = 68.5).$$

The z-values corresponding to $\bar{x}_1 = 67$ and $\bar{x}_2 = 69$ when $\mu = 68.5$ are

$$z_1 = \frac{67 - 68.5}{0.45} = -3.33 \quad \text{and} \quad z_2 = \frac{69 - 68.5}{0.45} = 1.11.$$

Therefore,

$$\beta = P(-3.33 < Z < 1.11) = P(Z < 1.11) - P(Z < -3.33)$$
$$= 0.8665 - 0.0004 = 0.8661.$$

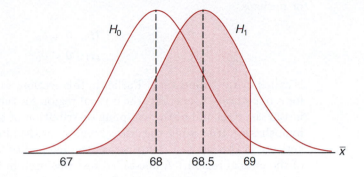

Figure 6.7: Type II error for testing $\mu = 68$ versus $\mu = 68.5$.

One very important concept that relates to error probabilities is the notion of the **power** of a test.

<table>
<tr><td>**Definition 6.4:**</td><td>The **power** of a test is the probability of rejecting H_0 given that a specific alternative is true.</td></tr>
</table>

The power of a test can be computed as $1 - \beta$. Often **different types of tests are compared by contrasting power properties**. Consider the previous illustration, in which we were testing H_0: $\mu = 68$ and H_1: $\mu \neq 68$. As before, suppose we are interested in assessing the sensitivity of the test. The test is governed by the rule that we do not reject H_0 if $67 \leq \bar{x} \leq 69$. We seek the capability of the test to properly reject H_0 when indeed $\mu = 68.5$. We have seen that the probability of a type II error is given by $\beta = 0.8661$. Thus, the **power** of the test is $1 - 0.8661 = 0.1339$. In a sense, the power is a more succinct measure of how sensitive the test is for detecting differences between a mean of 68 and a mean of 68.5. In this case, if μ is truly 68.5, the test as described will *properly reject H_0 only 13.39% of the time*. However, if $\mu = 70$, the power of the test is 0.99, a very satisfying value. From the foregoing, it is clear that to produce a desirable power (say, greater than 0.8), one must either increase α or increase the sample size.

So far in this chapter, much of the discussion of hypothesis testing has focused on foundations and definitions. In the sections that follow, we get more specific and put hypotheses in categories as well as discuss tests of hypotheses on various parameters of interest. We begin by drawing the distinction between a one-sided and a two-sided hypothesis.

One- and Two-Tailed Tests

A test of any statistical hypothesis where the alternative is **one sided**, such as

$$H_0: \ \theta = \theta_0,$$
$$H_1: \ \theta > \theta_0$$

or perhaps

$$H_0: \; \theta = \theta_0,$$
$$H_1: \; \theta < \theta_0,$$

is called a **one-tailed test**. Earlier in this section, we referred to the **test statistic** for a hypothesis. Generally, the critical region for the alternative hypothesis $\theta > \theta_0$ lies in the right tail of the sampling distribution of the test statistic when the null hypothesis is true, while the critical region for the alternative hypothesis $\theta < \theta_0$ lies entirely in the left tail. (In a sense, the inequality symbol points in the direction of the critical region.) A one-tailed test was used in the vaccine experiment to test the hypothesis $p = 1/4$ against the one-sided alternative $p > 1/4$ for the binomial distribution.

A test of any statistical hypothesis where the alternative is **two sided**, such as

$$H_0: \; \theta = \theta_0,$$
$$H_1: \; \theta \neq \theta_0,$$

is called a **two-tailed test**, since the critical region is split into two parts, often having equal probabilities, in each tail of the distribution of the test statistic. The alternative hypothesis $\theta \neq \theta_0$ states that either $\theta < \theta_0$ or $\theta > \theta_0$. A two-tailed test was used to test the null hypothesis that $\mu = 68$ kilograms against the two-sided alternative $\mu \neq 68$ kilograms in the example of the continuous population of student weights.

How Are the Null and Alternative Hypotheses Chosen?

The null hypothesis H_0 will often be stated using the *equality sign*. With this approach, it is clear how the probability of type I error is controlled: just as shown in the preceding examples. However, there are situations in which "do not reject H_0" implies that the parameter θ might be any value defined by the natural complement to the alternative hypothesis. For example, in the vaccine example, where the alternative hypothesis is $H_1: p > 1/4$, it is quite possible that nonrejection of H_0 cannot rule out a value of p less than $1/4$. Clearly though, in the case of one-tailed tests, the statement of the alternative is the most important consideration.

Whether one sets up a one-tailed or a two-tailed test will depend on the conclusion to be drawn if H_0 is rejected. The location of the critical region can be determined only after H_1 has been stated. For example, in testing a new drug, one sets up the hypothesis that it is no better than similar drugs now on the market and tests this against the alternative hypothesis that the new drug is superior. Such an alternative hypothesis will result in a one-tailed test with the critical region in the right tail. However, if we wish to compare a new teaching technique with the conventional classroom procedure, the alternative hypothesis should allow for the new approach to be either inferior or superior to the conventional procedure. Hence, the test is two-tailed with the critical region divided equally so as to fall in the extreme left and right tails of the distribution of our statistic.

Example 6.1: A manufacturer of a certain brand of rice cereal claims that the average saturated fat content does not exceed 1.5 grams per serving. State the null and alternative hypotheses to be used in testing this claim and determine where the critical region is located.

Solution: The manufacturer's claim should be rejected only if μ is greater than 1.5 milligrams and should not be rejected if μ is less than or equal to 1.5 milligrams. We test

$$H_0: \ \mu = 1.5,$$
$$H_1: \ \mu > 1.5.$$

Nonrejection of H_0 does not rule out values less than 1.5 milligrams. Since we have a one-tailed test, the greater than symbol indicates that the critical region lies entirely in the right tail of the distribution of our test statistic $\bar{X}$. ◢

Example 6.2: A real estate agent claims that 60% of all private residences being built today are 3-bedroom homes. To test this claim, a large sample of new residences is inspected; the proportion of these homes with 3 bedrooms is recorded and used as the test statistic. State the null and alternative hypotheses to be used in this test and determine the location of the critical region.

Solution: If the test statistic were substantially higher or lower than $p = 0.6$, we would reject the agent's claim. Hence, we should make the hypothesis

$$H_0: \ p = 0.6,$$
$$H_1: \ p \neq 0.6.$$

The alternative hypothesis implies a two-tailed test with the critical region divided equally in both tails of the distribution of $\widehat{P}$, our test statistic. ◢

6.3 The Use of *P*-Values for Decision Making in Testing Hypotheses

In testing hypotheses in which the test statistic is discrete, the critical region may be chosen arbitrarily and its size determined. If α is too large, it can be reduced by making an adjustment in the critical value. It may be necessary to increase the sample size to offset the decrease that occurs automatically in the power of the test.

Preselection of a Significance Level

The preselection of a significance level α has its roots in the philosophy that the maximum risk of making a type I error should be controlled. However, this approach does not account for values of test statistics that are "close" to the critical region. Suppose, for example, in the illustration with $H_0: \mu = 10$ versus $H_1: \mu \neq 10$, a value of $z = 1.87$ is observed; strictly speaking, with $\alpha = 0.05$, the value is not significant. But the risk of committing a type I error if one rejects H_0 in this case could hardly be considered severe. In fact, in a two-tailed scenario, one can quantify this risk as

$$P = 2P(Z > 1.87 \text{ when } \mu = 10) = 2(0.0307) = 0.0614.$$

As a result, 0.0614 is the probability of obtaining a value of z as large as or larger (in magnitude) than 1.87 when in fact $\mu = 10$. Although this evidence against H_0 is not as strong as that which would result from rejection at an $\alpha = 0.05$ level, it is important information to the user. Indeed, continued use of $\alpha = 0.05$ or 0.01 is only a result of what standards have been passed down through the generations. **The *P*-value approach has been adopted extensively by users of applied statistics**. The approach is designed to give the user an alternative (in terms of a probability) to a mere "reject" or "do not reject" conclusion. The P-value computation also gives the user important information when the z-value falls well *into the ordinary critical region*. For example, if z is 2.73, it is informative for the user to observe that

$$P = 2(0.0032) = 0.0064,$$

and thus the z-value is significant at a level considerably less than 0.05. It is important to know that under the condition of H_0, a value of $z = 2.73$ is an extremely rare event. That is, a value at least that large in magnitude would only occur 64 times in 10,000 experiments.

A Graphical Demonstration of a *P*-Value

One very simple way of explaining a P-value graphically is to consider two distinct samples. Suppose that two materials are being considered for coating a particular type of metal in order to inhibit corrosion. Specimens are obtained, and one collection is coated with material 1 and one collection coated with material 2. The sample sizes are $n_1 = n_2 = 10$, and corrosion is measured in percent of surface area affected. The hypothesis is that the samples came from common distributions with mean $\mu = 10$. Let us assume that the population variance is 1.0. Then we are testing

$$H_0: \ \mu_1 = \mu_2 = 10.$$

Let Figure 6.8 represent a point plot of the data; the data are placed on the distribution stated by the null hypothesis. Let us assume that the "×" data refer to material 1 and the "○" data refer to material 2. Now it seems clear that the data do refute the null hypothesis. But how can this be summarized in one number? **The *P*-value can be viewed as simply the probability of obtaining these data given that both samples come from the same distribution**. Clearly, this probability is quite small, say 0.00000001! Thus, the small P-value clearly refutes H_0, and the conclusion is that the population means are significantly different.

Use of the P-value approach as an aid in decision-making is quite natural, and nearly all computer packages that provide hypothesis-testing computation print out P-values along with values of the appropriate test statistic. The following is a formal definition of a P-value.

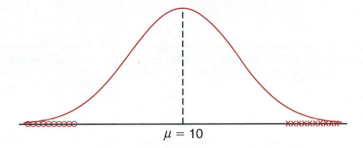

Figure 6.8: Data that are likely generated from populations having two different means.

Definition 6.5:	A **P-value** is the lowest level (of significance) at which the observed value of the test statistic is significant.

How Does the Use of *P*-Values Differ from Classic Hypothesis Testing?

It is tempting at this point to summarize the procedures associated with testing, say, H_0: $\theta = \theta_0$. However, the student who is a beginner in this area should understand that there are differences in approach and philosophy between the classic fixed α approach that is climaxed with either a "reject H_0" or a "do not reject H_0" conclusion and the *P*-value approach. In the latter, no fixed α is determined and conclusions are drawn on the basis of the size of the *P*-value in harmony with the subjective judgment of the engineer or scientist. While modern computer software will output *P*-values, nevertheless it is important that readers understand both approaches in order to appreciate the totality of the concepts. Thus, we offer a brief list of procedural steps for both the classical and the *P*-value approach.

Approach to Hypothesis Testing with Fixed Probability of Type I Error	**1.** State the null and alternative hypotheses. **2.** Choose a fixed significance level α. **3.** Choose an appropriate test statistic and establish the critical region based on α. **4.** Reject H_0 if the computed test statistic is in the critical region. Otherwise, do not reject. **5.** Draw scientific or engineering conclusions.
Significance Testing (*P*-Value Approach)	**1.** State null and alternative hypotheses. **2.** Choose an appropriate test statistic. **3.** Compute the *P*-value based on the computed value of the test statistic. **4.** Use judgment based on the *P*-value and knowledge of the scientific system in deciding whether or not to reject H_0.

In later sections of this chapter and chapters that follow, many examples and exercises emphasize the *P*-value approach to drawing scientific conclusions.

Exercises

6.1 Suppose that an allergist wishes to test the hypothesis that at least 30% of the public is allergic to some cheese products. Explain how the allergist could commit

(a) a type I error;

(b) a type II error.

6.2 A sociologist is concerned about the effectiveness of a training course designed to get more drivers to use seat belts in automobiles.

(a) What hypothesis is she testing if she commits a type I error by erroneously concluding that the training course is ineffective?

(b) What hypothesis is she testing if she commits a type II error by erroneously concluding that the training course is effective?

6.3 A large manufacturing firm is being charged with discrimination in its hiring practices.

(a) What hypothesis is being tested if a jury commits a type I error by finding the firm guilty?

(b) What hypothesis is being tested if a jury commits a type II error by finding the firm guilty?

6.4 A fabric manufacturer believes that the proportion of orders for raw material arriving late is $p = 0.6$. If a random sample of 10 orders shows that 3 or fewer arrived late, the hypothesis that $p = 0.6$ should be rejected in favor of the alternative $p < 0.6$. Use the binomial distribution.

(a) Find the probability of committing a type I error if the true proportion is $p = 0.6$.

(b) Find the probability of committing a type II error for the alternatives $p = 0.3$, $p = 0.4$, and $p = 0.5$.

6.5 Repeat Exercise 6.4 but assume that 50 orders are selected and the critical region is defined to be $x \leq 24$, where x is the number of orders in the sample that arrived late. Use the normal approximation.

6.6 The proportion of adults living in a small town who are college graduates is estimated to be $p = 0.6$. To test this hypothesis, a random sample of 15 adults is selected. If the number of college graduates in the sample is anywhere from 6 to 12, we shall not reject the null hypothesis that $p = 0.6$; otherwise, we shall conclude that $p \neq 0.6$.

(a) Evaluate α assuming that $p = 0.6$. Use the binomial distribution.

(b) Evaluate β for the alternatives $p = 0.5$ and $p = 0.7$.

(c) Is this a good test procedure?

6.7 Repeat Exercise 6.6 but assume that 200 adults are selected and the fail-to-reject region is defined to be $110 \leq x \leq 130$, where x is the number of college graduates in our sample. Use the normal approximation.

6.8 In *Relief from Arthritis* published by Thorsons Publishers, Ltd., John E. Croft claims that over 40% of those who suffer from osteoarthritis receive measurable relief from an ingredient produced by a particular species of mussel found off the coast of New Zealand. To test this claim, the mussel extract is to be given to a group of 7 osteoarthritic patients. If 3 or more of the patients receive relief, we shall not reject the null hypothesis that $p = 0.4$; otherwise, we shall conclude that $p < 0.4$.

(a) Evaluate α, assuming that $p = 0.4$.

(b) Evaluate β for the alternative $p = 0.3$.

6.9 A dry cleaning establishment claims that a new spot remover will remove more than 70% of the spots to which it is applied. To check this claim, the spot remover will be used on 12 spots chosen at random. If fewer than 11 of the spots are removed, we shall not reject the null hypothesis that $p = 0.7$; otherwise, we shall conclude that $p > 0.7$.

(a) Evaluate α, assuming that $p = 0.7$.

(b) Evaluate β for the alternative $p = 0.9$.

6.10 Repeat Exercise 6.9 but assume that 100 spots are treated and the critical region is defined to be $x > 82$, where x is the number of spots removed.

6.11 Repeat Exercise 6.8 but assume that 70 patients are given the mussel extract and the critical region is defined to be $x < 24$, where x is the number of osteoarthritic patients who receive relief.

6.12 A random sample of 400 voters in a certain city are asked if they favor an additional 4% gasoline sales tax to provide badly needed revenues for street repairs. If more than 220 but fewer than 260 favor the sales tax, we shall conclude that 60% of the voters are for it.

(a) Find the probability of committing a type I error if 60% of the voters favor the increased tax.

(b) What is the probability of committing a type II error using this test procedure if actually only 48% of the voters are in favor of the additional gasoline tax?

6.13 Suppose, in Exercise 6.12, we conclude that 60% of the voters favor the gasoline sales tax if more than 214 but fewer than 266 voters in our sample favor it. Show that this new critical region results in a smaller value for α at the expense of increasing β.

6.14 A manufacturer has developed a new fishing line, which the company claims has a mean breaking strength of 15 kilograms with a standard deviation of 0.5 kilogram. To test the hypothesis that $\mu = 15$ kilograms against the alternative that $\mu < 15$ kilograms, a random sample of 50 lines will be tested. The critical region is defined to be $\bar{x} < 14.9$.

(a) Find the probability of committing a type I error when H_0 is true.

(b) Evaluate β for the alternatives $\mu = 14.8$ and $\mu = 14.9$ kilograms.

6.15 A soft-drink machine at a steak house is regulated so that the amount of drink dispensed is approximately normally distributed with a mean of 200 milliliters and a standard deviation of 15 milliliters. The machine is checked periodically by taking a sample of 9 drinks and computing the average content. If $\bar{x}$ falls in the interval $191 < \bar{x} < 209$, the machine is thought to be operating satisfactorily; otherwise, we conclude that $\mu \neq 200$ milliliters.

(a) Find the probability of committing a type I error when $\mu = 200$ milliliters.

(b) Find the probability of committing a type II error when $\mu = 215$ milliliters.

6.16 Repeat Exercise 6.15 for samples of size $n = 25$.

Use the same critical region.

6.17 A new curing process developed for a certain type of cement results in a mean compressive strength of 5000 kilograms per square centimeter with a standard deviation of 120 kilograms per square centimeter. To test the hypothesis that $\mu = 5000$ against the alternative that $\mu < 5000$, a random sample of 50 pieces of cement is tested. The critical region is defined to be $\bar{x} < 4970$.

(a) Find the probability of committing a type I error when H_0 is true.

(b) Evaluate β for the alternatives $\mu = 4970$ and $\mu = 4960$.

6.18 If we plot the probabilities of failing to reject H_0 corresponding to various alternatives for μ (including the value specified by H_0) and connect all the points by a smooth curve, we obtain the **operating characteristic curve** of the test criterion, or the **OC curve**. Note that the probability of failing to reject H_0 when it is true is simply $1 - \alpha$. Operating characteristic curves are widely used in industrial applications to provide a visual display of the merits of the test criterion. With reference to Exercise 6.15, find the probabilities of failing to reject H_0 for the following 9 values of μ and plot the OC curve: 184, 188, 192, 196, 200, 204, 208, 212, and 216.

6.4 Single Sample: Tests Concerning a Single Mean

In this section, we formally consider tests of hypotheses on a single population mean. Many of the illustrations from previous sections involved tests on the mean, so the reader should already have insight into some of the details that are outlined here.

Tests on a Single Mean (Variance Known)

We should first describe the assumptions on which the experiment is based. The model for the underlying situation centers around an experiment with $X_1, X_2, \ldots,$ X_n representing a random sample from a distribution with mean μ and variance $\sigma^2 > 0$. Consider first the hypothesis

$$H_0\colon \mu = \mu_0,$$
$$H_1\colon \mu \neq \mu_0.$$

The appropriate test statistic should be based on the random variable $\bar{X}$. In Chapter 4, the Central Limit Theorem was introduced, which essentially states that despite the distribution of X, the random variable $\bar{X}$ has approximately a normal distribution with mean μ and variance σ^2/n for reasonably large sample sizes. So, $\mu_{\bar{X}} = \mu$ and $\sigma_{\bar{X}}^2 = \sigma^2/n$. We can then determine a critical region based on the computed sample average, $\bar{x}$. It should be clear to the reader by now that there will be a two-tailed critical region for the test.

Standardization of $\bar{X}$

It is convenient to standardize $\bar{X}$ and formally involve the **standard normal random variable** Z, where

$$Z = \frac{\bar{X} - \mu}{\sigma/\sqrt{n}}.$$

We know that *under H_0*, the Central Limit Theorem states that $\mu_{\bar{X}} = \mu_0$, $\sqrt{n}(\bar{X} - \mu_0)/\sigma$ follows an $n(x; 0, 1)$ distribution. Hence Z will fall between the critical values $-z_{\alpha/2}$ and $z_{\alpha/2}$ with a probability of $1 - \alpha$. The expression

$$P\left(-z_{\alpha/2} < \frac{\bar{X} - \mu_0}{\sigma/\sqrt{n}} < z_{\alpha/2}\right) = 1 - \alpha$$

can be used to write an appropriate nonrejection region. The reader should keep in mind that, formally, the critical region is designed to control α, the probability of a type I error. It should be obvious that a *two-tailed signal* of evidence is needed to support H_1. Thus, given a computed value $\bar{x}$, the formal test involves rejecting H_0 if the computed *test statistic z* falls in the critical region described next.

Test Procedure for a Single Mean (Variance Known)	$z = \dfrac{\bar{x} - \mu_0}{\sigma/\sqrt{n}} > z_{\alpha/2}$ or $z = \dfrac{\bar{x} - \mu_0}{\sigma/\sqrt{n}} < -z_{\alpha/2}$

If $-z_{\alpha/2} < z < z_{\alpha/2}$, do not reject H_0. Rejection of H_0, of course, implies acceptance of the alternative hypothesis $\mu \neq \mu_0$. With this definition of the critical region, it should be clear that there will be probability α of rejecting H_0 (falling into the critical region) when, indeed, $\mu = \mu_0$.

Although it is easier to understand the critical region written in terms of z, we can write the same critical region in terms of the computed average $\bar{x}$. The following can be written as an identical decision procedure:

$$\text{reject } H_0 \text{ if } \bar{x} < a \text{ or } \bar{x} > b,$$

where

$$a = \mu_0 - z_{\alpha/2}\frac{\sigma}{\sqrt{n}}, \qquad b = \mu_0 + z_{\alpha/2}\frac{\sigma}{\sqrt{n}}.$$

Hence, for a significance level α, the critical values of the random variable z and $\bar{x}$ are both depicted in Figure 6.9.

Tests of one-sided hypotheses on the mean involve the same statistic described in the two-sided case. The difference, of course, is that the critical region is only in one tail of the standard normal distribution. For example, suppose that we seek to test

$$H_0\!: \ \mu = \mu_0,$$
$$H_1\!: \ \mu > \mu_0.$$

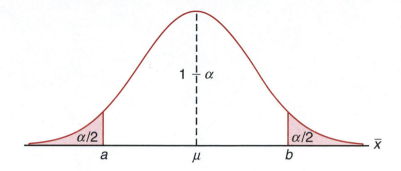

Figure 6.9: Critical region for the alternative hypothesis $\mu \neq \mu_0$.

The signal that favors H_1 comes from *large values* of z. Thus, rejection of H_0 results when the computed $z > z_\alpha$. Obviously, if the alternative is H_1: $\mu < \mu_0$, the critical region is entirely in the lower tail and thus rejection results from $z < -z_\alpha$. Although in a one-sided testing case the null hypothesis can be written as H_0: $\mu \leq \mu_0$ or H_0: $\mu \geq \mu_0$, it is usually written as H_0: $\mu = \mu_0$.

The following two examples illustrate tests on means for the case in which σ is known.

Example 6.3: A random sample of 100 recorded deaths in the United States during the past year showed an average life span of 71.8 years. Assuming a population standard deviation of 8.9 years, does this seem to indicate that the mean life span today is greater than 70 years? Use a 0.05 level of significance.

Solution: 1. H_0: $\mu = 70$ years.

2. H_1: $\mu > 70$ years.

3. $\alpha = 0.05$.

4. Critical region: $z > 1.645$, where $z = \frac{\bar{x}-\mu_0}{\sigma/\sqrt{n}}$.

5. Computations: $\bar{x} = 71.8$ years, $\sigma = 8.9$ years, and hence $z = \frac{71.8-70}{8.9/\sqrt{100}} = 2.02$.

6. Decision: Reject H_0 and conclude that the mean life span today is greater than 70 years.

The P-value corresponding to $z = 2.02$ is given by the area of the shaded region in Figure 6.10.

Using Table A.3, we have

$$P = P(Z > 2.02) = 0.0217.$$

As a result, the evidence in favor of H_1 is even stronger than that suggested by a 0.05 level of significance.

Example 6.4: A manufacturer of sports equipment has developed a new synthetic fishing line that the company claims has a mean breaking strength of 8 kilograms with a standard deviation of 0.5 kilogram. Test the hypothesis that $\mu = 8$ kilograms against the alternative that $\mu \neq 8$ kilograms if a random sample of 50 lines is tested and found to have a mean breaking strength of 7.8 kilograms. Use a 0.01 level of significance.

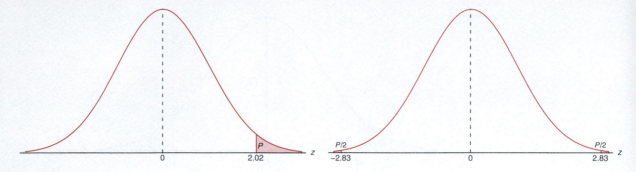

Figure 6.10: P-value for Example 6.3.   Figure 6.11: P-value for Example 6.4.

Solution: 1. H_0: $\mu = 8$ kilograms.

2. H_1: $\mu \neq 8$ kilograms.

3. $\alpha = 0.01$.

4. Critical region: $z < -2.575$ and $z > 2.575$, where $z = \frac{\bar{x} - \mu_0}{\sigma/\sqrt{n}}$.

5. Computations: $\bar{x} = 7.8$ kilograms, $n = 50$, and hence $z = \frac{7.8 - 8}{0.5/\sqrt{50}} = -2.83$.

6. Decision: Reject H_0 and conclude that the average breaking strength is not equal to 8.

Since the test in this example is two tailed, the desired P-value is twice the area of the shaded region in Figure 6.11 to the left of $z = -2.83$. Therefore, using Table A.3, we have

$$P = P(|Z| > 2.83) = 2P(Z < -2.83) = 0.0046,$$

which allows us to reject the null hypothesis that $\mu = 8$ kilograms at a level of significance smaller than 0.01.

Relationship to Confidence Interval Estimation

The reader should realize by now that the hypothesis-testing approach to statistical inference in this chapter is very closely related to the confidence interval approach in Chapter 5. Confidence interval estimation involves computation of bounds within which it is "reasonable" for the parameter in question to lie. For the case of a single population mean μ with σ^2 known, the structure of both hypothesis testing and confidence interval estimation is based on the random variable

$$Z = \frac{\bar{X} - \mu}{\sigma/\sqrt{n}}.$$

It turns out that the testing of H_0: $\mu = \mu_0$ against H_1: $\mu \neq \mu_0$ at a significance level α is equivalent to computing a $100(1 - \alpha)\%$ confidence interval on μ and rejecting H_0 if μ_0 is outside the confidence interval. If μ_0 is inside the confidence interval, the hypothesis is not rejected. The equivalence is very intuitive and quite simple to illustrate. Recall that with an observed value $\bar{x}$, failure to reject H_0 at significance level α implies that

$$-z_{\alpha/2} \leq \frac{\bar{x} - \mu_0}{\sigma/\sqrt{n}} \leq z_{\alpha/2},$$

which is equivalent to

$$\bar{x} - z_{\alpha/2}\frac{\sigma}{\sqrt{n}} \leq \mu_0 \leq \bar{x} + z_{\alpha/2}\frac{\sigma}{\sqrt{n}}.$$

The equivalence of confidence interval estimation to hypothesis testing extends to differences between two means, variances, ratios of variances, and so on. As a result, the student of statistics should not consider confidence interval estimation and hypothesis testing as separate forms of statistical inference.

Tests on a Single Sample (Variance Unknown)

One would certainly suspect that tests on a population mean μ with σ^2 unknown, like confidence interval estimation, should involve use of the Student t-distribution. Strictly speaking, the application of Student t for both confidence intervals and hypothesis testing is developed under the following assumptions. The random variables $X_1, X_2, \ldots, X_n$ represent a random sample from a normal distribution with unknown μ and σ^2. Then the random variable $\sqrt{n}(\bar{X} - \mu)/S$ has a Student t-distribution with $n-1$ degrees of freedom. The structure of the test is identical to that for the case of σ known, with the exception that the value σ in the test statistic is replaced by the computed estimate S and the standard normal distribution is replaced by a t-distribution.

The t-Statistic for a Test on a Single Mean (Variance Unknown)

For the two-sided hypothesis

$$H_0\!: \ \mu = \mu_0,$$
$$H_1\!: \ \mu \neq \mu_0,$$

we reject H_0 at significance level α when the computed t-statistic

$$t = \frac{\bar{x} - \mu_0}{s/\sqrt{n}}$$

exceeds $t_{\alpha/2,n-1}$ or is less than $-t_{\alpha/2,n-1}$.

The reader should recall from Chapters 4 and 5 that the t-distribution is symmetric around the value zero. Thus, this two-tailed critical region applies in a fashion similar to that for the case of known σ. For the two-sided hypothesis at significance level α, the two-tailed critical regions apply. For $H_1\!: \mu > \mu_0$, rejection results when $t > t_{\alpha,n-1}$. For $H_1\!: \mu < \mu_0$, the critical region is given by $t < -t_{\alpha,n-1}$.

Example 6.5: The Edison Electric Institute has published figures on the number of kilowatt hours used annually by various home appliances. It is claimed that a vacuum cleaner uses an average of 46 kilowatt hours per year. If a random sample of 12 homes included in a planned study indicates that vacuum cleaners use an average of 42 kilowatt hours per year with a sample standard deviation of 11.9 kilowatt hours, does this suggest at the 0.05 level of significance that vacuum cleaners use, on average, less than 46 kilowatt hours annually? Assume the population of kilowatt hours to be normal.

Solution: 1. H_0: $\mu = 46$ kilowatt hours.

2. H_1: $\mu < 46$ kilowatt hours.

3. $\alpha = 0.05$.

4. Critical region: $t < -1.796$, where $t = \frac{\bar{x}-\mu_0}{s/\sqrt{n}}$ with 11 degrees of freedom.

5. Computations: $\bar{x} = 42$ kilowatt hours, $s = 11.9$ kilowatt hours, and $n = 12$. Hence,

$$t = \frac{42-46}{11.9/\sqrt{12}} = -1.16, \qquad P = P(T < -1.16) \approx 0.135.$$

6. Decision: Do not reject H_0 and conclude that the average number of kilowatt hours used annually by home vacuum cleaners is not significantly less than 46.

Comment on the Single-Sample *t*-Test

Comments regarding the normality assumption are worth emphasizing at this point. We have indicated that when σ is known, the Central Limit Theorem allows for the use of a test statistic or a confidence interval which is based on Z, the standard normal random variable. Strictly speaking, of course, the Central Limit Theorem, and thus the use of the standard normal distribution, does not apply unless σ is known. In Chapter 4, the development of the t-distribution was given. There we pointed out that normality on $X_1, X_2, \ldots, X_n$ was an underlying assumption. Thus, *strictly speaking*, the Student's t-tables of percentage points for tests or confidence intervals should not be used unless it is known that the sample comes from a normal population. In practice, σ can rarely be assumed to be known. However, a very good estimate may be available from previous experiments. Many statistics textbooks suggest that one can safely replace σ by s in the test statistic

$$z = \frac{\bar{x}-\mu_0}{\sigma/\sqrt{n}}$$

when $n \geq 30$ with a bell-shaped population and still use the Z-tables for the appropriate critical region. The implication here is that the Central Limit Theorem is indeed being invoked and one is relying on the fact that $s \approx \sigma$. Obviously, when this is done, the results must be viewed as approximate. Thus, a computed P-value (from the Z-distribution) of 0.15 may be 0.12 or perhaps 0.17, or a computed confidence interval may be a 93% confidence interval rather than a 95% interval as desired. Now what about situations where $n \leq 30$? The user cannot rely on s being close to σ, and in order to take into account the inaccuracy of the estimate, the confidence interval should be wider or the critical value larger in magnitude. The t-distribution percentage points accomplish this but are correct only when the sample is from a normal distribution. Of course, normal probability plots can be used to ascertain some sense of the deviation from normality in a data set.

For small samples, it is often difficult to detect deviations from a normal distribution. (Goodness-of-fit tests are discussed in a later section of this chapter.) For bell-shaped distributions of the random variables $X_1, X_2, \ldots, X_n$, the use of the t-distribution for tests or confidence intervals is likely to produce quite good results. When in doubt, the user should resort to nonparametric procedures.

Annotated Computer Printout for Single-Sample t-Test

It should be of interest for the reader to see an annotated computer printout showing the result of a single-sample t-test. Suppose that an engineer is interested in testing the bias in a pH meter. Data are collected on a neutral substance (pH = 7.0). A sample of the measurements is taken, with the data as follows:

$$7.07 \quad 7.00 \quad 7.10 \quad 6.97 \quad 7.00 \quad 7.03 \quad 7.01 \quad 7.01 \quad 6.98 \quad 7.08$$

It is, then, of interest to test

$$H_0: \mu = 7.0,$$
$$H_1: \mu \neq 7.0.$$

In this illustration, we use the computer package *MINITAB* to illustrate the analysis of the data set above. Notice the key components of the printout shown in Figure 6.12. The mean $\bar{y}$ is 7.0250, StDev is simply the sample standard deviation $s = 0.044$, and SE Mean is the estimated standard error of the mean and is computed as $s/\sqrt{n} = 0.0139$. The t-value is the ratio

$$(7.0250 - 7)/0.0139 = 1.80.$$

```
pH-meter
   7.07    7.00    7.10    6.97    7.00    7.03    7.01    7.01    6.98    7.08
MTB > Onet 'pH-meter'; SUBC>    Test 7.

One-Sample T: pH-meter Test of mu = 7 vs not = 7
Variable  N     Mean    StDev   SE Mean       95% CI            T      P
pH-meter 10 7.02500  0.04403   0.01392  (6.99350, 7.05650)  1.80  0.106
```

Figure 6.12: *MINITAB* printout for one-sample t-test for pH meter.

The P-value of 0.106 suggests results that are inconclusive. There is no evidence suggesting a strong rejection of H_0 (based on an α of 0.05 or 0.10), **yet one certainly cannot confidently conclude that the pH meter is unbiased.** Notice that the sample size of 10 is rather small. An increase in sample size (perhaps another experiment) may sort things out. A discussion regarding appropriate sample size appears in Section 6.6.

6.5   Two Samples: Tests on Two Means

The reader should now understand the relationship between tests and confidence intervals, and can rely on details supplied by the confidence interval material in Chapter 5. Tests concerning two means represent a set of very important analytical tools for the scientist or engineer. The experimental setting is very much like that described in Section 5.8. Two independent random samples of sizes n_1 and n_2, respectively, are drawn from two populations with means μ_1 and μ_2 and variances σ_1^2 and σ_2^2. We know that the random variable

$$Z = \frac{(\bar{X}_1 - \bar{X}_2) - (\mu_1 - \mu_2)}{\sqrt{\sigma_1^2/n_1 + \sigma_2^2/n_2}}$$

has a standard normal distribution. Here we are assuming that n_1 and n_2 are sufficiently large that the Central Limit Theorem applies. Of course, if the two populations are normal, the statistic above has a standard normal distribution even for small n_1 and n_2. Obviously, if we can assume that $\sigma_1 = \sigma_2 = \sigma$, the statistic above reduces to

$$Z = \frac{(\bar{X}_1 - \bar{X}_2) - (\mu_1 - \mu_2)}{\sigma\sqrt{1/n_1 + 1/n_2}}.$$

The two statistics above serve as a basis for the development of the test procedures involving two means. The equivalence between tests and confidence intervals, along with the technical detail involving tests on one mean, allows a simple transition to tests on two means.

The two-sided hypothesis on two means can be written generally as

$$H_0\colon \mu_1 - \mu_2 = d_0.$$

Obviously, the alternative can be two sided or one sided. Again, the distribution used is the distribution of the test statistic under H_0. Values $\bar{x}_1$ and $\bar{x}_2$ are computed and, for σ_1 and σ_2 known, the test statistic is given by

$$z = \frac{(\bar{x}_1 - \bar{x}_2) - d_0}{\sqrt{\sigma_1^2/n_1 + \sigma_2^2/n_2}},$$

with a two-tailed critical region in the case of a two-sided alternative. That is, reject H_0 in favor of $H_1\colon \mu_1 - \mu_2 \neq d_0$ if $z > z_{\alpha/2}$ or $z < -z_{\alpha/2}$. One-tailed critical regions are used in the case of the one-sided alternatives. The reader should, as before, study the test statistic and be satisfied that for, say, $H_1\colon \mu_1 - \mu_2 > d_0$, the signal favoring H_1 comes from large values of z. Thus, the upper-tailed critical region applies.

Unknown but Equal Variances

The more prevalent situations involving tests on two means are those in which variances are unknown. If the scientist involved is willing to assume that both distributions are normal and that $\sigma_1 = \sigma_2 = \sigma$, the *pooled t-test* (often called the two-sample *t*-test) may be used. The test statistic (see Section 5.8) is given by the following test procedure.

| Two-Sample Pooled *t*-Test | For the two-sided hypothesis |

$$H_0: \ \mu_1 = \mu_2,$$
$$H_1: \ \mu_1 \neq \mu_2,$$

we reject H_0 at significance level α when the computed *t*-statistic

$$t = \frac{(\bar{x}_1 - \bar{x}_2) - d_0}{s_p\sqrt{1/n_1 + 1/n_2}},$$

where

$$s_p^2 = \frac{s_1^2(n_1 - 1) + s_2^2(n_2 - 1)}{n_1 + n_2 - 2},$$

exceeds $t_{\alpha/2, n_1+n_2-2}$ or is less than $-t_{\alpha/2, n_1+n_2-2}$.

Recall from Chapter 5 that the degrees of freedom for the *t*-distribution are a result of pooling of information from the two samples to estimate σ^2. One-sided alternatives suggest one-sided critical regions, as one might expect. For example, for $H_1: \mu_1 - \mu_2 > d_0$, reject $H_1: \mu_1 - \mu_2 = d_0$ when $t > t_{\alpha, n_1+n_2-2}$.

Example 6.6: An experiment was performed to compare the abrasive wear of two different laminated materials. Twelve pieces of material 1 were tested by exposing each piece to a machine measuring wear. Ten pieces of material 2 were similarly tested. In each case, the depth of wear was observed. The samples of material 1 gave an average (coded) wear of 85 units with a sample standard deviation of 4, while the samples of material 2 gave an average of 81 with a sample standard deviation of 5. Can we conclude at the 0.05 level of significance that the abrasive wear of material 1 exceeds that of material 2 by more than 2 units? Assume the populations to be approximately normal with equal variances.

Solution: Let μ_1 and μ_2 represent the population means of the abrasive wear for material 1 and material 2, respectively.

1. $H_0: \mu_1 - \mu_2 = 2.$

2. $H_1: \mu_1 - \mu_2 > 2.$

3. $\alpha = 0.05.$

4. Critical region: $t > 1.725$, where $t = \frac{(\bar{x}_1 - \bar{x}_2) - d_0}{s_p\sqrt{1/n_1 + 1/n_2}}$ with $v = 20$ degrees of freedom.

5. Computations:

$$\bar{x}_1 = 85, \quad s_1 = 4, \quad n_1 = 12,$$
$$\bar{x}_2 = 81, \quad s_2 = 5, \quad n_2 = 10.$$

Hence

$$s_p = \sqrt{\frac{(11)(16) + (9)(25)}{12 + 10 - 2}} = 4.478,$$

$$t = \frac{(85 - 81) - 2}{4.478\sqrt{1/12 + 1/10}} = 1.04,$$

$$P = P(T > 1.04) \approx 0.16. \qquad \text{(See Table A.4.)}$$

6. Decision: Do not reject H_0. We are unable to conclude that the abrasive wear of material 1 exceeds that of material 2 by more than 2 units. ⌐

Unknown but Unequal Variances

There are situations where the analyst is **not** able to assume that $\sigma_1 = \sigma_2$. Recall from Section 5.8 that, if the populations are normal, the statistic

$$T' = \frac{(\bar{X}_1 - \bar{X}_2) - d_0}{\sqrt{s_1^2/n_1 + s_2^2/n_2}}$$

has an approximate t-distribution with approximate degrees of freedom

$$v = \frac{(s_1^2/n_1 + s_2^2/n_2)^2}{(s_1^2/n_1)^2/(n_1 - 1) + (s_2^2/n_2)^2/(n_2 - 1)}.$$

As a result, the test procedure is to *not reject* H_0 when

$$-t_{\alpha/2,v} < t' < t_{\alpha/2,v},$$

with v given as above. Again, as in the case of the pooled t-test, one-sided alternatives suggest one-sided critical regions. The test procedures developed above are summarized in Table 6.3.

Paired Observations

A study of the two-sample t-test or confidence interval on the difference between means should suggest the need for an experimental design. Recall the discussion of experimental units in Chapter 5, where it was suggested that the conditions of the two populations (often referred to as the two treatments) should be assigned randomly to the experimental units. This is done to avoid biased results due to systematic differences between experimental units. In other words, in hypothesis-testing jargon, it is important that any significant difference found between means be due to the different conditions of the populations and not due to the experimental units in the study. For example, consider Exercise 5.28 in Section 5.9. The 20 seedlings play the role of the experimental units. Ten of them are to be treated with nitrogen and 10 not treated with nitrogen. It may be very important that this assignment to the "nitrogen" and "no-nitrogen" treatments be random to ensure that systematic differences between the seedlings do not interfere with a valid comparison between the means.

In Example 6.6, time of measurement is the most likely choice for the experimental unit. The 22 pieces of material should be measured in random order. We need to guard against the possibility that wear measurements made close together in time might tend to give similar results. **Systematic** (nonrandom) **differences in experimental units** *are not expected.* However, random assignments guard against the problem.

References to planning of experiments, randomization, choice of sample size, and so on, will continue to influence much of the development in Chapter 8. Any scientist or engineer whose interest lies in analysis of real data should study this material. The pooled t-test is extended in Chapter 8 to cover more than two means.

Testing of two means can be accomplished when data are in the form of paired observations, as discussed in Chapter 5. In this pairing structure, the conditions of the two populations (treatments) are assigned randomly within homogeneous units (pairs). Computation of the confidence interval for $\mu_1 - \mu_2$ in the situation with paired observations is based on the random variable

$$T = \frac{\bar{D} - \mu_D}{S_d/\sqrt{n}},$$

where $\bar{D}$ and S_d are random variables representing the sample mean and standard deviation of the differences of the observations in the experimental units. As in the case of the *pooled t-test*, the assumption is that the observations from each population are normal. This two-sample problem is essentially reduced to a one-sample problem by using the computed differences $d_1, d_2, \ldots, d_n$. Thus, the hypothesis reduces to

$$H_0: \ \mu_D = d_0.$$

The computed test statistic is then given by

$$t = \frac{\bar{d} - d_0}{s_d/\sqrt{n}}.$$

Critical regions are constructed using the t-distribution with $n - 1$ degrees of freedom.

Problem of Interaction in a Paired t-Test

Not only will the case study that follows illustrate the use of the paired t-test but the discussion will shed considerable light on the difficulties that arise when there is an interaction between the treatments and the experimental units in the paired t structure. The concept of interaction will be an important issue in Chapter 8.

There are some types of statistical tests in which the existence of interaction results in difficulty. The paired t-test is one such example. In Section 5.9, the paired structure was used in the computation of a confidence interval on the difference between two means, and the advantage in pairing was revealed for situations in which the experimental units are homogeneous. The pairing results in a reduction in σ_D, the standard deviation of a difference $D_i = X_{1i} - X_{2i}$, as discussed in Section 5.9. If interaction exists between treatments and experimental units, the advantage gained in pairing may be substantially reduced.

In order to demonstrate how interaction influences $\text{Var}(D)$ and hence the quality of the paired t-test, it is instructive to revisit the ith difference given by $D_i = X_{1i} - X_{2i} = (\mu_1 - \mu_2) + (\epsilon_1 - \epsilon_2)$, where X_{1i} and X_{2i} are taken on the ith experimental unit. If the pairing unit is homogeneous, the errors in X_{1i} and in X_{2i} should be similar and not independent. We noted in Chapter 5 that the positive covariance between the errors results in a reduced $\text{Var}(D)$. Thus, the size of the difference in the treatments and the relationship between the errors in X_{1i} and X_{2i} contributed by the experimental unit will tend to allow a significant difference to be detected.

Case Study 6.1: **Blood Sample Data**: In a study conducted in the Forestry and Wildlife Department at Virginia Tech, J. A. Wesson examined the influence of the drug succinylcholine on the circulation levels of androgens in the blood. Blood samples were taken from wild, free-ranging deer immediately after they had received an intramuscular injection of succinylcholine administered using darts and a capture gun. A second blood sample was obtained from each deer 30 minutes after the first sample, after which the deer was released. The levels of androgens at time of capture and 30 minutes later, measured in nanograms per milliliter (ng/mL), for 15 deer are given in Table 6.2.

Assuming that the populations of androgen levels at time of injection and 30 minutes later are normally distributed, test at the 0.05 level of significance whether the androgen concentrations are altered after 30 minutes.

Table 6.2: Data for Case Study 6.1

Deer	Androgen (ng/mL) At Time of Injection	30 Minutes after Injection	d_i
1	2.76	7.02	4.26
2	5.18	3.10	−2.08
3	2.68	5.44	2.76
4	3.05	3.99	0.94
5	4.10	5.21	1.11
6	7.05	10.26	3.21
7	6.60	13.91	7.31
8	4.79	18.53	13.74
9	7.39	7.91	0.52
10	7.30	4.85	−2.45
11	11.78	11.10	−0.68
12	3.90	3.74	−0.16
13	26.00	94.03	68.03
14	67.48	94.03	26.55
15	17.04	41.70	24.66

Solution: Let μ_1 and μ_2 be the average androgen concentration at the time of injection and 30 minutes later, respectively. We proceed as follows:

1. H_0: $\mu_1 = \mu_2$ or $\mu_D = \mu_1 - \mu_2 = 0$.
2. H_1: $\mu_1 \neq \mu_2$ or $\mu_D = \mu_1 - \mu_2 \neq 0$.
3. $\alpha = 0.05$.

4. Critical region: $t < -2.145$ and $t > 2.145$, where $t = \frac{\bar{d} - d_0}{s_D/\sqrt{n}}$ with $v = 14$ degrees of freedom.

5. Computations: The sample mean and standard deviation for the d_i are

$$\bar{d} = 9.848 \qquad \text{and} \qquad s_d = 18.474.$$

Therefore,

$$t = \frac{9.848 - 0}{18.474/\sqrt{15}} = 2.06.$$

6. Though the t-statistic is not significant at the 0.05 level, from Table A.4,

$$P = P(|T| > 2.06) \approx 0.06.$$

As a result, there is some evidence that there is a difference in mean circulating levels of androgen.

The assumption of no interaction would imply that the effect on androgen levels of the deer is roughly the same in the data for both treatments, i.e., at the time of injection of succinylcholine and 30 minutes following injection. This can be expressed with the two factors switching roles; for example, the difference in treatments is roughly the same across the units (i.e., the deer). There certainly are some deer/treatment combinations for which the no interaction assumption seems to hold, but there is hardly any strong evidence that the experimental units are homogeneous. However, the nature of the interaction and the resulting increase in $\text{Var}(\bar{D})$ appear to be dominated by a substantial difference in the treatments.

Annotated Computer Printout for Paired t-Test

Figure 6.13 displays a *SAS* computer printout for a paired t-test using the data of Case Study 6.1. Notice that the printout looks like that for a single sample t-test and, of course, that is exactly what is accomplished, since the test seeks to determine if $\bar{d}$ is significantly different from zero.

		Analysis Variable : Diff		
N	Mean	Std Error	t Value	Pr > |t|
15	9.8480000	4.7698699	2.06	0.0580

Figure 6.13: *SAS* printout of paired t-test for data of Case Study 6.1.

Summary of Test Procedures

As we complete the formal development of tests on population means, we offer Table 6.3, which summarizes the test procedure for the cases of a single mean and two means. Notice the approximate procedure when distributions are normal and variances are unknown but not assumed to be equal. This statistic was introduced in Chapter 5.

Table 6.3: Tests Concerning Means

H_0	Value of Test Statistic	H_1	Critical Region
$\mu = \mu_0$	$z = \dfrac{\bar{x} - \mu_0}{\sigma/\sqrt{n}};$ σ known	$\mu < \mu_0$ $\mu > \mu_0$ $\mu \neq \mu_0$	$z < -z_\alpha$ $z > z_\alpha$ $z < -z_{\alpha/2}$ or $z > z_{\alpha/2}$
$\mu = \mu_0$	$t = \dfrac{\bar{x} - \mu_0}{s/\sqrt{n}};$ $v = n-1,$ σ unknown	$\mu < \mu_0$ $\mu > \mu_0$ $\mu \neq \mu_0$	$t < -t_\alpha$ $t > t_\alpha$ $t < -t_{\alpha/2}$ or $t > t_{\alpha/2}$
$\mu_1 - \mu_2 = d_0$	$z = \dfrac{(\bar{x}_1 - \bar{x}_2) - d_0}{\sqrt{\sigma_1^2/n_1 + \sigma_2^2/n_2}};$ σ_1 and σ_2 known	$\mu_1 - \mu_2 < d_0$ $\mu_1 - \mu_2 > d_0$ $\mu_1 - \mu_2 \neq d_0$	$z < -z_\alpha$ $z > z_\alpha$ $z < -z_{\alpha/2}$ or $z > z_{\alpha/2}$
$\mu_1 - \mu_2 = d_0$	$t = \dfrac{(\bar{x}_1 - \bar{x}_2) - d_0}{s_p\sqrt{1/n_1 + 1/n_2}};$ $v = n_1 + n_2 - 2,$ $\sigma_1 = \sigma_2$ but unknown, $s_p^2 = \dfrac{(n_1-1)s_1^2 + (n_2-1)s_2^2}{n_1 + n_2 - 2}$	$\mu_1 - \mu_2 < d_0$ $\mu_1 - \mu_2 > d_0$ $\mu_1 - \mu_2 \neq d_0$	$t < -t_\alpha$ $t > t_\alpha$ $t < -t_{\alpha/2}$ or $t > t_{\alpha/2}$
$\mu_1 - \mu_2 = d_0$	$t' = \dfrac{(\bar{x}_1 - \bar{x}_2) - d_0}{\sqrt{s_1^2/n_1 + s_2^2/n_2}};$ $v = \dfrac{(s_1^2/n_1 + s_2^2/n_2)^2}{\frac{(s_1^2/n_1)^2}{n_1-1} + \frac{(s_2^2/n_2)^2}{n_2-1}},$ $\sigma_1 \neq \sigma_2$ and unknown	$\mu_1 - \mu_2 < d_0$ $\mu_1 - \mu_2 > d_0$ $\mu_1 - \mu_2 \neq d_0$	$t' < -t_\alpha$ $t' > t_\alpha$ $t' < -t_{\alpha/2}$ or $t' > t_{\alpha/2}$
$\mu_D = d_0$ paired observations	$t = \dfrac{\bar{d} - d_0}{s_d/\sqrt{n}};$ $v = n-1$	$\mu_D < d_0$ $\mu_D > d_0$ $\mu_D \neq d_0$	$t < -t_\alpha$ $t > t_\alpha$ $t < -t_{\alpha/2}$ or $t > t_{\alpha/2}$

6.6   Choice of Sample Size for Testing Means

In Section 6.2, we demonstrated how the analyst can exploit relationships among the sample size, the significance level α, and the power of the test to achieve a certain standard of quality. In most practical circumstances, the experiment should be planned, with a choice of sample size made prior to the data-taking process if possible. The sample size is usually determined to achieve good power for a fixed α and fixed specific alternative. This fixed alternative may be in the form of $\mu - \mu_0$ in the case of a hypothesis involving a single mean or $\mu_1 - \mu_2$ in the case of a problem involving two means. Specific cases will provide illustrations.

Suppose that we wish to test the hypothesis

$$H_0: \ \mu = \mu_0,$$
$$H_1: \ \mu > \mu_0,$$

with a significance level α, when the variance σ^2 is known. For a specific alternative, say $\mu = \mu_0 + \delta$, the power of our test is shown in Figure 6.14 to be

$$1 - \beta = P(\bar{X} > a \text{ when } \mu = \mu_0 + \delta).$$

Figure 6.14 shows that the α probability of a Type I error corresponds to a value (a in the figure) of the sample mean. If the sample mean is below this value, the null hypothesis cannot be rejected. But, there is a probability of β that the specific alternative hypothesis would give a sample mean less than a—this is the probability of a Type II error.

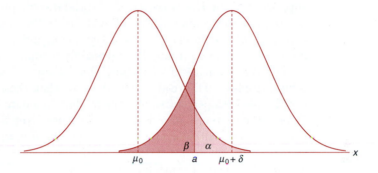

Figure 6.14: Testing $\mu = \mu_0$ versus $\mu = \mu_0 + \delta$.

Therefore,

$$\beta = P(\bar{X} < a \text{ when } \mu = \mu_0 + \delta)$$

$$= P\left[\frac{\bar{X} - (\mu_0 + \delta)}{\sigma/\sqrt{n}} < \frac{a - (\mu_0 + \delta)}{\sigma/\sqrt{n}} \text{ when } \mu = \mu_0 + \delta \right].$$

Under the alternative hypothesis $\mu = \mu_0 + \delta$, the statistic

$$\frac{\bar{X} - (\mu_0 + \delta)}{\sigma/\sqrt{n}}$$

is the standard normal variable Z. So

$$\beta = P\left(Z < \frac{a - \mu_0}{\sigma/\sqrt{n}} - \frac{\delta}{\sigma/\sqrt{n}} \right) = P\left(Z < z_\alpha - \frac{\delta}{\sigma/\sqrt{n}} \right),$$

from which we conclude that

$$-z_\beta = z_\alpha - \frac{\delta\sqrt{n}}{\sigma},$$

and hence

$$\text{Choice of sample size:} \quad n = \frac{(z_\alpha + z_\beta)^2 \sigma^2}{\delta^2},$$

a result that is also true when the alternative hypothesis is $\mu < \mu_0$.

In the case of a two-tailed test, we obtain the power $1 - \beta$ for a specified alternative when

$$n \approx \frac{(z_{\alpha/2} + z_\beta)^2 \sigma^2}{\delta^2}.$$

6.7   Graphical Methods for Comparing Means

In Chapter 4, considerable attention was directed to displaying data in graphical form, such as stem-and-leaf plots and box-and-whisker plots. Many computer software packages produce graphical displays. As we proceed to other forms of data analysis (e.g., regression analysis and analysis of variance), graphical methods become even more informative.

Graphical aids cannot be used as a replacement for the test procedure itself. Certainly, the value of the test statistic indicates the proper type of evidence in support of H_0 or H_1. However, a pictorial display provides a good illustration and is often a better communicator of evidence to the beneficiary of the analysis. Also, a picture will often clarify why a significant difference was found. Failure of an important assumption may be exposed by a summary type of graphical tool.

For the comparison of means, side-by-side box-and-whisker plots provide a telling display. The reader should recall that these plots display the 25th percentile, the 75th percentile, and the median in a data set. In addition, the whiskers display the extremes in a data set. Consider Exercise 6.40 at the end of this section. Plasma ascorbic acid levels were measured in two groups of pregnant women, smokers and nonsmokers. Figure 6.15 shows the box-and-whisker plots for both groups of women. Two things are very apparent. Taking into account variability, there appears to be a negligible difference in the sample means. In addition, the variability in the two groups appears to be somewhat different. Of course, the analyst must keep in mind the rather sizable differences between the sample sizes in this case.

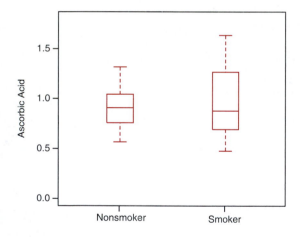

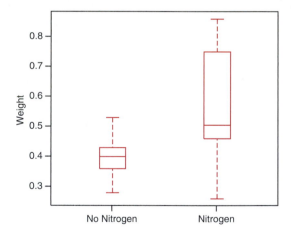

Figure 6.15: Two box-and-whisker plots of plasma ascorbic acid in smokers and nonsmokers.

Figure 6.16: Two box-and-whisker plots of seedling data.

Consider Exercise 5.28 in Section 5.9. Figure 6.16 shows the multiple box-and-whisker plot for the data on 20 seedlings, half given nitrogen and half not given nitrogen. The display reveals a smaller variability for the group that did not receive nitrogen. In addition, the lack of overlap of the box plots suggests a significant difference between the mean stem weights for the two groups. It would appear that the presence of nitrogen increases the stem weights and perhaps increases the variability in the weights.

There are no certain rules of thumb regarding when two box-and-whisker plots give evidence of significant difference between the means. However, a rough guideline is that if the 25th percentile line for one sample exceeds the median line for the other sample, there is strong evidence of a difference between means.

More emphasis is placed on graphical methods in a real-life case study presented later in this chapter.

Annotated Computer Printout for Two-Sample t-Test

Consider once again Exercise 5.28 on page 221, where seedling data under conditions of nitrogen and no nitrogen were collected. Test

$$H_0: \mu_{\text{NIT}} = \mu_{\text{NON}},$$
$$H_1: \mu_{\text{NIT}} > \mu_{\text{NON}},$$

where the population means indicate mean weights. Figure 6.17 is an annotated computer printout generated using the *SAS* package. Notice that sample standard deviation and standard error are shown for both samples. The t-statistics under the assumption of equal variance and unequal variance are both given. From the box-and-whisker plot of Figure 6.16 it would certainly appear that the equal variance assumption is violated. A P-value of 0.0229 suggests a conclusion of unequal means. This concurs with the diagnostic information given in Figure 6.17. Incidentally, notice that t and t' are equal in this case, since $n_1 = n_2$.

```
                    TTEST Procedure
Variable Weight
      Mineral     N    Mean   Std Dev Std Err
No nitrogen      10  0.3990    0.0728  0.0230
    Nitrogen     10  0.5650    0.1867  0.0591


Variances        DF    t Value     Pr > |t|
    Equal        18       2.62       0.0174
  Unequal      11.7       2.62       0.0229


          Test the Equality of Variances
Variable    Num DF    Den DF    F Value    Pr > F
  Weight         9         9       6.58    0.0098
```

Figure 6.17: *SAS* printout for two-sample t-test.

Exercises

6.19 In a research report, Richard H. Weindruch of the UCLA Medical School claims that mice with an average life span of 32 months will live to be about 40 months old when 40% of the calories in their diet are replaced by vitamins and protein. Is there any reason to believe that $\mu < 40$ if 64 mice that are placed on this diet have an average life of 38 months with a standard deviation of 5.8 months? Use a P-value in your conclusion.

6.20 A random sample of 64 bags of white cheddar popcorn weighed, on average, 5.23 ounces with a standard deviation of 0.24 ounce. Test the hypothesis that $\mu = 5.5$ ounces against the alternative hypothesis, $\mu < 5.5$ ounces, at the 0.05 level of significance.

6.21 An electrical firm manufactures light bulbs that have a lifetime that is approximately normally distributed with a mean of 800 hours and a standard deviation of 40 hours. Test the hypothesis that $\mu = 800$ hours against the alternative, $\mu \neq 800$ hours, if a random sample of 30 bulbs has an average life of 788 hours. Use a P-value in your answer.

6.22 In the American Heart Association journal *Hypertension*, researchers report that individuals who practice Transcendental Meditation (TM) lower their blood pressure significantly. If a random sample of 225 male TM practitioners meditate for 8.5 hours per week with a standard deviation of 2.25 hours, does that suggest that, on average, men who use TM meditate more than 8 hours per week? Quote a P-value in your conclusion.

6.23 Test the hypothesis that the average content of containers of a particular lubricant is 10 liters if the contents of a random sample of 10 containers are 10.2, 9.7, 10.1, 10.3, 10.1, 9.8, 9.9, 10.4, 10.3, and 9.8 liters. Use a 0.01 level of significance and assume that the distribution of contents is normal.

6.24 The average height of females in the freshman class of a certain college has historically been 162.5 centimeters with a standard deviation of 6.9 centimeters. Is there reason to believe that there has been a change in the average height if a random sample of 50 females in the present freshman class has an average height of 165.2 centimeters? Use a P-value in your conclusion. Assume the standard deviation remains the same.

6.25 It is claimed that automobiles are driven on average more than 20,000 kilometers per year. To test this claim, 100 randomly selected automobile owners are asked to keep a record of the kilometers they travel. Would you reject this claim if the random sample showed an average of 23,500 kilometers and a standard deviation of 3900 kilometers? Use a P-value in your conclusion.

6.26 According to a dietary study, high sodium intake may be related to ulcers, stomach cancer, and migraine headaches. The human requirement for salt is only 220 milligrams per day, which is surpassed in most single servings of ready-to-eat cereals. If a random sample of 20 similar servings of a certain cereal has a mean sodium content of 244 milligrams and a standard deviation of 24.5 milligrams, does this suggest at the 0.05 level of significance that the average sodium content for a single serving of such cereal is greater than 220 milligrams? Assume the distribution of sodium contents to be normal.

6.27 A study at the University of Colorado at Boulder shows that running increases the percent resting metabolic rate (RMR) in older women. The average RMR of 30 elderly women runners was 34.0% higher than the average RMR of 30 sedentary elderly women, and the standard deviations were reported to be 10.5 and 10.2%, respectively. Was there a significant increase in RMR of the women runners over the sedentary women? Assume the populations to be approximately normally distributed with equal variances. Use a P-value in your conclusions.

6.28 According to *Chemical Engineering*, an important property of fiber is its water absorbency. The average percent absorbency of 25 randomly selected pieces of cotton fiber was found to be 20 with a standard deviation of 1.5. A random sample of 25 pieces of acetate yielded an average percent of 12 with a standard deviation of 1.25. Is there strong evidence that the population mean percent absorbency is significantly higher for cotton fiber than for acetate? Assume that the percent absorbency is approximately normally distributed and that the population variances in percent absorbency for the two fibers are the same. Use a significance level of 0.05.

6.29 Past experience indicates that the time required for high school seniors to complete a standardized test is a normal random variable with a mean of 35 minutes. If a random sample of 20 high school seniors took an average of 33.1 minutes to complete this test with a standard deviation of 4.3 minutes, test the hypothesis, at the 0.05 level of significance, that $\mu = 35$ minutes against the alternative that $\mu < 35$ minutes.

6.30 A random sample of size $n_1 = 25$, taken from a normal population with a standard deviation $\sigma_1 = 5.2$, has a mean $\bar{x}_1 = 81$. A second random sample of size $n_2 = 36$, taken from a different normal population with a standard deviation $\sigma_2 = 3.4$, has a mean $\bar{x}_2 = 76$. Test the hypothesis that $\mu_1 = \mu_2$ against the alternative, $\mu_1 \neq \mu_2$. Quote a P-value in your conclusion.

6.31 A manufacturer claims that the average tensile strength of thread A exceeds the average tensile strength of thread B by at least 12 kilograms. To test this claim, 50 pieces of each type of thread were tested under similar conditions. Type A thread had an average tensile strength of 86.7 kilograms with a standard deviation of 6.28 kilograms, while type B thread had an average tensile strength of 77.8 kilograms with a standard deviation of 5.61 kilograms. Test the manufacturer's claim using a 0.05 level of significance.

6.32 *Amstat News* (December 2004) lists median salaries for associate professors of statistics at research institutions and at liberal arts and other institutions in the United States. Assume that a sample of 200 associate professors from research institutions has an average salary of $70,750 per year with a standard deviation of $6000. Assume also that a sample of 200 associate professors from other types of institutions has an average salary of $65,200 with a standard deviation of $5000. Test the hypothesis that the mean salary for associate professors in research institutions is $2000 higher than for those in other institutions. Use a 0.01 level of significance.

6.33 A study was conducted to see if increasing the substrate concentration has an appreciable effect on the velocity of a chemical reaction. With a substrate concentration of 1.5 moles per liter, the reaction was run 15 times, with an average velocity of 7.5 micromoles per 30 minutes and a standard deviation of 1.5. With a substrate concentration of 2.0 moles per liter, 12 runs were made, yielding an average velocity of 8.8 micromoles per 30 minutes and a sample standard deviation of 1.2. Is there any reason to believe that this increase in substrate concentration causes an increase in the mean velocity of the reaction of more than 0.5 micromole per 30 minutes? Use a 0.01 level of significance and assume the populations to be approximately normally distributed with equal variances.

6.34 A study was made to determine if the subject matter in a physics course is better understood when a lab constitutes part of the course. Students were randomly selected to participate in either a 3-semester-hour course without labs or a 4-semester-hour course with labs. In the section with labs, 11 students made an average grade of 85 with a standard deviation of 4.7, and in the section without labs, 17 students made an average grade of 79 with a standard deviation of 6.1. Would you say that the laboratory course increases the average grade by as much as 8 points? Use a P-value in your conclusion and assume the populations to be approximately normally distributed with equal variances.

6.35 To find out whether a new serum will arrest leukemia, 9 mice, all with an advanced stage of the disease, are selected. Five mice receive the treatment and 4 do not. Survival times, in years, from the time the experiment commenced are as follows:

Treatment	2.1	5.3	1.4	4.6	0.9
No Treatment	1.9	0.5	2.8	3.1	

At the 0.05 level of significance, can the serum be said to be effective? Assume the two populations to be normally distributed with equal variances.

6.36 Engineers at a large automobile manufacturing company are trying to decide whether to purchase brand A or brand B tires for the company's new models. To help them arrive at a decision, an experiment is conducted using 12 of each brand. The tires are run until they wear out. The results are as follows:

$$\text{Brand } A: \quad \bar{x}_1 = 37{,}900 \text{ kilometers},$$
$$s_1 = 5100 \text{ kilometers}.$$
$$\text{Brand } B: \quad \bar{x}_2 = 39{,}800 \text{ kilometers},$$
$$s_2 = 5900 \text{ kilometers}.$$

Test the hypothesis that there is no difference in the average wear of the two brands of tires. Assume the populations to be approximately normally distributed with equal variances. Use a P-value.

6.37 In Exercise 5.30 on page 222, test the hypothesis that the fuel economy of Volkswagen mini-trucks, on average, exceeds that of similarly equipped Toyota mini-trucks by 4 kilometers per liter. Use a 0.10 level of significance.

6.38 A UCLA researcher claims that the average life span of mice can be extended by as much as 8 months when the calories in their diet are reduced by approximately 40% from the time they are weaned. The restricted diets are enriched to normal levels by vitamins and protein. Suppose that a random sample of 10 mice is fed a normal diet and has an average life span of 32.1 months with a standard deviation of 3.2 months, while a random sample of 15 mice is fed the restricted diet and has an average life span of 37.6 months with a standard deviation of 2.8 months. Test the hypothesis, at the 0.05 level of significance, that the average life span of mice on this restricted diet is increased by 8 months against the alternative that the increase is less than 8 months. Assume the distributions of life spans for the regular and restricted diets are approximately normal with equal variances.

6.39 The following data represent the running times of films produced by two motion-picture companies:

Company	Time (minutes)						
1	102	86	98	109	92		
2	81	165	97	134	92	87	114

Test the hypothesis that the average running time of films produced by company 2 exceeds the average running time of films produced by company 1 by 10 minutes against the one-sided alternative that the difference is less than 10 minutes. Use a 0.1 level of significance and assume the distributions of times to be approximately normal with unequal variances.

6.40 In a study conducted at Virginia Tech, the plasma ascorbic acid levels of pregnant women were compared for smokers versus nonsmokers. Thirty-two women in the last three months of pregnancy, free of major health disorders and ranging in age from 15 to 32 years, were selected for the study. Prior to the collection of 20 mL of blood, the participants were told to avoid breakfast, forgo their vitamin supplements, and avoid foods high in ascorbic acid content. From the blood samples, the following plasma ascorbic acid values were determined, in milligrams per 100 milliliters:

Plasma Ascorbic Acid Values

Nonsmokers		Smokers
0.97	1.16	0.48
0.72	0.86	0.71
1.00	0.85	0.98
0.81	0.58	0.68
0.62	0.57	1.18
1.32	0.64	1.36
1.24	0.98	0.78
0.99	1.09	1.64
0.90	0.92	
0.74	0.78	
0.88	1.24	
0.94	1.18	

Is there sufficient evidence to conclude that there is a difference between plasma ascorbic acid levels of smokers and nonsmokers? Assume that the two sets of data came from normal populations with unequal variances. Use a P-value.

6.41 A study was conducted by the Department of Zoology at Virginia Tech to determine if there is a significant difference in the density of organisms at two different stations located on Cedar Run, a secondary stream in the Roanoke River drainage basin. Sewage from a sewage treatment plant and overflow from the Federal Mogul Corporation settling pond enter the stream near its headwaters. The following data give the density measurements, in number of organisms per square meter, at the two collecting stations:

Number of Organisms per Square Meter

Station 1		Station 2	
5030	4980	2800	2810
13,700	11,910	4670	1330
10,730	8130	6890	3320
11,400	26,850	7720	1230
860	17,660	7030	2130
2200	22,800	7330	2190
4250	1130		
15,040	1690		

Can we conclude, at the 0.05 level of significance, that the average densities at the two stations are equal? Assume that the observations come from normal populations with different variances.

6.42 Five samples of a ferrous-type substance were used to determine if there is a difference between a laboratory chemical analysis and an X-ray fluorescence analysis of the iron content. Each sample was split into two subsamples and the two types of analysis were applied. Following are the coded data showing the iron content analysis:

Analysis	Sample				
	1	2	3	4	5
X-ray	2.0	2.0	2.3	2.1	2.4
Chemical	2.2	1.9	2.5	2.3	2.4

Assuming that the populations are normal, test at the 0.05 level of significance whether the two methods of analysis give, on the average, the same result.

6.43 According to published reports, practice under fatigued conditions distorts mechanisms that govern performance. An experiment was conducted using 15 college males, who were trained to make a continuous horizontal right-to-left arm movement from a microswitch to a barrier, knocking over the barrier coincident with the arrival of a clock sweephand at the 6 o'clock position. The absolute value of the difference between the time, in milliseconds, that it took to knock over the barrier and the time for the sweephand to reach the 6 o'clock position (500 msec) was recorded. Each participant performed the task five times under prefatigue and postfatigue conditions, and the sums of the absolute differences for the five performances were recorded.

Absolute Time Differences

Subject	Prefatigue	Postfatigue
1	158	91
2	92	59
3	65	215
4	98	226
5	33	223
6	89	91
7	148	92
8	58	177
9	142	134
10	117	116
11	74	153
12	66	219
13	109	143
14	57	164
15	85	100

An increase in the mean absolute time difference when the task is performed under postfatigue conditions would support the claim that practice under fatigued conditions distorts mechanisms that govern performance. Assuming the populations to be normally distributed, test this claim.

6.44 In a study conducted by the Department of Human Nutrition and Foods at Virginia Tech, the following data were recorded on sorbic acid residuals, in parts per million, in ham immediately after being dipped in a sorbate solution and after 60 days of storage:

Sorbic Acid Residuals in Ham

Slice	Before Storage	After Storage
1	224	116
2	270	96
3	400	239
4	444	329
5	590	437
6	660	597
7	1400	689
8	680	576

Assuming the populations to be normally distributed, is there sufficient evidence, at the 0.05 level of significance, to say that the length of storage influences sorbic acid residual concentrations?

6.45 A taxi company manager is trying to decide whether the use of radial tires instead of regular belted tires improves fuel economy. Twelve cars were equipped with radial tires and driven over a prescribed test course. Without changing drivers, the same cars were then equipped with regular belted tires and driven once again over the test course. The gasoline consumption, in kilometers per liter, was recorded as follows:

Kilometers per Liter

Car	Radial Tires	Belted Tires
1	4.2	4.1
2	4.7	4.9
3	6.6	6.2
4	7.0	6.9
5	6.7	6.8
6	4.5	4.4
7	5.7	5.7
8	6.0	5.8
9	7.4	6.9
10	4.9	4.7
11	6.1	6.0
12	5.2	4.9

Can we conclude that cars equipped with radial tires give better fuel economy than those equipped with belted tires? Assume the populations to be normally distributed. Use a P-value in your conclusion.

6.46 In Review Exercise 5.60 on page 230, use the t-distribution to test the hypothesis that the diet reduces a woman's weight by 4.5 kilograms on average against the alternative hypothesis that the mean difference in weight is less than 4.5 kilograms. Use a P-value.

6.47 How large a sample is required in Exercise 6.20 if the power of the test is to be 0.90 when the true mean is 5.20? Assume that $\sigma = 0.24$.

6.48 If the distribution of life spans in Exercise 6.19 is approximately normal, how large a sample is required if the probability of committing a type II error is to be 0.1 when the true mean is 35.9 months? Assume that $\sigma = 5.8$ months.

6.49 How large a sample is required in Exercise 6.24 if the power of the test is to be 0.95 when the true average height differs from 162.5 by 3.1 centimeters? Use $\alpha = 0.02$.

6.50 How large should the samples be in Exercise 6.31 if the power of the test is to be 0.95 when the true difference between thread types A and B is 8 kilograms?

6.51 How large a sample is required in Exercise 6.22 if the power of the test is to be 0.8 when the true mean meditation time exceeds the hypothesized value by 1.2σ? Use $\alpha = 0.05$.

6.52 Nine subjects were used in an experiment to determine if exposure to carbon monoxide has an impact on breathing capability. The data were collected by personnel in the Health and Physical Education Department at Virginia Tech and were analyzed in the Statistics Consulting Center at Hokie Land. The subjects were exposed to breathing chambers, one of which contained a high concentration of CO. Breathing frequency measures were made for each subject for each chamber. The subjects were exposed to the breathing chambers in random sequence. The data give the breathing frequency, in number of breaths taken per minute. Make a one-sided test of the hypothesis that mean breathing frequency is the same for the two environments. Use $\alpha = 0.05$. Assume that breathing frequency is approximately normal.

Subject	With CO	Without CO
1	30	30
2	45	40
3	26	25
4	25	23
5	34	30
6	51	49
7	46	41
8	32	35
9	30	28

6.53 A study was conducted at the Department of Veterinary Medicine at Virginia Tech to determine if the "strength" of a wound from surgical incision is affected by the temperature of the knife. Eight dogs were used in the experiment. "Hot" and "cold" incisions were made on the abdomen of each dog, and the strength was measured. The resulting data appear below.

Dog	Knife	Strength
1	Hot	5120
1	Cold	8200
2	Hot	10,000
2	Cold	8600
3	Hot	10,000
3	Cold	9200
4	Hot	10,000
4	Cold	6200
5	Hot	10,000
5	Cold	10,000
6	Hot	7900
6	Cold	5200
7	Hot	510
7	Cold	885
8	Hot	1020
8	Cold	460

(a) Write an appropriate hypothesis to determine if there is a significant difference in strength between the hot and cold incisions.

(b) Test the hypothesis using a paired t-test. Use a P-value in your conclusion.

6.54 For testing

$$H_0: \ \mu = 14,$$
$$H_1: \ \mu \neq 14,$$

an $\alpha = 0.05$ level t-test is being considered. What sample size is necessary in order for the probability to be 0.1 of falsely failing to reject H_0 when the true population mean differs from 14 by 0.5? From a preliminary sample we estimate σ to be 1.25.

6.8   One Sample: Test on a Single Proportion

Tests of hypotheses concerning proportions are required in many areas. Politicians are certainly interested in knowing what fraction of the voters will favor them in the next election. All manufacturing firms are concerned about the proportion of defective items when a shipment is made. Gamblers depend on a knowledge of the proportion of outcomes that they consider favorable.

We shall consider the problem of testing the hypothesis that the proportion of successes in a binomial experiment equals some specified value. That is, we are testing the null hypothesis H_0 that $p = p_0$, where p is the parameter of the binomial distribution. The alternative hypothesis may be one of the usual one-sided or two-sided alternatives:

$$p < p_0, \quad p > p_0, \quad \text{or} \quad p \neq p_0.$$

The appropriate random variable on which we base our decision criterion is the binomial random variable X, although we could just as well use the statistic $\hat{p} = X/n$. Values of X that are far from the mean $\mu = np_0$ will lead to the rejection of the null hypothesis. Because X is a discrete binomial variable, it is unlikely that a critical region can be established whose size is *exactly* equal to a prespecified value of α. For this reason it is preferable, in dealing with small samples, to base our decisions on P-values. To test the hypothesis

$$H_0: \ p = p_0,$$
$$H_1: \ p < p_0,$$

we use the binomial distribution to compute the P-value

$$P = P(X \leq x \text{ when } p = p_0).$$

The value x is the number of successes in our sample of size n. If this P-value is less than or equal to α, our test is significant at the α level and we reject H_0 in favor of H_1. Similarly, to test the hypothesis

$$H_0: \ p = p_0,$$
$$H_1: \ p > p_0,$$

at the α-level of significance, we compute

$$P = P(X \geq x \text{ when } p = p_0)$$

and reject H_0 in favor of H_1 if this P-value is less than or equal to α. Finally, to test the hypothesis

$$H_0: \ p = p_0,$$
$$H_1: \ p \neq p_0,$$

at the α-level of significance, we compute

$$P = 2P(X \leq x \text{ when } p = p_0) \qquad \text{if } x < np_0$$

or

$$P = 2P(X \geq x \text{ when } p = p_0) \qquad \text{if } x > np_0$$

and reject H_0 in favor of H_1 if the computed P-value is less than or equal to α.

The steps for testing a null hypothesis about a proportion against various alternatives using the binomial probabilities of Table A.1 are as follows:

Testing a Proportion (Small Samples)	1. H_0: $p = p_0$. 2. One of the alternatives H_1: $p < p_0$, $p > p_0$, or $p \neq p_0$. 3. Choose a level of significance equal to α. 4. Test statistic: Binomial variable X with $p = p_0$. 5. Computations: Find x, the number of successes, and compute the appropriate P-value. 6. Decision: Draw appropriate conclusions based on the P-value.

Example 6.7: A builder claims that heat pumps are installed in 70% of all homes being constructed today in the city of Richmond, Virginia. Would you agree with this claim if a random survey of new homes in this city showed that 8 out of 15 had heat pumps installed? Use a 0.10 level of significance.

Solution: 1. H_0: $p = 0.7$.

2. H_1: $p \neq 0.7$.

3. $\alpha = 0.10$.

4. Test statistic: Binomial variable X with $p = 0.7$ and $n = 15$.

5. Computations: $x = 8$ and $np_0 = (15)(0.7) = 10.5$. Therefore, from Table A.1, the computed P-value is

$$P = 2P(X \leq 8 \text{ when } p = 0.7) = 2 \sum_{x=0}^{8} b(x; 15, 0.7) = 0.2622 > 0.10.$$

6. Decision: Do not reject H_0. Conclude that there is insufficient reason to doubt the builder's claim.　◢

In Section 3.2, we saw that binomial probabilities can be obtained from the actual binomial formula or from Table A.1 when n is small. For large n, approximation procedures are required. When the hypothesized value p_0 is very close to 0 or 1, the Poisson distribution, with parameter $\mu = np_0$, may be used. However, the normal curve approximation, with parameters $\mu = np_0$ and $\sigma^2 = np_0q_0$, is usually preferred for large n and is very accurate as long as p_0 is not extremely close to 0 or to 1. If we use the normal approximation, the **z-value for testing $p = p_0$** is given by

$$z = \frac{x - np_0}{\sqrt{np_0q_0}} = \frac{\hat{p} - p_0}{\sqrt{p_0q_0/n}},$$

which is a value of the standard normal variable Z. Hence, for a two-tailed test at the α-level of significance, the critical region is $z < -z_{\alpha/2}$ or $z > z_{\alpha/2}$. For the one-sided alternative $p < p_0$, the critical region is $z < -z_\alpha$, and for the alternative $p > p_0$, the critical region is $z > z_\alpha$.

6.9　Two Samples: Tests on Two Proportions

Situations often arise where we wish to test the hypothesis that two proportions are equal. For example, we might want to show evidence that the proportion of doctors who are pediatricians in one state is equal to the proportion in another state. A person may decide to give up smoking only if he or she is convinced that the proportion of smokers with lung cancer exceeds the proportion of nonsmokers with lung cancer.

In general, we wish to test the null hypothesis that two proportions, or binomial parameters, are equal. That is, we are testing $p_1 = p_2$ against one of the alternatives $p_1 < p_2$, $p_1 > p_2$, or $p_1 \neq p_2$. Of course, this is equivalent to testing the null hypothesis that $p_1 - p_2 = 0$ against one of the alternatives $p_1 - p_2 < 0$, $p_1 - p_2 > 0$, or $p_1 - p_2 \neq 0$. The statistic on which we base our decision is the random variable $\hat{P}_1 - \hat{P}_2$. Independent samples of sizes n_1 and n_2 are selected at random from two binomial populations and the proportions of successes $\hat{P}_1$ and $\hat{P}_2$ for the two samples are computed.

In our construction of confidence intervals for p_1 and p_2 we noted, for n_1 and n_2 sufficiently large, that the point estimator $\hat{P}_1$ minus $\hat{P}_2$ was approximately normally distributed with mean

$$\mu_{\hat{P}_1 - \hat{P}_2} = p_1 - p_2$$

and variance

$$\sigma^2_{\hat{P}_1 - \hat{P}_2} = \frac{p_1 q_1}{n_1} + \frac{p_2 q_2}{n_2}.$$

Therefore, our critical region(s) can be established by using the standard normal variable

$$Z = \frac{(\hat{P}_1 - \hat{P}_2) - (p_1 - p_2)}{\sqrt{p_1 q_1/n_1 + p_2 q_2/n_2}}.$$

When H_0 is true, we can substitute $p_1 = p_2 = p$ and $q_1 = q_2 = q$ (where p and q are the common values) in the preceding formula for Z to give the form

$$Z = \frac{\widehat{P}_1 - \widehat{P}_2}{\sqrt{pq(1/n_1 + 1/n_2)}}.$$

To compute a value of Z, however, we must estimate the parameters p and q that appear in the radical. Upon pooling the data from both samples, the **pooled estimate of the proportion** p is

$$\hat{p} = \frac{x_1 + x_2}{n_1 + n_2},$$

where x_1 and x_2 are the numbers of successes in each of the two samples. Substituting $\hat{p}$ for p and $\hat{q} = 1 - \hat{p}$ for q, the **z-value for testing $p_1 = p_2$** is determined from the formula

$$z = \frac{\hat{p}_1 - \hat{p}_2}{\sqrt{\hat{p}\hat{q}(1/n_1 + 1/n_2)}}.$$

The critical regions for the appropriate alternative hypotheses are set up as before, using critical points of the standard normal curve. Hence, for the alternative $p_1 \neq p_2$ at the α-level of significance, the critical region is $z < -z_{\alpha/2}$ or $z > z_{\alpha/2}$. For a test where the alternative is $p_1 < p_2$, the critical region is $z < -z_\alpha$, and when the alternative is $p_1 > p_2$, the critical region is $z > z_\alpha$.

Example 6.8: A vote is to be taken among the residents of a town and the surrounding county to determine whether a proposed chemical plant should be constructed. The construction site is within the town limits, and for this reason many voters in the county believe that the proposal will pass because of the large proportion of town voters who favor the construction. To determine if there is a significant difference in the proportions of town voters and county voters favoring the proposal, a poll is taken. If 120 of 200 town voters favor the proposal and 240 of 500 county residents favor it, would you agree that the proportion of town voters favoring the proposal is higher than the proportion of county voters? Use an $\alpha = 0.05$ level of significance.

Solution: Let p_1 and p_2 be the true proportions of voters in the town and county, respectively, favoring the proposal.

1. H_0: $p_1 = p_2$.

2. H_1: $p_1 > p_2$.

3. $\alpha = 0.05$.

4. Critical region: $z > 1.645$.

5. Computations:

$$\hat{p}_1 = \frac{x_1}{n_1} = \frac{120}{200} = 0.60, \quad \hat{p}_2 = \frac{x_2}{n_2} = \frac{240}{500} = 0.48, \quad \text{and}$$

$$\hat{p} = \frac{x_1 + x_2}{n_1 + n_2} = \frac{120 + 240}{200 + 500} = 0.51.$$

Therefore,

$$z = \frac{0.60 - 0.48}{\sqrt{(0.51)(0.49)(1/200 + 1/500)}} = 2.9,$$

$$P = P(Z > 2.9) = 0.0019.$$

6. Decision: Reject H_0 and agree that the proportion of town voters favoring the proposal is higher than the proportion of county voters.   ⏌

Exercises

6.55 A marketing expert for a pasta-making company believes that 40% of pasta lovers prefer lasagna. If 9 out of 20 pasta lovers choose lasagna over other pastas, what can be concluded about the expert's claim? Use a 0.05 level of significance.

6.56 Suppose that, in the past, 40% of all adults favored capital punishment. Do we have reason to believe that the proportion of adults favoring capital punishment has increased if, in a random sample of 15 adults, 8 favor capital punishment? Use a 0.05 level of significance.

6.57 A new radar device is being considered for a certain missile defense system. The system is checked by experimenting with aircraft in which a kill or a no kill is simulated. If, in 300 trials, 250 kills occur, accept or reject, at the 0.04 level of significance, the claim that the probability of a kill with the new system does not exceed the 0.8 probability of the existing device.

6.58 It is believed that at least 60% of the residents in a certain area favor an annexation suit by a neighboring city. What conclusion would you draw if only 110 in a sample of 200 voters favored the suit? Use a 0.05 level of significance.

6.59 A fuel oil company claims that one-fifth of the homes in a certain city are heated by oil. Do we have reason to believe that fewer than one-fifth are heated by oil if, in a random sample of 1000 homes in this city, 136 are heated by oil? Use a P-value in your conclusion.

6.60 At a certain college, it is estimated that at most 25% of the students ride bicycles to class. Does this seem to be a valid estimate if, in a random sample of 90 college students, 28 are found to ride bicycles to class? Use a 0.05 level of significance.

6.61 In a winter of an epidemic flu, the parents of 2000 babies were surveyed by researchers at a well-known pharmaceutical company to determine if the company's new medicine was effective after two days. Among 120 babies who had the flu and were given the medicine, 29 were cured within two days. Among 280 babies who had the flu but were not given the medicine, 56 recovered within two days. Is there any significant indication that supports the company's claim of the effectiveness of the medicine?

6.62 In a controlled laboratory experiment, scientists at the University of Minnesota discovered that 25% of a certain strain of rats subjected to a 20% coffee bean diet and then force-fed a powerful cancer-causing chemical later developed cancerous tumors. Would we have reason to believe that the proportion of rats developing tumors when subjected to this diet has increased if the experiment were repeated and 16 of 48 rats developed tumors? Use a 0.05 level of significance.

6.63 In a study to estimate the proportion of residents in a certain city and its suburbs who favor the construction of a nuclear power plant, it is found that 63 of 100 urban residents favor the construction while only 59 of 125 suburban residents are in favor. Is there a significant difference between the proportions of urban and suburban residents who favor construction of the nuclear plant? Make use of a P-value.

6.64 In a study on the fertility of married women conducted by Martin O'Connell and Carolyn C. Rogers for the Census Bureau in 1979, two groups of childless married women aged 25 to 29 were selected at random, and each was asked if she eventually planned to have a child. One group was selected from among women married less than two years and the other from among women married five years. Suppose that 240 of the 300 women married less than two years planned to have children some day compared to 288 of the 400 women married five years. Can we conclude that the proportion of women married less than two years who planned to have children is significantly higher than the proportion of women married five years? Make use of a *P*-value.

6.65 An urban community would like to show that the incidence of breast cancer is higher in their area than in a nearby rural area. (PCB levels were found to be higher in the soil of the urban community.) If it is found that 20 of 200 adult women in the urban community have breast cancer and 10 of 150 adult women in the rural community have breast cancer, can we conclude at the 0.05 level of significance that breast cancer is more prevalent in the urban community?

6.66 Group Project: The class should be divided into pairs of students for this project. Suppose it is conjectured that at least 25% of students at your university exercise for more than two hours a week. Collect data from a random sample of 50 students. Ask each student if he or she works out for at least two hours per week. Then do the computations that allow either rejection or nonrejection of the above conjecture. Show all work and quote a *P*-value in your conclusion.

6.10   Goodness-of-Fit Test

Throughout this chapter, we have been concerned with the testing of statistical hypotheses about single population parameters such as μ and p. Now we shall consider a test to determine if a population has a specified theoretical distribution. The test is based on how good a fit we have between the frequency of occurrence of observations in an observed sample and the expected frequencies obtained from the hypothesized distribution.

To illustrate, we consider the tossing of a die. We hypothesize that the die is honest, which is equivalent to testing the hypothesis that the distribution of outcomes is the discrete uniform distribution

$$f(x) = \frac{1}{6}, \quad x = 1, 2, \ldots, 6.$$

Suppose that the die is tossed 120 times and each outcome is recorded. Theoretically, if the die is balanced, we would expect each face to occur 20 times. The results are given in Table 6.4.

Table 6.4: Observed and Expected Frequencies of 120 Tosses of a Die

Face:	1	2	3	4	5	6
Observed	20	22	17	18	19	24
Expected	20	20	20	20	20	20

By comparing the observed frequencies with the corresponding expected frequencies, we must decide whether these discrepancies are likely to occur as a result of sampling fluctuations and the die is balanced or whether the die is not honest and the distribution of outcomes is not uniform. It is common practice to refer to each possible outcome of an experiment as a cell. In our illustration, we have 6 cells. The appropriate statistic on which we base our decision criterion for an experiment involving k cells is defined by the following.

Goodness-of-Fit Test

A **goodness-of-fit test** between observed and expected frequencies is based on the quantity

$$\chi^2 = \sum_{i=1}^{k} \frac{(o_i - e_i)^2}{e_i},$$

where χ^2 is a value of a random variable whose sampling distribution is approximated very closely by the chi-squared distribution with $v = k - 1$ degrees of freedom. The symbols o_i and e_i represent the observed and expected frequencies, respectively, for the ith cell.

The number of degrees of freedom associated with the chi-squared distribution used here is equal to $k - 1$, since there are only $k - 1$ freely determined cell frequencies. That is, once $k - 1$ cell frequencies are determined, so is the frequency for the kth cell.

If the observed frequencies are close to the corresponding expected frequencies, the χ^2-value will be small, indicating a good fit. If the observed frequencies differ considerably from the expected frequencies, the χ^2-value will be large and the fit is poor. A good fit leads to the acceptance of H_0, whereas a poor fit leads to its rejection. The critical region will, therefore, fall in the right tail of the chi-squared distribution. For a level of significance equal to α, we find the critical value χ_α^2 from Table A.5, and then $\chi^2 > \chi_\alpha^2$ constitutes the critical region. **The decision criterion described here should not be used unless each of the expected frequencies is at least equal to 5.** This restriction may require the combining of adjacent cells, resulting in a reduction in the number of degrees of freedom.

From Table 6.4, we find the χ^2-value to be

$$\chi^2 = \frac{(20 - 20)^2}{20} + \frac{(22 - 20)^2}{20} + \frac{(17 - 20)^2}{20}$$
$$+ \frac{(18 - 20)^2}{20} + \frac{(19 - 20)^2}{20} + \frac{(24 - 20)^2}{20} = 1.7.$$

Using Table A.5, we find $\chi_{0.05}^2 = 11.070$ for $v = 5$ degrees of freedom. Since 1.7 is less than the critical value, we fail to reject H_0. We conclude that there is insufficient evidence that the die is not balanced.

As a second illustration, let us test the hypothesis that the frequency distribution of battery lives given in Table 4.5 on page 187 may be approximated by a normal distribution with mean $\mu = 3.5$ and standard deviation $\sigma = 0.7$. The expected frequencies for the 7 classes (cells), listed in Table 6.5, are obtained by computing the areas under the hypothesized normal curve that fall between the various class boundaries.

For example, the z-values corresponding to the boundaries of the fourth class are

$$z_1 = \frac{2.95 - 3.5}{0.7} = -0.79 \quad \text{and} \quad z_2 = \frac{3.45 - 3.5}{0.7} = -0.07.$$

Table 6.5: Observed and Expected Frequencies of Battery Lives, Assuming Normality

Class Boundaries	o_i		e_i	
1.45–1.95	2 ⎫		0.5 ⎫	
1.95–2.45	1 ⎬ 7		2.1 ⎬ 8.5	
2.45–2.95	4 ⎭		5.9 ⎭	
2.95–3.45	15		10.3	
3.45–3.95	10		10.7	
3.95–4.45	5 ⎫ 8		7.0 ⎫ 10.5	
4.45–4.95	3 ⎭		3.5 ⎭	

From Table A.3 we find the area between $z_1 = -0.79$ and $z_2 = -0.07$ to be

$$\text{area} = P(-0.79 < Z < -0.07) = P(Z < -0.07) - P(Z < -0.79)$$
$$= 0.4721 - 0.2148 = 0.2573.$$

Hence, the expected frequency for the fourth class is

$$e_4 = (0.2573)(40) = 10.3.$$

It is customary to round these frequencies to one decimal.

The expected frequency for the first class interval is obtained by using the total area under the normal curve to the left of the boundary 1.95. For the last class interval, we use the total area to the right of the boundary 4.45. All other expected frequencies are determined by the method described for the fourth class. Note that we have combined adjacent classes in Table 6.5 where the expected frequencies are less than 5 (a rule of thumb in the goodness-of-fit test). Consequently, the total number of intervals is reduced from 7 to 4, resulting in $v = 3$ degrees of freedom. The χ^2-value is then given by

$$\chi^2 = \frac{(7 - 8.5)^2}{8.5} + \frac{(15 - 10.3)^2}{10.3} + \frac{(10 - 10.7)^2}{10.7} + \frac{(8 - 10.5)^2}{10.5} = 3.05.$$

Since the computed χ^2-value is less than $\chi^2_{0.05} = 7.815$ for 3 degrees of freedom, we have no reason to reject the null hypothesis and conclude that the normal distribution with $\mu = 3.5$ and $\sigma = 0.7$ provides a good fit for the distribution of battery lives.

The chi-squared goodness-of-fit test is an important resource, particularly since so many statistical procedures in practice depend, in a theoretical sense, on the assumption that the data gathered come from a specific type of distribution. As we have already seen, the normality assumption is often made. In the chapters that follow, we shall continue to make normality assumptions in order to provide a theoretical basis for certain tests and confidence intervals.

There are tests in the literature that are more powerful than the chi-squared test for testing normality. One such test is called **Geary's test**. This test is based on a very simple statistic which is a ratio of two estimators of the population standard deviation σ. Suppose that a random sample $X_1, X_2, \ldots, X_n$ is taken from a normal distribution, $N(\mu, \sigma)$. Consider the ratio

$$U = \frac{\sqrt{\pi/2} \sum\limits_{i=1}^{n} |X_i - \bar{X}|/n}{\sqrt{\sum\limits_{i=1}^{n} (X_i - \bar{X})^2/n}}.$$

The reader should recognize that the denominator is a reasonable estimator of σ whether the distribution is normal or not. The numerator is a good estimator of σ if the distribution is normal but may overestimate or underestimate σ when there are departures from normality. Thus, values of U differing considerably from 1.0 represent the signal that the hypothesis of normality should be rejected.

For large samples, a reasonable test is based on approximate normality of U. The test statistic is then a standardization of U, given by

$$Z = \frac{U - 1}{0.2661/\sqrt{n}}.$$

Of course, the test procedure involves the two-sided critical region. We compute a value of z from the data and do not reject the hypothesis of normality when

$$-z_{\alpha/2} < Z < z_{\alpha/2}.$$

A paper dealing with Geary's test is cited in the Bibliography (Geary, 1947).

6.11   Test for Independence (Categorical Data)

The chi-squared test procedure discussed in Section 6.10 can also be used to test the hypothesis of independence of two variables of classification. Suppose that we wish to determine whether the opinions of the voting residents of the state of Illinois concerning a new tax reform proposal are independent of their levels of income. Members of a random sample of 1000 registered voters from the state of Illinois are classified as to whether they are in a low, medium, or high income bracket and whether or not they favor the tax reform. The observed frequencies are presented in Table 6.6, which is known as a **contingency table**.

Table 6.6: 2 × 3 Contingency Table

Tax Reform	Income Level			Total
	Low	Medium	High	
For	182	213	203	598
Against	154	138	110	402
Total	336	351	313	1000

A contingency table with r rows and c columns is referred to as an $r \times c$ table ("$r \times c$" is read "r by c"). The row and column totals in Table 6.6 are called **marginal frequencies**. Our decision to accept or reject the null hypothesis, H_0, of independence between a voter's opinion concerning the tax reform and his or her level of income is based upon how good a fit we have between the observed frequencies in each of the 6 cells of Table 6.6 and the frequencies that we would expect for each cell under the assumption that H_0 is true. To find these expected frequencies, let us define the following events:

L: A person selected is in the low-income level.

M: A person selected is in the medium-income level.

H: A person selected is in the high-income level.

F: A person selected is for the tax reform.

A: A person selected is against the tax reform.

By using the marginal frequencies, we can list the following probability estimates:

$$P(L) = \frac{336}{1000}, \qquad P(M) = \frac{351}{1000}, \qquad P(H) = \frac{313}{1000},$$

$$P(F) = \frac{598}{1000}, \qquad P(A) = \frac{402}{1000}.$$

Now, if H_0 is true and the two variables are independent, we should have

$$P(L \cap F) \;=\; P(L)P(F) = \left(\frac{336}{1000}\right)\left(\frac{598}{1000}\right),$$

$$P(L \cap A) \;=\; P(L)P(A) = \left(\frac{336}{1000}\right)\left(\frac{402}{1000}\right),$$

$$P(M \cap F) = P(M)P(F) = \left(\frac{351}{1000}\right)\left(\frac{598}{1000}\right),$$

$$P(M \cap A) = P(M)P(A) = \left(\frac{351}{1000}\right)\left(\frac{402}{1000}\right),$$

$$P(H \cap F) \;=\; P(H)P(F) = \left(\frac{313}{1000}\right)\left(\frac{598}{1000}\right),$$

$$P(H \cap A) \;=\; P(H)P(A) = \left(\frac{313}{1000}\right)\left(\frac{402}{1000}\right).$$

The expected frequencies are obtained by multiplying each cell probability by the total number of observations. As before, we round these frequencies to one decimal. Thus, the expected number of low-income voters in our sample who favor the tax reform is estimated to be

$$\left(\frac{336}{1000}\right)\left(\frac{598}{1000}\right)(1000) = \frac{(336)(598)}{1000} = 200.9$$

when H_0 is true. The general rule for obtaining the expected frequency of any cell is given by the following formula:

$$\text{expected frequency} = \frac{\text{(column total)} \times \text{(row total)}}{\text{grand total}}.$$

The expected frequency for each cell is recorded in parentheses beside the actual observed value in Table 6.7. Note that the expected frequencies in any row or column add up to the appropriate marginal total. In our example, we need to compute only two expected frequencies in the top row of Table 6.7 and then find the others by subtraction. The number of degrees of freedom associated with the chi-squared test used here is equal to the number of cell frequencies that may be filled in freely when we are given the marginal totals and the grand total, and in this illustration that number is 2. A simple formula providing the correct number of degrees of freedom is

$$v = (r - 1)(c - 1).$$

Table 6.7: Observed and Expected Frequencies

Tax Reform	Income Level			Total
	Low	Medium	High	
For	182 (200.9)	213 (209.9)	203 (187.2)	598
Against	154 (135.1)	138 (141.1)	110 (125.8)	402
Total	336	351	313	1000

Hence, for our example, $v = (2 - 1)(3 - 1) = 2$ degrees of freedom. To test the null hypothesis of independence, we use the following decision criterion.

Test for Independence   Calculate

$$\chi^2 = \sum_i \frac{(o_i - e_i)^2}{e_i},$$

where the summation extends over all rc cells in the $r \times c$ contingency table.

If $\chi^2 > \chi_\alpha^2$ with $v = (r - 1)(c - 1)$ degrees of freedom, reject the null hypothesis of independence at the α-level of significance; otherwise, fail to reject the null hypothesis.

Applying this criterion to our example, we find that

$$\chi^2 = \frac{(182 - 200.9)^2}{200.9} + \frac{(213 - 209.9)^2}{209.9} + \frac{(203 - 187.2)^2}{187.2}$$
$$+ \frac{(154 - 135.1)^2}{135.1} + \frac{(138 - 141.1)^2}{141.1} + \frac{(110 - 125.8)^2}{125.8} = 7.85,$$
$$P \approx 0.02.$$

From Table A.5 we find that $\chi_{0.05}^2 = 5.991$ for $v = (2 - 1)(3 - 1) = 2$ degrees of freedom. The null hypothesis is rejected and we conclude that a voter's opinion concerning the tax reform and his or her level of income are not independent.

It is important to remember that the statistic on which we base our decision has a distribution that is only approximated by the chi-squared distribution. The computed χ^2-values depend on the cell frequencies and consequently are discrete. The continuous chi-squared distribution seems to approximate the discrete sampling distribution of χ^2 very well, provided that the number of degrees of freedom is greater than 1. In a 2×2 contingency table, where we have only 1 degree of freedom, a correction called **Yates' correction for continuity** is applied. The corrected formula then becomes

$$\chi^2(\text{corrected}) = \sum_i \frac{(|o_i - e_i| - 0.5)^2}{e_i}.$$

If the expected cell frequencies are large, the corrected and uncorrected results are almost the same. When the expected frequencies are between 5 and 10, Yates' correction should be applied. For expected frequencies less than 5, the Fisher-Irwin exact test should be used. A discussion of this test may be found in *Basic Concepts of Probability and Statistics* by Hodges and Lehmann (2005; see the Bibliography). The Fisher-Irwin test may be avoided, however, by choosing a larger sample.

6.12   Test for Homogeneity

When we tested for independence in Section 6.11, a random sample of 1000 voters was selected and the row and column totals for our contingency table were determined by chance. Another type of problem for which the method of Section 6.11 applies is one in which either the row or the column totals are predetermined. Suppose, for example, that we decide in advance to select 200 Democrats, 150 Republicans, and 150 Independents from the voters of the state of North Carolina and record whether they are for a proposed abortion law, against it, or undecided. The observed responses are given in Table 6.8.

Table 6.8: Observed Frequencies

| Abortion Law | Political Affiliation | | | |
	Democrat	Republican	Independent	Total
For	82	70	62	214
Against	93	62	67	222
Undecided	25	18	21	64
Total	200	150	150	500

Now, rather than test for independence, we test the hypothesis that the population proportions within each row are the same. That is, we test the hypothesis that the proportions of Democrats, Republicans, and Independents favoring the abortion law are the same; the proportions of each political affiliation against the law are the same; and the proportions of each political affiliation that are undecided are the same. We are basically interested in determining whether the three categories of voters are **homogeneous** with respect to their opinions concerning the proposed abortion law. Such a test is called a test for homogeneity.

Assuming homogeneity, we again find the expected cell frequencies by multiplying the corresponding row and column totals and then dividing by the grand total. The analysis then proceeds using the same chi-squared statistic as before. We illustrate this process for the data of Table 6.8 in the following example.

Example 6.9: Referring to the data of Table 6.8, test the hypothesis that opinions concerning the proposed abortion law are the same within each political affiliation. Use a 0.05 level of significance.

Solution: 1. H_0: For each opinion, the proportions of Democrats, Republicans, and Independents are the same.

2. H_1: For at least one opinion, the proportions of Democrats, Republicans, and Independents are not the same.

3. $\alpha = 0.05$.

4. Critical region: $\chi^2 > 9.488$ with $v = 4$ degrees of freedom.

5. Computations: Using the expected cell frequency formula on page 281, we need to compute 4 cell frequencies. All other frequencies are found by subtraction. The observed and expected cell frequencies are displayed in Table 6.9.

Table 6.9: Observed and Expected Frequencies

	Political Affiliation			
Abortion Law	**Democrat**	**Republican**	**Independent**	**Total**
For	82 (85.6)	70 (64.2)	62 (64.2)	214
Against	93 (88.8)	62 (66.6)	67 (66.6)	222
Undecided	25 (25.6)	18 (19.2)	21 (19.2)	64
Total	200	150	150	500

Now,

$$\chi^2 = \frac{(82 - 85.6)^2}{85.6} + \frac{(70 - 64.2)^2}{64.2} + \frac{(62 - 64.2)^2}{64.2}$$
$$+ \frac{(93 - 88.8)^2}{88.8} + \frac{(62 - 66.6)^2}{66.6} + \frac{(67 - 66.6)^2}{66.6}$$
$$+ \frac{(25 - 25.6)^2}{25.6} + \frac{(18 - 19.2)^2}{19.2} + \frac{(21 - 19.2)^2}{19.2}$$
$$= 1.53.$$

6. Decision: Do not reject H_0. There is insufficient evidence to conclude that the proportions of Democrats, Republicans, and Independents differ for each stated opinion.

Testing for Several Proportions

The chi-squared statistic for testing for homogeneity is also applicable when testing the hypothesis that k binomial parameters have the same value. This is, therefore, an extension of the test presented in Section 6.9 for determining differences between two proportions to a test for determining differences among k proportions. Hence, we are interested in testing the null hypothesis

$$H_0: \; p_1 = p_2 = \cdots = p_k$$

against the alternative hypothesis, H_1, that the population proportions are *not all equal*. To perform this test, we first observe independent random samples of size $n_1, n_2, \ldots, n_k$ from the k populations and arrange the data in a $2 \times k$ contingency table, Table 6.10.

Table 6.10: k Independent Binomial Samples

Sample:	1	2	$\cdots$	k
Successes	x_1	x_2	$\cdots$	x_k
Failures	$n_1 - x_1$	$n_2 - x_2$	$\cdots$	$n_k - x_k$

Depending on whether the sizes of the random samples were predetermined or occurred at random, the test procedure is identical to the test for homogeneity or the test for independence. Therefore, the expected cell frequencies are calculated as before and substituted, together with the observed frequencies, into the chi-squared statistic

$$\chi^2 = \sum_i \frac{(o_i - e_i)^2}{e_i},$$

with

$$v = (2 - 1)(k - 1) = k - 1$$

degrees of freedom.

By selecting the appropriate upper-tail critical region of the form $\chi^2 > \chi_\alpha^2$, we can now reach a decision concerning H_0.

Example 6.10: In a shop study, a set of data was collected to determine whether or not the proportion of defectives produced was the same for workers on the day, evening, and night shifts. The data collected are shown in Table 6.11.

Table 6.11: Data for Example 6.10

Shift:	Day	Evening	Night
Defectives	45	55	70
Nondefectives	905	890	870

Use a 0.025 level of significance to determine if the proportion of defectives is the same for all three shifts.

Solution: Let $p_1, p_2,$ and p_3 represent the true proportions of defectives for the day, evening, and night shifts, respectively.

1. H_0: $p_1 = p_2 = p_3$.

2. H_1: $p_1, p_2,$ and p_3 are not all equal.

3. $\alpha = 0.025$.

4. Critical region: $\chi^2 > 7.378$ for $v = 2$ degrees of freedom.

5. Computations: Corresponding to the observed frequencies $o_1 = 45$ and $o_2 = 55$, we find

$$e_1 = \frac{(950)(170)}{2835} = 57.0 \quad \text{and} \quad e_2 = \frac{(945)(170)}{2835} = 56.7.$$

Table 6.12: Observed and Expected Frequencies

Shift:	Day	Evening	Night	Total
Defectives	45 (57.0)	55 (56.7)	70 (56.3)	170
Nondefectives	905 (893.0)	890 (888.3)	870 (883.7)	2665
Total	950	945	940	2835

All other expected frequencies are found by subtraction and are displayed in Table 6.12.

Now

$$\chi^2 = \frac{(45-57.0)^2}{57.0} + \frac{(55-56.7)^2}{56.7} + \frac{(70-56.3)^2}{56.3}$$
$$+ \frac{(905-893.0)^2}{893.0} + \frac{(890-888.3)^2}{888.3} + \frac{(870-883.7)^2}{883.7} = 6.29,$$

$P \approx 0.04.$

6. Decision: We do not reject H_0 at $\alpha = 0.025$. Nevertheless, with the above P-value computed, it would certainly be dangerous to conclude that the proportion of defectives produced is the same for all shifts. ⌐

Often a complete study involving the use of statistical methods in hypothesis testing can be illustrated for the scientist or engineer using both test statistics, complete with P-values and statistical graphics. The graphics supplement the numerical diagnostics with pictures that show intuitively why the P-values appear as they do, as well as how reasonable (or not) the operative assumptions are.

6.13   Two-Sample Case Study

In this section, we consider a study involving a thorough graphical and formal analysis, along with annotated computer printout and conclusions. In a data analysis study conducted by personnel at the Statistics Consulting Center at Virginia Tech, two different materials, alloy A and alloy B, were compared in terms of breaking strength. Alloy B is more expensive, but it should certainly be adopted if it can be shown to be stronger than alloy A. The consistency of performance of the two alloys should also be taken into account.

Random samples of beams made from each alloy were selected, and strength was measured in units of 0.001-inch deflection as a fixed force was applied at both ends of the beam. Twenty specimens were used for each of the two alloys. The data are given in Table 6.13.

It is important that the engineer compare the two alloys. Of concern is average strength and reproducibility. Figure 6.18 shows two box-and-whisker plots on the same graph. The box-and-whisker plots suggest that there is similar, albeit somewhat different, variability of deflection for the two alloys. However, it seems that the mean deflection for alloy B is significantly smaller, suggesting, at least graphically, that alloy B is stronger. The sample means and standard deviations

Table 6.13: Data for Two-Sample Case Study

Alloy A			Alloy B		
88	82	87	75	81	80
79	85	90	77	78	81
84	88	83	86	78	77
89	80	81	84	82	78
81	85		80	80	
83	87		78	76	
82	80		83	85	
79	78		76	79	

are

$$\bar{y}_A = 83.55, \quad s_A = 3.663; \qquad \bar{y}_B = 79.70, \quad s_B = 3.097.$$

The *SAS* printout for the PROC TTEST is shown in Figure 6.19. The *F*-test suggests no significant difference in variances with $P = 0.4709$ (for the *F*-test in testing equal variances, the reader is referred to Walpole et al., 2011), and the two-sample *t*-statistic for testing

$$H_0: \ \mu_A = \mu_B,$$
$$H_1: \ \mu_A > \mu_B$$

($t = 3.59$, $P = 0.0009$) rejects H_0 in favor of H_1 and thus confirms what the graphical information suggests. Here we use the *t*-test that pools the two-sample variances together in light of the results of the *F*-test. On the basis of this analysis, the adoption of alloy *B* would seem to be in order.

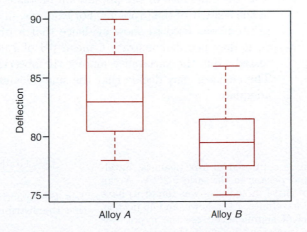

Figure 6.18: Box-and-whisker plots for both alloys.

```
                     The TTEST Procedure
        Alloy        N    Mean   Std Dev  Std Err
        Alloy A      20   83.55  3.6631   0.8191
        Alloy B      20   79.7   3.0967   0.6924


        Variances        DF    t Value    Pr > |t|
        Equal            38       3.59      0.0009
        Unequal          37       3.59      0.0010
                   Equality of Variances
        Num DF      Den DF     F Value    Pr > F
          19          19         1.40     0.4709
```

Figure 6.19: Annotated *SAS* printout for alloy data.

Statistical Significance and Engineering or Scientific Significance

While the statistician may feel quite comfortable with the results of the comparison between the two alloys in the case study above, a dilemma remains for the engineer. The analysis demonstrated a statistically significant improvement with the use of alloy B. However, is the difference found really worth pursuing, since alloy B is more expensive? This illustration highlights a very important issue often overlooked by statisticians and data analysts—*the distinction between statistical significance and engineering or scientific significance.* Here the average difference in deflection is $\bar{y}_A - \bar{y}_B = 0.00385$ inch. In a complete analysis, the engineer must determine if the difference is sufficient to justify the extra cost in the long run. This is an economic and engineering issue. The reader should understand that a statistically significant difference merely implies that the difference in the sample means found in the data could hardly have occurred by chance. It does not imply that the difference in the population means is profound or particularly significant in the context of the problem. For example, in Section 6.4, an annotated computer printout was used to show evidence that a pH meter was, in fact, biased. That is, it does not demonstrate a mean pH of 7.00 for the material on which it was tested. But the variability among the observations in the sample is very small. The engineer may decide that the small deviations from 7.0 render the pH meter adequate.

Exercises

6.67 A machine is supposed to mix peanuts, hazelnuts, cashews, and pecans in the ratio 5:2:2:1. A can containing 500 of these mixed nuts was found to have 269 peanuts, 112 hazelnuts, 74 cashews, and 45 pecans. At the 0.05 level of significance, test the hypothesis that the machine is mixing the nuts in the ratio 5:2:2:1.

6.68 The grades in a statistics course for a particular semester were as follows:

Grade	A	B	C	D	F
f	14	18	32	20	16

Test the hypothesis, at the 0.05 level of significance, that the distribution of grades is uniform.

6.69 A die is tossed 180 times with the following results:

x	1	2	3	4	5	6
f	28	36	36	30	27	23

Is this a balanced die? Use a 0.01 level of significance.

6.70 Three marbles are selected from an urn containing 5 red marbles and 3 green marbles. After the number X of red marbles is recorded, the marbles are replaced in the urn and the experiment repeated 112 times. The results obtained are as follows:

x	0	1	2	3
f	1	31	55	25

Test the hypothesis, at the 0.05 level of significance, that the recorded data may be fitted by the hypergeometric distribution $h(x; 8, 3, 5)$, $x = 0, 1, 2, 3$.

6.71 A coin is tossed until a head occurs and the number X of tosses recorded. After repeating the experiment 256 times, we obtained the following results:

x	1	2	3	4	5	6	7	8
f	136	60	34	12	9	1	3	1

Test the hypothesis, at the 0.05 level of significance, that the observed distribution of X may be fitted by the geometric distribution $g(x; 1/2)$, $x = 1, 2, 3, \ldots$.

6.72 The following scores represent the final examination grades for an elementary statistics course:

```
23  60  79  32  57  74  52  70  82
36  80  77  81  95  41  65  92  85
55  76  52  10  64  75  78  25  80
98  81  67  41  71  83  54  64  72
88  62  74  43  60  78  89  76  84
48  84  90  15  79  34  67  17  82
69  74  63  80  85  61
```

Test the goodness of fit between the observed class frequencies and the corresponding expected frequencies of a normal distribution with $\mu = 65$ and $\sigma = 21$, using a 0.05 level of significance.

6.73 For Exercise 4.49 on page 192, test the goodness of fit between the observed class frequencies and the corresponding expected frequencies of a normal distribution with $\mu = 1.8$ and $\sigma = 0.4$, using a 0.01 level of significance.

6.74 In an experiment to study the dependence of hypertension on smoking habits, the following data were taken on 180 individuals:

	Non-smokers	Moderate Smokers	Heavy Smokers
Hypertension	21	36	30
No hypertension	48	26	19

Test the hypothesis that the presence or absence of hypertension is independent of smoking habits. Use a 0.05 level of significance.

6.75 A random sample of 90 adults is classified according to gender and the number of hours of television watched during a week:

	Gender	
	Male	Female
Over 25 hours	15	29
Under 25 hours	27	19

Use a 0.01 level of significance and test the hypothesis that the time spent watching television is independent of whether the viewer is male or female.

6.76 A random sample of 200 married men, all retired, was classified according to education and number of children:

	Number of Children		
Education	0–1	2–3	Over 3
Elementary	14	37	32
Secondary	19	42	17
College	12	17	10

Test the hypothesis, at the 0.05 level of significance, that the size of a family is independent of the level of education attained by the father.

6.77 A criminologist conducted a survey to determine whether the incidence of certain types of crime varied from one part of a large city to another. The particular crimes of interest were assault, burglary, larceny, and homicide. The following table shows the numbers of crimes committed in four areas of the city during the past year.

	Type of Crime			
District	Assault	Burglary	Larceny	Homicide
1	162	118	451	18
2	310	196	996	25
3	258	193	458	10
4	280	175	390	19

Can we conclude from these data at the 0.01 level of significance that the occurrence of these types of crime is dependent on the city district?

6.78 According to a Johns Hopkins University study published in the *American Journal of Public Health*, widows live longer than widowers. Consider the following survival data collected on 100 widows and 100 widowers following the death of a spouse:

Years Lived	Widow	Widower
Less than 5	25	39
5 to 10	42	40
More than 10	33	21

Can we conclude at the 0.05 level of significance that the proportions of widows and widowers are equal with respect to the different time periods that a spouse survives after the death of his or her mate?

6.79 The following responses concerning the standard of living at the time of an independent opinion poll of 1000 households versus one year earlier seem to be in agreement with the results of a study published in *Across the Board* (June 1981):

	Standard of Living			
Period	Somewhat Better	Same	Not as Good	Total
1980: Jan.	72	144	84	300
May	63	135	102	300
Sept.	47	100	53	200
1981: Jan.	40	105	55	200

Test the hypothesis that the proportions of households within each standard of living category are the same for each of the four time periods. Use a *P*-value.

6.80 A college infirmary conducted an experiment to determine the degree of relief provided by three cough remedies. Each cough remedy was tried on 50 students and the following data recorded:

	Cough Remedy		
	NyQuil	Robitussin	Triaminic
No relief	11	13	9
Some relief	32	28	27
Total relief	7	9	14

Test the hypothesis that the three cough remedies are equally effective. Use a *P*-value in your conclusion.

6.81 To determine current attitudes about prayer in public schools, a survey was conducted in four Virginia counties. The following table gives the attitudes of 200 parents from Craig County, 150 parents from Giles County, 100 parents from Franklin County, and 100 parents from Montgomery County:

	County			
Attitude	Craig	Giles	Franklin	Mont.
Favor	65	66	40	34
Oppose	42	30	33	42
No opinion	93	54	27	24

Test for homogeneity of attitudes among the four counties concerning prayer in the public schools. Use a *P*-value in your conclusion.

6.82 A survey was conducted in Indiana, Kentucky, and Ohio to determine the attitude of voters concerning school busing. A poll of 200 voters from each of these states yielded the following results:

	Voter Attitude		
State	Support	Do Not Support	Undecided
Indiana	82	97	21
Kentucky	107	66	27
Ohio	93	74	33

At the 0.05 level of significance, test the null hypothesis that the proportions of voters within each attitude category are the same for each of the three states.

6.83 A survey was conducted in two Virginia cities to determine voter sentiment about two gubernatorial candidates in an upcoming election. Five hundred voters were randomly selected from each city and the following data were recorded:

	City	
Voter Sentiment	Richmond	Norfolk
Favor *A*	204	225
Favor *B*	211	198
Undecided	85	77

At the 0.05 level of significance, test the null hypothesis that proportions of voters favoring candidate *A*, favoring candidate *B*, and undecided are the same for each city.

6.84 In a study to estimate the proportion of married women who regularly watch soap operas, it is found that 52 of 200 married women in Denver, 31 of 150 married women in Phoenix, and 37 of 150 married women in Rochester watch at least one soap opera. Use a 0.05 level of significance to test the hypothesis that there is no difference among the true proportions of married women who watch soap operas in these three cities.

Review Exercises

6.85 State the null and alternative hypotheses to be used in testing the following claims and determine generally where the critical region is located:

(a) The mean snowfall at Lake George during the month of February is 21.8 centimeters.

(b) No more than 20% of the faculty at the local university contributed to the annual giving fund.

(c) On the average, children attend schools within 6.2 kilometers of their homes in suburban St. Louis.

(d) At least 70% of next year's new cars will be in the compact and subcompact category.

(e) The proportion of voters favoring the incumbent in the upcoming election is 0.58.

(f) The average rib-eye steak at the Longhorn Steak

house weighs at least 340 grams.

6.86 A geneticist is interested in the proportions of males and females in a population who have a certain minor blood disorder. In a random sample of 100 males, 31 are found to be afflicted, whereas only 24 of 100 females tested have the disorder. Can we conclude at the 0.01 level of significance that the proportion of men in the population afflicted with this blood disorder is significantly greater than the proportion of women afflicted?

6.87 A study was made to determine whether more Italians than Americans prefer white champagne to pink champagne at weddings. Of the 300 Italians selected at random, 72 preferred white champagne, and of the 400 Americans selected, 70 preferred white champagne. Can we conclude that a higher proportion of Italians than Americans prefer white champagne at weddings? Use a 0.05 level of significance.

6.88 Consider the situation of Exercise 6.52 on page 271. Along with breathing frequency, oxygen consumption in mL/kg/min was also measured.

Subject	With CO	Without CO
1	26.46	25.41
2	17.46	22.53
3	16.32	16.32
4	20.19	27.48
5	19.84	24.97
6	20.65	21.77
7	28.21	28.17
8	33.94	32.02
9	29.32	28.96

It is conjectured that oxygen consumption should be higher in an environment relatively free of CO. Do a significance test and discuss the conjecture.

6.89 In a study analyzed by the Statistics Consulting Center at Virginia Tech, a group of subjects was asked to complete a certain task on the computer. The response measured was the time to completion. The purpose of the experiment was to test a set of facilitation tools developed by the Department of Computer Science at the university. There were 10 subjects involved. With a random assignment, five were given a standard procedure using Fortran language for completion of the task. The other five were asked to do the task with the use of the facilitation tools. The data on the completion times for the task are given here.

Group 1 (Standard Procedure)	Group 2 (Facilitation Tool)
161	132
169	162
174	134
158	138
163	133

Assuming that the population distributions are normal and variances are the same for the two groups, support or refute the conjecture that the facilitation tools increase the speed with which the task can be accomplished.

6.90 State the null and alternative hypotheses to be used in testing the following claims, and determine generally where the critical region is located:

(a) At most, 20% of next year's wheat crop will be exported to the Soviet Union.

(b) On the average, American homemakers drink 3 cups of coffee per day.

(c) The proportion of college graduates in Virginia this year who majored in the social sciences is at least 0.15.

(d) The average donation to the American Lung Association is no more than $10.

(e) Residents in suburban Richmond commute, on the average, 15 kilometers to their place of employment.

6.91 If one can containing 500 nuts is selected at random from each of three different distributors of mixed nuts and there are, respectively, 345, 313, and 359 peanuts in each of the cans, can we conclude at the 0.01 level of significance that the mixed nuts of the three distributors contain equal proportions of peanuts?

6.92 A study was made to determine whether there is a difference between the proportions of parents in the states of Maryland (MD), Virginia (VA), Georgia (GA), and Alabama (AL) who favor placing Bibles in the elementary schools. The responses of 100 parents selected at random in each of these states are recorded in the following table:

Preference	State			
	MD	VA	GA	AL
Yes	65	71	78	82
No	35	29	22	18

Can we conclude that the proportions of parents who favor placing Bibles in the schools are the same for these four states? Use a 0.01 level of significance.

6.93 A study was conducted at the Virginia-Maryland Regional College of Veterinary Medicine Equine Center to determine if the performance of a certain type of surgery on young horses had any effect on certain kinds of blood cell types in the animal. Fluid samples were taken from each of six foals before and after surgery. The samples were analyzed for the number of postoperative white blood cell (WBC) leukocytes. A preoperative measure of WBC leukocytes was also taken. The data are given as follows:

Foal	Presurgery*	Postsurgery*
1	10.80	10.60
2	12.90	16.60
3	9.59	17.20
4	8.81	14.00
5	12.00	10.60
6	6.07	8.60

*All values $\times 10^{-3}$.

Use a paired sample t-test to determine if there is a significant change in WBC leukocytes with the surgery.

6.94 A study was conducted at the Department of Health and Physical Education at Virginia Tech to determine if 8 weeks of training truly reduces the cholesterol levels of the participants. A treatment group consisting of 15 people was given lectures twice a week on how to reduce cholesterol level. Another group of 18 people of similar age was randomly selected as a control group. All participants' cholesterol levels were recorded at the end of the 8-week program and are listed below.

Treatment:
    129 131 154 172 115 126 175 191
    122 238 159 156 176 175 126
Control:
    151 132 196 195 188 198 187 168 115
    165 137 208 133 217 191 193 140 146

Can we conclude, at the 5% level of significance, that the average cholesterol level has been reduced due to the program? Make the appropriate test on means.

6.95 In a study conducted by the Department of Mechanical Engineering and analyzed by the Statistics Consulting Center at Virginia Tech, steel rods supplied by two different companies were compared. Ten sample springs were made out of the steel rods supplied by each company, and the "bounciness" was studied. The data are as follows:

Company A:
 9.3 8.8 6.8 8.7 8.5 6.7 8.0 6.5 9.2 7.0
Company B:
11.0 9.8 9.9 10.2 10.1 9.7 11.0 11.1 10.2 9.6

Can you conclude that there is virtually no difference

in means between the steel rods supplied by the two companies? Use a P-value to reach your conclusion. Should variances be pooled here?

6.96 In a study conducted by the Water Resources Center and analyzed by the Statistics Consulting Center at Virginia Tech, two different wastewater treatment plants were compared. Plant A is located where the median household income is below $22,000 a year, and plant B is located where the median household income is above $60,000 a year. The amount of wastewater treated at each plant (thousands of gallons/day) was randomly sampled for 10 days. The data are as follows:

Plant A:
21 19 20 23 22 28 32 19 13 18

Plant B:
20 39 24 33 30 28 30 22 33 24

Can we conclude, at the 5% level of significance, that the average amount of wastewater treated at the plant in the high-income neighborhood is more than that treated at the plant in the low-income area? Assume normality.

6.97 The following data show the numbers of defects in 100,000 lines of code in two versions of a particular type of software program, one developed in the United States and one in Japan. Is there enough evidence to claim that there is a significant difference between the programs developed in the two countries? Test on means. Should variances be pooled?

U.S.	48 39 42 52 40 48 52 52
	54 48 52 55 43 46 48 52
Japan	50 48 42 40 43 48 50 46
	38 38 36 40 40 48 48 45

6.98 Studies show that the concentration of PCBs is much higher in malignant breast tissue than in normal breast tissue. If a study of 50 women with breast cancer reveals an average PCB concentration of 22.8×10^{-4} gram, with a standard deviation of 4.8×10^{-4} gram, is the mean concentration of PCBs less than 24×10^{-4} gram?

6.14   Potential Misconceptions and Hazards; Relationship to Material in Other Chapters

One of the easiest ways to misuse statistics relates to the final scientific conclusion drawn when the analyst does not reject the null hypothesis H_0. In this text, we have attempted to make clear what the null hypothesis means and what the alternative means, and to stress that, in a large sense, the alternative hypothesis is much more important. Put in the form of an example, if an engineer is attempt-ing to compare two gauges using a two-sample t-test, and H_0 is "the gauges are equivalent" while H_1 is "the gauges are not equivalent," not rejecting H_0 does not lead to the conclusion of equivalent gauges. In fact, a case can be made for never writing or saying "accept H_0"! Not rejecting H_0 merely implies insufficient evidence. Depending on the nature of the hypothesis, a lot of possibilities are still not ruled out.

In Chapter 5, we considered the case of the large-sample confidence interval using

$$z = \frac{\bar{x} - \mu}{s/\sqrt{n}}.$$

In hypothesis testing, replacing σ by s for $n < 30$ is risky. If $n \geq 30$ and the distribution is not normal but somehow close to normal, the Central Limit Theorem is being called upon and one is relying on the fact that with $n \geq 30$, $s \approx \sigma$. Of course, any t-test is accompanied by the concomitant assumption of normality. As in the case of confidence intervals, the t-test is relatively robust to normality.

Most of the chapters in this text include discussions whose purpose is to relate the chapter in question to other material that will follow. The topics of estimation and hypothesis testing are both used in a major way in nearly all of the techniques that fall under the umbrella of "statistical methods." This will be readily noted by students who advance to Chapters 7 through 9.

Chapter 7

Linear Regression

7.1 Introduction to Linear Regression

Often, in practice, one is called upon to solve problems involving sets of variables when it is known that there exists some inherent relationship among the variables. For example, in an industrial situation it may be known that the tar content in the outlet stream in a chemical process is related to the inlet temperature. It may be of interest to develop a method of prediction, that is, a procedure for estimating the tar content for various levels of the inlet temperature from experimental information. Now, of course, it is highly likely that for many example runs in which the inlet temperature is the same, say $130°C$, the outlet tar content will not be the same. This is much like what happens when we study several automobiles with the same engine volume. They will not all have the same gas mileage. Houses in the same part of the country that have the same square footage of living space will not all be sold for the same price. Tar content, gas mileage (mpg), and the price of houses (in thousands of dollars) are natural **dependent variables**, or responses, in these three scenarios. Inlet temperature, engine volume (cubic feet), and square feet of living space are, respectively, natural **independent variables**, or **regressors**. A reasonable form of a relationship between the **response Y** and the regressor x is the linear relationship

$$Y = \beta_0 + \beta_1 x,$$

where, of course, β_0 is the **intercept** and β_1 is the **slope**. The relationship is illustrated in Figure 7.1.

If the relationship is exact, then it is a **deterministic** relationship between two scientific variables and there is no random or probabilistic component to it. However, in the examples listed above, as well as in countless other scientific and engineering phenomena, the relationship is not deterministic (i.e., a given x does not always give the same value for Y). As a result, important problems here are probabilistic in nature since the relationship above cannot be viewed as being exact. The concept of **regression analysis** deals with finding the best relationship between Y and x, quantifying the strength of that relationship, and using methods that allow for prediction of the response values given values of the regressor x.

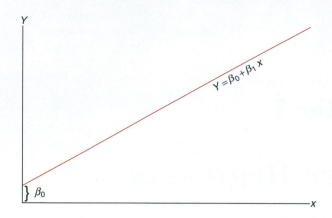

Figure 7.1: A linear relationship; β_0: intercept; β_1: slope.

In many applications, there will be more than one regressor (i.e., more than one independent variable **that helps to explain** Y). For example, in the case where the response is the price of a house, one would expect the age of the house to contribute to the explanation of the price, so in this case the multiple regression structure might be written

$$Y = \beta_0 + \beta_1 x_1 + \beta_2 x_2,$$

where Y is price, x_1 is square footage, and x_2 is age in years. The resulting analysis is termed **multiple regression**, while the analysis of the single regressor case is called **simple regression**. As a second illustration of multiple regression, a chemical engineer may be concerned with the amount of hydrogen lost from samples of a particular metal when the material is placed in storage. In this case, there may be two inputs, storage time x_1 in hours and storage temperature x_2 in degrees centigrade. The response would then be hydrogen loss Y in parts per million.

In this chapter, we first deal with the topic of **simple linear regression**, treating the case of a single regressor variable in which the relationship between y and x is linear. Then we turn to the case of more than one regressor variable. Denote a random sample of size n by the set $\{(x_i, y_i);\ i = 1, 2, \ldots, n\}$. If additional samples were taken using exactly the same values of x, we should expect the y values to vary. Hence, the value y_i in the ordered pair (x_i, y_i) is a value of some random variable Y_i.

7.2   The Simple Linear Regression (SLR) Model and the Least Squares Method

We have already confined the terminology *regression analysis* to situations in which relationships among variables are not deterministic (i.e., not exact). In other words, there must be a **random component** to the equation that relates the variables. This random component takes into account considerations that are not being measured or, in fact, are not understood by the scientists or engineers. Indeed, in most applications of regression, the linear equation, say $Y = \beta_0 + \beta_1 x$, is an approxima-tion that is a simplification of something unknown and much more complicated. For example, in our illustration involving the response $Y=$ tar content and $x =$ inlet temperature, $Y = \beta_0 + \beta_1 x$ is likely a reasonable approximation that may be operative within a confined range on x. More often than not, the models that are simplifications of more complicated and unknown structures are linear in nature (i.e., linear in the **parameters** β_0 and β_1 or, in the case of the model involving the price, size, and age of the house, linear in the **parameters** β_0, β_1, and β_2). These linear structures are simple and empirical in nature and are thus called **empirical models**.

An analysis of the relationship between Y and x requires the statement of a **statistical model**. A model is often used by a statistician as a representation of an **ideal** that essentially defines how we perceive that the data were generated by the system in question. The model must include the set $\{(x_i, y_i);\ i = 1, 2, \ldots, n\}$ of data involving n pairs of (x, y) values. One must bear in mind that the value y_i depends on x_i via a linear structure that also has the random component involved. The basis for the use of a statistical model relates to how the random variable Y moves with x and the random component. The model also includes what is assumed about the statistical properties of the random component. The statistical model for simple linear regression is given below.

Simple Linear Regression Model	The response Y is related to the independent variable x through the equation $$Y = \beta_0 + \beta_1 x + \epsilon.$$

In the above, β_0 and β_1 are unknown intercept and slope parameters, respectively, and ϵ is a random variable that is assumed to be distributed with $E(\epsilon) = 0$ and $\text{Var}(\epsilon) = \sigma^2$. The quantity σ^2 is often called the error variance or residual variance. Normal distribution of ϵ, $n(x; 0, \sigma^2)$, is also commonly assumed.

From the model above, several things become apparent. The quantity Y is a random variable since ϵ is random. The value x of the regressor variable is not random and, in fact, is measured with negligible error. The quantity ϵ, often called a **random error** or **random disturbance**, has constant variance. This portion of the assumptions is often called the **homogeneous variance assumption**. The presence of this random error, ϵ, keeps the model from becoming simply a deterministic equation. Now, the fact that $E(\epsilon) = 0$ implies that at a specific x the y-values are distributed around the **true**, or population, **regression line** $y = \beta_0 + \beta_1 x$. If the model is well chosen (i.e., there are no additional important regressors and the linear approximation is good within the ranges of the data), then positive and negative errors around the true regression are reasonable. We must keep in mind that in practice β_0 and β_1 are not known and must be estimated from data. In addition, the model described above is conceptual in nature. As a result, we never observe the actual ϵ values in practice and thus we can never draw the true regression line (but we assume it is there). We can only draw an estimated line. Figure 7.2 depicts the nature of hypothetical (x, y) data scattered around a true regression line for a case in which only $n = 5$ observations are available. Let us emphasize that what we see in Figure 7.2 is not the line that is used by the scientist or engineer. Rather, the picture merely describes what the assumptions mean! The regression that the user has at his or her disposal will now be described.

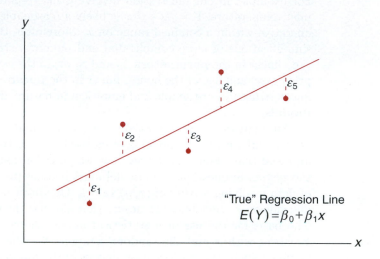

Figure 7.2: Hypothetical (x, y) data scattered around the true regression line for $n = 5$.

The Fitted Regression Line

An important aspect of regression analysis is, very simply, to estimate the parameters β_0 and β_1 (i.e., estimate the so-called **regression coefficients**). The method of estimation will be discussed in the next section. Suppose we denote the estimates b_0 for β_0 and b_1 for β_1. Then the estimated or **fitted regression** line is given by

$$\hat{y} = b_0 + b_1 x,$$

where $\hat{y}$ is the predicted or fitted value. Obviously, the fitted line is an estimate of the true regression line. We expect that the fitted line should be closer to the true regression line when a large amount of data are available. In the following example, we illustrate the fitted line for a real-life pollution study.

One of the more challenging problems confronting the water pollution control field is presented by the tanning industry. Tannery wastes are chemically complex. They are characterized by high values of chemical oxygen demand, volatile solids, and other pollution measures. Consider the experimental data in Table 7.1, which were obtained from 33 samples of chemically treated waste in a study conducted at Virginia Tech. Readings on x, the percent reduction in total solids, and y, the percent reduction in chemical oxygen demand, were recorded.

The data of Table 7.1 are plotted in a **scatter diagram** in Figure 7.3. From an inspection of this scatter diagram, it can be seen that the points closely follow a straight line, indicating that the assumption of linearity between the two variables appears to be reasonable.

Table 7.1: Measures of Reduction in Solids and Oxygen Demand

Solids Reduction, x (%)	Oxygen Demand Reduction, y (%)	Solids Reduction, x (%)	Oxygen Demand Reduction, y (%)
3	5	36	34
7	11	37	36
11	21	38	38
15	16	39	37
18	16	39	36
27	28	39	45
29	27	40	39
30	25	41	41
30	35	42	40
31	30	42	44
31	40	43	37
32	32	44	44
33	34	45	46
33	32	46	46
34	34	47	49
36	37	50	51
36	38		

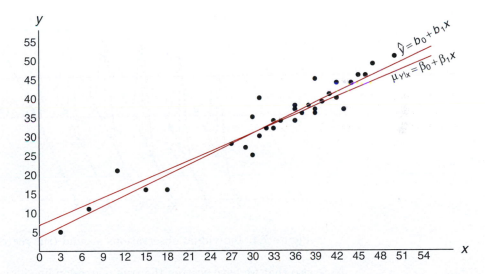

Figure 7.3: Scatter diagram with regression lines.

The fitted regression line and a *hypothetical true regression line* are shown on the scatter diagram of Figure 7.3.

Another Look at the Model Assumptions

It may be instructive to revisit the simple linear regression model presented previously and discuss in a graphical sense how it relates to the so-called true regression.

Let us expand on Figure 7.2 by illustrating not merely where the ϵ_i fall on a graph but also what the implication is of the normality assumption on the ϵ_i.

Suppose we have a simple linear regression with $n = 6$ evenly spaced values of x and a single y-value at each x. Consider the graph in Figure 7.4. This illustration should give the reader a clear representation of the model and the assumptions involved. The line in the graph is the true regression line. The points plotted are actual (y, x) points which are scattered about the line. Each point is on its own normal distribution with the center of the distribution (i.e., the mean of y) falling on the line. This is certainly expected since $E(Y) = \beta_0 + \beta_1 x$. As a result, the true regression line **goes through the means of the response**, and the actual observations are on the distribution around the means. Note also that all distributions have the same variance, which we referred to as σ^2. Of course, the deviation between an individual y and the point on the line will be its individual ϵ value. This is clear since

$$y_i - E(Y_i) = y_i - (\beta_0 + \beta_1 x_i) = \epsilon_i.$$

Thus, at a given x, Y and the corresponding ϵ both have variance σ^2.

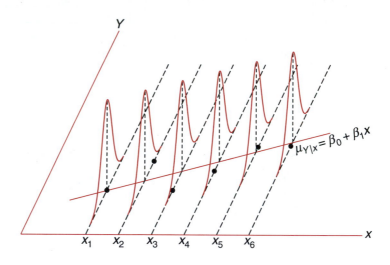

Figure 7.4: Individual observations around true regression line.

Note also that we have written the true regression line here as $\mu_{Y|x} = \beta_0 + \beta_1 x$ in order to reaffirm that the line goes through the mean of the Y random variable.

The Method of Least Squares

In this section, we discuss the method of fitting an estimated regression line to the data. This is tantamount to the determination of estimates b_0 for β_0 and b_1 for β_1. This of course allows for the computation of predicted values from the fitted line $\hat{y} = b_0 + b_1 x$ and other types of analyses and diagnostic information that will ascertain the strength of the relationship and the adequacy of the fitted model. Before we discuss the method of least squares estimation, it is important to introduce the concept of a **residual**. A residual is essentially an error in the fit of the model $\hat{y} = b_0 + b_1 x$.

Residual: Given a set of regression data $\{(x_i, y_i); i = 1, 2, \ldots, n\}$ and a fitted model, $\hat{y}_i = b_0 + b_1 x_i$, the ith residual e_i is given by

Error in Fit

$$e_i = y_i - \hat{y}_i, \quad i = 1, 2, \ldots, n.$$

Obviously, if a set of n absolute residuals is large, then the fit of the model is not good. Small residuals are a sign of a good fit. Another interesting relationship which is useful at times is the following:

$$y_i = b_0 + b_1 x_i + e_i.$$

The use of the above equation should result in clarification of the distinction between the residuals, e_i, and the conceptual model errors, ϵ_i. One must bear in mind that whereas the ϵ_i are not observed, the e_i not only are observed but also play an important role in the total analysis.

Figure 7.5 depicts the line fit to this set of data, namely $\hat{y} = b_0 + b_1 x$, and the line reflecting the model $\mu_{Y|x} = \beta_0 + \beta_1 x$. Now, of course, β_0 and β_1 are unknown parameters. The fitted line is an estimate of the line produced by the statistical model. Keep in mind that the line $\mu_{Y|x} = \beta_0 + \beta_1 x$ is not known.

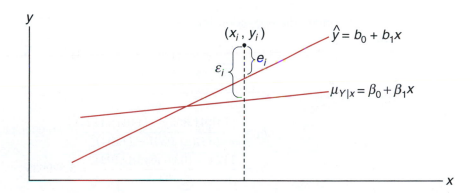

Figure 7.5: Comparing ϵ_i with the residual, e_i.

We shall find b_0 and b_1, the estimates of β_0 and β_1, so that the sum of the squares of the residuals is a minimum. The residual sum of squares is often called the sum of squares of the errors about the regression line and is denoted by SSE. This minimization procedure for estimating the parameters is called the **method of least squares**. Hence, we shall find b_0 and b_1 so as to minimize

$$SSE = \sum_{i=1}^{n} e_i^2 = \sum_{i=1}^{n} (y_i - \hat{y}_i)^2 = \sum_{i=1}^{n} (y_i - b_0 - b_1 x_i)^2.$$

Differentiating SSE with respect to b_0 and b_1, we have

$$\frac{\partial(SSE)}{\partial b_0} = -2 \sum_{i=1}^{n} (y_i - b_0 - b_1 x_i), \quad \frac{\partial(SSE)}{\partial b_1} = -2 \sum_{i=1}^{n} (y_i - b_0 - b_1 x_i) x_i.$$

Setting the partial derivatives equal to zero and rearranging the terms, we obtain the equations (called the **normal equations**)

$$nb_0 + b_1 \sum_{i=1}^{n} x_i = \sum_{i=1}^{n} y_i, \quad b_0 \sum_{i=1}^{n} x_i + b_1 \sum_{i=1}^{n} x_i^2 = \sum_{i=1}^{n} x_i y_i,$$

which may be solved simultaneously to yield computing formulas for b_0 and b_1.

Estimating the Regression Coefficients Given the sample $\{(x_i, y_i);\ i = 1, 2, \ldots, n\}$, the least squares estimates b_0 and b_1 of the regression coefficients β_0 and β_1 are computed from the formulas

$$b_1 = \frac{n \sum_{i=1}^{n} x_i y_i - \left(\sum_{i=1}^{n} x_i\right)\left(\sum_{i=1}^{n} y_i\right)}{n \sum_{i=1}^{n} x_i^2 - \left(\sum_{i=1}^{n} x_i\right)^2} = \frac{\sum_{i=1}^{n} (x_i - \bar{x})(y_i - \bar{y})}{\sum_{i=1}^{n} (x_i - \bar{x})^2} \quad \text{and}$$

$$b_0 = \frac{\sum_{i=1}^{n} y_i - b_1 \sum_{i=1}^{n} x_i}{n} = \bar{y} - b_1 \bar{x}.$$

The calculations of b_0 and b_1, using the data of Table 7.1, are illustrated by the following example.

Example 7.1: Estimate the regression line for the pollution data of Table 7.1.

Solution:
$$\sum_{i=1}^{33} x_i = 1104, \quad \sum_{i=1}^{33} y_i = 1124, \quad \sum_{i=1}^{33} x_i y_i = 41{,}355, \quad \sum_{i=1}^{33} x_i^2 = 41{,}086$$

Therefore,

$$b_1 = \frac{(33)(41{,}355) - (1104)(1124)}{(33)(41{,}086) - (1104)^2} = 0.903643 \text{ and}$$

$$b_0 = \frac{1124 - (0.903643)(1104)}{33} = 3.829633.$$

Thus, the estimated regression line is given by

$$\hat{y} = 3.8296 + 0.9036x.$$

Using the regression line of Example 7.1, we would predict a 31% reduction in the chemical oxygen demand when the reduction in the total solids is 30%. The 31% reduction in the chemical oxygen demand may be interpreted as an estimate of the population mean $\mu_{Y|30}$ or as an estimate of a new observation when the reduction in total solids is 30%. Such estimates, however, are subject to error. Even if the experiment were controlled so that the reduction in total solids was 30%, it is unlikely that we would measure a reduction in the chemical oxygen demand exactly equal to 31%. In fact, the original data recorded in Table 7.1 show that measurements of 25% and 35% were recorded for the reduction in oxygen demand when the reduction in total solids was kept at 30%.

What Is Good about Least Squares?

It should be noted that the least squares criterion is designed to provide a fitted line that results in a "closeness" between the line and the plotted points. One should remember that the residuals are the empirical counterpart to the ϵ values. Figure 7.6 illustrates a set of residuals. One should note that the fitted line has predicted values as points on the line and hence the residuals are vertical deviations from points to the line. As a result, the least squares procedure produces a line that **minimizes the sum of squares of vertical deviations** from the points to the line.

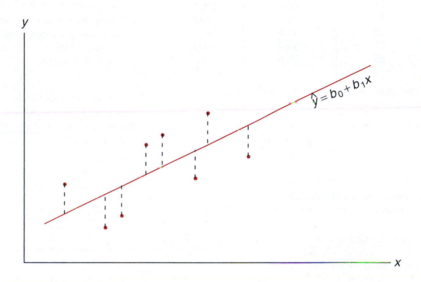

Figure 7.6: Residuals as vertical deviations.

Exercises

7.1 A study was conducted at Virginia Tech to determine if certain static arm-strength measures have an influence on the "dynamic lift" characteristics of an individual. Twenty-five individuals were subjected to strength tests and then were asked to perform a weight-lifting test in which weight was dynamically lifted overhead. The data are given here.

(a) Estimate β_0 and β_1 for the linear regression curve $\mu_{Y|x} = \beta_0 + \beta_1 x$.

(b) Find a point estimate of $\mu_{Y|30}$.

(c) Plot the residuals versus the x's (arm strength). Comment.

Individual	Arm Strength, x	Dynamic Lift, y
1	17.3	71.7
2	19.3	48.3
3	19.5	88.3
4	19.7	75.0
5	22.9	91.7
6	23.1	100.0
7	26.4	73.3
8	26.8	65.0
9	27.6	75.0
10	28.1	88.3

(cont.)

Individual	Arm Strength, x	Dynamic Lift, y
11	28.2	68.3
12	28.7	96.7
13	29.0	76.7
14	29.6	78.3
15	29.9	60.0
16	29.9	71.7
17	30.3	85.0
18	31.3	85.0
19	36.0	88.3
20	39.5	100.0
21	40.4	100.0
22	44.3	100.0
23	44.6	91.7
24	50.4	100.0
25	55.9	71.7

7.2 The grades of a class of 9 students on a midterm report (x) and on the final examination (y) are as follows:

x	77	50	71	72	81	94	96	99	67
y	82	66	78	34	47	85	99	99	68

(a) Estimate the linear regression line.

(b) Estimate the final examination grade of a student who received a grade of 85 on the midterm report.

7.3 The amounts of a chemical compound y that dissolved in 100 grams of water at various temperatures x were recorded as follows:

x (°C)	y (grams)		
0	8	6	8
15	12	10	14
30	25	21	24
45	31	33	28
60	44	39	42
75	48	51	44

(a) Find the equation of the regression line.

(b) Graph the line on a scatter diagram.

(c) Estimate the amount of chemical that will dissolve in 100 grams of water at 50°C.

7.4 The following data were collected to determine the relationship between pressure and the corresponding scale reading for the purpose of calibration.

Pressure, x (lb/sq in.)	Scale Reading, y
10	13
10	18
10	16
10	15
10	20
50	86
50	90
50	88
50	88
50	92

(a) Find the equation of the regression line.

(b) The purpose of calibration in this application is to estimate pressure from an observed scale reading. Estimate the pressure for a scale reading of 54 using $\hat{x} = (54 - b_0)/b_1$.

7.5 A study was made on the amount of converted sugar in a certain process at various temperatures. The data were coded and recorded.

(a) Estimate the linear regression line.

(b) Estimate the mean amount of converted sugar produced when the coded temperature is 1.75.

(c) Plot the residuals versus temperature. Comment.

Temperature, x	Converted Sugar, y
1.0	8.1
1.1	7.8
1.2	8.5
1.3	9.8
1.4	9.5
1.5	8.9
1.6	8.6
1.7	10.2
1.8	9.3
1.9	9.2
2.0	10.5

7.6 In a certain type of metal test specimen, the normal stress on a specimen is known to be functionally related to the shear resistance. A data set of coded experimental measurements on the two variables is given here.

(a) Estimate the regression line $\mu_{Y|x} = \beta_0 + \beta_1 x$.

(b) Estimate the shear resistance for a normal stress of 24.5.

Normal Stress, x	Shear Resistance, y
26.8	26.5
25.4	27.3
28.9	24.2
23.6	27.1
27.7	23.6
23.9	25.9
24.7	26.3
28.1	22.5
26.9	21.7
27.4	21.4
22.6	25.8
25.6	24.9

7.7 The following is a portion of a classic data set called the "pilot plot data" in *Fitting Equations to Data* by Daniel and Wood, published in 1971. The response y is the acid content of material produced by titration, whereas the regressor x is the organic acid content produced by extraction and weighing.

y	x	y	x
76	123	70	109
62	55	37	48
66	100	82	138
58	75	88	164
88	159	43	28

(a) Plot the data; does it appear that a simple linear regression will be a suitable model?

(b) Fit a simple linear regression; estimate a slope and intercept.

(c) Graph the regression line on the plot in (a).

7.8 A mathematics placement test is given to all entering freshmen at a small college. A student who receives a grade below 35 is denied admission to the regular mathematics course and placed in a remedial class. The placement test scores and the final grades for 20 students who took the regular course were recorded.

(a) Plot a scatter diagram.

(b) Find the equation of the regression line to predict course grades from placement test scores.

(c) Graph the line on the scatter diagram.

(d) If 60 is the minimum passing grade, which placement test score should be the cutoff below which students in the future should be denied admission to this course?

Placement Test	Course Grade
50	53
35	41
35	61
40	56
55	68
65	36
35	11
60	70
90	79
35	59
90	54
80	91
60	48
60	71
60	71
40	47
55	53
50	68
65	57
50	79

7.9 A study was made by a retail merchant to determine the relation between weekly advertising expenditures and sales.

Advertising Costs ($)	Sales ($)
40	385
20	400
25	395
20	365
30	475
50	440
40	490
20	420
50	560
40	525
25	480
50	510

(a) Plot a scatter diagram.

(b) Find the equation of the regression line to predict weekly sales from advertising expenditures.

(c) Estimate the weekly sales when advertising costs are $35.

(d) Plot the residuals versus advertising costs. Comment.

7.10 The following data are the selling prices z of a certain make and model of used car w years old. Fit a curve of the form $\mu_{z|w} = \gamma \delta^w$ by means of the nonlinear sample regression equation $\hat{z} = cd^w$. [*Hint*: Write $\ln \hat{z} = \ln c + (\ln d)w = b_0 + b_1 w$.]

w (years)	z (dollars)	w (years)	z (dollars)
1	6350	3	5395
2	5695	5	4985
2	5750	5	4895

7.11 The thrust of an engine (y) is a function of exhaust temperature (x) in °F when other important variables are held constant. Consider the following data.

y	x	y	x
4300	1760	4010	1665
4650	1652	3810	1550
3200	1485	4500	1700
3150	1390	3008	1270
4950	1820		

(a) Plot the data.

(b) Fit a simple linear regression to the data and plot the line through the data.

7.12 A study was done to study the effect of ambient temperature x on the electric power consumed by a chemical plant y. Other factors were held constant, and the data were collected from an experimental pilot plant.

y (BTU)	x (°F)	y (BTU)	x (°F)
250	27	265	31
285	45	298	60
320	72	267	34
295	58	321	74

(a) Plot the data.

(b) Estimate the slope and intercept in a simple linear regression model.

(c) Predict power consumption for an ambient temperature of 65°F.

7.13 A study of the amount of rainfall and the quantity of air pollution removed produced the data given here.

Daily Rainfall, x (0.01 cm)	Particulate Removed, y (μg/m^3)
4.3	126
4.5	121
5.9	116
5.6	118
6.1	114
5.2	118
3.8	132
2.1	141
7.5	108

(a) Find the equation of the regression line to predict the amount of particulate removed from the amount of daily rainfall.

(b) Estimate the amount of particulate removed when the daily rainfall is $x = 4.8$ units.

7.14 A professor in the School of Business in a university polled a dozen colleagues about the number of professional meetings they attended in the past five years (x) and the number of papers they submitted to refereed journals (y) during the same period. The summary data are given as follows:

$$n = 12, \quad \bar{x} = 4, \quad \bar{y} = 12,$$

$$\sum_{i=1}^{n} x_i^2 = 232, \quad \sum_{i=1}^{n} x_i y_i = 318.$$

Fit a simple linear regression model between x and y by finding out the estimates of intercept and slope. Comment on whether attending more professional meetings would result in publishing more papers.

7.3 Inferences Concerning the Regression Coefficients

In addition to the assumptions that the error term in the model

$$Y_i = \beta_0 + \beta_1 x_i + \epsilon_i$$

is a random variable with mean 0 and constant variance σ^2, suppose that we make the further assumption that $\epsilon_1, \epsilon_2, \ldots, \epsilon_n$ are independent from run to run in the experiment. This provides a foundation for finding the means and variances for the estimators of β_0 and β_1.

It is important to remember that our values of b_0 and b_1, based on a given sample of n observations, are only estimates of true parameters β_0 and β_1. If the experiment is repeated over and over again, each time using the same fixed values of x, the resulting estimates of β_0 and β_1 will most likely differ from experiment to experiment. These different estimates may be viewed as values assumed by the random variables B_0 and B_1, while b_0 and b_1 are specific realizations.

Since the values of x remain fixed, the values of B_0 and B_1 depend on the variations in the values of y or, more precisely, on the values of the random variables, $Y_1, Y_2, \ldots, Y_n$. The distributional assumptions imply that the Y_i, $i = 1, 2, \ldots, n$, are also independently distributed, with mean $\mu_{Y|x_i} = \beta_0 + \beta_1 x_i$ and equal variances σ^2; that is,

$$\sigma^2_{Y|x_i} = \sigma^2 \quad \text{for} \quad i = 1, 2, \ldots, n.$$

Mean and Variance of Estimators

In what follows, we show that the estimator B_1 is unbiased for β_1 and demonstrate the variances of both B_0 and B_1. This will begin a series of developments that lead to hypothesis testing and confidence interval estimation on the intercept and slope.

Since the estimator

$$B_1 = \frac{\sum_{i=1}^{n}(x_i - \bar{x})(Y_i - \bar{Y})}{\sum_{i=1}^{n}(x_i - \bar{x})^2} = \frac{\sum_{i=1}^{n}(x_i - \bar{x})Y_i}{\sum_{i=1}^{n}(x_i - \bar{x})^2}$$

is of the form $\sum_{i=1}^{n} c_i Y_i$, where

$$c_i = \frac{x_i - \bar{x}}{\sum_{i=1}^{n}(x_i - \bar{x})^2}, \quad i = 1, 2, \ldots, n,$$

we may conclude that B_1 has the mean

$$\mu_{B_1} = \frac{\sum_{i=1}^{n}(x_i - \bar{x})(\beta_0 + \beta_1 x_i)}{\sum_{i=1}^{n}(x_i - \bar{x})^2} = \beta_1,$$

and variance

$$\sigma^2_{B_1} = \frac{\sum_{i=1}^{n}(x_i - \bar{x})^2 \sigma^2_{Y_i}}{\left[\sum_{i=1}^{n}(x_i - \bar{x})^2\right]^2} = \frac{\sigma^2}{\sum_{i=1}^{n}(x_i - \bar{x})^2},$$

respectively.

It can also be shown that the random variable B_0 is normally distributed with

mean $\mu_{B_0} = \beta_0$ and variance $\sigma^2_{B_0} = \dfrac{\sum_{i=1}^{n} x_i^2}{n \sum_{i=1}^{n}(x_i - \bar{x})^2} \sigma^2.$

From the foregoing results, it is apparent that the **least squares estimators for** β_0 **and** β_1 **are both unbiased estimators.**

Partition of Total Variability and Estimation of σ^2

To draw inferences on β_0 and β_1, it becomes necessary to arrive at an estimate of the parameter σ^2 appearing in the two preceding variance formulas for B_0 and B_1. The parameter σ^2, the model error variance, reflects random variation or experimental error variation around the regression line. In much of what follows, it is advantageous to use the notation

$$S_{xx} = \sum_{i=1}^{n}(x_i - \bar{x})^2, \quad S_{yy} = \sum_{i=1}^{n}(y_i - \bar{y})^2, \quad S_{xy} = \sum_{i=1}^{n}(x_i - \bar{x})(y_i - \bar{y}).$$

Now we may write the error sum of squares as follows:

$$
\begin{aligned}
SSE &= \sum_{i=1}^{n}(y_i - b_0 - b_1 x_i)^2 = \sum_{i=1}^{n}[(y_i - \bar{y}) - b_1(x_i - \bar{x})]^2 \\
&= \sum_{i=1}^{n}(y_i - \bar{y})^2 - 2b_1 \sum_{i=1}^{n}(x_i - \bar{x})(y_i - \bar{y}) + b_1^2 \sum_{i=1}^{n}(x_i - \bar{x})^2 \\
&= S_{yy} - 2b_1 S_{xy} + b_1^2 S_{xx} = S_{yy} - b_1 S_{xy},
\end{aligned}
$$

the final step following from the fact that $b_1 = S_{xy}/S_{xx}$.

Theorem 7.1: An unbiased estimate of σ^2 is

$$s^2 = \frac{SSE}{n-2} = \sum_{i=1}^{n}\frac{(y_i - \hat{y}_i)^2}{n-2} = \frac{S_{yy} - b_1 S_{xy}}{n-2}.$$

The proof of Theorem 7.1 is left to the reader.

The Estimator of σ^2 as a Mean Squared Error

One should observe the result of Theorem 7.1 in order to gain some intuition about the estimator of σ^2. The parameter σ^2 measures variance or squared deviations between Y values and their mean given by $\mu_{Y|x}$ (i.e., squared deviations between Y and $\beta_0 + \beta_1 x$). Of course, $\beta_0 + \beta_1 x$ is estimated by $\hat{y} = b_0 + b_1 x$. Thus, it would make sense that the variance σ^2 is best depicted as a squared deviation of the typical observation y_i from the estimated mean, $\hat{y}_i$, which is the corresponding point on the fitted line. Thus, $(y_i - \hat{y}_i)^2$ values reveal the appropriate variance, much like the way $(y_i - \bar{y})^2$ values measure variance when one is sampling in a nonregression scenario. In other words, $\bar{y}$ estimates the mean in the latter simple situation, whereas $\hat{y}_i$ estimates the mean of y_i in a regression structure. Now, what about the divisor $n-2$? In future sections, we shall note that these are the degrees of freedom associated with the estimator s^2 of σ^2. Whereas in the standard normal i.i.d. scenario, one degree of freedom is subtracted from n in the denominator and a reasonable explanation is that one parameter is estimated, namely the mean μ by, say, $\bar{y}$, in the regression problem, **two parameters are estimated**, namely

β_0 and β_1 by b_0 and b_1. Thus, the important parameter σ^2, estimated by

$$s^2 = \sum_{i=1}^{n}(y_i - \hat{y}_i)^2/(n-2),$$

is called a **mean squared error**, depicting a type of mean (division by $n-2$) of the squared residuals.

Aside from merely estimating the linear relationship between x and Y for purposes of prediction, the experimenter may also be interested in drawing certain inferences about the slope and intercept. In order to allow for the testing of hypotheses and the construction of confidence intervals on β_0 and β_1, one must be willing to make the further assumption that each ϵ_i, $i = 1, 2, \ldots, n$, is normally distributed. This assumption implies that $Y_1, Y_2, \ldots, Y_n$ are also normally distributed, each with probability distribution $n(y_i; \beta_0 + \beta_1 x_i, \sigma)$.

Since B_1 follows a normal distribution, it turns out that under the normality assumption, a result very much analogous to that given in Theorem 4.4 allows us to conclude that $(n-2)S^2/\sigma^2$ is a chi-squared variable with $n-2$ degrees of freedom, independent of the random variable B_1. Theorem 4.5 then assures us that the statistic

$$T = \frac{(B_1 - \beta_1)/(\sigma/\sqrt{S_{xx}})}{S/\sigma} = \frac{B_1 - \beta_1}{S/\sqrt{S_{xx}}}$$

has a t-distribution with $n-2$ degrees of freedom. The statistic T can be used to construct a $100(1-\alpha)\%$ confidence interval for the coefficient β_1.

Confidence Interval for β_1 A $100(1-\alpha)\%$ confidence interval for the parameter β_1 in the regression line $\mu_{Y|x} = \beta_0 + \beta_1 x$ is

$$b_1 - t_{\alpha/2}\frac{s}{\sqrt{S_{xx}}} < \beta_1 < b_1 + t_{\alpha/2}\frac{s}{\sqrt{S_{xx}}},$$

where $t_{\alpha/2}$ is a value of the t-distribution with $n-2$ degrees of freedom.

Example 7.2: Find a 95% confidence interval for β_1 in the regression line $\mu_{Y|x} = \beta_0 + \beta_1 x$, based on the pollution data of Table 7.1.

Solution: From the results given in Example 7.1 we find that $S_{xx} = 4152.18$ and $S_{xy} = 3752.09$. In addition, we find that $S_{yy} = 3713.88$. Recall that $b_1 = 0.903643$. Hence,

$$s^2 = \frac{S_{yy} - b_1 S_{xy}}{n-2} = \frac{3713.88 - (0.903643)(3752.09)}{31} = 10.4299.$$

Therefore, taking the square root, we obtain $s = 3.2295$. Using Table A.4, we find $t_{0.025} \approx 2.045$ for 31 degrees of freedom. Therefore, a 95% confidence interval for β_1 is

$$0.903643 - \frac{(2.045)(3.2295)}{\sqrt{4152.18}} < \beta < 0.903643 + \frac{(2.045)(3.2295)}{\sqrt{4152.18}},$$

which simplifies to

$$0.8012 < \beta_1 < 1.0061.$$

Hypothesis Testing on the Slope

To test the null hypothesis H_0 that $\beta_1 = \beta_{10}$ against a suitable alternative, we again use the t-distribution with $n - 2$ degrees of freedom to establish a critical region and then base our decision on the value of

$$t = \frac{b_1 - \beta_{10}}{s/\sqrt{S_{xx}}}.$$

The method is illustrated by the following example.

Example 7.3: Using the estimated value $b_1 = 0.903643$ of Example 7.1, test the hypothesis that $\beta_1 = 1.0$ against the alternative that $\beta_1 < 1.0$.

Solution: The hypotheses are H_0: $\beta_1 = 1.0$ and H_1: $\beta_1 < 1.0$. So

$$t = \frac{0.903643 - 1.0}{3.2295/\sqrt{4152.18}} = -1.92,$$

with $n - 2 = 31$ degrees of freedom ($P \approx 0.03$).

Decision: The t-value is significant at the 0.03 level, suggesting strong evidence that $\beta_1 < 1.0$.

One important t-test on the slope is the test of the hypothesis

$$H_0: \ \beta_1 = 0 \text{ versus } H_1: \ \beta_1 \neq 0.$$

When the null hypothesis is not rejected, the conclusion is that there is no significant linear relationship between $E(y)$ and the independent variable x. The plot of the data for Example 7.1 would suggest that a linear relationship exists. However, in some applications in which σ^2 is large and thus considerable "noise" is present in the data, a plot, while useful, may not produce clear information for the researcher. Rejection of H_0 above implies that a significant linear regression exists.

Figure 7.7 displays a *MINITAB* printout showing the t-test for

$$H_0: \ \beta_1 = 0 \text{ versus } H_1: \ \beta_1 \neq 0,$$

for the data of Example 7.1. Note the regression coefficient (Coef), standard error (SE Coef), t-value (T), and P-value (P). The null hypothesis is rejected. Clearly, there is a significant linear relationship between mean chemical oxygen demand reduction and solids reduction. Note that the t-statistic is computed as

$$t = \frac{\text{coefficient}}{\text{standard error}} = \frac{b_1}{s/\sqrt{S_{xx}}}.$$

The failure to reject H_0: $\beta_1 = 0$ suggests that there is no linear relationship between Y and x. Figure 7.8 is an illustration of the implication of this result. It may mean that changing x has little impact on changes in Y, as seen in (a).

```
Regression Analysis: COD versus Per_Red
The regression equation is COD = 3.83 + 0.904 Per_Red

Predictor      Coef   SE Coef      T       P
Constant      3.830     1.768    2.17   0.038
Per_Red     0.90364   0.05012   18.03   0.000


S = 3.22954    R-Sq = 91.3%   R-Sq(adj) = 91.0%
Analysis of Variance
Source           DF       SS       MS       F       P
Regression        1   3390.6   3390.6  325.08   0.000
Residual Error   31    323.3     10.4
Total            32   3713.9
```

Figure 7.7: *MINITAB* printout for *t*-test for data of Example 7.1.

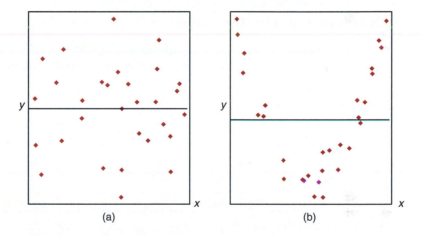

(a)   (b)

Figure 7.8: The hypothesis H_0: $\beta_1 = 0$ is not rejected.

However, it may also indicate that the true relationship is nonlinear, as indicated by (b).

When H_0: $\beta_1 = 0$ is rejected, there is an implication that the linear term in x residing in the model explains a significant portion of variability in Y. The two plots in Figure 7.9 illustrate possible scenarios. As depicted in (a) of the figure, rejection of H_0 may suggest that the relationship is, indeed, linear. As indicated in (b), it may suggest that while the model does contain a linear effect, a better representation may be found by including a polynomial (perhaps quadratic) term (i.e., terms that supplement the linear term).

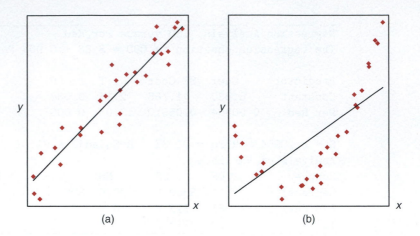

Figure 7.9: The hypothesis H_0: $\beta_1 = 0$ is rejected.

Statistical Inference on the Intercept

Confidence intervals and hypothesis testing on the coefficient β_0 may be established from the fact that B_0 is also normally distributed. It is not difficult to show that

$$T = \frac{B_0 - \beta_0}{S\sqrt{\sum_{i=1}^{n} x_i^2 / (nS_{xx})}}$$

has a t-distribution with $n - 2$ degrees of freedom from which we may construct a $100(1 - \alpha)\%$ confidence interval for β_0.

Confidence Interval   A $100(1 - \alpha)\%$ confidence interval for the parameter β_0 in the regression line
for β_0 $\mu_{Y|x} = \beta_0 + \beta_1 x$ is

$$b_0 - t_{\alpha/2}\frac{s}{\sqrt{nS_{xx}}}\sqrt{\sum_{i=1}^{n} x_i^2} < \beta_0 < b_0 + t_{\alpha/2}\frac{s}{\sqrt{nS_{xx}}}\sqrt{\sum_{i=1}^{n} x_i^2},$$

where $t_{\alpha/2}$ is a value of the t-distribution with $n - 2$ degrees of freedom.

To test the null hypothesis H_0 that $\beta_0 = \beta_{00}$ against a suitable alternative, we can use the t-distribution with $n - 2$ degrees of freedom to establish a critical region and then base our decision on the value of

$$t = \frac{b_0 - \beta_{00}}{s\sqrt{\sum_{i=1}^{n} x_i^2 / (nS_{xx})}}.$$

Example 7.4:   Using the estimated value $b_0 = 3.829633$ of Example 7.1, test the hypothesis that $\beta_0 = 0$ at the 0.05 level of significance against the alternative that $\beta_0 \neq 0$.

Solution: The hypotheses are H_0: $\beta_0 = 0$ and H_1: $\beta_0 \neq 0$. So

$$t = \frac{3.829633 - 0}{3.2295\sqrt{41{,}086/[(33)(4152.18)]}} = 2.17,$$

with 31 degrees of freedom. Thus, $P = P$-value ≈ 0.038 and we conclude that $\beta_0 \neq 0$. Note that this is merely Coef/StDev, as we see in the *MINITAB* printout in Figure 7.7. The SE Coef is the standard error of the estimated intercept. ◢

A Measure of Quality of Fit: Coefficient of Determination

Note in Figure 7.7 that an item denoted by R-Sq is given with a value of 91.3%. This quantity, R^2, is called the **coefficient of determination**. This quantity is a measure of the **proportion of variability in the response explained by the fitted model**. In Section 7.5, we shall introduce the notion of an analysis-of-variance approach to hypothesis testing in regression. The analysis-of-variance approach makes use of the error sum of squares $SSE = \sum_{i=1}^{n}(y_i - \hat{y}_i)^2$ and the **total corrected sum of squares** $SST = \sum_{i=1}^{n}(y_i - \bar{y}_i)^2$. The latter represents the variation in the response values that *ideally* would be explained by the model. The SSE value is the variation in the response due to error, or **variation unexplained**. Clearly, if $SSE = 0$, all variation is explained. The quantity that represents variation explained is $SST - SSE$. The R^2 is

$$\text{Coeff. of determination:} \quad R^2 = 1 - \frac{SSE}{SST}.$$

Note that if the fit is perfect, *all residuals are zero*, and thus $R^2 = 1.0$. But if SSE is only slightly smaller than SST, $R^2 \approx 0$. Note from the printout in Figure 7.7 that the coefficient of determination suggests that the model fit to the data explains 91.3% of the variability observed in the response, the reduction in chemical oxygen demand.

Figure 7.10 provides an illustration of a good fit ($R^2 \approx 1.0$) in plot (a) and a poor fit ($R^2 \approx 0$) in plot (b).

Pitfalls in the Use of R^2

Analysts quote values of R^2 quite often, perhaps due to its simplicity. However, there are pitfalls in its interpretation. The reliability of R^2 is a function of the size of the regression data set and the type of application. Clearly, $0 \leq R^2 \leq 1$ and the upper bound is achieved when the fit to the data is perfect (i.e., all of the residuals are zero). What is an acceptable value for R^2? This is a difficult question to answer. A chemist, charged with doing a linear calibration of a high-precision piece of equipment, certainly expects to experience a very high R^2-value (perhaps exceeding 0.99), while a behavioral scientist, dealing in data impacted by variability in human behavior, may feel fortunate to experience an R^2 as large as 0.70. An experienced model fitter senses when a value is large enough, given the situation confronted. Clearly, some scientific phenomena lend themselves to modeling with more precision than others.

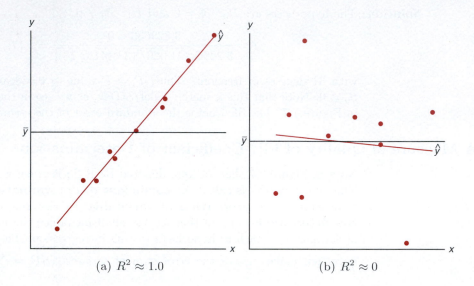

(a) $R^2 \approx 1.0$ (b) $R^2 \approx 0$

Figure 7.10: Plots depicting a very good fit and a poor fit.

The R^2 criterion is dangerous to use for comparing *competing models* for the same data set. Adding additional terms to the model (e.g., an additional regressor) decreases SSE and thus increases R^2 (or at least does not decrease it). This implies that R^2 can be made artificially high by an unwise practice of **overfitting** (i.e., the inclusion of too many model terms). Thus, the inevitable increase in R^2 enjoyed by adding an additional term does not imply the additional term was needed. In fact, the simple model may be superior for predicting response values. Suffice it to say at this point that one *should not subscribe to a model selection process that solely involves the consideration of R^2*.

7.4   Prediction

There are several reasons for building a linear regression. One, of course, is to predict response values at one or more values of the independent variable. In this section, the focus is on errors associated with prediction.

The equation $\hat{y} = b_0 + b_1 x$ may be used to predict or estimate the **mean response** $\mu_{Y|x_0}$ at $x = x_0$, where x_0 is not necessarily one of the prechosen values, or it may be used to predict a single value y_0 of the variable Y_0, when $x = x_0$. We would expect the error of prediction to be higher in the case of a single predicted value than in the case where a mean is predicted. This, then, will affect the width of our intervals for the values being predicted.

Suppose that the experimenter wishes to construct a confidence interval for $\mu_{Y|x_0}$. We shall use the point estimator $\hat{Y}_0 = B_0 + B_1 x_0$ to estimate $\mu_{Y|x_0} = \beta_0 + \beta_1 x$. It can be shown that the sampling distribution of $\hat{Y}_0$ is normal with mean

$$\mu_{Y|x_0} = E(\hat{Y}_0) = E(B_0 + B_1 x_0) = \beta_0 + \beta_1 x_0 = \mu_{Y|x_0}$$

and variance

$$\sigma^2_{\hat{Y}_0} = \sigma^2_{B_0 + B_1 x_0} = \sigma^2_{\bar{Y} + B_1(x_0 - \bar{x})} = \sigma^2 \left[\frac{1}{n} + \frac{(x_0 - \bar{x})^2}{S_{xx}} \right],$$

the latter following from the fact that $\text{Cov}(\bar{Y}, B_1) = 0$. Thus, a $100(1 - \alpha)\%$ confidence interval on the mean response $\mu_{Y|x_0}$ can now be constructed from the statistic

$$T = \frac{\hat{Y}_0 - \mu_{Y|x_0}}{S \sqrt{1/n + (x_0 - \bar{x})^2 / S_{xx}}},$$

which has a t-distribution with $n - 2$ degrees of freedom.

Confidence Interval
for $\mu_{Y|x_0}$

A $100(1 - \alpha)\%$ confidence interval for the mean response $\mu_{Y|x_0}$ is

$$\hat{y}_0 - t_{\alpha/2} s \sqrt{\frac{1}{n} + \frac{(x_0 - \bar{x})^2}{S_{xx}}} < \mu_{Y|x_0} < \hat{y}_0 + t_{\alpha/2} s \sqrt{\frac{1}{n} + \frac{(x_0 - \bar{x})^2}{S_{xx}}},$$

where $t_{\alpha/2}$ is a value of the t-distribution with $n - 2$ degrees of freedom.

Example 7.5: Using the data of Table 7.1 on page 299, construct 95% confidence limits for the mean response $\mu_{Y|x_0}$.

Solution: From the regression equation we find for, say, $x_0 = 20\%$ solids reduction,

$$\hat{y}_0 = 3.829633 + (0.903643)(20) = 21.9025.$$

In addition, $\bar{x} = 33.4545$, $S_{xx} = 4152.18$, $s = 3.2295$, and $t_{0.025} \approx 2.045$ for 31 degrees of freedom. Therefore, a 95% confidence interval for $\mu_{Y|20}$ is

$$21.9025 - (2.045)(3.2295) \sqrt{\frac{1}{33} + \frac{(20 - 33.4545)^2}{4152.18}} < \mu_{Y|20}$$

$$< 21.9025 + (2.045)(3.2295) \sqrt{\frac{1}{33} + \frac{(20 - 33.4545)^2}{4152.18}},$$

or simply $20.1071 < \mu_{Y|20} < 23.6979$.

Repeating the previous calculations for each of several different values of x_0, one can obtain the corresponding confidence limits on each $\mu_{Y|x_0}$. Figure 7.11 displays the data points, the estimated regression line, and the upper and lower confidence limits on the mean of $Y|x$.

In Example 7.5, we are 95% confident that the population mean reduction in chemical oxygen demand is between 20.1071% and 23.6979% when solids reduction is 20%.

Prediction Interval

Another type of interval that is often misinterpreted and confused with that given for $\mu_{Y|x}$ is the prediction interval for a future observed response. Actually, in

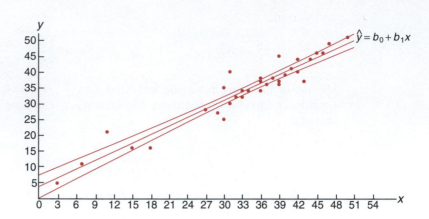

Figure 7.11: Confidence limits for the mean value of $Y|x$.

many instances, the prediction interval is more relevant to the scientist or engineer than the confidence interval on the mean. In the tar content and inlet temperature example cited in Section 7.1, there would certainly be interest not only in estimating the mean tar content at a specific temperature but also in constructing an interval that reflects the error in predicting a future observed amount of tar content at the given temperature.

To obtain a **prediction interval** for any single value y_0 of the variable Y_0, it is necessary to estimate the variance of the differences between the ordinates $\hat{y}_0$, obtained from the computed regression lines in repeated sampling when $x = x_0$, and the corresponding true ordinate y_0. We can think of the difference $\hat{y}_0 - y_0$ as a value of the random variable $\hat{Y}_0 - Y_0$, whose sampling distribution can be shown to be normal with mean

$$\mu_{\hat{Y}_0 - Y_0} = E(\hat{Y}_0 - Y_0) = E[B_0 + B_1 x_0 - (\beta_0 + \beta_1 x_0 + \epsilon_0)] = 0$$

and variance

$$\sigma^2_{\hat{Y}_0 - Y_0} = \sigma^2_{B_0 + B_1 x_0 - \epsilon_0} = \sigma^2_{\bar{Y} + B_1(x_0 - \bar{x}) - \epsilon_0} = \sigma^2 \left[1 + \frac{1}{n} + \frac{(x_0 - \bar{x})^2}{S_{xx}} \right].$$

Thus, a $100(1 - \alpha)\%$ prediction interval for a single predicted value y_0 can be constructed from the statistic

$$T = \frac{\hat{Y}_0 - Y_0}{S\sqrt{1 + 1/n + (x_0 - \bar{x})^2 / S_{xx}}},$$

which has a t-distribution with $n - 2$ degrees of freedom.

Prediction Interval for y_0   A $100(1 - \alpha)\%$ prediction interval for a single response y_0 is given by

$$\hat{y}_0 - t_{\alpha/2} s \sqrt{1 + \frac{1}{n} + \frac{(x_0 - \bar{x})^2}{S_{xx}}} < y_0 < \hat{y}_0 + t_{\alpha/2} s \sqrt{1 + \frac{1}{n} + \frac{(x_0 - \bar{x})^2}{S_{xx}}},$$

where $t_{\alpha/2}$ is a value of the t-distribution with $n - 2$ degrees of freedom.

Clearly, there is a distinction between the concept of a confidence interval and the prediction interval described previously. The interpretation of the confidence interval is identical to that described for all confidence intervals on population parameters discussed throughout the book. Indeed, $\mu_{Y|x_0}$ is a population parameter. The computed prediction interval, however, represents an interval that has a probability equal to $1 - \alpha$ of containing not a parameter but a future value y_0 of the random variable Y_0.

Example 7.6: Using the data of Table 7.1, construct a 95% prediction interval for y_0 when $x_0 = 20\%$.

Solution: We have $n = 33$, $x_0 = 20$, $\bar{x} = 33.4545$, $\hat{y}_0 = 21.9025$, $S_{xx} = 4152.18$, $s = 3.2295$, and $t_{0.025} \approx 2.045$ for 31 degrees of freedom. Therefore, a 95% prediction interval for y_0 is

$$21.9025 - (2.045)(3.2295)\sqrt{1 + \frac{1}{33} + \frac{(20 - 33.4545)^2}{4152.18}} < y_0$$
$$< 21.9025 + (2.045)(3.2295)\sqrt{1 + \frac{1}{33} + \frac{(20 - 33.4545)^2}{4152.18}},$$

which simplifies to $15.0585 < y_0 < 28.7464$.

Figure 7.12 shows another plot of the chemical oxygen demand reduction data, with both the confidence interval on the mean response and the prediction interval on an individual response plotted. The plot reflects a much tighter interval around the regression line in the case of the mean response.

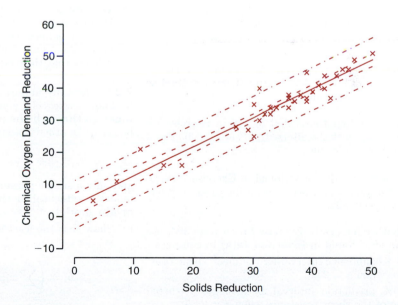

Figure 7.12: Confidence and prediction intervals for the chemical oxygen demand reduction data; inside bands indicate the confidence limits for the mean responses and outside bands indicate the prediction limits for the future responses.

Exercises

7.15 With reference to Exercise 7.1 on page 303,

(a) evaluate s^2;

(b) test the hypothesis that $\beta_1 = 0$ against the alternative that $\beta_1 \neq 0$ at the 0.05 level of significance and interpret the resulting decision.

7.16 With reference to Exercise 7.2 on page 304,

(a) evaluate s^2;

(b) construct a 95% confidence interval for β_0;

(c) construct a 95% confidence interval for β_1.

7.17 With reference to Exercise 7.5 on page 304,

(a) evaluate s^2;

(b) construct a 95% confidence interval for β_0;

(c) construct a 95% confidence interval for β_1.

7.18 With reference to Exercise 7.6 on page 304,

(a) evaluate s^2;

(b) construct a 99% confidence interval for β_0;

(c) construct a 99% confidence interval for β_1.

7.19 With reference to Exercise 7.3 on page 304,

(a) evaluate s^2;

(b) construct a 99% confidence interval for β_0;

(c) construct a 99% confidence interval for β_1.

7.20 Test the hypothesis that $\beta_0 = 10$ in Exercise 7.8 on page 305 against the alternative that $\beta_0 < 10$. Use a 0.05 level of significance.

7.21 Test the hypothesis that $\beta_1 = 6$ in Exercise 7.9 on page 305 against the alternative that $\beta_1 < 6$. Use a 0.025 level of significance.

7.22 Using the value of s^2 found in Exercise 7.16(a), construct a 95% confidence interval for $\mu_{Y|85}$ in Exercise 7.2 on page 304.

7.23 With reference to Exercise 7.6 on page 304, use the value of s^2 found in Exercise 7.18(a) to compute

(a) a 95% confidence interval for the mean shear resistance when $x = 24.5$;

(b) a 95% prediction interval for a single predicted value of the shear resistance when $x = 24.5$.

7.24 Using the value of s^2 found in Exercise 7.17(a), graph the regression line and the 95% confidence bands for the mean response $\mu_{Y|x}$ for the data of Exercise 7.5 on page 304.

7.25 Using the value of s^2 found in Exercise 7.17(a), construct a 95% confidence interval for the amount of converted sugar corresponding to $x = 1.6$ in Exercise 7.5 on page 304.

7.26 With reference to Exercise 7.3 on page 304, use the value of s^2 found in Exercise 7.19(a) to compute

(a) a 99% confidence interval for the average amount of chemical that will dissolve in 100 grams of water at $50°C$;

(b) a 99% prediction interval for the amount of chemical that will dissolve in 100 grams of water at $50°C$.

7.27 Consider the regression of mileage for certain automobiles, measured in miles per gallon (mpg) on their weight in pounds (wt). The data are from *Consumer Reports* (April 1997). Part of the *SAS* output from the procedure is shown in Figure 7.13.

(a) Estimate the mileage for a vehicle weighing 4000 pounds.

(b) Suppose that Honda engineers claim that, on average, the Civic (or any other model weighing 2440 pounds) gets more than 30 mpg. Based on the results of the regression analysis, would you believe that claim? Why or why not?

(c) The design engineers for the Lexus ES300 targeted 18 mpg as being ideal for this model (or any other model weighing 3390 pounds), although it is expected that some variation will be experienced. Is it likely that this target value is realistic? Discuss.

7.28 There are important applications in which, due to known scientific constraints, the regression line **must go through the origin** (i.e., the intercept must be zero). In other words, the model should read

$$Y_i = \beta_1 x_i + \epsilon_i, \quad i = 1, 2, \ldots, n,$$

and only a simple parameter requires estimation. The model is often called the **regression through the origin model**.

(a) Show that the least squares estimator of the slope is

$$b_1 = \left(\sum_{i=1}^{n} x_i y_i \right) \Big/ \left(\sum_{i=1}^{n} x_i^2 \right).$$

(b) Show that $\sigma_{B_1}^2 = \sigma^2 \Big/ \left(\sum_{i=1}^{n} x_i^2 \right)$.

(c) Show that b_1 in part (a) is an unbiased estimator for β_1. That is, show $E(B_1) = \beta_1$.

7.29 Use the data set

y	x
7	2
50	15
100	30
40	10
70	20

(a) Plot the data.

(b) Fit a regression line through the origin.

(c) Plot the regression line on the graph with the data.

(d) Give a general formula (in terms of the y_i and the slope b_1) for the estimator of σ^2.

(e) Give a formula for $\text{Var}(\hat{y}_i)$, $i = 1, 2, \ldots, n$, for this case.

(f) Plot 95% confidence limits for the mean response on the graph around the regression line.

7.30 For the data in Exercise 7.29, find a 95% prediction interval at $x = 25$.

	Root MSE		1.48794	R-Square		0.9509		
	Dependent Mean		21.50000	Adj R-Sq		0.9447		

Parameter Estimates

Variable	DF	Parameter Estimate	Standard Error	t Value	Pr > |t|
Intercept	1	44.78018	1.92919	23.21	<.0001
WT	1	-0.00686	0.00055133	-12.44	<.0001

MODEL	WT	MPG	Predict	LMean	UMean	Lpred	Upred	Residual
GMC	4520	15	13.7720	11.9752	15.5688	9.8988	17.6451	1.22804
Geo	2065	29	30.6138	28.6063	32.6213	26.6385	34.5891	-1.61381
Honda	2440	31	28.0412	26.4143	29.6681	24.2439	31.8386	2.95877
Hyundai	2290	28	29.0703	27.2967	30.8438	25.2078	32.9327	-1.07026
Infiniti	3195	23	22.8618	21.7478	23.9758	19.2543	26.4693	0.13825
Isuzu	3480	21	20.9066	19.8160	21.9972	17.3062	24.5069	0.09341
Jeep	4090	15	16.7219	15.3213	18.1224	13.0158	20.4279	-1.72185
Land	4535	13	13.6691	11.8570	15.4811	9.7888	17.5493	-0.66905
Lexus	3390	22	21.5240	20.4390	22.6091	17.9253	25.1227	0.47599
Lincoln	3930	18	17.8195	16.5379	19.1011	14.1568	21.4822	0.18051

Figure 7.13: *SAS* printout for Exercise 7.27.

7.5 Analysis-of-Variance Approach

Often the problem of analyzing the quality of the estimated regression line is handled by an **analysis-of-variance** (ANOVA) approach: a procedure whereby the total variation in the dependent variable is subdivided into meaningful components that are then observed and treated in a systematic fashion. The analysis of variance, discussed in Chapter 8, is a powerful resource that is used for many applications.

Suppose that we have n experimental data points in the usual form (x_i, y_i) and that the regression line is estimated. In our estimation of σ^2 in Section 7.3, we established the identity

$$S_{yy} = b_1 S_{xy} + SSE.$$

An alternative and perhaps more informative formulation is

$$\sum_{i=1}^{n} (y_i - \bar{y})^2 = \sum_{i=1}^{n} (\hat{y}_i - \bar{y})^2 + \sum_{i=1}^{n} (y_i - \hat{y}_i)^2.$$

We have achieved a partitioning of the **total corrected sum of squares of y** into two components that should convey particular meaning to the experimenter.

We shall indicate this partitioning symbolically as

$$SST = SSR + SSE.$$

The first component on the right, SSR, is called the **regression sum of squares**, and it reflects the amount of variation in the y-values **explained by the model**, in this case the postulated straight line. The second component is the familiar error sum of squares, which reflects variation about the regression line.

Suppose that we are interested in testing the hypothesis

$$H_0\colon \ \beta_1 = 0 \text{ versus } H_1\colon \ \beta_1 \neq 0,$$

where the null hypothesis says essentially that the model is $\mu_{Y|x} = \beta_0$. That is, the variation in Y results from chance or random fluctuations which are independent of the values of x. This condition is reflected in Figure 7.10(b). Under the conditions of this null hypothesis, it can be shown that SSR/σ^2 and SSE/σ^2 are values of independent chi-squared variables with 1 and $n-2$ degrees of freedom, respectively, and SST/σ^2 is also a value of a chi-squared variable with $n-1$ degrees of freedom. To test the hypothesis above, we compute

$$f = \frac{SSR/1}{SSE/(n-2)} = \frac{SSR}{s^2}$$

and reject H_0 at the α-level of significance when $f > f_\alpha(1, n-2)$.

The computations are usually summarized by means of an **analysis-of-variance table**, such as Table 7.2. It is customary to refer to the various sums of squares divided by their respective degrees of freedom as the **mean squares**.

Table 7.2: Analysis of Variance for Testing $\beta_1 = 0$

Source of Variation	Sum of Squares	Degrees of Freedom	Mean Square	Computed f
Regression	SSR	1	SSR	$\frac{SSR}{s^2}$
Error	SSE	$n-2$	$s^2 = \frac{SSE}{n-2}$	
Total	SST	$n-1$		

When the null hypothesis is rejected, that is, when the computed F-statistic exceeds the critical value $f_\alpha(1, n-2)$, we conclude that **there is a significant amount of variation in the response accounted for by the postulated model, the straight-line function.** If the F-statistic is in the fail-to-reject region, we conclude that the data did not reflect sufficient evidence to support the model postulated.

In Section 7.3, a procedure was given whereby the statistic

$$T = \frac{B_1 - \beta_{10}}{S/\sqrt{S_{xx}}}$$

is used to test the hypothesis

$$H_0\colon \ \beta_1 = \beta_{10} \text{ versus } H_1\colon \ \beta_1 \neq \beta_{10},$$

where T follows the t-distribution with $n - 2$ degrees of freedom. The hypothesis is rejected if $|t| > t_{\alpha/2}$ for an α-level of significance. It is interesting to note that in the special case in which we are testing

$$H_0\colon\ \beta_1 = 0 \text{ versus } H_1\colon\ \beta_1 \neq 0,$$

the value of our T-statistic becomes

$$t = \frac{b_1}{s/\sqrt{S_{xx}}},$$

and the hypothesis under consideration is identical to that being tested in Table 7.2. Namely, the null hypothesis states that the variation in the response is due merely to chance. The analysis of variance uses the F-distribution rather than the t-distribution. For the two-sided alternative, the two approaches are identical. This we can see by writing

$$t^2 = \frac{b_1^2 S_{xx}}{s^2} = \frac{b_1 S_{xy}}{s^2} = \frac{SSR}{s^2},$$

which is identical to the f-value used in the analysis of variance. The basic relationship between the t-distribution with v degrees of freedom and the F-distribution with 1 and v degrees of freedom is

$$t^2 = f(1, v).$$

Of course, the t-test allows for testing against a one-sided alternative while the F-test is restricted to testing against a two-sided alternative.

Annotated Computer Printout for Simple Linear Regression

Consider again the chemical oxygen demand reduction data of Table 7.1. Figures 7.14 and 7.15 show more complete annotated computer printouts, again with *MINITAB* software. The t-ratio column indicates tests for null hypotheses of zero values on the parameter. The term "Fit" denotes $\hat{y}$-values, often called **fitted values**. The term "SE Fit" is used in computing confidence intervals on mean response. The item R^2 is computed as $(SSR/SST) \times 100$ and signifies the proportion of variation in y explained by the straight-line regression. Also shown are confidence intervals on the mean response and prediction intervals on a new observation.

```
The regression equation is COD = 3.83 + 0.904 Per_Red
Predictor    Coef   SE Coef      T       P
  Constant   3.830    1.768    2.17   0.038
  Per_Red  0.90364  0.05012   18.03   0.000
S = 3.22954   R-Sq = 91.3%   R-Sq(adj) = 91.0%
              Analysis of Variance
Source          DF      SS      MS      F      P
Regression       1   3390.6  3390.6  325.08  0.000
Residual Error  31    323.3    10.4
Total           32   3713.9

Obs   Per_Red      COD      Fit   SE Fit   Residual   St Resid
  1       3.0    5.000    6.541    1.627     -1.541      -0.55
  2      36.0   34.000   36.361    0.576     -2.361      -0.74
  3       7.0   11.000   10.155    1.440      0.845       0.29
  4      37.0   36.000   37.264    0.590     -1.264      -0.40
  5      11.0   21.000   13.770    1.258      7.230       2.43
  6      38.0   38.000   38.168    0.607     -0.168      -0.05
  7      15.0   16.000   17.384    1.082     -1.384      -0.45
  8      39.0   37.000   39.072    0.627     -2.072      -0.65
  9      18.0   16.000   20.095    0.957     -4.095      -1.33
 10      39.0   36.000   39.072    0.627     -3.072      -0.97
 11      27.0   28.000   28.228    0.649     -0.228      -0.07
 12      39.0   45.000   39.072    0.627      5.928       1.87
 13      29.0   27.000   30.035    0.605     -3.035      -0.96
 14      40.0   39.000   39.975    0.651     -0.975      -0.31
 15      30.0   25.000   30.939    0.588     -5.939      -1.87
 16      41.0   41.000   40.879    0.678      0.121       0.04
 17      30.0   35.000   30.939    0.588      4.061       1.28
 18      42.0   40.000   41.783    0.707     -1.783      -0.57
 19      31.0   30.000   31.843    0.575     -1.843      -0.58
 20      42.0   44.000   41.783    0.707      2.217       0.70
 21      31.0   40.000   31.843    0.575      8.157       2.57
 22      43.0   37.000   42.686    0.738     -5.686      -1.81
 23      32.0   32.000   32.746    0.567     -0.746      -0.23
 24      44.0   44.000   43.590    0.772      0.410       0.13
 25      33.0   34.000   33.650    0.563      0.350       0.11
 26      45.0   46.000   44.494    0.807      1.506       0.48
 27      33.0   32.000   33.650    0.563     -1.650      -0.52
 28      46.0   46.000   45.397    0.843      0.603       0.19
 29      34.0   34.000   34.554    0.563     -0.554      -0.17
 30      47.0   49.000   46.301    0.881      2.699       0.87
 31      36.0   37.000   36.361    0.576      0.639       0.20
 32      50.0   51.000   49.012    1.002      1.988       0.65
 33      36.0   38.000   36.361    0.576      1.639       0.52
```

Figure 7.14: *MINITAB* printout of simple linear regression for chemical oxygen demand reduction data; part I.

Obs	Fit	SE Fit	95% CI	95% PI
1	6.541	1.627	(3.223, 9.858)	(-0.834, 13.916)
2	36.361	0.576	(35.185, 37.537)	(29.670, 43.052)
3	10.155	1.440	(7.218, 13.092)	(2.943, 17.367)
4	37.264	0.590	(36.062, 38.467)	(30.569, 43.960)
5	13.770	1.258	(11.204, 16.335)	(6.701, 20.838)
6	38.168	0.607	(36.931, 39.405)	(31.466, 44.870)
7	17.384	1.082	(15.177, 19.592)	(10.438, 24.331)
8	39.072	0.627	(37.793, 40.351)	(32.362, 45.781)
9	20.095	0.957	(18.143, 22.047)	(13.225, 26.965)
10	39.072	0.627	(37.793, 40.351)	(32.362, 45.781)
11	28.228	0.649	(26.905, 29.551)	(21.510, 34.946)
12	39.072	0.627	(37.793, 40.351)	(32.362, 45.781)
13	30.035	0.605	(28.802, 31.269)	(23.334, 36.737)
14	39.975	0.651	(38.648, 41.303)	(33.256, 46.694)
15	30.939	0.588	(29.739, 32.139)	(24.244, 37.634)
16	40.879	0.678	(39.497, 42.261)	(34.149, 47.609)
17	30.939	0.588	(29.739, 32.139)	(24.244, 37.634)
18	41.783	0.707	(40.341, 43.224)	(35.040, 48.525)
19	31.843	0.575	(30.669, 33.016)	(25.152, 38.533)
20	41.783	0.707	(40.341, 43.224)	(35.040, 48.525)
21	31.843	0.575	(30.669, 33.016)	(25.152, 38.533)
22	42.686	0.738	(41.181, 44.192)	(35.930, 49.443)
23	32.746	0.567	(31.590, 33.902)	(26.059, 39.434)
24	43.590	0.772	(42.016, 45.164)	(36.818, 50.362)
25	33.650	0.563	(32.502, 34.797)	(26.964, 40.336)
26	44.494	0.807	(42.848, 46.139)	(37.704, 51.283)
27	33.650	0.563	(32.502, 34.797)	(26.964, 40.336)
28	45.397	0.843	(43.677, 47.117)	(38.590, 52.205)
29	34.554	0.563	(33.406, 35.701)	(27.868, 41.239)
30	46.301	0.881	(44.503, 48.099)	(39.473, 53.128)
31	36.361	0.576	(35.185, 37.537)	(29.670, 43.052)
32	49.012	1.002	(46.969, 51.055)	(42.115, 55.908)
33	36.361	0.576	(35.185, 37.537)	(29.670, 43.052)

Figure 7.15: *MINITAB* printout of simple linear regression for chemical oxygen demand reduction data; part II.

7.6   Test for Linearity of Regression: Data with Repeated Observations

In certain kinds of experimental situations, the researcher has the capability of obtaining repeated observations on the response for each value of x. Although it is not necessary to have these repetitions in order to estimate β_0 and β_1, nevertheless repetitions enable the experimenter to obtain quantitative information concerning the appropriateness of the model. In fact, if repeated observations are generated, the experimenter can make a significance test to aid in determining whether or not the model is adequate.

Let us select a random sample of n observations using k distinct values of x, say $x_1, x_2, \ldots, x_n$, such that the sample contains n_1 observed values of the random variable Y_1 corresponding to x_1, n_2 observed values of Y_2 corresponding to $x_2, \ldots,$ n_k observed values of Y_k corresponding to x_k. Of necessity, $n = \sum_{i=1}^{k} n_i$.

We define

$$y_{ij} = \text{the } j\text{th value of the random variable } Y_i,$$

$$y_{i.} = T_{i.} = \sum_{j=1}^{n_i} y_{ij},$$

$$\bar{y}_{i.} = \frac{T_{i.}}{n_i}.$$

Hence, if $n_4 = 3$ measurements of Y were made corresponding to $x = x_4$, we would indicate these observations by $y_{41}, y_{42},$ and y_{43}. Then

$$T_{i.} = y_{41} + y_{42} + y_{43}.$$

Concept of Lack of Fit

The error sum of squares consists of two parts: the amount due to the variation between the values of Y within given values of x and a component that is normally called the **lack-of-fit** contribution. The first component reflects mere random variation, or **pure experimental error**, while the second component is a measure of the systematic variation brought about by higher-order terms. In our case, these are terms in x other than the linear, or first-order, contribution. Note that in choosing a linear model we are essentially assuming that this second component does not exist and hence our error sum of squares is completely due to random errors. If this should be the case, then $s^2 = SSE/(n-2)$ is an unbiased estimate of σ^2. However, if the model does not adequately fit the data, then the error sum of squares is inflated and produces a biased estimate of σ^2. Whether or not the model fits the data, an unbiased estimate of σ^2 can always be obtained when we have repeated observations simply by computing

$$s_i^2 = \frac{\sum_{j=1}^{n_i} (y_{ij} - \bar{y}_{i.})^2}{n_i - 1}, \quad i = 1, 2, \ldots, k,$$

for each of the k distinct values of x and then pooling these variances to get

$$s^2 = \frac{\sum_{i=1}^{k}(n_i - 1)s_i^2}{n-k} = \frac{\sum_{i=1}^{k}\sum_{j=1}^{n_i}(y_{ij} - \bar{y}_{i.})^2}{n-k}.$$

The numerator of s^2 is a **measure of the pure experimental error**. A computational procedure for separating the error sum of squares into the two components representing pure error and lack of fit is as follows:

Computation of Lack-of-Fit Sum of Squares

1. Compute the pure error sum of squares

$$\sum_{i=1}^{k}\sum_{j=1}^{n_i}(y_{ij} - \bar{y}_{i.})^2.$$

This sum of squares has $n - k$ degrees of freedom associated with it, and the resulting mean square is our unbiased estimate s^2 of σ^2.

2. Subtract the pure error sum of squares from the error sum of squares SSE, thereby obtaining the sum of squares due to lack of fit. The degrees of freedom for lack of fit are obtained by simply subtracting $(n - 2) - (n - k) = k - 2$.

The computations required for testing hypotheses in a regression problem with repeated measurements on the response may be summarized as shown in Table 7.3.

Table 7.3: Analysis of Variance for Testing Linearity of Regression

Source of Variation	Sum of Squares	Degrees of Freedom	Mean Square	Computed f
Regression	SSR	1	SSR	$\frac{SSR}{s^2}$
Error	SSE	$n - 2$		
Lack of fit	$\{$ $SSE - SSE$ (pure)	$\{$ $k - 2$	$\frac{SSE-SSE(\text{pure})}{k-2}$	$\frac{SSE-SSE(\text{pure})}{s^2(k-2)}$
Pure error	SSE (pure)	$n - k$	$s^2 = \frac{SSE(\text{pure})}{n-k}$	
Total	SST	$n - 1$		

Figures 7.16 and 7.17 display the sample points for the "correct model" and "incorrect model" situations. In Figure 7.16, where the $\mu_{Y|x}$ fall on a straight line, there is no lack of fit when a linear model is assumed, so the sample variation around the regression line is a pure error resulting from the variation that occurs among repeated observations. In Figure 7.17, where the $\mu_{Y|x}$ clearly do not fall on a straight line, the lack of fit from erroneously choosing a linear model accounts for a large portion of the variation around the regression line, supplementing the pure error.

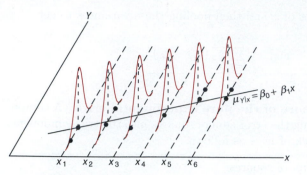

Figure 7.16: Correct linear model with no lack-of-fit component.

Figure 7.17: Incorrect linear model with lack-of-fit component.

What Is the Importance of Detecting Lack of Fit?

The concept of lack of fit is extremely important in applications of regression analysis. In fact, the need to construct or design an experiment that will account for lack of fit becomes more critical as the problem and the underlying mechanism involved become more complicated. Surely, one cannot always be certain that one's postulated structure, in this case the linear regression model, is correct or even an adequate representation. The following example shows how the error sum of squares is partitioned into the two components representing pure error and lack of fit. The adequacy of the model is tested at the α-level of significance by comparing the lack-of-fit mean square divided by s^2 with $f_\alpha(k - 2, n - k)$.

Example 7.7: Observations of the yield of a chemical reaction taken at various temperatures were recorded in Table 7.4. Estimate the linear model $\mu_{Y|x} = \beta_0 + \beta_1 x$ and test for lack of fit.

Table 7.4: Data for Example 7.7

y (%)	x (°C)	y (%)	x (°C)
77.4	150	88.9	250
76.7	150	89.2	250
78.2	150	89.7	250
84.1	200	94.8	300
84.5	200	94.7	300
83.7	200	95.9	300

Solution: Results of the computations are shown in Table 7.5. The partitioning of the total variation in this manner reveals a significant variation accounted for by the linear model and an insignificant amount of variation due to lack of fit. Thus, the experimental data do not seem to suggest the need to consider terms higher than first order in the model, and the null hypothesis is not rejected.

Table 7.5: Analysis of Variance on Yield-Temperature Data

Source of Variation	Sum of Squares	Degrees of Freedom	Mean Square	Computed f	P-Values
Regression	509.2507	1	509.2507	1531.58	<0.0001
Error	3.8660	10			
Lack of fit	1.2060	2	0.6030	1.81	0.2241
Pure error	2.6600	8	0.3325		
Total	513.1167	11			

Annotated Computer Printout for Test for Lack of Fit

Figure 7.18 is an annotated computer printout showing analysis of the data of Example 7.7 with *SAS*. Note the "LOF" with 2 degrees of freedom, representing the quadratic and cubic contribution to the model, and the *P*-value of 0.22, suggesting that the linear (first-order) model is adequate.

```
Dependent Variable: yield
                                  Sum of
Source                  DF        Squares      Mean Square    F Value    Pr > F
Model                    3     510.4566667    170.1522222     511.74    <.0001
Error                    8       2.6600000      0.3325000
Corrected Total         11     513.1166667

            R-Square      Coeff Var      Root MSE     yield Mean
            0.994816      0.666751       0.576628      86.48333

Source                  DF      Type I SS      Mean Square    F Value    Pr > F
temperature              1     509.2506667    509.2506667    1531.58    <.0001
LOF                      2       1.2060000      0.6030000       1.81    0.2241
```

Figure 7.18: *SAS* printout, showing analysis of data of Example 7.7.

Exercises

7.31 Test for linearity of regression in Exercise 7.3 on page 304. Use a 0.05 level of significance. Comment.

7.32 Test for linearity of regression in Exercise 7.8 on page 305. Comment.

7.33 Suppose we have a linear equation through the origin (Exercise 7.28) $\mu_{Y|x} = \beta x$.

(a) Estimate the regression line passing through the origin for the following data:

x	0.5	1.5	3.2	4.2	5.1	6.5
y	1.3	3.4	6.7	8.0	10.0	13.2

(b) Suppose it is not known whether the true regression should pass through the origin. Estimate the linear model $\mu_{Y|x} = \beta_0 + \beta_1 x$ and test the hypothesis that $\beta_0 = 0$, at the 0.10 level of significance, against the alternative that $\beta_0 \neq 0$.

7.34 Use an analysis-of-variance approach to test the hypothesis that $\beta_1 = 0$ against the alternative hypothesis $\beta_1 \neq 0$ in Exercise 7.5 on page 304 at the 0.05 level of significance.

7.35 Organophosphate (OP) compounds are used as pesticides. However, it is important to study their effect on species that are exposed to them. In the laboratory study *Some Effects of Organophosphate Pesticides on Wildlife Species*, by the Department of Fisheries and Wildlife at Virginia Tech, an experiment was conducted in which different dosages of a particular OP pesticide were administered to 5 groups of 5 mice (*Peromyscus leucopus*). The 25 mice were females of similar age and condition. One group received no chemical. The basic response y was a measure of activity in the brain. It was postulated that brain activity would decrease with an increase in OP dosage. Use the provided data to do the following.

(a) Find the least squares estimates of β_0 and β_1 using the model

$$Y_i = \beta_0 + \beta_1 x_i + \epsilon_i, \quad i = 1, 2, \dots, 25.$$

(b) Construct an analysis-of-variance table in which the lack of fit and pure error have been separated. Determine if the lack of fit is significant at the 0.05 level. Interpret the results.

Animal	Dose, x (mg/kg body weight)	Activity, y (moles/liter/min)
1	0.0	10.9
2	0.0	10.6
3	0.0	10.8
4	0.0	9.8
5	0.0	9.0
6	2.3	11.0
7	2.3	11.3
8	2.3	9.9
9	2.3	9.2
10	2.3	10.1
11	4.6	10.6
12	4.6	10.4
13	4.6	8.8
14	4.6	11.1
15	4.6	8.4
16	9.2	9.7
17	9.2	7.8
18	9.2	9.0
19	9.2	8.2
20	9.2	2.3
21	18.4	2.9
22	18.4	2.2
23	18.4	3.4
24	18.4	5.4
25	18.4	8.2

7.36 Transistor gain between emitter and collector in an integrated circuit device (hFE) is related to two variables (Myers, Montgomery, and Anderson-Cook, 2009) that can be controlled at the deposition process, emitter drive-in time (x_1, in minutes) and emitter dose (x_2, in ions $\times 10^{14}$). Fourteen samples were observed following deposition, and the resulting data are shown in the following table. We will consider linear regression models using gain as the response and emitter drive-in time or emitter dose as the regressor variable.

(a) Determine if emitter drive-in time influences gain in a linear relationship. That is, test H_0: $\beta_1 = 0$, where β_1 is the slope of the regressor variable.

(b) Do a lack-of-fit test to determine if the linear relationship is adequate. Draw conclusions.

(c) Determine if emitter dose influences gain in a linear relationship. Which regressor variable is the better predictor of gain?

Obs.	x_1 (drive-in time, min)	x_2 (dose, ions $\times 10^{14}$)	y (gain, or hFE)
1	195	4.00	1004
2	255	4.00	1636
3	195	4.60	852
4	255	4.60	1506
5	255	4.20	1272
6	255	4.10	1270
7	255	4.60	1269
8	195	4.30	903
9	255	4.30	1555
10	255	4.00	1260
11	255	4.70	1146
12	255	4.30	1276
13	255	4.72	1225
14	340	4.30	1321

7.37 The following data are a result of an investigation as to the effect of reaction temperature x on percent conversion of a chemical process y. (See Myers, Montgomery, and Anderson-Cook, 2009.) Fit a simple linear regression, and use a lack-of-fit test to determine if the model is adequate. Discuss.

Observation	Temperature (°C), x	Conversion (%), y
1	200	43
2	250	78
3	200	69
4	250	73
5	189.65	48
6	260.35	78
7	225	65
8	225	74
9	225	76
10	225	79
11	225	83
12	225	81

7.38 Heat treating is often used to carburize metal parts such as gears. The thickness of the carburized layer is considered an important feature of the gear, and it contributes to the overall reliability of the part. Because of the critical nature of this feature, a lab test is performed on each furnace load. The test is a destructive one, where an actual part is cross sectioned and soaked in a chemical for a period of time. This test involves running a carbon analysis on the surface of both the gear pitch (top of the gear tooth) and the gear root (between the gear teeth). The data provided are the results of the pitch carbon-analysis test for 19 parts.

(a) Fit a simple linear regression relating the pitch carbon analysis y against soak time. Test H_0: $\beta_1 = 0$.

(b) If the hypothesis in part (a) is rejected, determine if the linear model is adequate.

Soak Time	Pitch	Soak Time	Pitch
0.58	0.013	1.17	0.021
0.66	0.016	1.17	0.019
0.66	0.015	1.17	0.021
0.66	0.016	1.20	0.025
0.66	0.015	2.00	0.025
0.66	0.016	2.00	0.026
1.00	0.014	2.20	0.024
1.17	0.021	2.20	0.025
1.17	0.018	2.20	0.024
1.17	0.019		

7.39 A regression model is desired relating temperature and the proportion of impurities passing through solid helium. Temperature is listed in degrees centigrade. The data are as follows:

Temperature (°C)	Proportion of Impurities
−260.5	0.425
−255.7	0.224
−264.6	0.453
−265.0	0.475
−270.0	0.705
−272.0	0.860
−272.5	0.935
−272.6	0.961
−272.8	0.979
−272.9	0.990

(a) Fit a linear regression model.

(b) Does it appear that the proportion of impurities passing through helium increases as the temperature approaches −273 degrees centigrade?

(c) Find R^2.

(d) Based on the information above, does the linear model seem appropriate? What additional information would you need to better answer that question?

7.40 It is of interest to study the effect of population size in various cities in the United States on ozone concentrations. The data consist of the 1999 population in millions and the amount of ozone present per hour in ppb (parts per billion). The data are as follows.

Ozone (ppb/hour), y	Population, x
126	0.6
135	4.9
124	0.2
128	0.5
130	1.1
128	0.1
126	1.1
128	2.3
128	0.6
129	2.3

(a) Fit the linear regression model relating ozone concentration to population. Test H_0: $\beta_1 = 0$ using the ANOVA approach.

(b) Do a test for lack of fit. Is the linear model appropriate based on the results of your test?

(c) Test the hypothesis of part (a) using the pure mean square error in the F-test. Do the results change? Comment on the advantage of each test.

7.41 Evaluating nitrogen deposition from the atmosphere is a major role of the National Atmospheric Deposition Program (NADP), a partnership of many agencies. NADP is studying atmospheric deposition and its effect on agricultural crops, forest surface waters, and other resources. Nitrogen oxides may affect the ozone in the atmosphere and the amount of pure nitrogen in the air we breathe. The data are as follows:

Year	Nitrogen Oxide
1978	0.73
1979	2.55
1980	2.90
1981	3.83
1982	2.53
1983	2.77
1984	3.93
1985	2.03
1986	4.39
1987	3.04
1988	3.41
1989	5.07
1990	3.95
1991	3.14
1992	3.44
1993	3.63
1994	4.50
1995	3.95
1996	5.24
1997	3.30
1998	4.36
1999	3.33

(a) Plot the data.

(b) Fit a linear regression model and find R^2.

(c) What can you say about the trend in nitrogen oxide across time?

7.42 For a particular variety of plant, researchers wanted to develop a formula for predicting the quantity of seeds (in grams) as a function of the density of plants. They conducted a study with four levels of the factor x, the number of plants per plot. Four replications were used for each level of x. The data are as follows:

Plants per Plot, x	Quantity of Seeds, y (grams)			
10	12.6	11.0	12.1	10.9
20	15.3	16.1	14.9	15.6
30	17.9	18.3	18.6	17.8
40	19.2	19.6	18.9	20.0

Is a simple linear regression model adequate for analyzing this data set?

7.7   Diagnostic Plots of Residuals: Graphical Detection of Violation of Assumptions

Plots of the raw data can be extremely helpful in determining the nature of the model that should be fit to the data when there is a single independent variable. We have attempted to illustrate this in the foregoing. Detection of proper model form is, however, not the only benefit gained from diagnostic plotting. As in much of the material associated with significance testing in Chapter 6, plotting methods can illustrate and detect violation of assumptions. The reader should recall that much of what is illustrated in this chapter requires assumptions made on the model errors, the ϵ_i. In fact, we assume that the ϵ_i are independent $N(0, \sigma)$ random variables. Now, of course, the ϵ_i are not observed. However, the $e_i = y_i - \hat{y}_i$, the *residuals*, are the error in the fit of the regression line and thus serve to mimic the ϵ_i. Thus, the general complexion of these residuals can often highlight difficulties. Ideally, of course, the plot of the residuals is as depicted in Figure 7.19. That is, they should truly show random fluctuations around a value of zero.

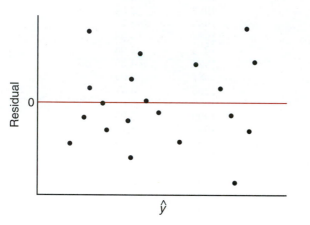

Figure 7.19: Ideal residual plot.

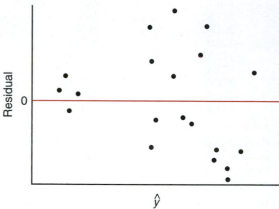

Figure 7.20: Residual plot depicting heterogeneous error variance.

Nonhomogeneous Variance

Homogeneous variance is an important assumption made in regression analysis. Violations can often be detected through the appearance of the residual plot. Increasing error variance with an increase in the regressor variable is a common condition in scientific data. Large error variance produces large residuals, and hence a residual plot like the one in Figure 7.20 is a signal of nonhomogeneous variance. More discussion regarding these residual plots and information regarding different types of residuals appears later in this chapter, where we deal with multiple linear regression.

Normal Probability Plotting

The assumption that the model errors are normal is made when the data analyst deals in either hypothesis testing or confidence interval estimation. Again, the numerical counterparts to the ϵ_i, namely the residuals, are subjects of diagnostic plotting to detect any extreme violations. To diagnose the normality of the residuals, normal quantile-quantile plots and normal probability plots can be used. These plots on residuals can be found in Walpole et al. (2011).

7.8 Correlation

The constant ρ (rho) is called the **population correlation coefficient** and plays a major role in many bivariate data analysis problems. It is important for the reader to understand the physical interpretation of this correlation coefficient and the distinction between correlation and regression. The term *regression* still has meaning here. In fact, the straight line given by $\mu_{Y|x} = \beta_0 + \beta_1 x$ is still called the regression line as before, and the estimates of β_0 and β_1 are identical to those given in Section 7.2. The value of ρ is 0 when $\beta_1 = 0$, which results when there essentially is no linear regression; that is, the regression line is horizontal and any knowledge of X is useless in predicting Y. Since $\sigma_Y^2 \geq \sigma^2$ as the variance in the response is the addition of the variance in X and the error, we must have $\rho^2 \leq 1$ and hence $-1 \leq \rho \leq 1$. Values of $\rho = \pm 1$ only occur when $\sigma^2 = 0$, in which case we have a perfect linear relationship between the two variables. Thus, a value of ρ equal to $+1$ implies a perfect linear relationship with a positive slope, while a value of ρ equal to -1 results from a perfect linear relationship with a negative slope. It might be said, then, that sample estimates of ρ close to unity in magnitude imply good correlation, or **linear association**, between X and Y, whereas values near zero indicate little or no correlation.

To obtain a sample estimate of ρ, recall from Section 7.3 that the error sum of squares is

$$SSE = S_{yy} - b_1 S_{xy}.$$

Dividing both sides of this equation by S_{yy} and replacing S_{xy} by $b_1 S_{xx}$, we obtain the relation

$$b_1^2 \frac{S_{xx}}{S_{yy}} = 1 - \frac{SSE}{S_{yy}}.$$

The value of $b_1^2 S_{xx}/S_{yy}$ is zero when $b_1 = 0$, which will occur when the sample points show no linear relationship. Since $S_{yy} \geq SSE$, we conclude that $b_1^2 S_{xx}/S_{xy}$ must be between 0 and 1. Consequently, $b_1 \sqrt{S_{xx}/S_{yy}}$ must range from -1 to $+1$, negative values corresponding to lines with negative slopes and positive values to lines with positive slopes. A value of -1 or $+1$ will occur when $SSE = 0$, but this is the case where all sample points lie in a straight line. Hence, a perfect linear relationship appears in the sample data when $b_1 \sqrt{S_{xx}/S_{yy}} = \pm 1$. Clearly, the quantity $b_1 \sqrt{S_{xx}/S_{yy}}$, which we shall henceforth designate as r, can be used as an estimate of the population correlation coefficient ρ. It is customary to refer to the estimate r as the **Pearson product-moment correlation coefficient** or simply the **sample correlation coefficient**.

Correlation Coefficient The measure ρ of linear association between two variables X and Y is estimated by the **sample correlation coefficient** r, where

$$r = b_1 \sqrt{\frac{S_{xx}}{S_{yy}}} = \frac{S_{xy}}{\sqrt{S_{xx} S_{yy}}}.$$

For values of r between -1 and $+1$ we must be careful in our interpretation. For example, values of r equal to 0.3 and 0.6 only mean that we have two positive correlations, one somewhat stronger than the other. It is wrong to conclude that $r = 0.6$ indicates a linear relationship twice as good as that indicated by the value $r = 0.3$. On the other hand, if we write

$$r^2 = \frac{S_{xy}^2}{S_{xx}S_{yy}} = \frac{SSR}{S_{yy}},$$

then r^2, which is usually referred to as the **sample coefficient of determination**, represents the proportion of the variation of S_{yy} explained by the regression of Y on x, namely SSR. That is, r^2 expresses the proportion of the total variation in the values of the variable Y that can be accounted for or explained by a linear relationship with the values of the random variable X. Thus, a correlation of 0.6 means that 0.36, or 36%, of the total variation of the values of Y in our sample is accounted for by a linear relationship with values of X.

Example 7.8: It is important that scientific researchers in the area of forest products be able to study correlation among the anatomy and mechanical properties of trees. For the study *Quantitative Anatomical Characteristics of Plantation Grown Loblolly Pine (Pinus taeda L.) and Cottonwood (Populus deltoides Bart. Ex Marsh.) and Their Relationships to Mechanical Properties*, conducted by the Department of Forestry and Forest Products at Virginia Tech, 29 loblolly pines were randomly selected for investigation. Table 7.6 shows the resulting data on the specific gravity in grams/cm^3 and the modulus of rupture in kilopascals (kPa). Compute and interpret the sample correlation coefficient.

Table 7.6: Data on 29 Loblolly Pines for Example 7.8

Specific Gravity, x (g/cm^3)	Modulus of Rupture, y (kPa)	Specific Gravity, x (g/cm^3)	Modulus of Rupture, y (kPa)
0.414	29,186	0.581	85,156
0.383	29,266	0.557	69,571
0.399	26,215	0.550	84,160
0.402	30,162	0.531	73,466
0.442	38,867	0.550	78,610
0.422	37,831	0.556	67,657
0.466	44,576	0.523	74,017
0.500	46,097	0.602	87,291
0.514	59,698	0.569	86,836
0.530	67,705	0.544	82,540
0.569	66,088	0.557	81,699
0.558	78,486	0.530	82,096
0.577	89,869	0.547	75,657
0.572	77,369	0.585	80,490
0.548	67,095		

Solution: From the data we find that

$$S_{xx} = 0.11273, \quad S_{yy} = 11,807,324,805, \quad S_{xy} = 34,422.27572.$$

Therefore,

$$r = \frac{34{,}422.27572}{\sqrt{(0.11273)(11{,}807{,}324{,}805)}} = 0.9435.$$

A correlation coefficient of 0.9435 indicates a good linear relationship between X and Y. Since $r^2 = 0.8902$, we can say that approximately 89% of the variation in the values of Y is accounted for by a linear relationship with X.

A test of the special hypothesis $\rho = 0$ versus an appropriate alternative is equivalent to testing $\beta_1 = 0$ for the simple linear regression model, and therefore the procedures of Section 7.5 using either the t-distribution with $n - 2$ degrees of freedom or the F-distribution with 1 and $n - 2$ degrees of freedom are applicable. However, if one wishes to avoid the analysis-of-variance procedure and compute only the sample correlation coefficient, it can be verified that the t-value

$$t = \frac{b_1}{s/\sqrt{S_{xx}}}$$

can also be written as

$$t = \frac{r\sqrt{n - 2}}{\sqrt{1 - r^2}},$$

which, as before, is a value of the statistic T having a t-distribution with $n - 2$ degrees of freedom.

Example 7.9: For the data of Example 7.8, test the hypothesis that there is no linear association among the variables.

Solution: 1. H_0: $\rho = 0$.

2. H_1: $\rho \neq 0$.

3. $\alpha = 0.05$.

4. Critical region: $t < -2.052$ or $t > 2.052$.

5. Computations: $t = \frac{0.9435\sqrt{27}}{\sqrt{1 - 0.9435^2}} = 14.79$, $P < 0.0001$.

6. Decision: Reject the hypothesis of no linear association.

7.9 Simple Linear Regression Case Study

In the manufacture of commercial wood products, it is important to estimate the relationship between the density of a wood product and its stiffness. A relatively new type of particleboard is being considered that can be formed with considerably more ease than the accepted commercial product. It is necessary to know at what density the stiffness is comparable to that of the well-known, well-documented commercial product. A study of these properties was done by Terrance E. Conners: *Investigation of Certain Mechanical Properties of a Wood-Foam Composite* (Master's Thesis, Department of Forestry and Wildlife Management, University of Massachusetts). Thirty particleboards were produced at densities ranging from roughly 8 to 26 pounds per cubic foot, and the stiffness was measured in pounds per square inch. Table 7.7 shows the data.

It is necessary for the data analyst to focus on an appropriate fit to the data and use inferential methods discussed in this chapter. Hypothesis testing on the slope of the regression, as well as confidence or prediction interval estimation, may

Table 7.7: Density and Stiffness for 30 Particleboards

Density, x	Stiffness, y	Density, x	Stiffness, y
9.50	14,814.00	8.40	17,502.00
9.80	14,007.00	11.00	19,443.00
8.30	7573.00	9.90	14,191.00
8.60	9714.00	6.40	8076.00
7.00	5304.00	8.20	10,728.00
17.40	43,243.00	15.00	25,319.00
15.20	28,028.00	16.40	41,792.00
16.70	49,499.00	15.40	25,312.00
15.00	26,222.00	14.50	22,148.00
14.80	26,751.00	13.60	18,036.00
25.60	96,305.00	23.40	104,170.00
24.40	72,594.00	23.30	49,512.00
19.50	32,207.00	21.20	48,218.00
22.80	70,453.00	21.70	47,661.00
19.80	38,138.00	21.30	53,045.00

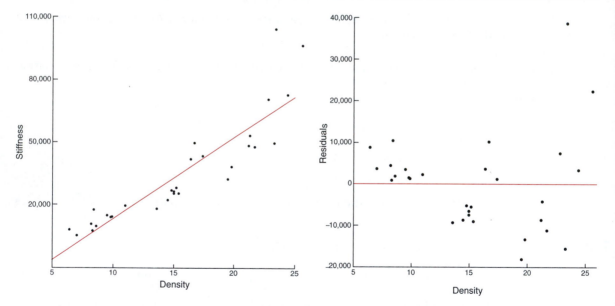

Figure 7.21: Scatter plot of the wood density data.

Figure 7.22: Residual plot for the wood density data.

well be appropriate. We begin by demonstrating a simple scatter plot of the raw data with a simple linear regression superimposed. Figure 7.21 shows this plot. The simple linear regression fit to the data produced the fitted model

$$\hat{y} = -25{,}433.739 + 3884.976x \quad (R^2 = 0.7975),$$

and the residuals were computed. Figure 7.22 shows the residuals plotted against the measurements of density. This is hardly an ideal or healthy set of residuals.

They do not show a random scatter around a value of zero. In fact, clusters of positive and negative values suggest that a curvilinear trend in the data should be investigated.

Exercises

7.43 Compute and interpret the correlation coefficient for the following grades of 6 students selected at random:

Mathematics grade	70 92 80 74 65 83
English grade	74 84 63 87 78 90

7.44 Use the data of Exercise 7.1 on page 303 to do the followings.

(a) Calculate r.

(b) Test the hypothesis that $\rho = 0$ against the alternative that $\rho \neq 0$ at the 0.05 level of significance.

7.45 With reference to Exercise 7.13 on page 306, assume a bivariate normal distribution for x and y.

(a) Calculate r.

(b) Test the null hypothesis that $\rho = -0.5$ against the alternative that $\rho < -0.5$ at the 0.025 level of significance.

(c) Determine the percentage of the variation in the amount of particulate removed that is due to changes in the daily amount of rainfall.

7.46 Test the hypothesis that $\rho = 0$ in Exercise 7.43 against the alternative that $\rho \neq 0$. Use a 0.05 level of significance.

7.47 The following data were obtained in a study of the relationship between the weight and chest size of infants at birth.

Weight (kg)	Chest Size (cm)
2.75	29.5
2.15	26.3
4.41	32.2
5.52	36.5
3.21	27.2
4.32	27.7
2.31	28.3
4.30	30.3
3.71	28.7

(a) Calculate r.

(b) Test the null hypothesis that $\rho = 0$ against the alternative that $\rho > 0$ at the 0.01 level of significance.

(c) What percentage of the variation in infant chest sizes is explained by difference in weight?

7.10 Multiple Linear Regression and Estimation of the Coefficients

In most research problems where regression analysis is applied, more than one independent variable is needed in the regression model. The complexity of most scientific mechanisms is such that in order to be able to predict an important response, a **multiple regression model** is needed. When this model is linear in the coefficients, it is called a **multiple linear regression model**. For the case of k independent variables $x_1, x_2, \ldots, x_k$, the mean of $Y|x_1, x_2, \ldots, x_k$ is given by the multiple linear regression model

$$\mu_{Y|x_1,x_2,\ldots,x_k} = \beta_0 + \beta_1 x_1 + \cdots + \beta_k x_k,$$

and the estimated response is obtained from the sample regression equation

$$\hat{y} = b_0 + b_1 x_1 + \cdots + b_k x_k,$$

where each regression coefficient β_i is estimated by b_i from the sample data using the method of least squares. As in the case of a single independent variable, the multiple linear regression model can often be an adequate representation of a more complicated structure within certain ranges of the independent variables.

Similar least squares techniques can also be applied for estimating the coefficients when the linear model involves, say, powers and products of the independent variables. For example, when $k = 1$, the experimenter may believe that the means $\mu_{Y|x}$ do not fall on a straight line but are more appropriately described by the **polynomial regression model**

$$\mu_{Y|x} = \beta_0 + \beta_1 x + \beta_2 x^2 + \cdots + \beta_r x^r,$$

and the estimated response is obtained from the polynomial regression equation

$$\hat{y} = b_0 + b_1 x + b_2 x^2 + \cdots + b_r x^r.$$

Confusion arises occasionally when we speak of a polynomial model as a linear model. However, statisticians normally refer to a linear model as one in which the parameters occur linearly, regardless of how the independent variables enter the model. An example of a nonlinear model is the **exponential relationship**

$$\mu_{Y|x} = \alpha \beta^x,$$

whose response is estimated by the regression equation

$$\hat{y} = ab^x.$$

There are many phenomena in science and engineering that are inherently nonlinear in nature, and when the true structure is known, an attempt should certainly be made to fit the actual model. The literature on estimation by least squares of nonlinear models is voluminous. The nonlinear models discussed in this chapter deal with nonideal conditions in which the analyst is certain that the response and hence the response model error are not normally distributed but, rather, have a binomial or Poisson distribution. These situations do occur extensively in practice.

A student who wants a more general account of nonlinear regression should consult *Classical and Modern Regression with Applications* by Myers (1990; see the Bibliography).

In this section, we obtain the least squares estimators of the parameters $\beta_0, \beta_1, \ldots,$ by fitting the multiple linear regression model

$$\mu_{Y|x_1,x_2,\ldots,x_k} = \beta_0 + \beta_1 x_1 + \cdots + \beta_k x_k$$

to the data points

$$\{(x_{1i}, x_{2i}, \ldots, x_{ki}, y_i); \quad i = 1, 2, \ldots, n \text{ and } n > k\},$$

where y_i is the observed response to the values $x_{1i}, x_{2i}, \ldots, x_{ki}$ of the k independent variables $x_1, x_2, \ldots, x_k$. Each observation $(x_{1i}, x_{2i}, \ldots, x_{ki}, y_i)$ is assumed to satisfy the following equation.

Multiple Linear Regression Model

$$y_i = \beta_0 + \beta_1 x_{1i} + \beta_2 x_{2i} + \cdots + \beta_k x_{ki} + \epsilon_i$$

or

$$y_i = \hat{y}_i + e_i = b_0 + b_1 x_{1i} + b_2 x_{2i} + \cdots + b_k x_{ki} + e_i,$$

where ϵ_i and e_i are the random error and residual, respectively, associated with the response y_i and fitted value $\hat{y}_i$.

As in the case of simple linear regression, it is assumed that the ϵ_i are independent and identically distributed with mean 0 and common variance σ^2.

In using the concept of least squares to arrive at estimates $b_0, b_1, \ldots, b_k$, we minimize the expression

$$SSE = \sum_{i=1}^{n} e_i^2 = \sum_{i=1}^{n} (y_i - b_0 - b_1 x_{1i} - b_2 x_{2i} - \cdots - b_k x_{ki})^2.$$

Differentiating SSE in turn with respect to $b_0, b_1, \ldots, b_k$ and equating to zero, we generate the set of $k + 1$ **normal equations for multiple linear regression**.

Normal Estimation Equations for Multiple Linear Regression	$$nb_0 + b_1 \sum_{i=1}^{n} x_{1i} \;\; + b_2 \sum_{i=1}^{n} x_{2i} \;\; + \;\cdots\; + b_k \sum_{i=1}^{n} x_{ki} \;\; = \sum_{i=1}^{n} y_i$$ $$b_0 \sum_{i=1}^{n} x_{1i} + b_1 \sum_{i=1}^{n} x_{1i}^2 \;\; + b_2 \sum_{i=1}^{n} x_{1i} x_{2i} + \;\cdots\; + b_k \sum_{i=1}^{n} x_{1i} x_{ki} = \sum_{i=1}^{n} x_{1i} y_i$$ $$\vdots \qquad\qquad \vdots \qquad\qquad\quad \vdots \qquad\qquad\qquad \vdots \qquad\qquad \vdots$$ $$b_0 \sum_{i=1}^{n} x_{ki} + b_1 \sum_{i=1}^{n} x_{ki} x_{1i} + b_2 \sum_{i=1}^{n} x_{ki} x_{2i} + \;\cdots\; + b_k \sum_{i=1}^{n} x_{ki}^2 \;\; = \sum_{i=1}^{n} x_{ki} y_i$$

These equations can be solved for $b_0, b_1, b_2, \ldots, b_k$ by any appropriate method for solving systems of linear equations. Most statistical software can be used to obtain numerical solutions of the above equations.

Example 7.10: A study was done on a diesel-powered light-duty pickup truck to see if humidity, air temperature, and barometric pressure influence emission of nitrous oxide (in ppm). Emission measurements were taken at different times, with varying experimental conditions. The data are given in Table 7.8. The model is

$$\mu_{Y|x_1,x_2,x_3} = \beta_0 + \beta_1 x_1 + \beta_2 x_2 + \beta_3 x_3,$$

or, equivalently,

$$y_i = \beta_0 + \beta_1 x_{1i} + \beta_2 x_{2i} + \beta_3 x_{3i} + \epsilon_i, \quad i = 1, 2, \ldots, 20.$$

Fit this multiple linear regression model to the given data and then estimate the amount of nitrous oxide emitted for the conditions where humidity is 50%, temperature is 76°F, and barometric pressure is 29.30.

Solution: The solution of the set of estimating equations yields the unique estimates

$$b_0 = -3.507778, \;\; b_1 = -0.002625, \;\; b_2 = 0.000799, \;\; b_3 = 0.154155.$$

Therefore, the regression equation is

$$\hat{y} = -3.507778 - 0.002625\, x_1 + 0.000799\, x_2 + 0.154155\, x_3.$$

For 50% humidity, a temperature of 76°F, and a barometric pressure of 29.30, the estimated amount of nitrous oxide emitted is

$$\hat{y} = -3.507778 - 0.002625(50.0) + 0.000799(76.0) + 0.1541553(29.30)$$
$$= 0.9384 \text{ ppm.}$$

Table 7.8: Data for Example 7.10

Nitrous Oxide, y	Humidity, x_1	Temp., x_2	Pressure, x_3	Nitrous Oxide, y	Humidity, x_1	Temp., x_2	Pressure, x_3
0.90	72.4	76.3	29.18	1.07	23.2	76.8	29.38
0.91	41.6	70.3	29.35	0.94	47.4	86.6	29.35
0.96	34.3	77.1	29.24	1.10	31.5	76.9	29.63
0.89	35.1	68.0	29.27	1.10	10.6	86.3	29.56
1.00	10.7	79.0	29.78	1.10	11.2	86.0	29.48
1.10	12.9	67.4	29.39	0.91	73.3	76.3	29.40
1.15	8.3	66.8	29.69	0.87	75.4	77.9	29.28
1.03	20.1	76.9	29.48	0.78	96.6	78.7	29.29
0.77	72.2	77.7	29.09	0.82	107.4	86.8	29.03
1.07	24.0	67.7	29.60	0.95	54.9	70.9	29.37

Source: Charles T. Hare, "Light-Duty Diesel Emission Correction Factors for Ambient Conditions," EPA-600/2-77-116. U.S. Environmental Protection Agency.

Polynomial Regression

Now suppose that we wish to fit the polynomial equation

$$\mu_{Y|x} = \beta_0 + \beta_1 x + \beta_2 x^2 + \cdots + \beta_r x^r$$

to the n pairs of observations $\{(x_i, y_i); \ i = 1, 2, \ldots, n\}$. Each observation, y_i, satisfies the equation

$$y_i = \beta_0 + \beta_1 x_i + \beta_2 x_i^2 + \cdots + \beta_r x_i^r + \epsilon_i$$

or

$$y_i = \hat{y}_i + e_i = b_0 + b_1 x_i + b_2 x_i^2 + \cdots + b_r x_i^r + e_i,$$

where r is the degree of the polynomial and ϵ_i and e_i are again the random error and residual associated with the response y_i and fitted value $\hat{y}_i$, respectively. Here, the number of pairs, n, must be at least as large as $r+1$, the number of parameters to be estimated.

Notice that the polynomial model can be considered a special case of the more general multiple linear regression model, where we set $x_1 = x, x_2 = x^2, \ldots, x_r = x^r$. The normal equations assume the same form as those given on page 337. They are then solved for $b_0, b_1, b_2, \ldots, b_r$.

Example 7.11: Given the data

x	0	1	2	3	4	5	6	7	8	9
y	9.1	7.3	3.2	4.6	4.8	2.9	5.7	7.1	8.8	10.2

fit a regression curve of the form $\mu_{Y|x} = \beta_0 + \beta_1 x + \beta_2 x^2$ and then estimate $\mu_{Y|2}$.

Solution: From the data given, we find that

$$10b_0 + \quad 45b_1 + \quad 285b_2 = \quad 63.7,$$
$$45b_0 + \quad 285b_1 + \quad 2025b_2 = \quad 307.3,$$
$$285b_0 + 2025b_1 + 15{,}333b_2 = 2153.3.$$

Solving these normal equations, we obtain

$$b_0 = 8.698, \quad b_1 = -2.341, \quad b_2 = 0.288.$$

Therefore,

$$\hat{y} = 8.698 - 2.341x + 0.288x^2.$$

When $x = 2$, our estimate of $\mu_{Y|2}$ is

$$\hat{y} = 8.698 - (2.341)(2) + (0.288)(2^2) = 5.168.$$

Example 7.12: The data in Table 7.9 represent the percent of impurities that resulted for various temperatures and sterilizing times during a reaction associated with the manufacturing of a certain beverage. Estimate the regression coefficients in the polynomial model

$$y_i = \beta_0 + \beta_1 x_{1i} + \beta_2 x_{2i} + \beta_{11} x_{1i}^2 + \beta_{22} x_{2i}^2 + \beta_{12} x_{1i} x_{2i} + \epsilon_i,$$

for $i = 1, 2, \ldots, 18$.

Table 7.9: Data for Example 7.12

Sterilizing	Temperature, x_1 (°C)		
Time, x_2 (min)	75	100	125
15	14.05	10.55	7.55
	14.93	9.48	6.59
20	16.56	13.63	9.23
	15.85	11.75	8.78
25	22.41	18.55	15.93
	21.66	17.98	16.44

Solution: Using the normal equations, we obtain

$$b_0 = 56.4411, \quad b_1 = -0.36190, \quad b_2 = -2.75299,$$
$$b_{11} = 0.00081, \quad b_{22} = 0.08173, \quad b_{12} = 0.00314,$$

and our estimated regression equation is

$$\hat{y} = 56.4411 - 0.36190x_1 - 2.75299x_2 + 0.00081x_1^2 + 0.08173x_2^2 + 0.00314x_1 x_2.$$

Many of the principles and procedures associated with the estimation of polynomial regression functions fall into the category of **response surface methodology**, a collection of techniques that have been used quite successfully by scientists and engineers in many fields. The x_i^2 are called **pure quadratic terms**, and the $x_i x_j$ $(i \neq j)$ are called **interaction terms**. Such problems as selecting a proper experimental design, particularly in cases where a large number of variables are in the model, and choosing optimum operating conditions for $x_1, x_2, \ldots, x_k$ are often approached through the use of these methods. For an extensive discussion, the reader is referred to *Response Surface Methodology: Process and Product Optimization Using Designed Experiments* by Myers, Montgomery, and Anderson-Cook (2009; see the Bibliography).

Exercises

7.48 In *Applied Spectroscopy*, the infrared reflectance spectra properties of a viscous liquid used in the electronics industry as a lubricant were studied. The designed experiment consisted of testing the effect of band frequency x_1 and film thickness x_2 on optical density y using a Perkin-Elmer Model 621 infrared spectrometer.

y	x_1	x_2
0.231	740	1.10
0.107	740	0.62
0.053	740	0.31
0.129	805	1.10
0.069	805	0.62
0.030	805	0.31
1.005	980	1.10
0.559	980	0.62
0.321	980	0.31
2.948	1235	1.10
1.633	1235	0.62
0.934	1235	0.31

(From Pacansky, England, and Wattman, 1986.) Estimate the multiple linear regression equation

$$\hat{y} = b_0 + b_1 x_1 + b_2 x_2.$$

7.49 A set of experimental runs was made to determine a way of predicting cooking time y at various values of oven width x_1 and flue temperature x_2. The coded data were recorded as follows:

y	x_1	x_2
6.40	1.32	1.15
15.05	2.69	3.40
18.75	3.56	4.10
30.25	4.41	8.75
44.85	5.35	14.82
48.94	6.20	15.15
51.55	7.12	15.32
61.50	8.87	18.18
100.44	9.80	35.19
111.42	10.65	40.40

Estimate the multiple linear regression equation

$$\mu_{Y|x_1,x_2} = \beta_0 + \beta_1 x_1 + \beta_2 x_2.$$

7.50 An experiment was conducted to determine if the weight of an animal can be predicted after a given period of time on the basis of the initial weight of the animal and the amount of feed that was eaten. Use the data, measured in kilograms, to do the followings.

(a) Fit a multiple regression equation of the form

$$\mu_{Y|x_1,x_2} = \beta_0 + \beta_1 x_1 + \beta_2 x_2.$$

(b) Predict the final weight of an animal having an initial weight of 35 kilograms that is given 250 kilograms of feed.

Final Weight, y	Initial Weight, x_1	Feed Weight, x_2
95	42	272
77	33	226
80	33	259
100	45	292
97	39	311
70	36	183
50	32	173
80	41	236
92	40	230
84	38	235

7.51 The following data represent the chemistry grades for a random sample of 12 freshmen at a certain college along with their scores on an intelligence test administered while they were still seniors in high school. The number of class periods missed is also given.

Student	Chemistry Grade, y	Test Score, x_1	Classes Missed, x_2
1	85	65	1
2	74	50	7
3	76	55	5
4	90	65	2
5	85	55	6
6	87	70	3
7	94	65	2
8	98	70	5
9	81	55	4
10	91	70	3
11	76	50	1
12	74	55	4

(a) Fit a multiple linear regression equation of the form $\hat{y} = b_0 + b_1 x_1 + b_2 x_2$.

(b) Estimate the chemistry grade for a student who has an intelligence test score of 60 and missed 4 classes.

7.52 An experiment was conducted on a new model of a particular make of automobile to determine the stopping distance at various speeds. The following data were recorded.

Speed, v (km/hr)	35 50 65 80 95 110
Stopping Distance, d (m)	16 26 41 62 88 119

(a) Fit a multiple regression curve of the form $\mu_{D|v} = \beta_0 + \beta_1 v + \beta_2 v^2$.

(b) Estimate the stopping distance when the car is traveling at 70 kilometers per hour.

7.53 The electric power consumed each month by a chemical plant is thought to be related to the average ambient temperature x_1, the number of days in the month x_2, the average product purity x_3, and the tons of product produced x_4. The past year's historical data are available and are presented here.

y	x_1	x_2	x_3	x_4
240	25	24	91	100
236	31	21	90	95
290	45	24	88	110
274	60	25	87	88
301	65	25	91	94
316	72	26	94	99
300	80	25	87	97
296	84	25	86	96
267	75	24	88	110
276	60	25	91	105
288	50	25	90	100
261	38	23	89	98

(a) Fit a multiple linear regression model using the above data set.

(b) Predict power consumption for a month in which $x_1 = 75°F$, $x_2 = 24$ days, $x_3 = 90\%$, and $x_4 = 98$ tons.

7.54 The following is a set of coded experimental data on the compressive strength of a particular alloy at various values of the concentration of some additive:

Concentration, x	Compressive Strength, y		
10.0	25.2	27.3	28.7
15.0	29.8	31.1	27.8
20.0	31.2	32.6	29.7
25.0	31.7	30.1	32.3
30.0	29.4	30.8	32.8

(a) Estimate the quadratic regression equation $\mu_{Y|x} = \beta_0 + \beta_1 x + \beta_2 x^2$.

(b) Test for lack of fit of the model.

7.55 An experiment was conducted in order to determine if cerebral blood flow in human beings can be predicted from arterial oxygen tension (millimeters of mercury). Fifteen patients participated in the study, and the following data were collected:

Blood Flow, y	Arterial Oxygen Tension, x
84.33	603.40
87.80	582.50
82.20	556.20
78.21	594.60
78.44	558.90
80.01	575.20
83.53	580.10
79.46	451.20
75.22	404.00
76.58	484.00
77.90	452.40

(cont.)

Blood Flow, y	Arterial Oxygen Tension, x
78.80	448.40
80.67	334.80
86.60	320.30
78.20	350.30

Estimate the quadratic regression equation

$$\mu_{Y|x} = \beta_0 + \beta_1 x + \beta_2 x^2.$$

7.56 The following data are given:

x	0	1	2	3	4	5	6
y	1	4	5	3	2	3	4

(a) Fit the cubic model $\mu_{Y|x} = \beta_0 + \beta_1 x + \beta_2 x^2 + \beta_3 x^3$.

(b) Predict Y when $x = 2$.

7.57 (a) Fit a multiple regression equation of the form $\mu_{Y|x} = \beta_0 + \beta_1 x_1 + \beta_2 x^2$ to the data of Example 7.7 on page 326.

(b) Estimate the yield of the chemical reaction for a temperature of $225°C$.

7.58 The following data reflect information from 17 U.S. Navy hospitals at various sites around the world. The regressors are workload variables, that is, items that result in the need for personnel in a hospital. A brief description of the variables is as follows:

$y =$ monthly labor-hours,

$x_1 =$ average daily patient load,

$x_2 =$ monthly X-ray exposures,

$x_3 =$ monthly occupied bed-days,

$x_4 =$ eligible population in the area/1000,

$x_5 =$ average length of patient's stay, in days.

Site	x_1	x_2	x_3	x_4	x_5	y
1	15.57	2463	472.92	18.0	4.45	566.52
2	44.02	2048	1339.75	9.5	6.92	696.82
3	20.42	3940	620.25	12.8	4.28	1033.15
4	18.74	6505	568.33	36.7	3.90	1003.62
5	49.20	5723	1497.60	35.7	5.50	1611.37
6	44.92	11,520	1365.83	24.0	4.60	1613.27
7	55.48	5779	1687.00	43.3	5.62	1854.17
8	59.28	5969	1639.92	46.7	5.15	2160.55
9	94.39	8461	2872.33	78.7	6.18	2305.58
10	128.02	20,106	3655.08	180.5	6.15	3503.93
11	96.00	13,313	2912.00	60.9	5.88	3571.59
12	131.42	10,771	3921.00	103.7	4.88	3741.40
13	127.21	15,543	3865.67	126.8	5.50	4026.52
14	252.90	36,194	7684.10	157.7	7.00	10,343.81
15	409.20	34,703	12,446.33	169.4	10.75	11,732.17
16	463.70	39,204	14,098.40	331.4	7.05	15,414.94
17	510.22	86,533	15,524.00	371.6	6.35	18,854.45

The goal here is to produce an empirical equation that will estimate (or predict) personnel needs for naval hospitals. Estimate the multiple linear regression equation

$$\mu_{Y|x_1,x_2,x_3,x_4,x_5} = \beta_0 + \beta_1 x_1 + \beta_2 x_2 + \beta_3 x_3 + \beta_4 x_4 + \beta_5 x_5.$$

7.59 An experiment was conducted to study the size of squid eaten by sharks and tuna. The regressor variables are characteristics of the beaks of the squid. The data are given as follows:

x_1	x_2	x_3	x_4	x_5	y
1.31	1.07	0.44	0.75	0.35	1.95
1.55	1.49	0.53	0.90	0.47	2.90
0.99	0.84	0.34	0.57	0.32	0.72
0.99	0.83	0.34	0.54	0.27	0.81
1.01	0.90	0.36	0.64	0.30	1.09
1.09	0.93	0.42	0.61	0.31	1.22
1.08	0.90	0.40	0.51	0.31	1.02
1.27	1.08	0.44	0.77	0.34	1.93
0.99	0.85	0.36	0.56	0.29	0.64
1.34	1.13	0.45	0.77	0.37	2.08
1.30	1.10	0.45	0.76	0.38	1.98
1.33	1.10	0.48	0.77	0.38	1.90
1.86	1.47	0.60	1.01	0.65	8.56
1.58	1.34	0.52	0.95	0.50	4.49
1.97	1.59	0.67	1.20	0.59	8.49
1.80	1.56	0.66	1.02	0.59	6.17
1.75	1.58	0.63	1.09	0.59	7.54
1.72	1.43	0.64	1.02	0.63	6.36
1.68	1.57	0.72	0.96	0.68	7.63
1.75	1.59	0.68	1.08	0.62	7.78
2.19	1.86	0.75	1.24	0.72	10.15
1.73	1.67	0.64	1.14	0.55	6.88

In the study, the regressor variables and response considered are

x_1 = rostral length, in inches,

x_2 = wing length, in inches,

x_3 = rostral to notch length, in inches,

x_4 = notch to wing length, in inches,

x_5 = width, in inches,

y = weight, in pounds.

Estimate the multiple linear regression equation

$$\mu_{Y|x_1,x_2,x_3,x_4,x_5} = \beta_0 + \beta_1 x_1 + \beta_2 x_2 + \beta_3 x_3 + \beta_4 x_4 + \beta_5 x_5.$$

7.60 An engineer at a semiconductor company wants to model the relationship between the gain or hFE of a device (y) and three parameters: emitter-RS (x_1), base-RS (x_2), and emitter-to-base-RS (x_3). The data are presented here are from Myers, Montgomery, and Anderson-Cook, 2009.

(a) Fit a multiple linear regression to the data.

(b) Predict hFE when $x_1 = 14$, $x_2 = 220$, and $x_3 = 5$.

x_1, Emitter-RS	x_2, Base-RS	x_3, E-B-RS	y, hFE
14.62	226.0	7.000	128.40
15.63	220.0	3.375	52.62
14.62	217.4	6.375	113.90
15.00	220.0	6.000	98.01
14.50	226.5	7.625	139.90
15.25	224.1	6.000	102.60
16.12	220.5	3.375	48.14
15.13	223.5	6.125	109.60
15.50	217.6	5.000	82.68
15.13	228.5	6.625	112.60
15.50	230.2	5.750	97.52
16.12	226.5	3.750	59.06
15.13	226.6	6.125	111.80
15.63	225.6	5.375	89.09
15.38	234.0	8.875	171.90
15.50	230.0	4.000	66.80

7.61 A study was performed on a type of bearing to find the relationship of amount of wear y to $x_1 =$ oil viscosity and $x_2 =$ load. The following data were obtained.

y	x_1	x_2	y	x_1	x_2
193	1.6	851	230	15.5	816
172	22.0	1058	91	43.0	1201
113	33.0	1357	125	40.0	1115

(Data from Myers, Montgomery, and Anderson-Cook, 2009.)

(a) Estimate the unknown parameters of the multiple linear regression equation

$$\mu_{Y|x_1,x_2} = \beta_0 + \beta_1 x_1 + \beta_2 x_2.$$

(b) Predict wear when oil viscosity is 20 and load is 1200.

7.62 Eleven student teachers took part in an evaluation program designed to measure teacher effectiveness and determine what factors are important. The response measure was a quantitative evaluation of the teacher. The regressor variables were scores on four standardized tests given to each teacher. The data are as follows:

y	x_1	x_2	x_3	x_4
410	69	125	59.00	55.66
569	57	131	31.75	63.97
425	77	141	80.50	45.32
344	81	122	75.00	46.67
324	0	141	49.00	41.21
505	53	152	49.35	43.83
235	77	141	60.75	41.61
501	76	132	41.25	64.57
400	65	157	50.75	42.41
584	97	166	32.25	57.95
434	76	141	54.50	57.90

Estimate the multiple linear regression equation

$$\mu_{Y|x_1,x_2,x_3,x_4} = \beta_0 + \beta_1 x_1 + \beta_2 x_2 + \beta_3 x_3 + \beta_4 x_4.$$

7.63 The personnel department of a certain industrial firm used 12 subjects in a study to determine the relationship between job performance rating (y) and scores on four tests. Use the data provided here to estimate the regression coefficients in the model

$$\hat{y} = b_0 + b_1 x_1 + b_2 x_2 + b_3 x_3 + b_4 x_4.$$

y	x_1	x_2	x_3	x_4
11.2	56.5	71.0	38.5	43.0
14.5	59.5	72.5	38.2	44.8
17.2	69.2	76.0	42.5	49.0
17.8	74.5	79.5	43.4	56.3
19.3	81.2	84.0	47.5	60.2
24.5	88.0	86.2	47.4	62.0
21.2	78.2	80.5	44.5	58.1
16.9	69.0	72.0	41.8	48.1
14.8	58.1	68.0	42.1	46.0
20.0	80.5	85.0	48.1	60.3
13.2	58.3	71.0	37.5	47.1
22.5	84.0	87.2	51.0	65.2

7.11   Inferences in Multiple Linear Regression

The means and variances of the estimators $b_0, b_1, \ldots, b_k$ are readily obtained under certain assumptions on the random errors $\epsilon_1, \epsilon_2, \ldots, \epsilon_k$ that are identical to those made in the case of simple linear regression. When we assume these errors to be independent, each with mean 0 and variance σ^2, it can be shown that $b_0, b_1, \ldots, b_k$ are, respectively, unbiased estimators of the regression coefficients $\beta_0, \beta_1, \ldots, \beta_k$. In addition, the variances of the b's are obtained through the elements of the inverse of the $\mathbf{A}$ matrix. Note that the off-diagonal elements of $\mathbf{A} = \mathbf{X}'\mathbf{X}$ represent sums of products of elements in the columns of $\mathbf{X}$, while the diagonal elements of $\mathbf{A}$ represent sums of squares of elements in the columns of $\mathbf{X}$. The inverse matrix, $\mathbf{A}^{-1}$, apart from the multiplier σ^2, represents the **variance-covariance matrix** of the estimated regression coefficients. That is, the elements of the matrix $\mathbf{A}^{-1}\sigma^2$ display the variances of $b_0, b_1, \ldots, b_k$ on the main diagonal and covariances on the off-diagonal. For example, in a $k = 2$ multiple linear regression problem, we might write

$$(\mathbf{X}'\mathbf{X})^{-1} = \begin{bmatrix} c_{00} & c_{01} & c_{02} \\ c_{10} & c_{11} & c_{12} \\ c_{20} & c_{21} & c_{22} \end{bmatrix}$$

with the elements below the main diagonal determined through the symmetry of the matrix. Then we can write

$$\sigma_{b_i}^2 = c_{ii}\sigma^2, \qquad i = 0, 1, 2,$$
$$\sigma_{b_i b_j} = \mathrm{Cov}(b_i, b_j) = c_{ij}\sigma^2, \quad i \neq j.$$

Of course, the estimates of the variances and hence the standard errors of these estimators are obtained by replacing σ^2 with the appropriate estimate obtained through experimental data. An unbiased estimate of σ^2 is once again defined in terms of the error sum of squares, which is computed using the formula established in Theorem 7.2. In the theorem, we are making the assumptions on the ϵ_i described above.

Theorem 7.2: For the linear regression equation

$$\mathbf{y} = \mathbf{X}\boldsymbol{\beta} + \boldsymbol{\epsilon},$$

an unbiased estimate of σ^2 is given by the error or residual mean square

$$s^2 = \frac{SSE}{n-k-1}, \quad \text{where} \quad SSE = \sum_{i=1}^{n} e_i^2 = \sum_{i=1}^{n}(y_i - \hat{y}_i)^2.$$

We can see that Theorem 7.2 represents a generalization of Theorem 7.1 for the simple linear regression case. The proof is left for the reader. As in the simpler linear regression case, the estimate s^2 is a measure of the variation in the prediction errors or residuals.

The error and regression sums of squares take on the same form and play the same role as in the simple linear regression case. In fact, the sum-of-squares identity

$$\sum_{i=1}^{n}(y_i - \bar{y})^2 = \sum_{i=1}^{n}(\hat{y}_i - \bar{y})^2 + \sum_{i=1}^{n}(y_i - \hat{y}_i)^2$$

continues to hold, and we retain our previous notation, namely

$$SST = SSR + SSE,$$

with

$$SST = \sum_{i=1}^{n}(y_i - \bar{y})^2 = \text{total sum of squares}$$

and

$$SSR = \sum_{i=1}^{n}(\hat{y}_i - \bar{y})^2 = \text{regression sum of squares}.$$

There are k degrees of freedom associated with SSR, and, as always, SST has $n-1$ degrees of freedom. Therefore, after subtraction, SSE has $n-k-1$ degrees of freedom. Thus, our estimate of σ^2 is again given by the error sum of squares divided by its degrees of freedom. All three of these sums of squares will appear on the printouts of most multiple regression computer packages. Note that the condition $n > k$ in Section 7.10 guarantees that the degrees of freedom of SSE cannot be negative.

Analysis of Variance in Multiple Regression

The partition of the total sum of squares into its components, the regression and error sums of squares, plays an important role. An **analysis of variance** can be conducted to shed light on the quality of the regression equation. A useful hypothesis that determines if a significant amount of variation is explained by the model is

$$H_0: \ \beta_1 = \beta_2 = \beta_3 = \cdots = \beta_k = 0.$$

The analysis of variance involves an F-test which can be summarized via a table as follows:

Source	Sum of Squares	Degrees of Freedom	Mean Squares	F
Regression	SSR	k	$MSR = \frac{SSR}{k}$	$f = \frac{MSR}{MSE}$
Error	SSE	$n - (k+1)$	$MSE = \frac{SSE}{n-(k+1)}$	
Total	SST	$n - 1$		

This test is an **upper-tailed test**. Rejection of H_0 implies that the **regression equation differs from a constant**. That is, at least one regressor variable is important. However, this and other tests of hypotheses considered later require that the errors ϵ_i be normally distributed in the multiple linear regression model.

Further utility of the mean square error (or residual mean square) lies in its use in hypothesis testing and confidence interval estimation. In addition, the mean square error plays an important role in situations where the scientist is searching for the best from a set of competing models. Many model-building criteria involve the statistic s^2.

A knowledge of the distributions of the individual coefficient estimators enables the experimenter to construct confidence intervals for the coefficients and to test hypotheses about them. Since the b_j $(j = 0, 1, 2, \ldots, k)$ are normally distributed with mean β_j and variance $c_{jj}\sigma^2$, we can use the statistic

$$t = \frac{b_j - \beta_{j0}}{s\sqrt{c_{jj}}}$$

with $n - k - 1$ degrees of freedom to test hypotheses and construct confidence intervals on β_j. For example, if we wish to test

$$H_0\colon\ \beta_j = \beta_{j0},$$
$$H_1\colon\ \beta_j \neq \beta_{j0},$$

we compute the above t-statistic and do not reject H_0 if $-t_{\alpha/2} < t < t_{\alpha/2}$, where $t_{\alpha/2}$ has $n - k - 1$ degrees of freedom.

Individual t-Tests for Variable Screening

The t-test most often used in multiple regression is the one that tests the importance of individual coefficients (i.e., $H_0\colon \beta_j = 0$ against the alternative $H_1\colon \beta_j \neq 0$). These tests often contribute to what is termed **variable screening**, where the analyst attempts to arrive at the most useful model (i.e., the choice of which regressors to use). It should be emphasized here that if a coefficient is found insignificant (i.e., the hypothesis $H_0\colon \beta_j = 0$ **is not rejected**), the conclusion drawn is that the **variable** is insignificant (i.e., explains an insignificant amount of variation in y), **in the presence of the other regressors in the model**.

Exercises

7.64 For the data of Exercise 7.49 on page 340, estimate σ^2.

7.65 For the data of Exercise 7.48 on page 340, estimate σ^2.

7.66 Obtain estimates of the variances and the covariance of the estimators b_1 and b_2 of Exercise 7.48 on page 340.

7.67 For the data of Exercise 7.53 on page 340, estimate σ^2.

7.68 For the model of Exercise 7.55 on page 341, test the hypothesis that $\beta_2 = 0$ at the 0.05 level of significance against the alternative that $\beta_2 \neq 0$.

7.69 Referring to Exercise 7.53 on page 340, find the estimate of

(a) $\sigma_{b_2}^2$;

(b) $\text{Cov}(b_1, b_4)$.

7.70 For the model of Exercise 7.49 on page 340, test the hypothesis that $\beta_1 = 2$ against the alternative that $\beta_1 \neq 2$. Use a P-value in your conclusion.

7.71 For the model of Exercise 7.48 on page 340, test the hypothesis that $\beta_1 = 0$ at the 0.05 level of significance against the alternative that $\beta_1 \neq 0$.

7.72 For Exercise 7.54 on page 341, construct a 90% confidence interval for the mean compressive strength when the concentration is $x = 19.5$ and a quadratic model is used.

7.73 Using the data of Exercise 7.48 on page 340 and the estimate of σ^2 from Exercise 7.65, compute 95% confidence intervals for the predicted response and the mean response when $x_1 = 900$ and $x_2 = 1.00$.

7.74 Consider the following data from Exercise 7.61 on page 342.

y (wear)	x_1 (oil viscosity)	x_2 (load)
193	1.6	851
230	15.5	816
172	22.0	1058
91	43.0	1201
113	33.0	1357
125	40.0	1115

(a) Estimate σ^2 using multiple regression of y on x_1 and x_2.

(b) Compute predicted values, a 95% confidence interval for mean wear, and a 95% prediction interval for observed wear if $x_1 = 20$ and $x_2 = 1000$.

7.75 Using the data of Exercise 7.53 on page 340 and the estimate of σ^2 from Exercise 7.67, compute 95% confidence intervals for the predicted response and the mean response when $x_1 = 75$, $x_2 = 24$, $x_3 = 90$, and $x_4 = 98$.

7.76 Use the data from Exercise 7.60 on page 342.

(a) Estimate σ^2 using the multiple regression of y on x_1, x_2, and x_3.

(b) Compute a 95% prediction interval for the observed gain with the three regressors at $x_1 = 15.0$, $x_2 = 220.0$, and $x_3 = 6.0$.

7.77 Using the data from Exercise 7.74, test the following at the 0.05 level.

(a) H_0: $\beta_1 = 0$ versus H_1: $\beta_1 \neq 0$;

(b) H_0: $\beta_2 = 0$ versus H_1: $\beta_2 \neq 0$.

(c) Do you have any reason to believe that the model in Exercise 7.74 should be changed? Why or why not?

Review Exercises

7.78 With reference to Exercise 7.8 on page 305, construct

(a) a 95% confidence interval for the average course grade of students who score 35 on the placement test;

(b) a 95% prediction interval for the course grade of a student who scored 35 on the placement test.

7.79 The Statistics Consulting Center at Virginia Tech analyzed data on normal woodchucks for the Department of Veterinary Medicine. The variables of interest were body weight in grams and heart weight in grams. It was desired to develop a linear regression equation in order to determine if there is a significant linear relationship between heart weight and total body weight. Use heart weight as the independent variable and body weight as the dependent variable and fit a simple linear regression using the following data. In addition, test the hypothesis H_0: $\beta_1 = 0$ versus H_1: $\beta_1 \neq 0$. Draw conclusions.

Body Weight (grams)	Heart Weight (grams)
4050	11.2
2465	12.4
3120	10.5
5700	13.2
2595	9.8
3640	11.0
2050	10.8
4235	10.4
2935	12.2
4975	11.2
3690	10.8
2800	14.2
2775	12.2
2170	10.0
2370	12.3
2055	12.5
2025	11.8
2645	16.0
2675	13.8

7.80 The amounts of solids removed from a particular material when exposed to drying periods of different lengths are as shown.

x (hours)	y (grams)	
4.4	13.1	14.2
4.5	9.0	11.5
4.8	10.4	11.5
5.5	13.8	14.8
5.7	12.7	15.1
5.9	9.9	12.7
6.3	13.8	16.5
6.9	16.4	15.7
7.5	17.6	16.9
7.8	18.3	17.2

(a) Estimate the linear regression line.

(b) Test at the 0.05 level of significance whether the linear model is adequate.

7.81 With reference to Exercise 7.9 on page 305, construct

(a) a 95% confidence interval for the average weekly sales when $45 is spent on advertising;

(b) a 95% prediction interval for the weekly sales when $45 is spent on advertising.

7.82 An experiment was designed for the Department of Materials Engineering at Virginia Tech to study hydrogen embrittlement properties based on electrolytic hydrogen pressure measurements. The solution used was 0.1 N NaOH, and the material was a certain type of stainless steel. The cathodic charging current density was controlled and varied at four levels. The effective hydrogen pressure was observed as the response. The data follow.

Run	Charging Current Density, x (mA/cm^2)	Effective Hydrogen Pressure, y (atm)
1	0.5	86.1
2	0.5	92.1
3	0.5	64.7
4	0.5	74.7
5	1.5	223.6
6	1.5	202.1
7	1.5	132.9
8	2.5	413.5
9	2.5	231.5
10	2.5	466.7
11	2.5	365.3
12	3.5	493.7
13	3.5	382.3
14	3.5	447.2
15	3.5	563.8

(a) Run a simple linear regression of y against x.

(b) Compute the pure error sum of squares and make a test for lack of fit.

(c) Does the information in part (b) indicate a need for a model in x beyond a first-order regression? Explain.

7.83 The following data represent the chemistry grades for a random sample of 12 freshmen at a certain college along with their scores on an intelligence test administered while they were still seniors in high school.

Student	Test Score, x	Chemistry Grade, y
1	65	85
2	50	74
3	55	76
4	65	90
5	55	85
6	70	87
7	65	94
8	70	98
9	55	81
10	70	91
11	50	76
12	55	74

(a) Compute and interpret the sample correlation coefficient.

(b) State necessary assumptions on random variables.

(c) Test the hypothesis that $\rho = 0.5$ against the alternative that $\rho > 0.5$. Use a P-value in the conclusion.

7.84 The *Washington Times* business section in March of 1997 listed 21 different used computers and printers and their sale prices. Also listed was the average hover bid. Partial results from regression analysis using *SAS* software are shown in Figure 7.23 on page 350.

(a) Explain the difference between the confidence interval on the mean and the prediction interval.

(b) Explain why the standard errors of prediction vary from observation to observation.

(c) Which observation has the lowest standard error of prediction? Why?

7.85 Consider the vehicle data from *Consumer Reports* in Figure 7.24 on page 351. The data include weight (in tons), mileage (in miles per gallon), and drive ratio. A regression model was fitted relating weight x to mileage y. A partial *SAS* printout in Figure 7.24 shows some of the results of that regression analysis, and Figure 7.25 on page 352 gives a plot of the residuals and weight for each vehicle.

(a) From the analysis and the residual plot, does it appear that an improved model might be found by using a transformation? Explain.

(b) Fit the model by replacing weight with log weight. Comment on the results.

(c) Fit a model by replacing mpg with gallons per 100 miles traveled, as mileage is often reported in other countries. Which of the three models is preferable? Explain.

7.86 Observations on the yield of a chemical reaction taken at various temperatures were recorded as follows:

x (°C)	y (%)	x (°C)	y (%)
150	75.4	150	77.7
150	81.2	200	84.4
200	85.5	200	85.7
250	89.0	250	89.4
250	90.5	300	94.8
300	96.7	300	95.3

(a) Plot the data.

(b) Does it appear from the plot as if the relationship is linear?

(c) Fit a simple linear regression and test for lack of fit.

(d) Draw conclusions based on your result in (c).

7.87 Physical fitness testing is an important aspect of athletic training. A common measure of the magnitude of cardiovascular fitness is the maximum volume of oxygen uptake during strenuous exercise. A study was conducted on 24 middle-aged men to determine the influence on oxygen uptake of the time required to complete a two-mile run. Oxygen uptake was measured with standard laboratory methods as the subjects performed on a treadmill. The work was published in "Maximal Oxygen Intake Prediction in Young and Middle Aged Males," *Journal of Sports Medicine*

9, 1969, 17–22. The data are as follows:

Subject	y, Maximum Volume of O_2	x, Time in Seconds
1	42.33	918
2	53.10	805
3	42.08	892
4	50.06	962
5	42.45	968
6	42.46	907
7	47.82	770
8	49.92	743
9	36.23	1045
10	49.66	810
11	41.49	927
12	46.17	813
13	46.18	858
14	43.21	860
15	51.81	760
16	53.28	747
17	53.29	743
18	47.18	803
19	56.91	683
20	47.80	844
21	48.65	755
22	53.67	700
23	60.62	748
24	56.73	775

(a) Estimate the parameters in a simple linear regression model.

(b) Does the time it takes to run two miles have a significant influence on maximum oxygen uptake? Use H_0: $\beta_1 = 0$ versus H_1: $\beta_1 \neq 0$.

(c) Plot the residuals on a graph against x and comment on the appropriateness of the simple linear model.

7.88 Consider the fictitious set of data shown below, where the line through the data is the fitted simple linear regression line. Sketch a residual plot.

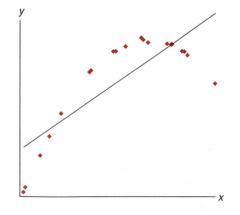

7.89 Consider a 2nd-order response surface model that contains the linear, pure quadratic and cross-product terms as follows:

$$y_i = \beta_0 + \beta_1 x_{1i} + \beta_2 x_{2i} + \beta_{11} x_{1i}^2$$
$$+ \beta_{22} x_{2i}^2 + \beta_{12} x_{1i} x_{2i} + \epsilon_i.$$

Fit the model above to the following data, and suggest any model editing that may be needed.

x_1	x_2	y
-1	-1	35.38
-1	0	38.42
-1	1	42.15
0	-1	36.52
0	0	45.27
0	1	53.22
1	-1	37.18
1	0	48.25
1	1	55.07

7.90 A study was conducted to determine whether lifestyle changes could replace medication in reducing blood pressure among hypertensives. The factors considered were a healthy diet with an exercise program, the typical dosage of medication for hypertension, and no intervention. The pretreatment body mass index (BMI) was also calculated because it is known to affect blood pressure. The response considered in this study was change in blood pressure. The variable "group" had the following levels.

1 = Healthy diet and an exercise program

2 = Medication

3 = No intervention

(a) Fit an appropriate model using the data below. Does it appear that exercise and diet could be effectively used to lower blood pressure? Explain your answer from the results.

(b) Would exercise and diet be an effective alternative to medication?

(*Hint:* You may wish to form the model in more than one way to answer both of these questions.)

Change in Blood Pressure	Group	BMI
-32	1	27.3
-21	1	22.1
-26	1	26.1
-16	1	27.8
-11	2	19.2
-19	2	26.1
-23	2	28.6
-5	2	23.0
-6	3	28.1
5	3	25.3
-11	3	26.7
14	3	22.3

7.91 Show that in choosing the so-called best subset model from a series of candidate models, choosing the model with the smallest s^2 is equivalent to choosing the model with the smallest R_{adj}^2.

7.92 Case Study: Consider the data set for Exercise 7.58, page 341 (hospital data), repeated here.

Site	x_1	x_2	x_3	x_4	x_5	y
1	15.57	2463	472.92	18.0	4.45	566.52
2	44.02	2048	1339.75	9.5	6.92	696.82
3	20.42	3940	620.25	12.8	4.28	1033.15
4	18.74	6505	568.33	36.7	3.90	1003.62
5	49.20	5723	1497.60	35.7	5.50	1611.37
6	44.92	11,520	1365.83	24.0	4.60	1613.27
7	55.48	5779	1687.00	43.3	5.62	1854.17
8	59.28	5969	1639.92	46.7	5.15	2160.55
9	94.39	8461	2872.33	78.7	6.18	2305.58
10	128.02	20,106	3655.08	180.5	6.15	3503.93
11	96.00	13,313	2912.00	60.9	5.88	3571.59
12	131.42	10,771	3921.00	103.7	4.88	3741.40
13	127.21	15,543	3865.67	126.8	5.50	4026.52
14	252.90	36,194	7684.10	157.7	7.00	10,343.81
15	409.20	34,703	12,446.33	169.4	10.75	11,732.17
16	463.70	39,204	14,098.40	331.4	7.05	15,414.94
17	510.22	86,533	15,524.00	371.6	6.35	18,854.45

(a) The *SAS* PROC REG outputs provided in Figures 7.26 and 7.27 on pages 353 and 354 supply a considerable amount of information. Goals are to do outlier detection and eventually determine which model terms are to be used in the final model.

(b) Often the role of a single regressor variable is not apparent when it is studied in the presence of several other variables. This is due to multicollinearity. With this in mind, comment on the importance of x_2 and x_3 in the full model as opposed to their importance in a model in which they are the only variables.

(c) Comment on what other analyses should be run.

(d) Run appropriate analyses and write your conclusions concerning the final model.

7.93 Project: This project can be done in groups or by individuals. Each person or group must find a set of data, preferably from (but not restricted to) their field of study. The data need to fit the regression framework with a regression variable x and a response variable y. Carefully make the assignment as to which variable is x and which y. It may be necessary to consult a journal or periodical from your field if you do not have other research data available.

(a) Plot y versus x. Comment on the relationship as seen from the plot.

(b) Fit an appropriate regression model from the data. Use simple linear regression or fit a polynomial model to the data. Comment on measures of quality.

(c) Plot residuals as illustrated in the text. Check possible violation of assumptions. Show graphically a plot of confidence intervals on a mean response plotted against x. Comment.

R-Square	Coeff Var	Root MSE	Price Mean
0.967472	7.923338	70.83841	894.0476

Parameter	Estimate	Standard Error	t Value	Pr > |t|
Intercept	59.93749137	38.34195754	1.56	0.1345
Buyer	1.04731316	0.04405635	23.77	<.0001

product	Buyer	Price	Predict Value	Std Err Predict	Lower 95% Mean	Upper 95% Mean	Lower 95% Predict	Upper 95% Predict
IBM PS/1 486/66 420MB	325	375	400.31	25.8906	346.12	454.50	242.46	558.17
IBM ThinkPad 500	450	625	531.23	21.7232	485.76	576.70	376.15	686.31
IBM Think-Dad 755CX	1700	1850	1840.37	42.7041	1750.99	1929.75	1667.25	2013.49
AST Pentium 90 540MB	800	875	897.79	15.4590	865.43	930.14	746.03	1049.54
Dell Pentium 75 1GB	650	700	740.69	16.7503	705.63	775.75	588.34	893.05
Gateway 486/75 320MB	700	750	793.06	16.0314	759.50	826.61	641.04	945.07
Clone 586/133 1GB	500	600	583.59	20.2363	541.24	625.95	429.40	737.79
Compaq Contura 4/25 120MB	450	600	531.23	21.7232	485.76	576.70	376.15	686.31
Compaq Deskpro P90 1.2GB	800	850	897.79	15.4590	865.43	930.14	746.03	1049.54
Micron P75 810MB	800	675	897.79	15.4590	865.43	930.14	746.03	1049.54
Micron P100 1.2GB	900	975	1002.52	16.1176	968.78	1036.25	850.46	1154.58
Mac Quadra 840AV 500MB	450	575	531.23	21.7232	485.76	576.70	376.15	686.31
Mac Performer 6116 700MB	700	775	793.06	16.0314	759.50	826.61	641.04	945.07
PowerBook 540c 320MB	1400	1500	1526.18	30.7579	1461.80	1590.55	1364.54	1687.82
PowerBook 5300 500MB	1350	1575	1473.81	28.8747	1413.37	1534.25	1313.70	1633.92
Power Mac 7500/100 1GB	1150	1325	1264.35	21.9454	1218.42	1310.28	1109.13	1419.57
NEC Versa 486 340MB	800	900	897.79	15.4590	865.43	930.14	746.03	1049.54
Toshiba 1960CS 320MB	700	825	793.06	16.0314	759.50	826.61	641.04	945.07
Toshiba 4800VCT 500MB	1000	1150	1107.25	17.8715	1069.85	1144.66	954.34	1260.16
HP Laser jet III	350	475	426.50	25.0157	374.14	478.86	269.26	583.74
Apple Laser Writer Pro 63	750	800	845.42	15.5930	812.79	878.06	693.61	997.24

Figure 7.23: *SAS* printout, showing partial analysis of data of Review Exercise 7.84.

Obs	Model	WT	MPG	DR_RATIO
1	Buick Estate Wagon	4.360	16.9	2.73
2	Ford Country Squire Wagon	4.054	15.5	2.26
3	Chevy Ma libu Wagon	3.605	19.2	2.56
4	Chrysler LeBaron Wagon	3.940	18.5	2.45
5	Chevette	2.155	30.0	3.70
6	Toyota Corona	2.560	27.5	3.05
7	Datsun 510	2.300	27.2	3.54
8	Dodge Omni	2.230	30.9	3.37
9	Audi 5000	2.830	20.3	3.90
10	Volvo 240 CL	3.140	17.0	3.50
11	Saab 99 GLE	2.795	21.6	3.77
12	Peugeot 694 SL	3.410	16.2	3.58
13	Buick Century Special	3.380	20.6	2.73
14	Mercury Zephyr	3.070	20.8	3.08
15	Dodge Aspen	3.620	18.6	2.71
16	AMC Concord D/L	3.410	18.1	2.73
17	Chevy Caprice Classic	3.840	17.0	2.41
18	Ford LTP	3.725	17.6	2.26
19	Mercury Grand Marquis	3.955	16.5	2.26
20	Dodge St Regis	3.830	18.2	2.45
21	Ford Mustang 4	2.585	26.5	3.08
22	Ford Mustang Ghia	2.910	21.9	3.08
23	Macda GLC	1.975	34.1	3.73
24	Dodge Colt	1.915	35.1	2.97
25	AMC Spirit	2.670	27.4	3.08
26	VW Scirocco	1.990	31.5	3.78
27	Honda Accord LX	2.135	29.5	3.05
28	Buick Skylark	2.570	28.4	2.53
29	Chevy Citation	2.595	28.8	2.69
30	Olds Omega	2.700	26.8	2.84
31	Pontiac Phoenix	2.556	33.5	2.69
32	Plymouth Horizon	2.200	34.2	3.37
33	Datsun 210	2.020	31.8	3.70
34	Fiat Strada	2.130	37.3	3.10
35	VW Dasher	2.190	30.5	3.70
36	Datsun 810	2.815	22.0	3.70
37	BMW 320i	2.600	21.5	3.64
38	VW Rabbit	1.925	31.9	3.78

R-Square	Coeff Var	Root MSE	MPG Mean
0.817244	11.46010	2.837580	24.76053

Parameter	Estimate	Standard Error	t Value	Pr > |t|
Intercept	48.67928080	1.94053995	25.09	<.0001
WT	-8.36243141	0.65908398	-12.69	<.0001

Figure 7.24: *SAS* printout, showing partial analysis of data of Review Exercise 7.85.

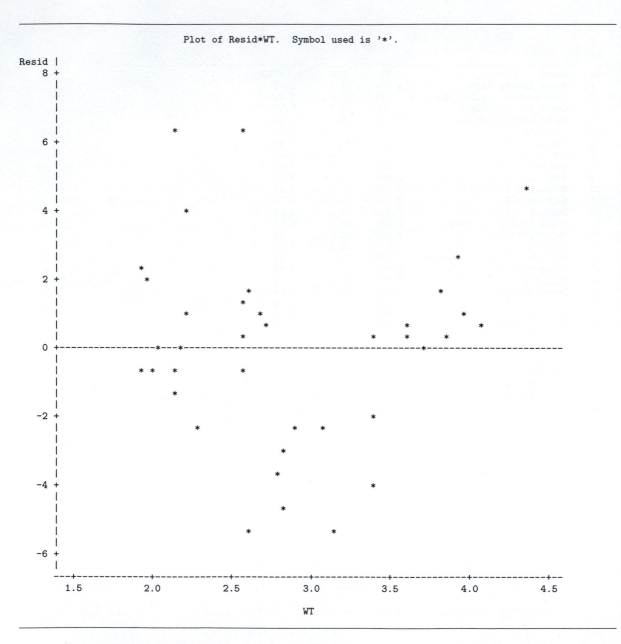

Figure 7.25: *SAS* printout, showing residual plot of Review Exercise 7.85.

```
Dependent Variable: y
                   Analysis of Variance
                              Sum of          Mean
Source            DF          Squares         Square      F Value    Pr > F
Model             5          490177488      98035498      237.79     <.0001
Error            11            4535052        412277
Corrected Total  16          494712540
               Root MSE              642.08838   R-Square    0.9908
               Dependent Mean       4978.48000   Adj R-Sq    0.9867
               Coeff Var              12.89728
                            Parameter Estimates
                                    Parameter    Standard
Variable   Label              DF     Estimate       Error  t Value  Pr > |t|
Intercept  Intercept          1    1962.94816  1071.36170    1.83     0.0941
x1         Average Daily Patient Load  1   -15.85167    97.65299   -0.16     0.8740
x2         Monthly X-Ray Exposure      1     0.05593     0.02126    2.63     0.0234
x3         Monthly Occupied Bed Days   1     1.58962     3.09208    0.51     0.6174
x4         Eligible Population in the  1    -4.21867     7.17656   -0.59     0.5685
           Area/100
x5         Average Length of Patients  1  -394.31412   209.63954   -1.88     0.0867
           Stay in Days
```

Figure 7.26: *SAS* output for Review Exercise 7.92; part I.

Obs	Dependent Variable	Predicted Value	Std Error Mean Predict	95% CL Mean		95% CL Predict	
1	566.5200	775.0251	241.2323	244.0765	1306	-734.6494	2285
2	696.8200	740.6702	331.1402	11.8355	1470	-849.4275	2331
3	1033	1104	278.5116	490.9234	1717	-436.5244	2644
4	1604	1240	268.1298	650.3459	1831	-291.0028	2772
5	1611	1564	211.2372	1099	2029	76.6816	3052
6	1613	2151	279.9293	1535	2767	609.5796	3693
7	1854	1690	218.9976	1208	2172	196.5345	3183
8	2161	1736	468.9903	703.9948	2768	-13.8306	3486
9	2306	2737	290.4749	2098	3376	1186	4288
10	3504	3682	585.2517	2394	4970	1770	5594
11	3572	3239	189.0989	2823	3655	1766	4713
12	3741	4353	328.8507	3630	5077	2766	5941
13	4027	4257	314.0481	3566	4948	2684	5830
14	10344	8768	252.2617	8213	9323	7249	10286
15	11732	12237	573.9168	10974	13500	10342	14133
16	15415	15038	585.7046	13749	16328	13126	16951
17	18854	19321	599.9780	18000	20641	17387	21255

Obs	Residual	Std Error Residual	Student Residual	-2-1 0 1 2
1	-208.5051	595.0	-0.350	| | |
2	-43.8502	550.1	-0.0797	| | |
3	-70.7734	578.5	-0.122	| | |
4	363.1244	583.4	0.622	| |* |
5	46.9483	606.3	0.0774	| | |
6	-538.0017	577.9	-0.931	| *| |
7	164.4696	603.6	0.272	| | |
8	424.3145	438.5	0.968	| |* |
9	-431.4090	572.6	-0.753	| *| |
10	-177.9234	264.1	-0.674	| *| |
11	332.6011	613.6	0.542	| |* |
12	-611.9330	551.5	-1.110	| **| |
13	-230.5684	560.0	-0.412	| | |
14	1576	590.5	2.669	| |***** |
15	-504.8574	287.9	-1.753	| ***| |
16	376.5491	263.1	1.431	| |** |
17	-466.2470	228.7	-2.039	| ****| |

Figure 7.27: *SAS* output for Review Exercise 7.92; part II.

Chapter 8

One-Factor Experiments: General

8.1 Analysis-of-Variance Technique and the Strategy of Experimental Design

In the estimation and hypothesis testing material covered in Chapters 5 and 6, we were restricted in each case to considering no more than two population parameters. Such was the case, for example, in testing for the equality of two population means using independent samples from normal populations with common but unknown variance, where it was necessary to obtain a pooled estimate of σ^2. This material dealing in two-sample inference represents a special case of what we call the *one-factor problem*. For example, in Exercise 6.35 on page 269, the survival time was measured for two samples of mice, where one sample received a new serum for leukemia treatment and the other sample received no treatment. In this case, we say that there is *one factor*, namely *treatment*, and the factor is at *two levels*. If several competing treatments were being used in the sampling process, more samples of mice would be necessary. In this case, the problem would involve one factor with more than two levels and thus more than two samples.

In the $k > 2$ sample problem, it will be assumed that there are k samples from k populations. One very common procedure used to deal with testing population means is called the **analysis of variance**, or **ANOVA**.

The analysis of variance is certainly not a new technique to the reader who has followed the material on regression theory. We used the analysis-of-variance approach to partition the total sum of squares into a portion due to regression and a portion due to error.

Suppose in an industrial experiment that an engineer is interested in how the mean absorption of moisture in concrete varies among 5 different concrete aggregates. The samples are exposed to moisture for 48 hours. It is decided that 6 samples are to be tested for each aggregate, requiring a total of 30 samples to be tested. The data are recorded in Table 8.1.

The model for this situation may be set up as follows. There are 6 observations taken from each of 5 populations with means $\mu_1, \mu_2, \ldots, \mu_5$, respectively. We may

Table 8.1: Absorption of Moisture in Concrete Aggregates

Aggregate:	1	2	3	4	5	
	551	595	639	417	563	
	457	580	615	449	631	
	450	508	511	517	522	
	731	583	573	438	613	
	499	633	648	415	656	
	632	517	677	555	679	
Total	3320	3416	3663	2791	3664	16,854
Mean	553.33	569.33	610.50	465.17	610.67	561.80

wish to test

$$H_0: \ \mu_1 = \mu_2 = \cdots = \mu_5,$$

$$H_1: \ \text{At least two of the means are not equal.}$$

In addition, we may be interested in making individual comparisons among these 5 population means.

In the analysis-of-variance procedure, it is assumed that whatever variation exists among the aggregate averages is attributed to (1) variation in absorption among observations *within* aggregate types and (2) variation *among* aggregate types, that is, due to differences in the chemical composition of the aggregates. The **within-aggregate variation** is, of course, brought about by various causes. Perhaps humidity and temperature conditions were not kept entirely constant throughout the experiment. It is possible that there was a certain amount of heterogeneity in the batches of raw materials that were used. At any rate, we shall consider the within-sample variation to be **chance or random variation**. Part of the goal of the analysis of variance is to determine if the differences among the 5 sample means are what we would expect due to random variation alone or, rather, due to variation beyond merely random effects, i.e., differences in the chemical composition of the aggregates.

Many pointed questions appear at this stage concerning the preceding problem. For example, how many samples must be tested for each aggregate? This is a question that continually haunts the practitioner. In addition, what if the within-sample variation is so large that it is difficult for a statistical procedure to detect the systematic differences? Can we systematically control extraneous sources of variation and thus remove them from the portion we call random variation? We shall attempt to answer these and other questions in the following sections.

In Chapters 5 and 6, the notions of estimation and testing for the two-sample case were covered against the important backdrop of the way the experiment is conducted. This discussion falls into the broad category of design of experiments. For example, for the **pooled *t*-test** discussed in Chapter 6, it is assumed that the factor levels (treatments in the mice example) are assigned randomly to the experimental units (mice). The notion of experimental units was discussed in Chapters 5 and 6 and illustrated through examples. Simply put, experimental units are the units (mice, patients, concrete specimens, time) that **provide the heterogeneity that leads to experimental error** in a scientific investigation. The random assignment eliminates bias that could result with systematic assignment. The goal is to distribute uniformly among the factor levels the risks brought about by the heterogeneity of the experimental units. Random assignment best simulates the conditions that are assumed by the model. In Section 8.5, we discuss **blocking** in experiments. The notion of blocking was presented in Chapters 5 and 6, when comparisons between means were accomplished with **pairing**, that is, the division of the experimental units into homogeneous pairs called **blocks**. The factor levels or treatments are then assigned randomly within blocks. The purpose of blocking is to reduce the effective experimental error. In this chapter, we naturally extend the pairing to larger block sizes, with analysis of variance being the primary analytical tool.

8.2   One-Way Analysis of Variance (One-Way ANOVA): Completely Randomized Design

Random samples of size n are selected from each of k populations. The k different populations are classified on the basis of a single criterion such as different treatments or groups. Today the term **treatment** is used generally to refer to the various classifications, whether they be different aggregates, different analysts, different fertilizers, or different regions of the country.

Assumptions and Hypotheses in One-Way ANOVA

It is assumed that the k populations are independent and normally distributed with means $\mu_1, \mu_2, \ldots, \mu_k$ and common variance σ^2. As indicated in Section 8.1, these assumptions are made more palatable by randomization. We wish to derive appropriate methods for testing the hypothesis

$$H_0:\ \mu_1 = \mu_2 = \cdots = \mu_k,$$
$$H_1:\ \text{At least two of the means are not equal.}$$

Let y_{ij} denote the jth observation from the ith treatment and arrange the data as in Table 8.2. Here, $Y_{i.}$ is the total of all observations in the sample from the ith treatment, $\bar{y}_{i.}$ is the mean of all observations in the sample from the ith treatment, $Y_{..}$ is the total of all nk observations, and $\bar{y}_{..}$ is the mean of all nk observations.

Model for One-Way ANOVA

Each observation may be written in the form

$$Y_{ij} = \mu_i + \epsilon_{ij},$$

where ϵ_{ij} measures the deviation of the jth observation of the ith sample from the corresponding treatment mean. The ϵ_{ij}-term represents random error and plays the same role as the error terms in the regression models. An alternative and

Table 8.2: k Random Samples

Treatment:	1	2	$\cdots$	i	$\cdots$	k	
	y_{11}	y_{21}	$\cdots$	y_{i1}	$\cdots$	y_{k1}	
	y_{12}	y_{22}	$\cdots$	y_{i2}	$\cdots$	y_{k2}	
	$\vdots$	$\vdots$		$\vdots$		$\vdots$	
	y_{1n}	y_{2n}	$\cdots$	y_{in}	$\cdots$	y_{kn}	
Total	$Y_{1.}$	$Y_{2.}$	$\cdots$	$Y_{i.}$	$\cdots$	$Y_{k.}$	$Y_{..}$
Mean	$\bar{y}_{1.}$	$\bar{y}_{2.}$	$\cdots$	$\bar{y}_{i.}$	$\cdots$	$\bar{y}_{k.}$	$\bar{y}_{..}$

preferred form of this equation is obtained by substituting $\mu_i = \mu + \alpha_i$, subject to the constraint $\sum_{i=1}^{k} \alpha_i = 0$. Hence, we may write

$$Y_{ij} = \mu + \alpha_i + \epsilon_{ij},$$

where μ is just the **grand mean** of all the μ_i, that is,

$$\mu = \frac{1}{k} \sum_{i=1}^{k} \mu_i,$$

and α_i is called the **effect** of the ith treatment.

The null hypothesis that the k population means are equal, against the alternative that at least two of the means are unequal, may now be replaced by the equivalent hypothesis

$$H_0: \ \alpha_1 = \alpha_2 = \cdots = \alpha_k = 0,$$
$$H_1: \ \text{At least one of the } \alpha_i \text{ is not equal to zero.}$$

Resolution of Total Variability into Components

Our test will be based on a comparison of two independent estimates of the common population variance σ^2. These estimates will be obtained by partitioning the total variability of our data, designated by the double summation

$$\sum_{i=1}^{k} \sum_{j=1}^{n} (y_{ij} - \bar{y}_{..})^2,$$

into two components.

Theorem 8.1:

Sum-of-Squares Identity
$$\sum_{i=1}^{k} \sum_{j=1}^{n} (y_{ij} - \bar{y}_{..})^2 = n \sum_{i=1}^{k} (\bar{y}_{i.} - \bar{y}_{..})^2 + \sum_{i=1}^{k} \sum_{j=1}^{n} (y_{ij} - \bar{y}_{i.})^2$$

It will be convenient in what follows to identify the terms of the sum-of-squares identity by the following notation:

<table>
<tr><td>

Three Important
Measures of
Variability

</td><td>

$$SST = \sum_{i=1}^{k}\sum_{j=1}^{n}(y_{ij} - \bar{y}_{..})^2 = \text{total sum of squares,}$$

$$SSA = n\sum_{i=1}^{k}(\bar{y}_{i.} - \bar{y}_{..})^2 = \text{treatment sum of squares,}$$

$$SSE = \sum_{i=1}^{k}\sum_{j=1}^{n}(y_{ij} - \bar{y}_{i.})^2 = \text{error sum of squares.}$$

</td></tr>
</table>

The sum-of-squares identity can then be represented symbolically by the equation

$$SST = SSA + SSE.$$

The identity above expresses how between-treatment and within-treatment variation add to the total sum of squares. However, much insight can be gained by investigating the **expected value of both *SSA* and *SSE***. Eventually, we shall develop variance estimates that formulate the ratio to be used to test the equality of population means.

Theorem 8.2:

$$E(SSA) = (k-1)\sigma^2 + n\sum_{i=1}^{k}\alpha_i^2$$

The proof of the theorem is left to the reader.

If H_0 is true, an estimate of σ^2, based on $k-1$ degrees of freedom, is provided by this expression:

Treatment Mean
Square

$$s_1^2 = \frac{SSA}{k-1}$$

If H_0 is true and thus each α_i in Theorem 8.2 is equal to zero, we see that

$$E\left(\frac{SSA}{k-1}\right) = \sigma^2,$$

and s_1^2 is an unbiased estimate of σ^2. However, if H_1 is true, we have

$$E\left(\frac{SSA}{k-1}\right) = \sigma^2 + \frac{n}{k-1}\sum_{i=1}^{k}\alpha_i^2,$$

and s_1^2 estimates σ^2 plus an additional positive term, which measures variation due to the systematic effects.

A second and independent estimate of σ^2, based on $k(n-1)$ degrees of freedom, is the following formula:

Error Mean
Square

$$s^2 = \frac{SSE}{k(n-1)}$$

It is instructive to point out the importance of the expected values of the mean squares. In the next section, we discuss the use of an **F-ratio** with the treatment mean square residing in the numerator. It turns out that when H_1 is true, the presence of the condition $E(s_1^2) > E(s^2)$ suggests that the F-ratio be used in the context of a **one-sided upper-tailed test**. That is, when H_1 is true, we would expect the numerator s_1^2 to exceed the denominator.

Use of *F*-Test in ANOVA

The estimate s^2 is unbiased regardless of the truth or falsity of the null hypothesis. It is important to note that the sum-of-squares identity has partitioned not only the total variability of the data, but also the total number of degrees of freedom. That is,

$$nk - 1 = k - 1 + k(n - 1).$$

F-Ratio for Testing Equality of Means

When H_0 is true, the ratio $f = s_1^2/s^2$ is a value of the random variable F having the F-distribution with $k - 1$ and $k(n - 1)$ degrees of freedom (see Theorem 4.8). Since s_1^2 overestimates σ^2 when H_0 is false, we have a one-tailed test with the critical region entirely in the right tail of the distribution.

The null hypothesis H_0 is rejected at the α-level of significance when

$$f > f_\alpha[k - 1, k(n - 1)].$$

Another approach, the *P*-value approach, suggests that the evidence in favor of or against H_0 is

$$P = P\{f[k - 1, k(n - 1)] > f\}.$$

The computations for an analysis-of-variance problem are usually summarized in tabular form, as shown in Table 8.3.

Table 8.3: Analysis of Variance for the One-Way ANOVA

Source of Variation	Sum of Squares	Degrees of Freedom	Mean Square	Computed f
Treatments	SSA	$k - 1$	$s_1^2 = \dfrac{SSA}{k - 1}$	$\dfrac{s_1^2}{s^2}$
Error	SSE	$k(n - 1)$	$s^2 = \dfrac{SSE}{k(n - 1)}$	
Total	SST	$kn - 1$		

Example 8.1: Test the hypothesis $\mu_1 = \mu_2 = \cdots = \mu_5$ at the 0.05 level of significance for the data of Table 8.1 on absorption of moisture by various types of cement aggregates.

Solution: The hypotheses are

$$H_0: \ \mu_1 = \mu_2 = \cdots = \mu_5,$$
$$H_1: \ \text{At least two of the means are not equal,}$$

with $\alpha = 0.05$.

Critical region: $f > 2.76$ with $v_1 = 4$ and $v_2 = 25$ degrees of freedom. The sum-of-squares computations give

$$SST = 209{,}377, \ \ SSA = 85{,}356,$$
$$SSE = 209{,}377 - 85{,}356 = 124{,}021.$$

These results and the remaining computations are exhibited in Figure 8.1 in the *SAS* ANOVA procedure.

```
                        The GLM Procedure
Dependent Variable: moisture
```

Source	DF	Sum of Squares	Mean Square	F Value	Pr > F
Model	4	85356.4667	21339.1167	4.30	0.0088
Error	25	124020.3333	4960.8133		
Corrected Total	29	209376.8000			

R-Square	Coeff Var	Root MSE	moisture Mean
0.407669	12.53703	70.43304	561.8000

Source	DF	Type I SS	Mean Square	F Value	Pr > F
aggregate	4	85356.46667	21339.11667	4.30	0.0088

Figure 8.1: *SAS* output for the analysis-of-variance procedure.

Decision: Reject H_0 and conclude that the aggregates do not have the same mean absorption. The *P*-value for $f = 4.30$ is 0.0088, which is smaller than 0.05.

In addition to the ANOVA, a box plot was constructed for each aggregate. The plots are shown in Figure 8.2. From these plots it is evident that the absorption is not the same for all aggregates. In fact, it appears as if aggregate 4 stands out from the rest. A more formal analysis showing this result will appear in Exercise 8.21 on page 372.

During experimental work, one often loses some of the desired observations. Experimental animals may die, experimental material may be damaged, or human subjects may drop out of a study. The previous analysis for equal sample size will still be valid if we slightly modify the sum of squares formulas. We now assume the k random samples to be of sizes $n_1, n_2, \ldots, n_k$, respectively.

Sum of Squares, Unequal Sample Sizes

$$SST = \sum_{i=1}^{k} \sum_{j=1}^{n_i} (y_{ij} - \bar{y}_{..})^2, \ \ SSA = \sum_{i=1}^{k} n_i (\bar{y}_{i.} - \bar{y}_{..})^2, \ \ SSE = SST - SSA$$

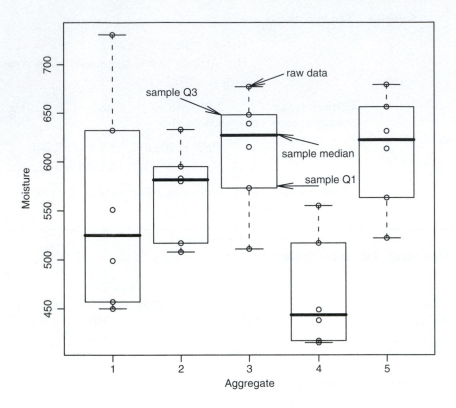

Figure 8.2: Box plots for the absorption of moisture in concrete aggregates.

The degrees of freedom are then partitioned as before: $N - 1$ for SST, $k - 1$ for SSA, and $N - 1 - (k - 1) = N - k$ for SSE, where $N = \sum_{i=1}^{k} n_i$.

Example 8.2: Part of a study conducted at Virginia Tech was designed to measure serum alkaline phosphatase activity levels (in Bessey-Lowry units) in children with seizure disorders who were receiving anticonvulsant therapy under the care of a private physician. Forty-five subjects were found for the study and categorized into four drug groups:

G-1: Control (not receiving anticonvulsants and having no history of seizure disorders)

G-2: Phenobarbital

G-3: Carbamazepine

G-4: Other anticonvulsants

From blood samples collected from each subject, the serum alkaline phosphatase activity level was determined and recorded as shown in Table 8.4. Test the hypothesis at the 0.05 level of significance that the average serum alkaline phosphatase activity level is the same for the four drug groups.

Table 8.4: Serum Alkaline Phosphatase Activity Level

G-1		G-2	G-3	G-4
49.20	97.50	97.07	62.10	110.60
44.54	105.00	73.40	94.95	57.10
45.80	58.05	68.50	142.50	117.60
95.84	86.60	91.85	53.00	77.71
30.10	58.35	106.60	175.00	150.00
36.50	72.80	0.57	79.50	82.90
82.30	116.70	0.79	29.50	111.50
87.85	45.15	0.77	78.40	
105.00	70.35	0.81	127.50	
95.22	77.40			

Solution: With the level of significance at 0.05, the hypotheses are

$$H_0: \ \mu_1 = \mu_2 = \mu_3 = \mu_4,$$
$$H_1: \ \text{At least two of the means are not equal.}$$

Critical region: $f > 2.836$, from interpolating in Table A.6.
Computations: $Y_{1.} = 1460.25$, $Y_{2.} = 440.36$, $Y_{3.} = 842.45$, $Y_{4.} = 707.41$, and $Y_{..} = 3450.47$. The analysis of variance is shown in the *MINITAB* output of Figure 8.3.

```
One-way ANOVA: G-1, G-2, G-3, G-4

Source   DF    SS     MS     F     P
Factor    3   13939  4646   3.57  0.022
Error    41   53376  1302
Total    44   67315

S = 36.08    R-Sq = 20.71%    R-Sq(adj) = 14.90%

                        Individual 95% CIs For Mean Based on
                        Pooled StDev
Level   N    Mean   StDev   --+---------+---------+---------+-------
G-1    20   73.01   25.75              (----*-----)
G-2     9   48.93   47.11   (-------*-------)
G-3     9   93.61   46.57                 (-------*-------)
G-4     7  101.06   30.76                   (--------*--------)
                            --+---------+---------+---------+-------
                            30        60        90       120

Pooled StDev = 36.08
```

Figure 8.3: *MINITAB* analysis of data in Table 8.4.

Decision: Reject H_0 and conclude that the average serum alkaline phosphatase activity levels for the four drug groups are not all the same. The calculated P-value is 0.022.

In concluding our discussion on the analysis of variance for the one-way classification, we state the advantages of choosing equal sample sizes rather than unequal sample sizes. The first advantage is that the f-ratio is insensitive to slight departures from the assumption of equal variances for the k populations when the samples are of equal size. Second, the choice of equal sample sizes minimizes the probability of committing a type II error.

8.3   Tests for the Equality of Several Variances

Although the f-ratio obtained from the analysis-of-variance procedure is insensitive to departures from the assumption of equal variances for the k normal populations when the samples are of equal size, we may still prefer to exercise caution and run a preliminary test for homogeneity of variances. Such a test would certainly be advisable in the case of unequal sample sizes if there was a reasonable doubt concerning the homogeneity of the population variances. Suppose, therefore, that we wish to test the null hypothesis

$$H_0\colon\ \sigma_1^2 = \sigma_2^2 = \cdots = \sigma_k^2$$

against the alternative

$$H_1\colon\ \text{The variances are not all equal.}$$

The test that we shall use, called **Bartlett's test**, is based on a statistic whose sampling distribution provides exact critical values when the sample sizes are equal. These critical values for equal sample sizes can also be used to yield highly accurate approximations to the critical values for unequal sample sizes.

First, we compute the k sample variances $s_1^2, s_2^2, \ldots, s_k^2$ from samples of size $n_1, n_2, \ldots, n_k$, with $\sum_{i=1}^{k} n_i = N$. Second, we combine the sample variances to give the pooled estimate

$$s_p^2 = \frac{1}{N-k} \sum_{i=1}^{k} (n_i - 1)s_i^2.$$

Now

$$b = \frac{[(s_1^2)^{n_1-1}(s_2^2)^{n_2-1} \cdots (s_k^2)^{n_k-1}]^{1/(N-k)}}{s_p^2}$$

is a value of a random variable B having the **Bartlett distribution**. For the special case where $n_1 = n_2 = \cdots = n_k = n$, we reject H_0 at the α-level of significance if

$$b < b_k(\alpha; n),$$

where $b_k(\alpha; n)$ is the critical value leaving an area of size α in the left tail of the Bartlett distribution. Table A.8 gives the critical values, $b_k(\alpha; n)$, for $\alpha = 0.01$ and 0.05; $k = 2, 3, \ldots, 10$; and selected values of n from 3 to 100.

When the sample sizes are unequal, the null hypothesis is rejected at the α-level of significance if

$$b < b_k(\alpha; n_1, n_2, \ldots, n_k),$$

where

$$b_k(\alpha; n_1, n_2, \ldots, n_k) \approx \frac{n_1 b_k(\alpha; n_1) + n_2 b_k(\alpha; n_2) + \cdots + n_k b_k(\alpha; n_k)}{N}.$$

As before, all the $b_k(\alpha; n_i)$ for sample sizes $n_1, n_2, \ldots, n_k$ are obtained from Table A.8.

Example 8.3: Use Bartlett's test to test the hypothesis at the 0.01 level of significance that the population variances of the four drug groups of Example 8.2 are equal.

Solution: We have the hypotheses

$$H_0: \sigma_1^2 = \sigma_2^2 = \sigma_3^2 = \sigma_4^2,$$
$$H_1: \text{The variances are not equal},$$
$$\text{with } \alpha = 0.01.$$

Critical region: Referring to Example 8.2, we have $n_1 = 20$, $n_2 = 9$, $n_3 = 9$, $n_4 = 7$, $N = 45$, and $k = 4$. Therefore, we reject when

$$b < b_4(0.01; 20, 9, 9, 7)$$
$$\approx \frac{(20)(0.8586) + (9)(0.6892) + (9)(0.6892) + (7)(0.6045)}{45}$$
$$= 0.7513.$$

Computations: First compute

$$s_1^2 = 662.862, \quad s_2^2 = 2219.781, \quad s_3^2 = 2168.434, \quad s_4^2 = 946.032,$$

and then

$$s_p^2 = \frac{(19)(662.862) + (8)(2219.781) + (8)(2168.434) + (6)(946.032)}{41}$$
$$= 1301.861.$$

Now

$$b = \frac{[(662.862)^{19}(2219.781)^8(2168.434)^8(946.032)^6]^{1/41}}{1301.861} = 0.8557.$$

Decision: Do not reject the hypothesis, and conclude that the population variances of the four drug groups are not significantly different.

Although Bartlett's test is most often used for testing of homogeneity of variances, other methods are available. A method due to Cochran provides a computationally simple procedure, but it is restricted to situations in which the sample sizes are equal. **Cochran's test** is particularly useful for detecting if one variance is much larger than the others. The statistic that is used is

$$G = \frac{\text{largest } S_i^2}{\sum\limits_{i=1}^{k} S_i^2},$$

and the hypothesis of equality of variances is rejected if $g > g_\alpha$, where the value of g_α is obtained from Table A.9.

To illustrate Cochran's test, let us refer again to the data of Table 8.1 on moisture absorption in concrete aggregates. Were we justified in assuming equal variances when we performed the analysis of variance in Example 8.1? We find that

$$s_1^2 = 12{,}134, \ s_2^2 = 2303, \ s_3^2 = 3594, \ s_4^2 = 3319, \ s_5^2 = 3455.$$

Therefore,

$$g = \frac{12{,}134}{24{,}805} = 0.4892,$$

which does not exceed the table value $g_{0.05} = 0.5065$. Hence, we conclude that the assumption of equal variances is reasonable.

Exercises

8.1 Six different machines are being considered for use in manufacturing rubber seals. The machines are being compared with respect to tensile strength of the product. A random sample of four seals from each machine is used to determine whether the mean tensile strength varies from machine to machine. The following are the tensile-strength measurements in kilograms per square centimeter $\times 10^{-1}$:

Machine					
1	**2**	**3**	**4**	**5**	**6**
17.5	16.4	20.3	14.6	17.5	18.3
16.9	19.2	15.7	16.7	19.2	16.2
15.8	17.7	17.8	20.8	16.5	17.5
18.6	15.4	18.9	18.9	20.5	20.1

Perform the analysis of variance at the 0.05 level of significance and indicate whether or not the mean tensile strengths differ significantly for the six machines.

8.2 The data in the following table represent the number of hours of relief provided by five different brands of headache tablets administered to 25 subjects experiencing fevers of $38°C$ or more. Perform the analysis of variance and test the hypothesis at the 0.05 level of significance that the mean number of hours of relief provided by the tablets is the same for all five brands. Discuss the results.

Tablet				
A	**B**	**C**	**D**	**E**
5.2	9.1	3.2	2.4	7.1
4.7	7.1	5.8	3.4	6.6
8.1	8.2	2.2	4.1	9.3
6.2	6.0	3.1	1.0	4.2
3.0	9.1	7.2	4.0	7.6

8.3 In an article "Shelf-Space Strategy in Retailing," published in *Proceedings: Southern Marketing Association*, the effect of shelf height on the supermarket sales of canned dog food is investigated. An experiment was conducted at a small supermarket for a period of 8 days on the sales of a single brand of dog food, referred to as Arf dog food, involving three levels of shelf height: knee level, waist level, and eye level. During each day, the shelf height of the canned dog food was randomly changed on three different occasions. The remaining sections of the gondola that housed the given brand were filled with a mixture of dog food brands that were both familiar and unfamiliar to customers in this particular geographic area. Sales, in hundreds of dollars, of Arf dog food per day for the three shelf heights are given. Based on the data, is there a significant difference in the average daily sales of this dog food based on shelf height? Use a 0.01 level of significance.

Shelf Height		
Knee Level	**Waist Level**	**Eye Level**
77	88	85
82	94	85
86	93	87
78	90	81
81	91	80
86	94	79
77	90	87
81	87	93

8.4 Immobilization of free-ranging white-tailed deer by drugs allows researchers the opportunity to closely examine the deer and gather valuable physiological information. In the study *Influence of Physical Restraint and Restraint Facilitating Drugs on Blood Measurements of White-Tailed Deer and Other Selected Mammals*, conducted at Virginia Tech, wildlife biologists tested the "knockdown" time (time from injection to immobilization) of three different immobilizing drugs. Immobilization, in this case, is defined as the point where the animal no longer has enough muscle control to remain standing. Thirty male white-tailed deer were randomly assigned to each of three treatments. Group *A* received 5 milligrams of liquid succinylcholine chloride (SCC); group *B* received 8 milligrams of powdered SCC; and group *C* received 200 milligrams of phencyclidine hydrochloride. Knockdown times, in minutes, were recorded. Perform an analysis of variance at the 0.01 level of significance and determine whether or not the average knockdown time for the three drugs is the same.

Group		
A	*B*	*C*
11	10	4
5	7	4
14	16	6
7	7	3
10	7	5
7	5	6
23	10	8
4	10	3
11	6	7
11	12	3

8.5 The mitochondrial enzyme NADPH:NAD transhydrogenase of the common rat tapeworm (*Hymenolepiasis diminuta*) catalyzes hydrogen in the transfer from NADPH to NAD, producing NADH. This enzyme is known to serve a vital role in the tapeworm's anaerobic metabolism, and it has recently been hypothesized that it may serve as a proton exchange pump, transferring protons across the mitochondrial membrane. A study on *Effect of Various Substrate Concentrations on the Conformational Variation of the NADPH:NAD Transhydrogenase of Hymenolepiasis diminuta*, conducted at Bowling Green State University, was designed to assess the ability of this enzyme to undergo conformation or shape changes. Changes in the specific activity of the enzyme caused by variations in the concentration of NADP could be interpreted as supporting the theory of conformational change. The enzyme in question is located in the inner membrane of the tapeworm's mitochondria. Tapeworms were homogenized, and through a series of centrifugations, the enzyme was isolated. Various concentrations of NADP were then added to the isolated enzyme solution, and the mixture was then incubated in a water bath at 56°C for 3 minutes. The enzyme was then analyzed on a dual-beam spectrophotometer, and the results shown were calculated, with the specific activity of the enzyme given in nanomoles per minute per milligram of protein. Test the hypothesis at the 0.01 level that the average specific activity is the same for the four concentrations.

NADP Concentration (nm)				
0	**80**	**160**	**360**	
11.01	11.38	11.02	6.04	10.31
12.09	10.67	10.67	8.65	8.30
10.55	12.33	11.50	7.76	9.48
11.26	10.08	10.31	10.13	8.89
			9.36	

8.6 A study measured the sorption (either absorption or adsorption) rates of three different types of organic chemical solvents. These solvents are used to clean industrial fabricated-metal parts and are potential hazardous waste. Independent samples from each type of solvent were tested, and their sorption rates were recorded as a mole percentage. (See McClave, Dietrich, and Sincich, 1997.)

Aromatics		Chloroalkanes		Esters		
1.06	0.95	1.58	1.12	0.29	0.43	0.06
0.79	0.65	1.45	0.91	0.06	0.51	0.09
0.82	1.15	0.57	0.83	0.44	0.10	0.17
0.89	1.12	1.16	0.43	0.55	0.53	0.17
1.05				0.61	0.34	0.60

Is there a significant difference in the mean sorption rates for the three solvents? Use a *P*-value for your conclusions. Which solvent would you use?

8.7 It has been shown that the fertilizer magnesium ammonium phosphate, $MgNH_4PO_4$, is an effective supplier of the nutrients necessary for plant growth. The compounds supplied by this fertilizer are highly soluble in water, allowing the fertilizer to be applied directly on the soil surface or mixed with the growth substrate during the potting process. A study on the *Effect of Magnesium Ammonium Phosphate on Height of Chrysanthemums* was conducted at George Mason University to determine a possible optimum level of fertilization, based on the enhanced vertical growth response of the chrysanthemums. Forty chrysanthemum seedlings were divided into four groups, each containing 10 plants. Each was planted in a similar pot containing a uniform growth medium. To each group of plants an increasing concentration of $MgNH_4PO_4$, measured in grams per bushel, was added. The four groups of plants were grown under uniform conditions in a greenhouse for a period of four weeks. The treatments and the respective changes in heights, measured in centimeters, are shown next.

Treatment

50 g/bu		100 g/bu		200 g/bu		400 g/bu	
13.2	12.4	16.0	12.6	7.8	14.4	21.0	14.8
12.8	17.2	14.8	13.0	20.0	15.8	19.1	15.8
13.0	14.0	14.0	23.6	17.0	27.0	18.0	26.0
14.2	21.6	14.0	17.0	19.6	18.0	21.1	22.0
15.0	20.0	22.2	24.4	20.2	23.2	25.0	18.2

Can we conclude at the 0.05 level of significance that different concentrations of $MgNH_4PO_4$ affect the average attained height of chrysanthemums? How much $MgNH_4PO_4$ appears to be best?

8.8 For the data set in Exercise 8.7, use Bartlett's test to check whether the variances are equal. Use $\alpha = 0.05$.

8.9 Use Bartlett's test at the 0.01 level of significance to test for homogeneity of variances in Exercise 8.5 on page 367.

8.10 Use Cochran's test at the 0.01 level of significance to test for homogeneity of variances in Exercise 8.4 on page 367.

8.11 Use Bartlett's test at the 0.05 level of significance to test for homogeneity of variances in Exercise 8.6 on page 367.

8.4   Multiple Comparisons

The analysis of variance is a powerful procedure for testing the homogeneity of a set of means. However, if we reject the null hypothesis and accept the stated alternative—that the means are not all equal—we still do not know which of the population means are equal and which are different.

Often it is of interest to make several (perhaps all possible) **paired comparisons** among the treatments. Actually, a paired comparison may be viewed as a simple contrast, namely, a test of

$$H_0: \ \mu_i - \mu_j = 0,$$
$$H_1: \ \mu_i - \mu_j \neq 0,$$

for all $i \neq j$. Making all possible paired comparisons among the means can be very beneficial when particular complex contrasts are not known *a priori*. For example, in the aggregate data of Table 8.1, suppose that we wish to test

$$H_0: \ \mu_1 - \mu_5 = 0,$$
$$H_1: \ \mu_1 - \mu_5 \neq 0.$$

The test is developed through use of an F, t, or confidence interval approach. Using t, we have

$$t = \frac{\bar{y}_{1.} - \bar{y}_{5.}}{s\sqrt{2/n}},$$

where s is the square root of the mean square error and $n = 6$ is the sample size per treatment. Using the mean square error from Figure 8.1 as the common variance, we have

$$t = \frac{553.33 - 610.67}{\sqrt{4960.8133}\sqrt{1/3}} = -1.41.$$

The P-value for the t-test with 25 degrees of freedom is 0.17. Thus, there is not sufficient evidence to reject H_0.

Relationship between T and F

In the foregoing, we displayed the use of a pooled t-test along the lines of that discussed in Chapter 6. The pooled estimate was taken from the mean squared error in order to enjoy the degrees of freedom that are pooled across all five samples. In addition, we have tested a contrast. The reader should note that if the t-value is squared, the result is exactly of the same form as the value of f for a test on a contrast, discussed in the preceding section. In fact,

$$f = \frac{(\bar{y}_{1.} - \bar{y}_{5.})^2}{s^2(1/6 + 1/6)} = \frac{(553.33 - 610.67)^2}{4961(1/3)} = 1.988,$$

which, of course, is t^2.

Confidence Interval Approach to a Paired Comparison

It is straightforward to solve the same problem of a paired comparison (or a contrast) using a confidence interval approach. Clearly, if we compute a $100(1 - \alpha)\%$ confidence interval on $\mu_1 - \mu_5$, we have

$$\bar{y}_{1.} - \bar{y}_{5.} \pm t_{\alpha/2} s \sqrt{\frac{2}{6}},$$

where $t_{\alpha/2}$ is the upper $100(1 - \alpha/2)\%$ point of a t-distribution with 25 degrees of freedom (degrees of freedom coming from s^2). This straightforward connection between hypothesis testing and confidence intervals should be obvious from discussions in Chapters 5 and 6. The test of the simple contrast $\mu_1 - \mu_5$ involves no more than observing whether or not the confidence interval above covers zero. Substituting the numbers, we have as the 95% confidence interval

$$(553.33 - 610.67) \pm 2.060\sqrt{4961}\sqrt{\frac{1}{3}} = -57.34 \pm 83.77.$$

Thus, since the interval covers zero, the contrast is not significant. In other words, we do not find a significant difference between the means of aggregates 1 and 5.

Experiment-wise Error Rate

Serious difficulties occur when the analyst attempts to make many or all possible paired comparisons. For the case of k means, there will be, of course, $r = k(k - 1)/2$ possible paired comparisons. Assuming independent comparisons, the **experiment-wise error rate** or **family error rate** (i.e., the probability of false rejection of at least one of the hypotheses) is given by $1 - (1 - \alpha)^r$, where α is the selected probability of a type I error for a specific comparison. Clearly, this measure of experiment-wise type I error can be quite large. For example, even if there are only 6 comparisons, say, in the case of 4 means, and $\alpha = 0.05$, the experiment-wise rate is

$$1 - (0.95)^6 \approx 0.26.$$

When many paired comparisons are being tested, there is usually a need to make the effective contrast on a single comparison more conservative. That is, with the confidence interval approach, the confidence intervals would be much wider than the $\pm t_{\alpha/2} s \sqrt{2/n}$ used for the case where only a single comparison is being made.

Tukey's Test

There are several standard methods for making paired comparisons that sustain the credibility of the type I error rate. We shall discuss and illustrate two of them here. The first one, called **Tukey's procedure**, allows formation of simultaneous $100(1 - \alpha)\%$ confidence intervals for all paired comparisons. The method is based on the *studentized* range distribution. The appropriate percentile point is a function of α, k, and v = degrees of freedom for s^2. A list of upper percentage points for $\alpha = 0.05$ is shown in Table A.10. The method of paired comparisons by Tukey involves finding a significant difference between means i and j $(i \neq j)$ if $|\bar{y}_{i.} - \bar{y}_{j.}|$ exceeds $q(\alpha, k, v) \sqrt{\frac{s^2}{n}}$.

Tukey's procedure is easily illustrated. Consider a hypothetical example where we have 6 treatments in a one-factor completely randomized design, with 5 observations taken per treatment. Suppose that the mean square error taken from the analysis-of-variance table is $s^2 = 2.45$ (24 degrees of freedom). The sample means are in ascending order:

$\bar{y}_{2.}$	$\bar{y}_{5.}$	$\bar{y}_{1.}$	$\bar{y}_{3.}$	$\bar{y}_{6.}$	$\bar{y}_{4.}$
14.50	16.75	19.84	21.12	22.90	23.20

With $\alpha = 0.05$, the value of $q(0.05, 6, 24)$ is 4.37. Thus, all absolute differences are to be compared to

$$4.37 \sqrt{\frac{2.45}{5}} = 3.059.$$

As a result, the following represent means found to be significantly different using Tukey's procedure:

4 and 1,	4 and 5,	4 and 2,	6 and 1,	6 and 5,
6 and 2,	3 and 5,	3 and 2,	1 and 5,	1 and 2.

Where Does the α-Level Come From in Tukey's Test?

We briefly alluded to the concept of **simultaneous confidence intervals** being employed for Tukey's procedure. The reader will gain a useful insight into the notion of multiple comparisons if he or she gains an understanding of what is meant by simultaneous confidence intervals.

In Chapter 5, we saw that if we compute a 95% confidence interval on, say, a mean μ, then the probability that the interval covers the true mean μ is 0.95. However, as we have discussed, for the case of multiple comparisons, the effective probability of interest is tied to the experiment-wise error rate, and it should be emphasized that the confidence intervals of the type $\bar{y}_{i.} - \bar{y}_{j.} \pm q(\alpha, k, v) s \sqrt{1/n}$ are not independent since they all involve s and many involve the use of the same averages, the $\bar{y}_{i.}$. Despite the difficulties, if we use $q(0.05, k, v)$, the simultaneous confidence level is controlled at 95%. The same holds for $q(0.01, k, v)$; namely, the confidence level is controlled at 99%. In the case of $\alpha = 0.05$, there is a probability of 0.05 that at least one pair of measures will be falsely found to be different (false rejection of at least one null hypothesis). In the $\alpha = 0.01$ case, the corresponding probability will be 0.01.

Although we only discussed Tukey's test in multiple comparison procedures, there are many other tests available.

Exercises

8.12 The study *Loss of Nitrogen Through Sweat by Preadolescent Boys Consuming Three Levels of Dietary Protein* was conducted by the Department of Human Nutrition and Foods at Virginia Tech to determine perspiration nitrogen loss at various dietary protein levels. Twelve preadolescent boys ranging in age from 7 years, 8 months to 9 years, 8 months, all judged to be clinically healthy, were used in the experiment. Each boy was subjected to one of three controlled diets in which 29, 54, or 84 grams of protein were consumed per day. The following data represent the body perspiration nitrogen loss, in milligrams, during the last two days of the experimental period:

Protein Level		
29 Grams	**54 Grams**	**84 Grams**
190	318	390
266	295	321
270	271	396
	438	399
	402	

(a) Perform an analysis of variance at the 0.05 level of significance to show that the mean perspiration nitrogen losses at the three protein levels are different.

(b) Use Tukey's test to determine which protein levels are significantly different from each other in mean nitrogen loss.

8.13 The purpose of the study *The Incorporation of a Chelating Agent into a Flame Retardant Finish of a Cotton Flannelette and the Evaluation of Selected Fabric Properties*, conducted at Virginia Tech, was to evaluate the use of a chelating agent as part of the flame-retardant finish of cotton flannelette by determining its effects upon flammability after the fabric is laundered under specific conditions. Two baths were prepared, one with carboxymethyl cellulose and one without. Twelve pieces of fabric were laundered 5 times in bath I, and 12 other pieces of fabric were laundered 10 times in bath I. This procedure was repeated using 24 additional pieces of cloth in bath II. After the washings the lengths of fabric that burned and the burn

times were measured. For convenience, let us define the following treatments:

Treatment 1: 5 launderings in bath I,

Treatment 2: 5 launderings in bath II,

Treatment 3: 10 launderings in bath I,

Treatment 4: 10 launderings in bath II.

Burn times, in seconds, were recorded as follows:

	Treatment		
1	**2**	**3**	**4**
13.7	6.2	27.2	18.2
23.0	5.4	16.8	8.8
15.7	5.0	12.9	14.5
25.5	4.4	14.9	14.7
15.8	5.0	17.1	17.1
14.8	3.3	13.0	13.9
14.0	16.0	10.8	10.6
29.4	2.5	13.5	5.8
9.7	1.6	25.5	7.3
14.0	3.9	14.2	17.7
12.3	2.5	27.4	18.3
12.3	7.1	11.5	9.9

(a) Perform an analysis of variance, using a 0.01 level of significance, and determine whether there are any significant differences among the treatment means.

(b) Use Tukey's test, with a 0.05 level of significance, to test whether treatment levels are different from each other.

8.14 An investigation was conducted to determine the source of reduction in yield of a certain chemical product. It was known that the loss in yield occurred in the mother liquor, that is, the material removed at the filtration stage. It was thought that different blends of the original material might result in different yield reductions at the mother liquor stage. The data below on the percent reductions for 3 batches of each of 4 preselected blends were obtained.

(a) Perform an analysis of variance at the $\alpha = 0.05$ level of significance.

(b) Use Tukey's test to determine which blends differ.

Blend			
1	2	3	4
25.6	25.2	20.8	31.6
24.3	28.6	26.7	29.8
27.9	24.7	22.2	34.3

8.15 Use Tukey's test, with a 0.05 level of significance, to analyze the means of the five different brands of headache tablets in Exercise 8.2 on page 366.

8.16 The following data are values of pressure (psi) in a torsion spring for several settings of the angle between the legs of the spring in a free position:

Angle (°)					
67	71	75		79	83
83	84	86	87	89	90
85	85	87	87	90	92
	85	88	88	90	
	86	88	88	91	
	86	88	89		
	87	90			

Compute a one-way analysis of variance for this experiment and state your conclusion concerning the effect of angle on the pressure in the spring. (From Hicks and Turner, 1999.)

8.17 In the study *An Evaluation of the Removal Method for Estimating Benthic Populations and Diversity*, conducted by Virginia Tech on the Jackson River, 5 different sampling procedures were used to determine the species counts. Twenty samples were selected at random, and each of the 5 sampling procedures was repeated 4 times. The species counts were recorded as follows:

Sampling Procedure				
Deple-	Modified		Substrate Removal	Kick-
tion	Hess	Surber	Kicknet	net
85	75	31	43	17
55	45	20	21	10
40	35	9	15	8
77	67	37	27	15

(a) Is there a significant difference in the average species counts for the different sampling procedures? Use a *P*-value in your conclusion.

(b) Use Tukey's test with $\alpha = 0.05$ to find which sam-

pling procedures differ.

8.18 The following table gives tensile strengths (in deviations from 340) for wires taken from nine cables to be used for a high-voltage network. Each cable is made from 12 wires. We want to know whether the mean strengths of the wires in the nine cables are the same. If the cables are different, which ones differ? Use a *P*-value in your analysis of variance.

Cable	Tensile Strength											
1	5	−13	−5	−2	−10	−6	−5	0	−3	2	−7	−5
2	−11	−13	−8	8	−3	−12	−12	−10	5	−6	−12	−10
3	0	−10	−15	−12	−2	−8	−5	0	−4	−1	−5	−11
4	−12	4	2	10	−5	−8	−12	0	−5	−3	−3	0
5	7	1	5	0	10	6	5	2	0	−1	−10	−2
6	1	0	−5	−4	−1	0	2	5	1	−2	6	7
7	−1	0	2	1	−4	2	7	5	1	0	−4	2
8	−1	0	7	5	10	8	1	2	−3	6	0	5
9	2	6	7	8	15	11	−7	7	10	7	8	1

(From A. Hald, *Statistical Theory with Engineering Applications*, John Wiley & Sons, New York, 1952)

8.19 It is suspected that the environmental temperature at which batteries are activated affects their life. Thirty homogeneous batteries were tested, six at each of five temperatures, and the data are shown below (activated life in seconds). Analyze and interpret the data.

Temperature (°C)				
0	25	50	75	100
55	60	70	72	65
55	61	72	72	66
57	60	72	72	60
54	60	68	70	64
54	60	77	68	65
56	60	77	69	65

(From Hicks and Turner, 1999.)

8.20 Do Tukey's test for paired comparisons for the data of Exercise 8.6 on page 367. Discuss the results.

8.21 The printout in Figure 8.4 on page 373 gives information on Tukey's test, using PROC GLM in *SAS*, for the aggregate data in Example 8.1. Give conclusions regarding paired comparisons using Tukey's test results.

8.5 Concept of Blocks and the Randomized Complete Block Design

In Section 8.1, we discussed the idea of blocking, that is, isolating sets of experimental units that are reasonably homogeneous and randomly assigning treatments

---

The GLM Procedure
Tukey's Studentized Range (HSD) Test for moisture

NOTE: This test controls the Type I experimentwise error rate, but
it generally has a higher Type II error rate than REGWQ.

```
        Alpha                                      0.05
        Error Degrees of Freedom                     25
        Error Mean Square                      4960.813
        Critical Value of Studentized Range     4.15336
        Minimum Significant Difference          119.43
```

Means with the same letter are not significantly different.

Tukey Grouping		Mean	N	aggregate
	A	610.67	6	5
	A			
	A	610.50	6	3
	A			
B	A	569.33	6	2
B	A			
B	A	553.33	6	1
B				
B		465.17	6	4

Figure 8.4: *SAS* printout for Exercise 8.21.

to these units. This is an extension of the "pairing" concept discussed in Chapters 5 and 6, and it is done to reduce experimental error, since the units in a block have more common characteristics than units in different blocks.

The reader should not view blocks as a second factor, although this is a tempting way of visualizing the design. In fact, the main factor (treatments) still carries the major thrust of the experiment. Experimental units are still the source of error, just as in the completely randomized design. We merely treat sets of these units more systematically when blocking is accomplished. In this way, we say there are restrictions in randomization. Before we turn to a discussion of blocking, let us look at two examples of a **completely randomized design**. The first example is a chemical experiment designed to determine if there is a difference in mean reaction yield among four catalysts. The data are shown in Table 8.5. Samples of materials to be tested are drawn from the same batches of raw materials, while other conditions, such as temperature and concentration of reactants, are held constant. In this case, the time of day for the experimental runs might represent the experimental units, and if the experimenter believed that there could possibly be a slight time effect, he or she would randomize the assignment of the catalysts to the runs to counteract the possible trend. As a second example of such a design, consider an experiment to compare four methods of measuring a particular physical property of a fluid substance. Suppose the sampling process is destructive; that is, once a sample of the substance has been measured by one method, it cannot be measured again by any of the other methods. If it is decided that five measurements are to be taken for each method, then 20 samples of the material are selected from a large batch at random and are used in the experiment to compare the four measuring methods. The experimental units are the randomly selected samples. Any variation from sample to sample will appear in the error variation, as measured by s^2 in the analysis.

Table 8.5: Yield of Reaction

Control	Catalyst 1	Catalyst 2	Catalyst 3
50.7	54.1	52.7	51.2
51.5	53.8	53.9	50.8
49.2	53.1	57.0	49.7
53.1	52.5	54.1	48.0
52.7	54.0	52.5	47.2
$\bar{y}_{0.} = 51.44$	$\bar{y}_{1.} = 53.50$	$\bar{y}_{2.} = 54.04$	$\bar{y}_{3.} = 49.38$

What Is the Purpose of Blocking?

If the variation due to heterogeneity in experimental units is so large that the sensitivity with which treatment differences are detected is reduced due to an inflated value of s^2, a better plan might be to "block off" variation due to these units and thus reduce the extraneous variation to that accounted for by smaller or more homogeneous blocks. For example, suppose that in the previous catalyst illustration it is known *a priori* that there definitely is a significant day-to-day effect on the yield and that we can measure the yield for four catalysts on a given day. Rather than assign the four catalysts to the 20 test runs completely at random, we choose, say, five days and run each of the four catalysts on each day, randomly assigning the catalysts to the runs within days. In this way, the day-to-day variation is removed from the analysis, and consequently the experimental error, which still includes any time trend *within days*, more accurately represents chance variation. Each day is referred to as a **block**.

The most straightforward of the randomized block designs is one in which we randomly assign each treatment once to every block. Such an experimental layout is called a **randomized complete block (RCB) design**, each block constituting a single replication of the treatments.

A typical layout for the randomized complete block design using 3 measurements in 4 blocks is as follows:

Block 1	Block 2	Block 3	Block 4
t_2	t_1	t_3	t_2
t_1	t_3	t_2	t_1
t_3	t_2	t_1	t_3

The t's denote the assignment to blocks of each of the 3 treatments. Of course, the true allocation of treatments to units within blocks is done at random. Once the experiment has been completed, the data can be recorded in the following 3 × 4 array:

Treatment Block:	1	2	3	4
1	y_{11}	y_{12}	y_{13}	y_{14}
2	y_{21}	y_{22}	y_{23}	y_{24}
3	y_{31}	y_{32}	y_{33}	y_{34}

where y_{11} represents the response obtained by using treatment 1 in block l, y_{12} represents the response obtained by using treatment 1 in block 2, ..., and y_{34} represents the response obtained by using treatment 3 in block 4.

Let us now generalize and consider the case of k treatments assigned to b blocks. The data may be summarized as shown in the $k \times b$ rectangular array of Table 8.6. It will be assumed that the y_{ij}, $i = 1, 2, \ldots, k$ and $j = 1, 2, \ldots, b$, are values of independent random variables having normal distributions with mean μ_{ij} and common variance σ^2.

Table 8.6: $k \times b$ Array for the RCB Design

Treatment	Block					Total	Mean
	1	2	$\cdots$	j	$\cdots$	b	
1	y_{11}	y_{12}	$\cdots$	y_{1j}	$\cdots$	y_{1b}	$T_{1.}$	$\bar{y}_{1.}$
2	y_{21}	y_{22}	$\cdots$	y_{2j}	$\cdots$	y_{2b}	$T_{2.}$	$\bar{y}_{2.}$
$\vdots$	$\vdots$	$\vdots$		$\vdots$		$\vdots$	$\vdots$	$\vdots$
i	y_{i1}	y_{i2}	$\cdots$	y_{ij}	$\cdots$	y_{ib}	$T_{i.}$	$\bar{y}_{i.}$
$\vdots$	$\vdots$	$\vdots$		$\vdots$		$\vdots$	$\vdots$	$\vdots$
k	y_{k1}	y_{k2}	$\cdots$	y_{kj}	$\cdots$	y_{kb}	$T_{k.}$	$\bar{y}_{k.}$
Total	$T_{.1}$	$T_{.2}$	$\cdots$	$T_{.j}$	$\cdots$	$T_{.b}$	$T_{..}$	
Mean	$\bar{y}_{.1}$	$\bar{y}_{.2}$	$\cdots$	$\bar{y}_{.j}$	$\cdots$	$\bar{y}_{.b}$		$\bar{y}_{..}$

Let $\mu_{i.}$ represent the average (rather than the total) of the b population means for the ith treatment. That is,

$$\mu_{i.} = \frac{1}{b} \sum_{j=1}^{b} \mu_{ij}, \text{ for } i = 1, \ldots, k.$$

Similarly, the average of the population means for the jth block, $\mu_{.j}$, is defined by

$$\mu_{.j} = \frac{1}{k} \sum_{i=1}^{k} \mu_{ij}, \text{ for } j = 1, \ldots, b$$

and the average of the bk population means, μ, is defined by

$$\mu = \frac{1}{bk} \sum_{i=1}^{k} \sum_{j=1}^{b} \mu_{ij}.$$

To determine if part of the variation in our observations is due to differences among the treatments, we consider the following test:

**Hypothesis of
Equal Treatment
Means**

H_0: $\mu_{1.} = \mu_{2.} = \cdots = \mu_{k.} = \mu$,

H_1: The $\mu_{i.}$ are not all equal.

Model for the RCB Design

Each observation may be written in the form

$$y_{ij} = \mu_{ij} + \epsilon_{ij},$$

where ϵ_{ij} measures the deviation of the observed value y_{ij} from the population mean μ_{ij}. The preferred form of this equation is obtained by substituting

$$\mu_{ij} = \mu + \alpha_i + \beta_j,$$

where α_i is, as before, the effect of the ith treatment and β_j is the effect of the jth block. It is assumed that the treatment and block effects are additive. Hence, we may write

$$y_{ij} = \mu + \alpha_i + \beta_j + \epsilon_{ij}.$$

Notice that the model resembles that of the one-way classification, the essential difference being the introduction of the block effect β_j. The basic concept is much like that of the one-way classification except that we must account in the analysis for the additional effect due to blocks, since we are now systematically controlling variation *in two directions*. If we now impose the restrictions that

$$\sum_{i=1}^{k} \alpha_i = 0 \qquad \text{and} \qquad \sum_{j=1}^{b} \beta_j = 0,$$

then

$$\mu_{i.} = \frac{1}{b} \sum_{j=1}^{b} (\mu + \alpha_i + \beta_j) = \mu + \alpha_i, \text{ for } i = 1, \ldots, k,$$

and

$$\mu_{.j} = \frac{1}{k} \sum_{i=1}^{k} (\mu + \alpha_i + \beta_j) = \mu + \beta_j, \text{ for } j = 1, \ldots, b.$$

The null hypothesis that the k treatment means $\mu_{i.}$ are equal, and therefore equal to μ, is now **equivalent to testing the hypothesis**

$$H_0: \ \alpha_1 = \alpha_2 = \cdots = \alpha_k = 0,$$
$$H_1: \ \text{At least one of the } \alpha_i \text{ is not equal to zero.}$$

Each of the tests on treatments will be based on a comparison of independent estimates of the common population variance σ^2. These estimates will be obtained by splitting the total sum of squares of our data into three components by means of the following identity. The proof is left to the reader.

Theorem 8.3: | **Sum-of-Squares Identity**

$$\sum_{i=1}^{k}\sum_{j=1}^{b}(y_{ij}-\bar{y}_{..})^2 = b\sum_{i=1}^{k}(\bar{y}_{i.}-\bar{y}_{..})^2 + k\sum_{j=1}^{b}(\bar{y}_{.j}-\bar{y}_{..})^2$$

$$+ \sum_{i=1}^{k}\sum_{j=1}^{b}(y_{ij}-\bar{y}_{i.}-\bar{y}_{.j}+\bar{y}_{..})^2$$

Using Theorem 8.3, we can denote the following terms.

The sum-of-squares identity may be presented symbolically by the equation

$$SST = SSA + SSB + SSE,$$

where

$$SST = \sum_{i=1}^{k}\sum_{j=1}^{b}(y_{ij}-\bar{y}_{..})^2 \qquad = \text{total sum of squares,}$$

$$SSA = b\sum_{i=1}^{k}(\bar{y}_{i.}-\bar{y}_{..})^2 \qquad = \text{treatment sum of squares,}$$

$$SSB = k\sum_{j=1}^{b}(\bar{y}_{.j}-\bar{y}_{..})^2 \qquad = \text{block sum of squares,}$$

$$SSE = \sum_{i=1}^{k}\sum_{j=1}^{b}(y_{ij}-\bar{y}_{i.}-\bar{y}_{.j}+\bar{y}_{..})^2 = \text{error sum of squares.}$$

Following the procedure outlined in Theorem 8.2, where we interpreted the sums of squares as functions of the independent random variables $Y_{11}, Y_{12}, \ldots, Y_{kb}$, we can show that the expected values of the treatment, block, and error sums of squares are given by

$$E(SSA) = (k-1)\sigma^2 + b\sum_{i=1}^{k}\alpha_i^2, \quad E(SSB) = (b-1)\sigma^2 + k\sum_{j=1}^{b}\beta_j^2,$$

$$E(SSE) = (b-1)(k-1)\sigma^2.$$

As in the case of the one-factor problem, we have the treatment mean square

$$s_1^2 = \frac{SSA}{k-1}.$$

If the treatment effects $\alpha_1 = \alpha_2 = \cdots = \alpha_k = 0$, s_1^2 is an unbiased estimate of σ^2. However, if the treatment effects are not all zero, we have the following:

Expected Treatment Mean Square

$$E\left(\frac{SSA}{k-1}\right) = \sigma^2 + \frac{b}{k-1}\sum_{i=1}^{k}\alpha_i^2$$

In this case, s_1^2 overestimates σ^2. A second estimate of σ^2, based on $b-1$ degrees of freedom, is

$$s_2^2 = \frac{SSB}{b-1}.$$

The estimate s_2^2 is an unbiased estimate of σ^2 if the block effects $\beta_1 = \beta_2 = \cdots = \beta_b = 0$. If the block effects are not all zero, then

$$E\left(\frac{SSB}{b-1}\right) = \sigma^2 + \frac{k}{b-1}\sum_{j=1}^{b}\beta_j^2,$$

and s_2^2 will overestimate σ^2. A third estimate of σ^2, based on $(k-1)(b-1)$ degrees of freedom and independent of s_1^2 and s_2^2, is

$$s^2 = \frac{SSE}{(k-1)(b-1)},$$

which is unbiased regardless of the truth or falsity of either null hypothesis.

To test the null hypothesis that the treatment effects are all equal to zero, we compute the ratio $f_1 = s_1^2/s^2$, which is a value of the random variable F_1 having an F-distribution with $k-1$ and $(k-1)(b-1)$ degrees of freedom when the null hypothesis is true. The null hypothesis is rejected at the α-level of significance when

$$f_1 > f_\alpha[k-1, (k-1)(b-1)].$$

In practice, we first compute SST, SSA, and SSB and then, using the sum-of-squares identity, obtain SSE by subtraction. The degrees of freedom associated with SSE are also usually obtained by subtraction; that is,

$$(k-1)(b-1) = kb - 1 - (k-1) - (b-1).$$

The computations in an analysis-of-variance problem for a randomized complete block design may be summarized as shown in Table 8.7.

Table 8.7: Analysis of Variance for the Randomized Complete Block Design

Source of Variation	Sum of Squares	Degrees of Freedom	Mean Square	Computed f
Treatments	SSA	$k-1$	$s_1^2 = \dfrac{SSA}{k-1}$	$f_1 = \dfrac{s_1^2}{s^2}$
Blocks	SSB	$b-1$	$s_2^2 = \dfrac{SSB}{b-1}$	
Error	SSE	$(k-1)(b-1)$	$s^2 = \dfrac{SSE}{(k-1)(b-1)}$	
Total	SST	$kb-1$		

Example 8.4: Four different machines, M_1, M_2, M_3, and M_4, are being considered for the assembling of a particular product. It was decided that six different operators would be used in a randomized block experiment to compare the machines. The machines were assigned in a random order to each operator. The operation of the machines requires physical dexterity, and it was anticipated that there would be a difference among the operators in the speed with which they operated the machines. The amounts of time (in seconds) required to assemble the product are shown in Table 8.8.

Table 8.8: Time, in Seconds, to Assemble Product

Machine	Operator 1	2	3	4	5	6	Total
1	42.5	39.3	39.6	39.9	42.9	43.6	247.8
2	39.8	40.1	40.5	42.3	42.5	43.1	248.3
3	40.2	40.5	41.3	43.4	44.9	45.1	255.4
4	41.3	42.2	43.5	44.2	45.9	42.3	259.4
Total	163.8	162.1	164.9	169.8	176.2	174.1	1010.9

Test the hypothesis H_0, at the 0.05 level of significance, that the machines perform at the same mean rate of speed.

Solution: The hypotheses are

$$H_0: \alpha_1 = \alpha_2 = \alpha_3 = \alpha_4 = 0 \quad \text{(machine effects are zero)},$$
$$H_1: \text{At least one of the } \alpha_i \text{ is not equal to zero.}$$

The sum-of-squares formulas shown on page 377 and the degrees of freedom are used to produce the analysis of variance in Table 8.9. The value $f = 3.34$ is significant at $P = 0.048$. If we use $\alpha = 0.05$ as at least an approximate yardstick, we conclude that the machines do not perform at the same mean rate of speed. ◾

Table 8.9: Analysis of Variance for the Data of Table 8.8

Source of Variation	Sum of Squares	Degrees of Freedom	Mean Square	Computed f
Machines	15.93	3	5.31	3.34
Operators	42.09	5	8.42	
Error	23.84	15	1.59	
Total	81.86	23		

Interaction between Blocks and Treatments

Another important assumption that is implicit in writing the model for a randomized complete block design is that the treatment and block effects are additive.

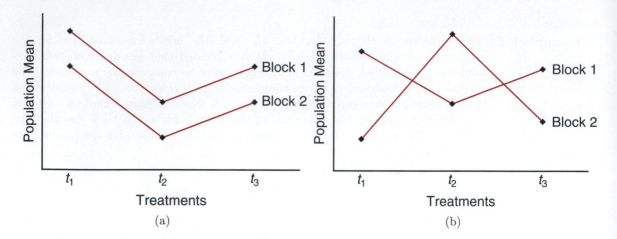

Figure 8.5: Population means for (a) additive results and (b) interacting effects.

This is equivalent to stating that

$$\mu_{ij} - \mu_{ij'} = \mu_{i'j} - \mu_{i'j'} \qquad \text{or} \qquad \mu_{ij} - \mu_{i'j} = \mu_{ij'} - \mu_{i'j'},$$

for every value of i, i', j, and j'. That is, the difference between the population means for blocks j and j' is the same for every treatment and the difference between the population means for treatments i and i' is the same for every block. The parallel lines of Figure 8.5(a) illustrate a set of mean responses for which the treatment and block effects are additive, whereas the intersecting lines of Figure 8.5(b) show a situation in which treatment and block effects are said to **interact**.

Exercises

8.22 Three varieties of potatoes are being compared for yield. The experiment is conducted by assigning each variety at random to one of 3 equal-size plots at each of 4 different locations. The following yields for varieties A, B, and C, in 100 kilograms per plot, were recorded:

Location 1	Location 2	Location 3	Location 4
B: 13	C: 21	C: 9	A: 11
A: 18	A: 20	B: 12	C: 10
C: 12	B: 23	A: 14	B: 17

Perform a randomized complete block analysis of variance to test the hypothesis that there is no difference in the yielding capabilities of the 3 varieties of potatoes. Use a 0.05 level of significance. Draw conclusions.

8.23 Four kinds of fertilizer f_1, f_2, f_3, and f_4 are used to study the yield of beans. The soil is divided into 3 blocks, each containing 4 homogeneous plots. The yields in kilograms per plot and the corresponding treatments are as follows:

Block 1	Block 2	Block 3
$f_1 = 42.7$	$f_3 = 50.9$	$f_4 = 51.1$
$f_3 = 48.5$	$f_1 = 50.0$	$f_2 = 46.3$
$f_4 = 32.8$	$f_2 = 38.0$	$f_1 = 51.9$
$f_2 = 39.3$	$f_4 = 40.2$	$f_3 = 53.5$

Conduct an analysis of variance at the 0.05 level of significance using the randomized complete block model.

8.24 The following data represent the final grades obtained by 5 students in mathematics, English, French, and biology:

	Subject			
Student	Math	English	French	Biology
1	68	57	73	61
2	83	94	91	86
3	72	81	63	59
4	55	73	77	66
5	92	68	75	87

Test the hypothesis that the courses are of equal difficulty. Use a *P*-value in your conclusions and discuss your findings.

8.25 The following data are the percents of foreign additives measured by 5 analysts for 3 similar brands of strawberry jam, *A*, *B*, and *C*:

Analyst 1	Analyst 2	Analyst 3	Analyst 4	Analyst 5
B: 2.7	C: 7.5	B: 2.8	A: 1.7	C: 8.1
C: 3.6	A: 1.6	A: 2.7	B: 1.9	A: 2.0
A: 3.8	B: 5.2	C: 6.4	C: 2.6	B: 4.8

Perform a randomized complete block analysis of variance to test the hypothesis, at the 0.05 level of significance, that the percent of foreign additives is the same for all 3 brands of jam. Which brand of jam appears to have fewer additives?

8.26 A nuclear power facility produces a vast amount of heat, which is usually discharged into aquatic systems. This heat raises the temperature of the aquatic system, resulting in a greater concentration of chlorophyll *a*, which in turn extends the growing season. To study this effect, water samples were collected monthly at 3 stations for a period of 12 months. Station *A* is located closest to a potential heated water discharge, station *C* is located farthest away from the discharge, and station *B* is located halfway between stations *A* and *C*. The concentrations of chlorophyll *a* were recorded. Perform an analysis of variance and test the hypothesis, at the 0.05 level of significance, that there is no difference in the mean concentrations of chlorophyll *a* at the 3 stations.

	Station		
Month	*A*	*B*	*C*
January	9.867	3.723	4.410
February	14.035	8.416	11.100
March	10.700	20.723	4.470
April	13.853	9.168	8.010
May	7.067	4.778	34.080
June	11.670	9.145	8.990
July	7.357	8.463	3.350
August	3.358	4.086	4.500
September	4.210	4.233	6.830
October	3.630	2.320	5.800
November	2.953	3.843	3.480
December	2.640	3.610	3.020

8.27 In a study on *The Periphyton of the South River, Virginia: Mercury Concentration, Productivity, and Autotropic Index Studies*, conducted by the Department of Environmental Sciences and Engineering at Virginia Tech, the total mercury concentration in periphyton total solids was measured at 6 different stations on 6 different days. Determine whether the mean mercury content is significantly different between the stations by using the following recorded data. Use a *P*-value and discuss your findings.

			Station			
Date	*CA*	*CB*	*E1*	*E2*	*E3*	*E4*
April 8	0.45	3.24	1.33	2.04	3.93	5.93
June 23	0.10	0.10	0.99	4.31	9.92	6.49
July 1	0.25	0.25	1.65	3.13	7.39	4.43
July 8	0.09	0.06	0.92	3.66	7.88	6.24
July 15	0.15	0.16	2.17	3.50	8.82	5.39
July 23	0.17	0.39	4.30	2.91	5.50	4.29

8.28 Organic arsenicals are used by forestry personnel as silvicides. The amount of arsenic that the body takes in when exposed to these silvicides is a major health problem. It is important that the amount of exposure be determined quickly so that a field worker with a high level of arsenic can be removed from the job. In an experiment reported in the paper "A Rapid Method for the Determination of Arsenic Concentrations in Urine at Field Locations," published in the *American Industrial Hygiene Association Journal* (Vol. 37, 1976), urine specimens from 4 forest service personnel were divided equally into 3 samples each so that each individual's urine could be analyzed for arsenic by a university laboratory, by a chemist using a portable system, and by a forest-service employee after a brief orientation. The arsenic levels, in parts per million, were recorded for this problem. Perform an analysis of variance and test the hypothesis, at the 0.05 level of significance, that there is no difference in the arsenic levels for the 3 methods of analysis.

		Analyst	
Individual	Employee	Chemist	Laboratory
1	0.05	0.05	0.04
2	0.05	0.05	0.04
3	0.04	0.04	0.03
4	0.15	0.17	0.10

8.29 In a study conducted by the Department of Health and Physical Education at Virginia Tech, 3 diets were assigned for a period of 3 days to each of 6 subjects in a randomized complete block design. The subjects, playing the role of blocks, were assigned the following 3 diets in a random order:

> Diet 1:   mixed fat and carbohydrates,
> Diet 2:   high fat,
> Diet 3:   high carbohydrates.

At the end of the 3-day period, each subject was put on a treadmill and the time to exhaustion, in seconds, was measured.

			Subject			
Diet	1	2	3	4	5	6
1	84	35	91	57	56	45
2	91	48	71	45	61	61
3	122	53	110	71	91	122

Perform an analysis of variance, separating out the diet, subject, and error sum of squares. Use a *P*-value to determine if there are significant differences among the diets, using the preceeding recorded data.

8.30 In the paper "Self-Control and Therapist Control in the Behavioral Treatment of Overweight Women," published in *Behavioral Research and Therapy* (Volume 10, 1972), two reduction treatments and a control treatment were studied for their effects on the weight change of obese women. The two reduction treatments were a self-induced weight reduction program and a therapist-controlled reduction program. Each of 10 subjects was assigned to one of the 3 treatment programs in a random order and measured for weight loss. The following weight changes were recorded:

Subject	Control	Self-induced	Therapist
1	1.00	−2.25	−10.50
2	3.75	−6.00	−13.50
3	0.00	−2.00	0.75
4	−0.25	−1.50	−4.50
5	−2.25	−3.25	−6.00
6	−1.00	−1.50	4.00
7	−1.00	−10.75	−12.25
8	3.75	−0.75	−2.75
9	1.50	0.00	−6.75
10	0.50	−3.75	−7.00

Perform an analysis of variance and test the hypothesis, at the 0.01 level of significance, that there is no difference in the mean weight losses for the 3 treatments. Which treatment was best?

8.31 Scientists in the Department of Plant Pathology at Virginia Tech devised an experiment in which 5 different treatments were applied to 6 different locations in an apple orchard to determine if there were significant differences in growth among the treatments. Treatments 1 through 4 represent different herbicides and treatment 5 represents a control. The growth period was from May to November in 1982, and the amounts of new growth, measured in centimeters, for samples selected from the 6 locations in the orchard were recorded as follows:

Treatment	1	2	3	4	5	6
1	455	72	61	215	695	501
2	622	82	444	170	437	134
3	695	56	50	443	701	373
4	607	650	493	257	490	262
5	388	263	185	103	518	622

Perform an analysis of variance, separating out the treatment, location, and error sum of squares. De-

termine if there are significant differences among the treatment means. Quote a *P*-value.

8.32 An experiment was conducted to compare three types of coating materials for copper wire. The purpose of the coating is to eliminate flaws in the wire. Ten different specimens of length 5 millimeters were randomly assigned to receive each coating, and the thirty specimens were subjected to an abrasive type process. The number of flaws was measured for each, and the results are as follows:

Material

1				2				3			
6	8	4	5	3	3	5	4	12	8	7	14
7	7	9	6	2	4	4	5	18	6	7	18
7	8			4	3			8	5		

Suppose it is assumed that the Poisson process applies and thus the model is $Y_{ij} = \mu_i + \epsilon_{ij}$, where μ_i is the mean of a Poisson distribution and $\sigma^2_{Y_{ij}} = \mu_i$.

(a) Do an appropriate transformation on the data and perform an analysis of variance.

(b) Determine whether or not there is sufficient evidence to choose one coating material over the other. Show whatever findings suggest a conclusion.

(c) Do a plot of the residuals and comment.

(d) Give the purpose of your data transformation.

(e) What additional assumption is made here that may not have been completely satisfied by your transformation?

(f) Comment on (e) after doing a normal probability plot of the residuals.

8.33 In the book *Design of Experiments for the Quality Improvement*, published by the Japanese Standards Association (1989), a study on the amount of dye needed to get the best color for a certain type of fabric was reported. The three amounts of dye, $(1/3)\%$ wof $((1/3)\%$ of the weight of a fabric), 1% wof, and 3% wof, were each administered at two different plants. The color density of the fabric was then observed four times for each level of dye at each plant.

Amount of Dye

	(1/3)%		1%		3%	
Plant 1	5.2	6.0	12.3	10.5	22.4	17.8
	5.9	5.9	12.4	10.9	22.5	18.4
Plant 2	6.5	5.5	14.5	11.8	29.0	23.2
	6.4	5.9	16.0	13.6	29.7	24.0

Perform an analysis of variance to test the hypothesis, at the 0.05 level of significance, that there is no difference in the color density of the fabric for the three levels of dye. Consider plants to be blocks.

8.6   Random Effects Models

Throughout this chapter, we deal with analysis-of-variance procedures in which the primary goal is to study the effect on some response of certain fixed or predetermined treatments. Experiments in which the treatments or treatment levels are preselected by the experimenter as opposed to being chosen randomly are called **fixed effects experiments**. For the fixed effects model, inferences are made only on those particular treatments used in the experiment.

It is often important that the experimenter be able to draw inferences about a population of treatments by means of an experiment in which the treatments used are chosen randomly from the population. For example, a biologist may be interested in whether or not there is significant variance in some physiological characteristic due to animal type. The animal types actually used in the experiment are then chosen randomly and represent the treatment effects. A chemist may be interested in studying the effect of analytical laboratories on the chemical analysis of a substance. She is not concerned with particular laboratories but rather with a large population of laboratories. She might then select a group of laboratories at random and allocate samples to each for analysis. The statistical inference would then involve (1) testing whether or not the laboratories contribute a nonzero variance to the analytical results and (2) estimating the variance between and within laboratories.

Model and Assumptions for Random Effects Model

The one-way **random effects model** is written like the fixed effects model but with the terms taking on different meanings. The response $y_{ij} = \mu + \alpha_i + \epsilon_{ij}$ is now a value of the random variable

$$Y_{ij} = \mu + A_i + \epsilon_{ij}, \text{ with } i = 1, 2, \ldots, k \text{ and } j = 1, 2, \ldots, n,$$

where the A_i are independently and normally distributed with mean 0 and variance σ_α^2 and are independent of the ϵ_{ij}. As for the fixed effects model, the ϵ_{ij} are also independently and normally distributed with mean 0 and variance σ^2. Note that for a random effects experiment, the constraint that $\sum_{i=1}^{k} \alpha_i = 0$ no longer applies.

Theorem 8.4: For the one-way random effects analysis-of-variance model,

$$E(SSA) = (k-1)\sigma^2 + n(k-1)\sigma_\alpha^2 \qquad \text{and} \qquad E(SSE) = k(n-1)\sigma^2.$$

Table 8.10 shows the expected mean squares for both a fixed effects and a random effects experiment. The computations for a random effects experiment are carried out in exactly the same way as for a fixed effects experiment. That is, the sum-of-squares, degrees-of-freedom, and mean-square columns in an analysis-of-variance table are the same for both models.

For the random effects model, the hypothesis that the treatment effects are all zero is written as follows:

Table 8.10: Expected Mean Squares for the One-Factor Experiment

Source of Variation	Degrees of Freedom	Mean Squares	Expected Mean Squares Fixed Effects	Random Effects
Treatments	$k-1$	s_1^2	$\sigma^2 + \dfrac{n}{k-1}\sum_i \alpha_i^2$	$\sigma^2 + n\sigma_\alpha^2$
Error	$k(n-1)$	s^2	σ^2	σ^2
Total	$nk-1$			

Hypothesis for a
Random Effects
Experiment

$$H_0:\ \sigma_\alpha^2 = 0,$$
$$H_1:\ \sigma_\alpha^2 \neq 0.$$

This hypothesis says that the different treatments contribute nothing to the variability of the response. It is obvious from Table 8.10 that s_1^2 and s^2 are both estimates of σ^2 when H_0 is true and that the ratio

$$f = \frac{s_1^2}{s^2}$$

is a value of the random variable F having the F-distribution with $k-1$ and $k(n-1)$ degrees of freedom. The null hypothesis is rejected at the α-level of significance when

$$f > f_\alpha[k-1, k(n-1)].$$

In many scientific and engineering studies, interest is not centered on the F-test. The scientist knows that the random effect is, indeed, significant. What is more important is estimation of the various variance components. This produces a *ranking* in terms of what factors produce the most variability and by how much. In the present context, it may be of interest to quantify how much larger the *single-factor variance component* is than that produced by chance (random variation).

Estimation of Variance Components

Table 8.10 can also be used to estimate the **variance components** σ^2 and σ_α^2. Since s_1^2 estimates $\sigma^2 + n\sigma_\alpha^2$ and s^2 estimates σ^2,

$$\hat{\sigma}^2 = s^2, \qquad \hat{\sigma}_\alpha^2 = \frac{s_1^2 - s^2}{n}.$$

Example 8.5: The data in Table 8.11 are coded observations on the yield of a chemical process, using five batches of raw material selected randomly. Show that the batch variance component is significantly greater than zero and obtain its estimate.

Solution: The total, batch, and error sums of squares are, respectively,

$$SST = 194.64, \ \ SSA = 72.60, \ \text{and} \ SSE = 194.64 - 72.60 = 122.04.$$

Table 8.11: Data for Example 8.5

Batch:	1	2	3	4	5	
	9.7	10.4	15.9	8.6	9.7	
	5.6	9.6	14.4	11.1	12.8	
	8.4	7.3	8.3	10.7	8.7	
	7.9	6.8	12.8	7.6	13.4	
	8.2	8.8	7.9	6.4	8.3	
	7.7	9.2	11.6	5.9	11.7	
	8.1	7.6	9.8	8.1	10.7	
Total	55.6	59.7	80.7	58.4	75.3	329.7

Table 8.12: Analysis of Variance for Example 8.5

Source of Variation	Sum of Squares	Degrees of Freedom	Mean Square	Computed f
Batches	72.60	4	18.15	4.46
Error	122.04	30	4.07	
Total	194.64	34		

These results, with the remaining computations, are shown in Table 8.12.

The f-ratio is significant at the $\alpha = 0.05$ level, indicating that the hypothesis of a zero batch component is rejected. An estimate of the batch variance component is

$$\hat{\sigma}_\alpha^2 = \frac{18.15 - 4.07}{7} = 2.01.$$

Note that while the **batch variance component** is significantly different from zero, when gauged against the estimate of σ^2, namely $\hat{\sigma}^2 = MSE = 4.07$, it appears as if the batch variance component is not appreciably large.

If the result using the formula for σ_α^2 appears negative (i.e., when s_1^2 is smaller than s^2), $\hat{\sigma}_\alpha^2$ is then set to zero. This is a biased estimator. In order to have a better estimator of σ_α^2, a method called **restricted** (or **residual**) **maximum likelihood (REML)** is commonly used (see Harville, 1977, in the Bibliography). Such an estimator can be found in many statistical software packages. The details for this estimation procedure are beyond the scope of this text.

8.7 Case Study for One-Way Experiment

Case Study 8.1: **Chemical Analysis**: Personnel in the Chemistry Department of Virginia Tech were called upon to analyze a data set that was produced to compare 4 different methods of analysis of aluminum in a certain solid igniter mixture. To get a broad range of analytical laboratories involved, 5 laboratories were used in the experiment as blocks. These laboratories were selected because they are generally adept in doing these types of analyses. Twenty samples of igniter material containing 2.70%

aluminum were assigned randomly, 4 to each laboratory, and directions were given on how to carry out the chemical analysis using all 4 methods. The data retrieved are shown in Table 8.13.

Table 8.13: Data Set for Case Study 8.1

| Method | Laboratory | | | | | Mean |
	1	2	3	4	5	
A	2.67	2.69	2.62	2.66	2.70	2.668
B	2.71	2.74	2.69	2.70	2.77	2.722
C	2.76	2.76	2.70	2.76	2.81	2.758
D	2.65	2.69	2.60	2.64	2.73	2.662

The laboratories are not considered as random effects since they were not selected randomly from a larger population of laboratories. The data were analyzed as a randomized complete block design. Plots of the data were sought to determine if an additive model of the type

$$y_{ij} = \mu + m_i + l_j + \epsilon_{ij}$$

is appropriate: in other words, a model with additive effects. The randomized block is not appropriate when interaction between laboratories and methods exists. Consider the plot shown in Figure 8.6. Although this plot is a bit difficult to interpret because each point is a single observation, there appears to be no appreciable interaction between methods and laboratories. A more complete discussion of the concept of interaction will be given in Chapter 9.

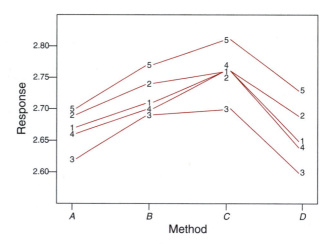

Figure 8.6: Interaction plot for data of Case Study 8.1.

Residual Plots

Residual plots were used as diagnostic indicators regarding the homogeneous variance assumption. Figure 8.7 shows a plot of residuals against analytical methods. The variability depicted in the residuals seems to be remarkably homogeneous.

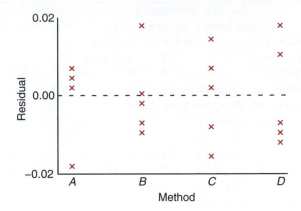

Figure 8.7: Plot of residuals against method for the data of Case Study 8.1.

The residual plots show no difficulty with either the assumption of normal errors or the assumption of homogeneous variance. *SAS* PROC GLM was used to conduct the analysis of variance. Figure 8.8 on page 388 shows the annotated computer printout.

The computed *f*- and *P*-values do indicate a significant difference between analytical methods. This analysis can be followed by a multiple comparison analysis to determine where the differences are among the methods.

Exercises

8.34 An experiment is conducted in which 4 treatments are to be compared in 5 blocks. The data are given below.

Treatment	Block 1	2	3	4	5
1	12.8	10.6	11.7	10.7	11.0
2	11.7	14.2	11.8	9.9	13.8
3	11.5	14.7	13.6	10.7	15.9
4	12.6	16.5	15.4	9.6	17.1

(a) Assuming a random effects model, test the hypothesis, at the 0.05 level of significance, that there is no difference between treatment means.

(b) Compute estimates of the treatment and block variance components.

8.35 Testing patient blood samples for HIV antibodies, a spectrophotometer determines the optical density of each sample. Optical density is measured as the absorbance of light at a particular wavelength. The blood sample is positive if it exceeds a certain cutoff value that is determined by the control samples for that run. Researchers are interested in comparing the laboratory variability for the positive control values. The data represent positive control values for 10 different runs at 4 randomly selected laboratories.

Run	Laboratory 1	2	3	4
1	0.888	1.065	1.325	1.232
2	0.983	1.226	1.069	1.127
3	1.047	1.332	1.219	1.051
4	1.087	0.958	0.958	0.897
5	1.125	0.816	0.819	1.222
6	0.997	1.015	1.140	1.125
7	1.025	1.071	1.222	0.990
8	0.969	0.905	0.995	0.875
9	0.898	1.140	0.928	0.930
10	1.018	1.051	1.322	0.775

(a) Write an appropriate model for this experiment.

(b) Estimate the laboratory variance component and the variance within laboratories.

8.36 Five pours of metals have had 5 core samples each analyzed for the amount of a trace element. The data for the 5 randomly selected pours are given here.

(a) The intent is that the pours be identical. Thus, test that the "pour" variance component is zero. Draw conclusions.

```
                         The GLM Procedure
                     Class Level Information
               Class          Levels    Values
               Method            4      A B C D
               Lab               5      1 2 3 4 5
               Number of Observations Read        20
               Number of Observations Used        20
Dependent Variable: Response
                         Sum of
Source              DF     Squares   Mean Square   F Value   Pr > F
Model                7   0.05340500   0.00762929    42.19    <.0001
Error               12   0.00217000   0.00018083
Corrected Total     19   0.05557500


R-Square    Coeff Var      Root MSE     Response Mean
0.960954    0.497592       0.013447        2.702500


Source              DF   Type III SS   Mean Square   F Value   Pr > F
Method               3    0.03145500    0.01048500    57.98    <.0001
Lab                  4    0.02195000    0.00548750    30.35    <.0001


Observation       Observed         Predicted           Residual
          1     2.67000000        2.66300000         0.00700000
          2     2.71000000        2.71700000        -0.00700000
          3     2.76000000        2.75300000         0.00700000
          4     2.65000000        2.65700000        -0.00700000
          5     2.69000000        2.68550000         0.00450000
          6     2.74000000        2.73950000         0.00050000
          7     2.76000000        2.77550000        -0.01550000
          8     2.69000000        2.67950000         0.01050000
          9     2.62000000        2.61800000         0.00200000
         10     2.69000000        2.67200000         0.01800000
         11     2.70000000        2.70800000        -0.00800000
         12     2.60000000        2.61200000        -0.01200000
         13     2.66000000        2.65550000         0.00450000
         14     2.70000000        2.70950000        -0.00950000
         15     2.76000000        2.74550000         0.01450000
         16     2.64000000        2.64950000        -0.00950000
         17     2.70000000        2.71800000        -0.01800000
         18     2.77000000        2.77200000        -0.00200000
         19     2.81000000        2.80800000         0.00200000
         20     2.73000000        2.71200000         0.01800000
```

Figure 8.8: *SAS* printout for data of Case Study 8.1.

(b) Show a complete ANOVA along with an estimate of the within-pour variance.

Pour

Core	1	2	3	4	5
1	0.98	0.85	1.12	1.21	1.00
2	1.02	0.92	1.68	1.19	1.21
3	1.57	1.16	0.99	1.32	0.93
4	1.25	1.43	1.26	1.08	0.86
5	1.16	0.99	1.05	0.94	1.41

8.37 A data set shows the effect of 4 operators, chosen randomly, on the output of a particular machine.

Operator

1	2	3	4
175.4	168.5	170.1	175.2
171.7	162.7	173.4	175.7
173.0	165.0	175.7	180.1
170.5	164.1	170.7	183.7

(a) Perform a random effects analysis of variance at the 0.05 level of significance.

(b) Compute an estimate of the operator variance component and the experimental error variance component.

8.38 A textile company weaves a certain fabric on a large number of looms. The managers would like the looms to be homogeneous so that their fabric is of uniform strength. It is suspected that there may be significant variation in strength among looms. Consider the following data for 4 randomly selected looms. Each observation is a determination of strength of the fabric in pounds per square inch.

Loom

1	2	3	4
99	97	94	93
97	96	95	94
97	92	90	90
96	98	92	92

(a) Write a model for the experiment.

(b) Does the loom variance component differ significantly from zero?

(c) Comment on the managers' suspicion.

Review Exercises

8.39 An analysis was conducted by the Statistics Consulting Center at Virginia Tech in conjunction with the Department of Forestry. A certain treatment was applied to a set of tree stumps in which the chemical Garlon was used with the purpose of regenerating the roots of the stumps. A spray was used with four levels of Garlon concentration. After a period of time, the height of the shoots was observed. Perform a one-factor analysis of variance on the following data. Test to see if the concentration of Garlon has a significant impact on the height of the shoots. Use $\alpha = 0.05$.

Garlon Level

1		2		3		4	
2.87	2.31	3.27	2.66	2.39	1.91	3.05	0.91
3.91	2.04	3.15	2.00	2.89	1.89	2.43	0.01

8.40 Consider the aggregate data of Example 8.1. Perform Bartlett's test, at level $\alpha = 0.1$, to determine if there is heterogeneity of variance among the aggregates.

8.41 A company that stamps gaskets out of sheets of rubber, plastic, and cork wants to compare the mean number of gaskets produced per hour for the three types of material. Two randomly selected stamping machines are chosen as blocks. The data represent the number of gaskets (in thousands) produced per hour. The data are given below. In addition, the printout

analysis is given in Figure 8.9 on page 390.

Material

Machine	Cork			Rubber			Plastic		
A	4.31	4.27	4.40	3.36	3.42	3.48	4.01	3.94	3.89
B	3.94	3.81	3.99	3.91	3.80	3.85	3.48	3.53	3.42

(a) Why would the stamping machines be chosen as blocks?

(b) Plot the six means for machine and material combinations.

(c) Is there a single material that is best?

(d) Is there an interaction between treatments and blocks? If so, is the interaction causing any serious difficulty in arriving at a proper conclusion? Explain.

8.42 Four laboratories are being used to perform chemical analysis. Samples of the same material are sent to the laboratories for analysis as part of a study to determine whether or not they give, on the average, the same results. Use the data set provided to do the following.

Laboratory

A	B	C	D
58.7	62.7	55.9	60.7
61.4	64.5	56.1	60.3
60.9	63.1	57.3	60.9
59.1	59.2	55.2	61.4
58.2	60.3	58.1	62.3

(a) Run Bartlett's test to show that the within-laboratory variances are not significantly different at the $\alpha = 0.05$ level of significance.

(b) Perform the analysis of variance and give conclusions concerning the laboratories.

(c) Do a normal probability plot of residuals.

```
                        The GLM Procedure
Dependent Variable: gasket
                             Sum of
Source               DF      Squares    Mean Square  F Value   Pr > F
Model                 5   1.68122778     0.33624556    76.52   <.0001
Error                12   0.05273333     0.00439444
Corrected Total      17   1.73396111
R-Square      Coeff Var       Root MSE    gasket Mean
0.969588      1.734095        0.066291       3.822778

Source               DF  Type III SS   Mean Square  F Value   Pr > F
material              2   0.81194444    0.40597222    92.38   <.0001
machine               1   0.10125000    0.10125000    23.04   0.0004
material*machine      2   0.76803333    0.38401667    87.39   <.0001
 Level of     Level of        -----------gasket-----------
 material     machine     N          Mean            Std Dev
 cork         A           3     4.32666667        0.06658328
 cork         B           3     3.91333333        0.09291573
 plastic      A           3     3.94666667        0.06027714
 plastic      B           3     3.47666667        0.05507571
 rubber       A           3     3.42000000        0.06000000
 rubber       B           3     3.85333333        0.05507571

 Level of               ------------gasket-----------
 material     N               Mean            Std Dev
 cork         6         4.12000000        0.23765521
 plastic      6         3.71166667        0.26255793
 rubber       6         3.63666667        0.24287171

 Level of               ------------gasket-----------
 machine      N               Mean            Std Dev
 A            9         3.89777778        0.39798800
 B            9         3.74777778        0.21376259
```

Figure 8.9: *SAS* printout for Review Exercise 8.41.

8.43 A study is conducted to compare gas mileage for 3 competing brands of gasoline. Four different automobile models of varying size are randomly selected. The data, in miles per gallon, follow. The order of testing is random for each model.

(a) Discuss the need for the use of more than a single model of car.

(b) Consider the ANOVA from the *SAS* printout in Figure 8.10 on page 391. Does brand of gasoline matter?

(c) Which brand of gasoline would you select? Consult the result of Tukey's test.

Model	Gasoline Brand		
	A	*B*	*C*
A	32.4	35.6	38.7
B	28.8	28.6	29.9
C	36.5	37.6	39.1
D	34.4	36.2	37.9

```
                         The GLM Procedure
Dependent Variable: MPG
                          Sum of
   Source          DF     Squares    Mean Square   F Value   Pr > F
   Model            5   153.2508333   30.6501667    24.66    0.0006
   Error            6     7.4583333    1.2430556
   Corrected Total 11   160.7091667


   R-Square     Coeff Var       Root MSE      MPG Mean
   0.953591     3.218448        1.114924      34.64167


   Source          DF   Type III SS   Mean Square  F Value   Pr > F
   Model            3   130.3491667   43.4497222    34.95    0.0003
   Brand            2    22.9016667   11.4508333     9.21    0.0148


             Tukey's Studentized Range (HSD) Test for MPG

NOTE: This test controls the Type I experimentwise error rate, but
it generally has a higher Type II error rate than REGWQ.
              Alpha                                  0.05
              Error Degrees of Freedom                  6
              Error Mean Square                   1.243056
              Critical Value of Studentized Range 4.33920
              Minimum Significant Difference       2.4189


       Means with the same letter are not significantly different.
           Tukey Grouping          Mean     N    Brand
                          A       36.4000    4    C
                          A
                      B   A       34.5000    4    B
                      B
                      B           33.0250    4    A
```

Figure 8.10: *SAS* printout for Review Exercise 8.43.

8.44 In a study that was analyzed for personnel in the Department of Biochemistry at Virginia Tech, three diets were given to groups of rats in order to study the effect of each on dietary residual zinc in the bloodstream. Five pregnant rats were randomly assigned to each diet group, and each was given the diet on day 22 of pregnancy. The amount of zinc in parts per million was measured. Use the data to determine if there is a significant difference in residual dietary zinc among the three diets. Use $\alpha = 0.05$. Perform a one-way analysis of variance.

Diet 1:	1	0.50	0.42	0.65	0.47	0.44
Diet 2:	2	0.42	0.40	0.73	0.47	0.69
Diet 3:	3	1.06	0.82	0.72	0.72	0.82

8.45 An experiment was conducted to compare three types of paint for evidence of differences in their wearing qualities. They were exposed to abrasive action and the time in hours until abrasion was noticed was observed. Six specimens were used for each type of paint. The data are as follows.

Paint Type		
1	2	3
158 97 282	515 264 544	317 662 213
315 220 115	525 330 525	536 175 614

(a) Do an analysis of variance to determine if the evidence suggests that wearing quality differs for the three paints. Use a *P*-value in your conclusion.

(b) If significant differences are found, characterize what they are. Is there one paint that stands out? Discuss your findings.

(c) Do whatever graphical analysis you need to determine if assumptions used in (a) are valid. Discuss your findings.

(d) Suppose it is determined that the data for each treatment follow an exponential distribution. Does this suggest an alternative analysis? If so, do the alternative analysis and give findings.

8.46 Four different locations in the northeast were used for collecting ozone measurements in parts per million. Amounts of ozone were collected for 5 samples at each location.

(a) Is there sufficient information here to suggest that there are differences in the mean ozone levels across locations? Be guided by a P-value.

(b) If significant differences are found in (a), characterize the nature of the differences. Use whatever methods you have learned.

Location			
1	2	3	4
0.09	0.15	0.10	0.10
0.10	0.12	0.13	0.07
0.08	0.17	0.08	0.05
0.08	0.18	0.08	0.08
0.11	0.14	0.09	0.09

8.47 Group Project: It is of interest to determine which type of sports ball can be thrown the longest distance. The competition involves a tennis ball, a baseball, and a softball. Divide the class into teams of five individuals. Each team should design and conduct a separate experiment. Each team should also analyze the data from its own experiment. For a given team, each of the five individuals will throw each ball (after sufficient arm warmup). The experimental response will be the distance (in feet) that the ball is thrown. The data for each team will involve 15 observations. Important points:

(a) This is not a competition among teams. The competition is among the three types of sports balls. One would expect that the conclusion drawn by each team would be similar.

(b) Each team should be gender mixed.

(c) The experimental design for each team should be a randomized complete block design. The five individuals throwing are the blocks.

(d) Be sure to incorporate the appropriate randomization in conducting the experiment.

(e) The results should contain a description of the experiment with an ANOVA table complete with a P-value and appropriate conclusions. Use graphical techniques where appropriate. Use multiple comparisons where appropriate. Draw practical conclusions concerning differences between the ball types. Be thorough.

8.8 Potential Misconceptions and Hazards; Relationship to Material in Other Chapters

As in other procedures covered in previous chapters, the analysis of variance is reasonably robust to the normality assumption but less robust to the homogeneous variance assumption. Also we note here that Bartlett's test for equal variance is extremely nonrobust to normality.

This is a pivotal chapter in that it is essentially an "entry level" point for important topics such as design of experiments and analysis of variance. Chapter 9 will concern itself with the same topics, but the expansion will be to more than one factor, with the total analysis further complicated by the interpretation of interaction among factors. There are times when the role of interaction in a scientific experiment is more important than the role of the main factors (main effects). The presence of interaction results in even more emphasis placed on graphical displays.

Chapter 9

Factorial Experiments
(Two or More Factors)

9.1   Introduction

Consider a situation where it is of interest to study the effects of **two factors**, A and B, on some response. For example, in a chemical experiment, we would like to vary simultaneously the reaction pressure and reaction time and study the effect of each on the yield. In a biological experiment, it is of interest to study the effects of drying time and temperature on the amount of solids (percent by weight) left in samples of yeast. As in Chapter 8, the term **factor** is used in a general sense to denote any feature of the experiment such as temperature, time, or pressure that may be varied from trial to trial. We define the **levels** of a factor to be the actual values used in the experiment.

For each of these cases, it is important to determine not only if each of the two factors has an influence on the response, but also if there is a significant interaction between the two factors. As far as terminology is concerned, the experiment described here is a two-factor experiment and the experimental design may be either a completely randomized design, in which the various treatment combinations are assigned randomly to all the experimental units, or a randomized complete block design, in which factor combinations are assigned randomly within blocks. In the case of the yeast example, the various treatment combinations of temperature and drying time would be assigned randomly to the samples of yeast if we were using a completely randomized design.

Many of the concepts studied in Chapter 8 are extended in this chapter to two and three factors. The main thrust of this material is the use of the completely randomized design with a *factorial experiment*. A factorial experiment in two factors involves experimental trials (or a single trial) with all factor combinations. For example, in the temperature-drying-time example with, say, 3 levels of each and $n = 2$ runs at each of the 9 combinations, we have a *two-factor factorial experiment in a completely randomized design*. Neither factor is a blocking factor; we are interested in how each influences percent solids in the samples and whether or not they interact. The biologist would have available 18 physical samples of material which are experimental units. These would then be assigned randomly to the 18 combinations (9 treatment combinations, each duplicated).

Before we launch into analytical details, sums of squares, and so on, it may be of interest for the reader to observe the obvious connection between what we have described and the situation with the one-factor problem. Consider the yeast experiment. Explanation of degrees of freedom aids the reader or the analyst in visualizing the extension. We should initially view the 9 treatment combinations as if they represented one factor with 9 levels (8 degrees of freedom). Thus, an initial look at degrees of freedom gives

Treatment combinations	8
Error	9
Total	17

Main Effects and Interaction

The experiment could be analyzed as described in the above table. However, the *F*-test for combinations would probably not give the analyst the information he or she desires, namely, that which considers the role of temperature and drying time. Three drying times have 2 associated degrees of freedom; three temperatures have 2 degrees of freedom. The main factors, temperature and drying time, are called **main effects**. The main effects represent 4 of the 8 degrees of freedom for *factor combinations*. The additional 4 degrees of freedom are associated with *interaction* between the two factors. As a result, the analysis involves

Combinations	8
Temperature	2
Drying time	2
Interaction	4
Error	9
Total	17

Recall from Chapter 8 that factors in an analysis of variance may be viewed as fixed or random, depending on the type of inference desired and how the levels were chosen. Here we must consider fixed effects, random effects, and even cases where effects are mixed. Most attention will be directed toward expected mean squares when we advance to these topics. In the following section, we focus on the concept of interaction.

9.2 Interaction in the Two-Factor Experiment

In the randomized block model discussed previously, it was assumed that one observation on each treatment is taken in each block. If the model assumption is correct, that is, if blocks and treatments are the only real effects and interaction does not exist, the expected value of the mean square error is the experimental error variance σ^2. Suppose, however, that there is interaction occurring between treatments and blocks as indicated by the model

$$y_{ij} = \mu + \alpha_i + \beta_j + (\alpha\beta)_{ij} + \epsilon_{ij}$$

of Section 8.5. The expected value of the mean square error is then given as

$$E\left[\frac{SSE}{(b-1)(k-1)}\right] = \sigma^2 + \frac{1}{(b-1)(k-1)}\sum_{i=1}^{k}\sum_{j=1}^{b}(\alpha\beta)_{ij}^2.$$

The treatment and block effects do not appear in the expected mean square error, but the interaction effects do. Thus, if there is interaction in the model, the mean square error reflects variation due to experimental error plus an interaction contribution, and for this experimental plan, there is no way of separating them.

Interaction and the Interpretation of Main Effects

From an experimenter's point of view it should seem necessary to arrive at a significance test on the existence of interaction by separating true error variation from that due to interaction. The main effects, A and B, take on a different meaning in the presence of interaction. In the previous biological example, the effect that drying time has on the amount of solids left in the yeast might very well depend on the temperature to which the samples are exposed. In general, there could be experimental situations in which factor A has a positive effect on the response at one level of factor B, while at a different level of factor B the effect of A is negative. We use the term **positive effect** here to indicate that the yield or response increases as the levels of a given factor increase according to some defined order. In the same sense, a **negative effect** corresponds to a decrease in response for increasing levels of the factor.

Consider, for example, the following data on temperature (factor A at levels t_1, t_2, and t_3 in increasing order) and drying time d_1, d_2, and d_3 (also in increasing order). The response is percent solids. These data are completely hypothetical and given to illustrate a point.

	d_1	d_2	d_3	Total
t_1	4.4	8.8	5.2	18.4
t_2	7.5	8.5	2.4	18.4
t_3	9.7	7.9	0.8	18.4
Total	21.6	25.2	8.4	55.2

with a header of *B* spanning the d_1, d_2, d_3 columns and *A* labeling the row header.

Clearly the effect of temperature on percent solids is positive at the low drying time d_1 but negative for high drying time d_3. This **clear interaction** between temperature and drying time is obviously of interest to the biologist, but, based on the totals of the responses for temperatures t_1, t_2, and t_3, the temperature sum of squares, SSA, will yield a value of zero. We say then that the presence of interaction is **masking** the effect of temperature. Thus, if we consider the average effect of temperature, averaged over drying time, **there is no effect**. This then defines the main effect. But, of course, this is likely not what is pertinent to the biologist.

Before drawing any final conclusions resulting from tests of significance on the main effects and interaction effects, the **experimenter should first observe whether or not the test for interaction is significant**. If interaction is not significant, then the results of the tests on the main effects are meaningful. However, if interaction should be significant, then only those tests on the main effects that turn out to be significant are meaningful. Nonsignificant main effects in the presence of interaction might well be a result of masking and dictate the need to observe the influence of each factor at fixed levels of the other.

A Graphical Look at Interaction

The presence of interaction as well as its scientific impact can be interpreted nicely through the use of **interaction plots**. The plots clearly give a pictorial view of the tendency in the data to show the effect of changing one factor as one moves from one level to another of a second factor. Figure 9.1 illustrates the strong temperature by drying time interaction. The interaction is revealed in nonparallel lines.

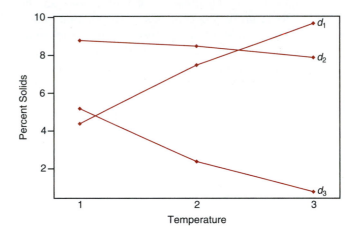

Figure 9.1: Interaction plot for temperature–drying time data.

The relatively strong *temperature effect* on percent solids at the lower drying time is reflected in the steep slope at d_1. At the middle drying time d_2 the temperature has very little effect, while at the high drying time d_3 the negative slope illustrates a negative effect of temperature. Interaction plots such as this set give the scientist a quick and meaningful interpretation of the interaction that is present. It should be apparent that **parallelism** in the plots signals an **absence of interaction**.

Need for Multiple Observations

Interaction and experimental error are separated in the two-factor experiment only if multiple observations are taken at the various treatment combinations. For maximum efficiency, there should be the same number n of observations at each combination. These should be true replications, not just repeated measurements. For example, in the yeast illustration, if we take $n = 2$ observations at each combination of temperature and drying time, there should be two separate samples and not merely repeated measurements on the same sample. This allows variability due to experimental units to appear in "error," so the variation is not merely measurement error.

9.3 Two-Factor Analysis of Variance

To present general formulas for the analysis of variance of a two-factor experiment using repeated observations in a completely randomized design, we shall consider the case of n replications of the treatment combinations determined by a levels of factor A and b levels of factor B. The observations may be classified by means of a rectangular array where the rows represent the levels of factor A and the columns represent the levels of factor B. Each treatment combination defines a cell in our array. Thus, we have ab cells, each cell containing n observations. Denoting the kth observation taken at the ith level of factor A and the jth level of factor B by y_{ijk}, Table 9.1 shows the abn observations.

Table 9.1: Two-Factor Experiment with n Replications

A	B 1	2	$\cdots$	b	Total	Mean
1	y_{111}	y_{121}	$\cdots$	y_{1b1}	$Y_{1..}$	$\bar{y}_{1..}$
	y_{112}	y_{122}	$\cdots$	y_{1b2}		
	$\vdots$	$\vdots$		$\vdots$		
	y_{11n}	y_{12n}	$\cdots$	y_{1bn}		
2	y_{211}	y_{221}	$\cdots$	y_{2b1}	$Y_{2..}$	$\bar{y}_{2..}$
	y_{212}	y_{222}	$\cdots$	y_{2b2}		
	$\vdots$	$\vdots$		$\vdots$		
	y_{21n}	y_{22n}	$\cdots$	y_{2bn}		
$\vdots$	$\vdots$	$\vdots$		$\vdots$	$\vdots$	$\vdots$
a	y_{a11}	y_{a21}	$\cdots$	y_{ab1}	$Y_{a..}$	$\bar{y}_{a..}$
	y_{a12}	y_{a22}	$\cdots$	y_{ab2}		
	$\vdots$	$\vdots$		$\vdots$		
	y_{a1n}	y_{a2n}	$\cdots$	y_{abn}		
Total	$Y_{.1.}$	$Y_{.2.}$	$\cdots$	$Y_{.b.}$	$Y_{...}$	
Mean	$\bar{y}_{.1.}$	$\bar{y}_{.2.}$	$\cdots$	$\bar{y}_{.b.}$		$\bar{y}_{...}$

The observations in the (ij)th cell constitute a random sample of size n from a population that is assumed to be normally distributed with mean μ_{ij} and variance σ^2. All ab populations are assumed to have the same variance σ^2. Let us define the following useful symbols, some of which are used in Table 9.1:

$$Y_{ij.} = \text{sum of the observations in the } (ij)\text{th cell,}$$
$$Y_{i..} = \text{sum of the observations for the } i\text{th level of factor } A,$$
$$Y_{.j.} = \text{sum of the observations for the } j\text{th level of factor } B,$$
$$Y_{...} = \text{sum of all } abn \text{ observations,}$$
$$\bar{y}_{ij.} = \text{mean of the observations in the } (ij)\text{th cell,}$$
$$\bar{y}_{i..} = \text{mean of the observations for the } i\text{th level of factor } A,$$
$$\bar{y}_{.j.} = \text{mean of the observations for the } j\text{th level of factor } B,$$
$$\bar{y}_{...} = \text{mean of all } abn \text{ observations.}$$

Unlike in the one-factor situation covered at length in Chapter 8, here we are assuming that the **populations**, where n independent identically distributed observations are taken, are **combinations** of factors. Also we will assume throughout that an equal number (n) of observations are taken at each factor combination. In cases in which the sample sizes per combination are unequal, the computations are more complicated but the concepts are transferable.

Model and Hypotheses for the Two-Factor Problem

Each observation in Table 9.1 may be written in the form

$$y_{ijk} = \mu_{ij} + \epsilon_{ijk},$$

where ϵ_{ijk} measures the deviations of the observed y_{ijk} values in the (ij)th cell from the population mean μ_{ij}. If we let $(\alpha\beta)_{ij}$ denote the interaction effect of the ith level of factor A and the jth level of factor B, α_i the effect of the ith level of factor A, β_j the effect of the jth level of factor B, and μ the overall mean, we can write

$$\mu_{ij} = \mu + \alpha_i + \beta_j + (\alpha\beta)_{ij},$$

and then

$$y_{ijk} = \mu + \alpha_i + \beta_j + (\alpha\beta)_{ij} + \epsilon_{ijk},$$

on which we impose the restrictions

$$\sum_{i=1}^{a} \alpha_i = 0, \qquad \sum_{j=1}^{b} \beta_j = 0, \qquad \sum_{i=1}^{a} (\alpha\beta)_{ij} = 0, \qquad \sum_{j=1}^{b} (\alpha\beta)_{ij} = 0.$$

The three hypotheses to be tested are as follows:

1. H_0': $\alpha_1 = \alpha_2 = \cdots = \alpha_a = 0,$
   H_1': At least one of the α_i is not equal to zero.
2. H_0'': $\beta_1 = \beta_2 = \cdots = \beta_b = 0,$
   H_1'': At least one of the β_j is not equal to zero.

3. $H_0''': (\alpha\beta)_{11} = (\alpha\beta)_{12} = \cdots = (\alpha\beta)_{ab} = 0$,

$H_1''':$ At least one of the $(\alpha\beta)_{ij}$ is not equal to zero.

We warned the reader about the problem of masking of main effects when interaction is a heavy contributor in the model. It is recommended that the interaction test result be considered first. The interpretation of the main effect test follows, and the nature of the scientific conclusion depends on whether interaction is found. If interaction is ruled out, then hypotheses 1 and 2 above can be tested and the interpretation is quite simple. However, if interaction is found to be present, the interpretation can be more complicated, as we have seen from the discussion of the drying time and temperature in the previous section. In what follows, the structure of the tests of hypotheses 1, 2, and 3 will be discussed. Interpretation of results will be incorporated in the discussion of the analysis in Example 9.1.

The tests of the hypotheses above will be based on a comparison of independent estimates of σ^2 provided by splitting the total sum of squares of our data into four components by means of the following identity.

Partitioning of Variability in the Two-Factor Case

Theorem 9.1: **Sum-of-Squares Identity**

$$\sum_{i=1}^{a}\sum_{j=1}^{b}\sum_{k=1}^{n}(y_{ijk} - \bar{y}_{...})^2 = bn\sum_{i=1}^{a}(\bar{y}_{i..} - \bar{y}_{...})^2 + an\sum_{j=1}^{b}(\bar{y}_{.j.} - \bar{y}_{...})^2$$

$$+ n\sum_{i=1}^{a}\sum_{j=1}^{b}(\bar{y}_{ij.} - \bar{y}_{i..} - \bar{y}_{.j.} + \bar{y}_{...})^2 + \sum_{i=1}^{a}\sum_{j=1}^{b}\sum_{k=1}^{n}(y_{ijk} - \bar{y}_{ij.})^2$$

Symbolically, we write the sum-of-squares identity as

$$SST = SSA + SSB + SS(AB) + SSE,$$

where SSA and SSB are called the sums of squares for the main effects A and B, respectively, $SS(AB)$ is called the interaction sum of squares for A and B, and SSE is the error sum of squares. The degrees of freedom are partitioned according to the identity

$$abn - 1 = (a - 1) + (b - 1) + (a - 1)(b - 1) + ab(n - 1).$$

Formation of Mean Squares

If we divide each of the sums of squares on the right side of the sum-of-squares identity by its corresponding number of degrees of freedom, we obtain the four statistics

$$S_1^2 = \frac{SSA}{a-1}, \qquad S_2^2 = \frac{SSB}{b-1}, \qquad S_3^2 = \frac{SS(AB)}{(a-1)(b-1)}, \qquad S^2 = \frac{SSE}{ab(n-1)}.$$

All of these variance estimates are independent estimates of σ^2 under the condition that there are no effects α_i, β_j, and, of course, $(\alpha\beta)_{ij}$. If we interpret the sums of squares as functions of the independent random variables $y_{111}, y_{112}, \ldots, y_{abn}$, it is not difficult to verify that

$$E(S_1^2) = E\left[\frac{SSA}{a-1}\right] = \sigma^2 + \frac{nb}{a-1}\sum_{i=1}^{a}\alpha_i^2,$$

$$E(S_2^2) = E\left[\frac{SSB}{b-1}\right] = \sigma^2 + \frac{na}{b-1}\sum_{j=1}^{b}\beta_j^2,$$

$$E(S_3^2) = E\left[\frac{SS(AB)}{(a-1)(b-1)}\right] = \sigma^2 + \frac{n}{(a-1)(b-1)}\sum_{i=1}^{a}\sum_{j=1}^{b}(\alpha\beta)_{ij}^2,$$

$$E(S^2) = E\left[\frac{SSE}{ab(n-1)}\right] = \sigma^2,$$

from which we immediately observe that all four estimates of σ^2 are unbiased when H_0', H_0'', and H_0''' are true.

To test the hypothesis H_0', that the effects of factors A are all equal to zero, we compute the following ratio:

F-Test for Factor A

$$f_1 = \frac{s_1^2}{s^2},$$

---

which is a value of the random variable F_1 having the F-distribution with $a-1$ and $ab(n-1)$ degrees of freedom when H_0' is true. The null hypothesis is rejected at the α-level of significance when $f_1 > f_\alpha[a-1, ab(n-1)]$.

Similarly, to test the hypothesis H_0'' that the effects of factor B are all equal to zero, we compute the following ratio:

F-Test for Factor B

$$f_2 = \frac{s_2^2}{s^2},$$

---

which is a value of the random variable F_2 having the F-distribution with $b-1$ and $ab(n-1)$ degrees of freedom when H_0'' is true. This hypothesis is rejected at the α-level of significance when $f_2 > f_\alpha[b-1, ab(n-1)]$.

Finally, to test the hypothesis H_0''', that the interaction effects are all equal to zero, we compute the following ratio:

F-Test for Interaction

$$f_3 = \frac{s_3^2}{s^2},$$

---

which is a value of the random variable F_3 having the F-distribution with $(a-1)(b-1)$ and $ab(n-1)$ degrees of freedom when H_0''' is true. We conclude that, at the α-level of significance, interaction is present when $f_3 > f_\alpha[(a-1)(b-1), ab(n-1)]$.

As indicated in Section 9.2, it is advisable to interpret the test for interaction before attempting to draw inferences on the main effects. If interaction is not significant, there is certainly evidence that the tests on main effects are interpretable. Rejection of hypothesis 1 on page 398 implies that the response means at the levels of factor A are significantly different, while rejection of hypothesis 2 implies a similar condition for the means at levels of factor B. However, a significant interaction could very well imply that the data should be analyzed in a somewhat different manner—**perhaps observing the effect of factor A at fixed levels of factor B**, and so forth.

The computations in an analysis-of-variance problem, for a two-factor experiment with n replications, are usually summarized as in Table 9.2.

Table 9.2: Analysis of Variance for the Two-Factor Experiment with n Replications

Source of Variation	Sum of Squares	Degrees of Freedom	Mean Square	Computed f
Main effect:				
A	SSA	$a-1$	$s_1^2 = \frac{SSA}{a-1}$	$f_1 = \frac{s_1^2}{s^2}$
B	SSB	$b-1$	$s_2^2 = \frac{SSB}{b-1}$	$f_2 = \frac{s_2^2}{s^2}$
Two-factor interactions:				
AB	$SS(AB)$	$(a-1)(b-1)$	$s_3^2 = \frac{SS(AB)}{(a-1)(b-1)}$	$f_3 = \frac{s_3^2}{s^2}$
Error	SSE	$ab(n-1)$	$s^2 = \frac{SSE}{ab(n-1)}$	
Total	SST	$abn-1$		

Example 9.1: In an experiment conducted to determine which of 3 different missile systems is preferable, the propellant burning rate for 24 static firings was measured. Four different propellant types were used. The experiment yielded duplicate observations of burning rates at each combination of the treatments.

The data, after coding, are given in Table 9.3. Test the following hypotheses: (a) H_0': there is no difference in the mean propellant burning rates when different missile systems are used, (b) H_0'': there is no difference in the mean propellant burning rates of the 4 propellant types, (c) H_0''': there is no interaction between the different missile systems and the different propellant types.

Table 9.3: Propellant Burning Rates

Missile System	Propellant Type			
	b_1	b_2	b_3	b_4
a_1	34.0	30.1	29.8	29.0
	32.7	32.8	26.7	28.9
a_2	32.0	30.2	28.7	27.6
	33.2	29.8	28.1	27.8
a_3	28.4	27.3	29.7	28.8
	29.3	28.9	27.3	29.1

Solution: 1. (a) H_0': $\alpha_1 = \alpha_2 = \alpha_3 = 0$.

    (b) H_0'': $\beta_1 = \beta_2 = \beta_3 = \beta_4 = 0$.

    (c) H_0''': $(\alpha\beta)_{11} = (\alpha\beta)_{12} = \cdots = (\alpha\beta)_{34} = 0$.

2. (a) H_1': At least one of the α_i is not equal to zero.

    (b) H_1'': At least one of the β_j is not equal to zero.

    (c) H_1''': At least one of the $(\alpha\beta)_{ij}$ is not equal to zero.

The sum-of-squares formula is used as described in Theorem 9.1. The analysis of variance is shown in Table 9.4.

Table 9.4: Analysis of Variance for the Data of Table 9.3

Source of Variation	Sum of Squares	Degrees of Freedom	Mean Square	Computed f
Missile system	14.52	2	7.26	5.84
Propellant type	40.08	3	13.36	10.75
Interaction	22.16	6	3.69	2.97
Error	14.91	12	1.24	
Total	91.68	23		

The reader is directed to a *SAS* GLM Procedure (General Linear Models) for analysis of the burning rate data in Figure 9.2. Note how the "model" (11 degrees of freedom) is initially tested and the system, type, and system by type interaction are tested separately. The F-test on the model ($P = 0.0030$) is testing the accumulation of the two main effects and the interaction.

(a) Reject H_0' and conclude that different missile systems result in different mean propellant burning rates. The P-value is approximately 0.0169.

(b) Reject H_0'' and conclude that the mean propellant burning rates are not the same for the four propellant types. The P-value is approximately 0.0010.

(c) Interaction is barely insignificant at the 0.05 level, but the P-value of approximately 0.0513 would indicate that interaction must be taken seriously.

At this point we should draw some type of interpretation of the interaction. It should be emphasized that statistical significance of a main effect merely implies that *marginal means are significantly different*. However, consider the two-way table of averages in Table 9.5.

Table 9.5: Interpretation of Interaction

	b_1	b_2	b_3	b_4	**Average**
a_1	33.35	31.45	28.25	28.95	30.50
a_2	32.60	30.00	28.40	27.70	29.68
a_3	28.85	28.10	28.50	28.95	28.60
Average	31.60	29.85	28.38	28.53	

It is apparent that more important information exists in the body of the table—trends that are inconsistent with the trend depicted by marginal averages. Table

```
                           The GLM Procedure
   Dependent Variable: rate
                               Sum of
   Source                 DF      Squares   Mean Square   F Value   Pr > F
   Model                  11   76.76833333    6.97893939      5.62   0.0030
   Error                  12   14.91000000    1.24250000
   Corrected Total        23   91.67833333


   R-Square     Coeff Var      Root MSE      rate Mean
   0.837366      3.766854      1.114675      29.59167


   Source          DF    Type III SS   Mean Square   F Value   Pr > F
   system           2   14.52333333    7.26166667      5.84   0.0169
   type             3   40.08166667   13.36055556     10.75   0.0010
   system*type      6   22.16333333    3.69388889      2.97   0.0512
```

Figure 9.2: *SAS* printout of the analysis of the propellant rate data of Table 9.3.

9.5 certainly suggests that the effect of propellant type depends on the system being used. For example, for system 3 the propellant-type effect does not appear to be important, although it does have a large effect if either system 1 or system 2 is used. This explains the "significant" interaction between these two factors. More will be revealed subsequently concerning this interaction.   ◢

Graphical Analysis for the Two-Factor Problem of Example 9.1

Many of the same types of graphical displays that were suggested in the one-factor problems certainly apply in the two-factor case. Two-dimensional plots of cell means or treatment combination means can provide insight into the presence of interactions between the two factors. In addition, a plot of residuals against fitted values may well provide an indication of whether or not the homogeneous variance assumption holds. Often, of course, a violation of the homogeneous variance assumption involves an increase in the error variance as *the response numbers get larger*. As a result, this plot may point out the violation.

Figure 9.3 shows the plot of cell means in the case of the missile system propellant illustration in Example 9.1. Notice how graphically (in this case) the lack of parallelism shows through. Note the flatness of the part of the figure showing the propellant effect for system 3. This illustrates interaction among the factors. Figure 9.4 shows the plot of residuals against fitted values for the same data. There is no apparent sign of difficulty with the homogeneous variance assumption.

Example 9.2: An electrical engineer is investigating a plasma etching process used in semiconductor manufacturing. It is of interest to study the effects of two factors, the C_2F_6 gas flow rate (A) and the power applied to the cathode (B). The response is the etch rate. Each factor is run at 3 levels, and 2 experimental runs on etch rate are made for each of the 9 combinations. The setup is that of a completely randomized

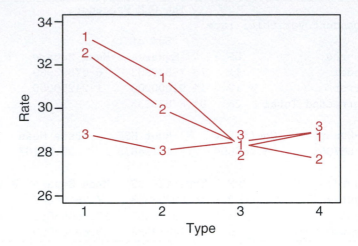

Figure 9.3: Plot of cell means for data of Example 9.1. Numbers represent missile systems.

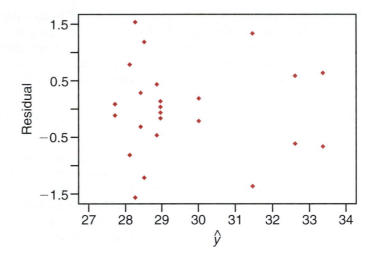

Figure 9.4: Residual plot of data of Example 9.1.

design. The data are given in Table 9.6. The etch rate is in A°/min.

The levels of the factors are in ascending order, with level 1 being the lowest level and level 3 being the highest.

   (a) Show an analysis of variance table and draw conclusions, beginning with the test on interaction.

   (b) Do tests on main effects and draw conclusions.

Solution: A *SAS* output is given in Figure 9.5. From the output we learn the following.

   (a) The *P*-value for the test of interaction is 0.4485. We can conclude that there is no significant interaction.

Table 9.6: Data for Example 9.2

C$_2$F$_6$ Flow Rate	Power Supplied 1	2	3
1	288	488	670
	360	465	720
2	385	482	692
	411	521	724
3	488	595	761
	462	612	801

```
                    The GLM Procedure
Dependent Variable: etchrate
                        Sum of
Source            DF     Squares   Mean Square   F Value   Pr > F
Model              8  379508.7778   47438.5972     61.00   <.0001
Error              9    6999.5000     777.7222
Corrected Total   17  386508.2778


R-Square    Coeff Var      Root MSE     etchrate Mean
0.981890    5.057714      27.88767         551.3889


Source            DF   Type III SS   Mean Square   F Value   Pr > F
c2f6               2    46343.1111    23171.5556     29.79   0.0001
power              2   330003.4444   165001.7222    212.16   <.0001
c2f6*power         4     3162.2222     790.5556      1.02   0.4485
```

Figure 9.5: *SAS* printout for Example 9.2.

(b) There is a significant difference in mean etch rate based on the level of power to the cathode. Tukey's test revealed that the etch rate for level 3 is significantly higher than that for level 2 and the rate for level 2 is significantly higher than that for level 1. See Figure 9.6(a). There is also a significant difference in mean etch rate for the 3rd C$_2$F$_6$ flow rate from the 1st and 2nd C$_2$F$_6$ flow rates. Tukey's test shows that the mean etch rate for level 3 is significantly higher than those for levels 1 and 2 and there is no significant difference in the mean etch rate between the flow rate levels 1 and 2. See Figure 9.6(b).∎

Tukey Grouping	Mean	N	power		Tukey Grouping	Mean	N	c2f6
A	728.00	6	3		A	619.83	6	3
B	527.17	6	2		B	535.83	6	2
					B			
C	399.00	6	1		C	498.50	6	1
(a)					(b)			

Figure 9.6: *SAS* output for Example 9.2. (a) Tukey's test on power; (b) Tukey's test on gas flow rate.

Exercises

9.1 An experiment was conducted to study the effects of temperature and type of oven on the life of a particular component. Four types of ovens and 3 temperature levels were used in the experiment. Twenty-four pieces were assigned randomly, two to each combination of treatments, and the following results recorded.

	Oven			
Temperature (°F)	O_1	O_2	O_3	O_4
500	227	214	225	260
	221	259	236	229
550	187	181	232	246
	208	179	198	273
600	174	198	178	206
	202	194	213	219

Using a 0.05 level of significance, test the hypothesis that

(a) different temperatures have no effect on the life of the component;

(b) different ovens have no effect on the life of the component;

(c) the type of oven and temperature do not interact.

9.2 To ascertain the stability of vitamin C in reconstituted frozen orange juice concentrate stored in a refrigerator for a period of up to one week, the study *Vitamin C Retention in Reconstituted Frozen Orange Juice* was conducted by the Department of Human Nutrition and Foods at Virginia Tech. Three types of frozen orange juice concentrate were tested using 3 different time periods. The time periods refer to the number of days from when the orange juice was blended until it was tested. The results, in milligrams of ascorbic acid per liter, were recorded. Use a 0.05 level of significance to test the hypothesis that

(a) there is no difference in ascorbic acid contents among the different brands of orange juice concentrate;

(b) there is no difference in ascorbic acid contents for the different time periods;

(c) the brands of orange juice concentrate and the number of days from the time the juice was blended until it was tested do not interact.

Brand	Time (days)					
	0		3		7	
Richfood	52.6	54.2	49.4	49.2	42.7	48.8
	49.8	46.5	42.8	53.2	40.4	47.6
Sealed-Sweet	56.0	48.0	48.8	44.0	49.2	44.0
	49.6	48.4	44.0	42.4	42.0	43.2
Minute Maid	52.5	52.0	48.0	47.0	48.5	43.3
	51.8	53.6	48.2	49.6	45.2	47.6

9.3 Three strains of rats were studied under 2 environmental conditions for their performance in a maze test. The error scores for the 48 rats were recorded.

Environment	Strain					
	Bright		Mixed		Dull	
Free	28	12	33	83	101	94
	22	23	36	14	33	56
	25	10	41	76	122	83
	36	86	22	58	35	23
Restricted	72	32	60	89	136	120
	48	93	35	126	38	153
	25	31	83	110	64	128
	91	19	99	118	87	140

Use a 0.01 level of significance to test the hypothesis that

(a) there is no difference in error scores for different environments;

(b) there is no difference in error scores for different strains;

(c) the environments and strains of rats do not interact.

9.4 Corrosion fatigue in metals has been defined as the simultaneous action of cyclic stress and chemical attack on a metal structure. A widely used technique for minimizing corrosion fatigue damage in aluminum involves the application of a protective coating. A study conducted by the Department of Mechanical Engineering at Virginia Tech used 3 different levels of humidity (Low: 20–25% relative humidity; Medium: 55–60% relative humidity; High: 86–91% relative humidity) and 3 types of surface coatings (Uncoated: no coating; Anodized: sulfuric acid anodic oxide coating; Conversion: chromate chemical conversion coating). The corrosion fatigue data, expressed in thousands of cycles to failure, were recorded as follows:

	Relative Humidity					
Coating	**Low**		**Medium**		**High**	
Uncoated	361	469	314	522	1344	1216
	466	937	244	739	1027	1097
	1069	1357	261	134	1011	1011
Anodized	114	1032	322	471	78	466
	1236	92	306	130	387	107
	533	211	68	398	130	327
Conversion	130	1482	252	874	586	524
	841	529	105	755	402	751
	1595	754	847	573	846	529

(a) Perform an analysis of variance with $\alpha = 0.05$ to test for significant main and interaction effects.

(b) Use Tukey's multiple-range test at the 0.05 level of significance to determine which humidity levels result in different corrosion fatigue damage.

9.5 To determine which muscles need to be subjected to a conditioning program in order to improve one's performance on the flat serve used in tennis, a study was conducted by the Department of Health, Physical Education and Recreation at Virginia Tech. Five different muscles

> 1: anterior deltoid   4: middle deltoid
>
> 2: pectoralis major   5: triceps
>
> 3: posterior deltoid

were tested on each of 3 subjects, and the experiment was carried out 3 times for each treatment combination. The electromyographic data, recorded during the serve, are presented here.

	Muscle				
Subject	**1**	**2**	**3**	**4**	**5**
1	32	5	58	10	19
	59	1.5	61	10	20
	38	2	66	14	23
2	63	10	64	45	43
	60	9	78	61	61
	50	7	78	71	42
3	43	41	26	63	61
	54	43	29	46	85
	47	42	23	55	95

Use a 0.01 level of significance to test the hypothesis that

(a) different subjects have equal electromyographic measurements;

(b) different muscles have no effect on electromyographic measurements;

(c) subjects and types of muscle do not interact.

9.6 An experiment was conducted to determine whether additives increase the adhesiveness of rubber products. Sixteen products were made with the new additive and another 16 without the new additive. The observed adhesiveness was as recorded below.

	Temperature (°C)			
	50	**60**	**70**	**80**
Without Additive	2.3	3.4	3.8	3.9
	2.9	3.7	3.9	3.2
	3.1	3.6	4.1	3.0
	3.2	3.2	3.8	2.7
With Additive	4.3	3.8	3.9	3.5
	3.9	3.8	4.0	3.6
	3.9	3.9	3.7	3.8
	4.2	3.5	3.6	3.9

Perform an analysis of variance to test for significant main and interaction effects.

9.7 The extraction rate of a certain polymer is known to depend on the reaction temperature and the amount of catalyst used. An experiment was conducted at four levels of temperature and five levels of the catalyst, and the extraction rate was recorded in the following table.

	Amount of Catalyst				
	0.5%	**0.6%**	**0.7%**	**0.8%**	**0.9%**
50°C	38	45	57	59	57
	41	47	59	61	58
60°C	44	56	70	73	61
	43	57	69	72	58
70°C	44	56	70	73	61
	47	60	67	61	59
80°C	49	62	70	62	53
	47	65	55	69	58

Perform an analysis of variance. Test for significant main and interaction effects.

9.8 In Myers, Montgomery, and Anderson-Cook (2009), a scenario is discussed involving an auto bumper plating process. The response is the thickness of the material. Factors that may impact the thickness include amount of nickel (A) and pH (B). A two-factor experiment is designed. The plan is a completely randomized design in which the individual bumpers are assigned randomly to the factor combinations. Three levels of pH and two levels of nickel content are involved in the experiment. The thickness data, in cm $\times 10^{-3}$, are as follows:

Nickel Content (grams)	pH		
	5	5.5	6
18	250	211	221
	195	172	150
	188	165	170
10	115	88	69
	165	112	101
	142	108	72

(a) Display the analysis-of-variance table with tests for both main effects and interaction. Show *P*-values.

(b) Give engineering conclusions. What have you learned from the analysis of the data?

(c) Show a plot that depicts either a presence or an absence of interaction.

9.9 An engineer is interested in the effects of cutting speed and tool geometry on the life in hours of a machine tool. Two cutting speeds and two different geometries are used. Three experimental tests are accomplished at each of the four combinations. The data are as follows.

Tool Geometry	Cutting Speed	
	Low	High
1	22 28 20	34 37 29
2	18 15 16	11 10 10

(a) Show an analysis-of-variance table with tests on interaction and main effects.

(b) Comment on the effect that interaction has on the test on cutting speed.

(c) Do secondary tests that will allow the engineer to learn the true impact of cutting speed.

(d) Show a plot that graphically displays the interaction effect.

9.10 Two factors in a manufacturing process for an integrated circuit are studied in a two-factor experiment. The purpose of the experiment is to learn their effect on the resistivity of the wafer. The factors are implant dose (2 levels) and furnace position (3 levels). Experimentation is costly so only one experimental run is made at each combination. The data are as follows.

Dose	Position		
1	15.5	14.8	21.3
2	27.2	24.9	26.1

It is to be assumed that no interaction exists between these two factors.

(a) Write the model and explain terms.

(b) Show the analysis-of-variance table.

(c) Explain the 2 "error" degrees of freedom.

(d) Use Tukey's test to do multiple-comparison tests on furnace position. Explain what the results show.

9.11 A study was done to determine the impact of two factors, method of analysis and the laboratory doing the analysis, on the level of sulfur content in coal. Twenty-eight coal specimens were randomly assigned to 14 factor combinations, the structure of the experimental units represented by combinations of seven laboratories and two methods of analysis with two specimens per factor combination. The data, expressed in percent of sulfur, are as follows:

Laboratory	Method			
	1		2	
1	0.109	0.105	0.105	0.108
2	0.129	0.122	0.127	0.124
3	0.115	0.112	0.109	0.111
4	0.108	0.108	0.117	0.118
5	0.097	0.096	0.110	0.097
6	0.114	0.119	0.116	0.122
7	0.155	0.145	0.164	0.160

(The data are taken from G. Taguchi, "Signal to Noise Ratio and Its Applications to Testing Material," *Reports of Statistical Application Research*, Union of Japanese Scientists and Engineers, Vol. 18, No. 4, 1971.)

(a) Do an analysis of variance and show results in an analysis-of-variance table.

(b) Is interaction significant? If so, discuss what it means to the scientist. Use a *P*-value in your conclusion.

(c) Are the individual main effects, laboratory, and method of analysis statistically significant? Discuss what is learned and let your answer be couched in the context of any significant interaction.

(d) Do an interaction plot that illustrates the effect of interaction.

(e) Do a test comparing methods 1 and 2 at laboratory 1 and do the same test at laboratory 7. Comment on what these results illustrate.

9.12 In an experiment conducted in the Civil Engineering Department at Virginia Tech, growth of a certain type of algae in water was observed as a function of time and the dosage of copper added to the water. The data are as follows. Response is in units of algae.

Copper	Time in Days		
	5	12	18
1	0.30	0.37	0.25
	0.34	0.36	0.23
	0.32	0.35	0.24
2	0.24	0.30	0.27
	0.23	0.32	0.25
	0.22	0.31	0.25
3	0.20	0.30	0.27
	0.28	0.31	0.29
	0.24	0.30	0.25

(a) Do an analysis of variance and show the analysis-of-variance table.

(b) Comment concerning whether the data are sufficient to show a time effect on algae concentration.

(c) Do the same for copper content. Does the level of copper impact algae concentration?

(d) Comment on the results of the test for interaction. How is the effect of copper content influenced by time?

9.13 In Myers, *Classical and Modern Regression with Applications* (1990), an experiment is described in which the Environmental Protection Agency seeks to determine the effect of two water treatment methods on magnesium uptake. Magnesium levels in grams per cubic centimeter (cc) are measured, and two different time levels are incorporated into the experiment. The data are as follows:

Time (hr)	Treatment 1			Treatment 2		
1	2.19	2.15	2.16	2.03	2.01	2.04
2	2.01	2.03	2.04	1.88	1.86	1.91

(a) Do an interaction plot. What is your impression?

(b) Do an analysis of variance and show tests for the main effects and interaction.

(c) Give your findings regarding how time and treatment influence magnesium uptake.

(d) Fit the appropriate regression model with treatment as a categorical variable. Include interaction in the model.

(e) Is interaction significant in the regression model?

9.14 Consider the data set in Exercise 9.12 and answer the following questions.

(a) Both factors, copper and time, are quantitative in nature. As a result, a regression model may be of interest. Describe what might be an appropriate model using x_1 = copper content and x_2 = time.

Fit the model to the data, showing regression coefficients and a *t*-test on each.

(b) Fit the model

$$Y = \beta_0 + \beta_1 x_1 + \beta_2 x_2 + \beta_{12} x_1 x_2$$
$$+ \beta_{11} x_1^2 + \beta_{22} x_2^2 + \epsilon,$$

and compare it to the one you chose in (a). Which is more appropriate? Use R_{adj}^2 as a criterion.

9.15 The purpose of the study *The Incorporation of a Chelating Agent into a Flame Retardant Finish of a Cotton Flannelette and the Evaluation of Selected Fabric Properties*, conducted at Virginia Tech, was to evaluate the use of a chelating agent as part of the flame retardant finish of cotton flannelette by determining its effect upon flammability after the fabric is laundered under specific conditions. There were two treatments at two levels. Two baths were prepared, one with carboxymethyl cellulose (bath I) and one without (bath II). Half of the fabric was laundered 5 times and half was laundered 10 times. There were 12 pieces of fabric in each bath/number of launderings combination. After the washings, the lengths of fabric that burned and the burn times were measured. Burn times (in seconds) were recorded as follows:

Launderings	Bath I			Bath II		
5	13.7	23.0	15.7	6.2	5.4	5.0
	25.5	15.8	14.8	4.4	5.0	3.3
	14.0	29.4	9.7	16.0	2.5	1.6
	14.0	12.3	12.3	3.9	2.5	7.1
10	27.2	16.8	12.9	18.2	8.8	14.5
	14.9	17.1	13.0	14.7	17.1	13.9
	10.8	13.5	25.5	10.6	5.8	7.3
	14.2	27.4	11.5	17.7	18.3	9.9

(a) Perform an analysis of variance. Is there a significant interaction term?

(b) Are there main effect differences? Discuss.

9.4 Three-Factor Experiments

In this section, we consider an experiment with three factors, A, B, and C, at a, b, and c levels, respectively, in a completely randomized experimental design. Assume again that we have n observations for each of the abc treatment combinations. We shall proceed to outline significance tests for the three main effects and interactions involved. It is hoped that the reader can then use the description given here to generalize the analysis to $k > 3$ factors.

<table>
<tr><td>Model for the
Three-Factor
Experiment</td><td>The model for the three-factor experiment is</td></tr>
</table>

$$y_{ijkl} = \mu + \alpha_i + \beta_j + \gamma_k + (\alpha\beta)_{ij} + (\alpha\gamma)_{ik} + (\beta\gamma)_{jk} + (\alpha\beta\gamma)_{ijk} + \epsilon_{ijkl},$$

$i = 1, 2, \ldots, a$; $j = 1, 2, \ldots, b$; $k = 1, 2, \ldots, c$; and $l = 1, 2, \ldots, n$, where α_i, β_j, and γ_k are the main effects and $(\alpha\beta)_{ij}$, $(\alpha\gamma)_{ik}$, and $(\beta\gamma)_{jk}$ are the two-factor interaction effects that have the same interpretation as in the two-factor experiment.

The term $(\alpha\beta\gamma)_{ijk}$ is called the **three-factor interaction effect**, a term that represents a nonadditivity of the $(\alpha\beta)_{ij}$ over the different levels of the factor C. As before, the sum of all main effects is zero and the sum over any subscript of the two- and three-factor interaction effects is zero. In many experimental situations, these higher-order interactions are insignificant and their mean squares reflect only random variation, but we shall outline the analysis in its most general form.

Again, in order that valid significance tests can be made, we must assume that the errors are values of independent and normally distributed random variables, each with mean 0 and common variance σ^2.

The general philosophy concerning the analysis is the same as that discussed for the one- and two-factor experiments. The sum of squares is partitioned into eight terms, each representing a source of variation from which we obtain independent estimates of σ^2 when all the main effects and interaction effects are zero. If the effects of any given factor or interaction are not all zero, then the mean square will estimate the error variance plus a component due to the systematic effect in question.

<table>
<tr><td>Sum of Squares
for a
Three-Factor
Experiment</td><td></td></tr>
</table>

$$SSA = bcn\sum_{i=1}^{a}(\bar{y}_{i\ldots} - \bar{y}_{\ldots})^2 \quad SS(AB) = cn\sum_{i}\sum_{j}(\bar{y}_{ij\ldots} - \bar{y}_{i\ldots} - \bar{y}_{\cdot j\ldots} + \bar{y}_{\ldots})^2$$

$$SSB = acn\sum_{j=1}^{b}(\bar{y}_{\cdot j\ldots} - \bar{y}_{\ldots})^2 \quad SS(AC) = bn\sum_{i}\sum_{k}(\bar{y}_{i\cdot k\cdot} - \bar{y}_{i\ldots} - \bar{y}_{\cdot\cdot k\cdot} + \bar{y}_{\ldots})^2$$

$$SSC = abn\sum_{k=1}^{c}(\bar{y}_{\cdot\cdot k\cdot} - \bar{y}_{\ldots})^2 \quad SS(BC) = an\sum_{j}\sum_{k}(\bar{y}_{\cdot jk\cdot} - \bar{y}_{\cdot j\ldots} - \bar{y}_{\cdot\cdot k\cdot} + \bar{y}_{\ldots})^2$$

$$SS(ABC) = n\sum_{i}\sum_{j}\sum_{k}(\bar{y}_{ijk\cdot} - \bar{y}_{ij\ldots} - \bar{y}_{i\cdot k\cdot} - \bar{y}_{\cdot jk\cdot} + \bar{y}_{i\ldots} + \bar{y}_{\cdot j\ldots} + \bar{y}_{\cdot\cdot k\cdot} - \bar{y}_{\ldots})^2$$

$$SST = \sum_{i}\sum_{j}\sum_{k}\sum_{l}(y_{ijkl} - \bar{y}_{\ldots})^2 \quad SSE = \sum_{i}\sum_{j}\sum_{k}\sum_{l}(y_{ijkl} - \bar{y}_{ijk\cdot})^2$$

Although we emphasize interpretation of annotated computer printout in this section rather than being concerned with laborious computation of sums of squares, we do offer the following as the sums of squares for the three main effects and interactions. Notice the obvious extension from the two- to three-factor problem. The averages in the formulas follow from what is given in the two-factor problem.

The computations in an analysis-of-variance table for a three-factor problem with n replicated runs at each factor combination are summarized in Table 9.7.

For the three-factor experiment with a single experimental run per combina-

Table 9.7: ANOVA for the Three-Factor Experiment with n Replications

Source of Variation	Sum of Squares	Degrees of Freedom	Mean Square	Computed f
Main effect:				
A	SSA	$a-1$	s_1^2	$f_1 = \frac{s_1^2}{s^2}$
B	SSB	$b-1$	s_2^2	$f_2 = \frac{s_2^2}{s^2}$
C	SSC	$c-1$	s_3^2	$f_3 = \frac{s_3^2}{s^2}$
Two-factor interaction:				
AB	$SS(AB)$	$(a-1)(b-1)$	s_4^2	$f_4 = \frac{s_4^2}{s^2}$
AC	$SS(AC)$	$(a-1)(c-1)$	s_5^2	$f_5 = \frac{s_5^2}{s^2}$
BC	$SS(BC)$	$(b-1)(c-1)$	s_6^2	$f_6 = \frac{s_6^2}{s^2}$
Three-factor interaction:				
ABC	$SS(ABC)$	$(a-1)(b-1)(c-1)$	s_7^2	$f_7 = \frac{s_7^2}{s^2}$
Error	SSE	$abc(n-1)$	s^2	
Total	SST	$abcn-1$		

tion, we may use the analysis of Table 9.7 by setting $n = 1$ and using the ABC interaction sum of squares for SSE. In this case, we are assuming that the $(\alpha\beta\gamma)_{ijk}$ interaction effects are all equal to zero so that

$$E\left[\frac{SS(ABC)}{(a-1)(b-1)(c-1)}\right] = \sigma^2 + \frac{n}{(a-1)(b-1)(c-1)}\sum_{i=1}^{a}\sum_{j=1}^{b}\sum_{k=1}^{c}(\alpha\beta\gamma)_{ijk}^2 = \sigma^2.$$

That is, $SS(ABC)$ represents variation due only to experimental error. Its mean square thereby provides an unbiased estimate of the error variance. With $n = 1$ and $SSE = SS(ABC)$, the error sum of squares is found by subtracting the sums of squares of the main effects and two-factor interactions from the total sum of squares.

Example 9.3: In the production of a particular material, three variables are of interest: A, the operator effect (three operators): B, the catalyst used in the experiment (three catalysts); and C, the washing time of the product following the cooling process (15 minutes and 20 minutes). Three runs were made at each combination of factors. It was felt that all interactions among the factors should be studied. The coded yields are in Table 9.8. Perform an analysis of variance to test for significant effects.

Solution: Table 9.9 shows an analysis of variance of the data given above. None of the interactions show a significant effect at the $\alpha = 0.05$ level. However, the P-value for BC is 0.0610; thus, it should not be ignored. The operator and catalyst effects are significant, while the effect of washing time is not significant. ⌐

Table 9.8: Data for Example 9.3

Operator, A	Washing Time, C					
	15 Minutes			20 Minutes		
	Catalyst, B			Catalyst, B		
	1	2	3	1	2	3
1	10.7	10.3	11.2	10.9	10.5	12.2
	10.8	10.2	11.6	12.1	11.1	11.7
	11.3	10.5	12.0	11.5	10.3	11.0
2	11.4	10.2	10.7	9.8	12.6	10.8
	11.8	10.9	10.5	11.3	7.5	10.2
	11.5	10.5	10.2	10.9	9.9	11.5
3	13.6	12.0	11.1	10.7	10.2	11.9
	14.1	11.6	11.0	11.7	11.5	11.6
	14.5	11.5	11.5	12.7	10.9	12.2

Table 9.9: ANOVA for a Three-Factor Experiment in a Completely Randomized Design

Source	df	Sum of Squares	Mean Square	F-Value	P-Value
A	2	13.98	6.99	11.64	0.0001
B	2	10.18	5.09	8.48	0.0010
AB	4	4.77	1.19	1.99	0.1172
C	1	1.19	1.19	1.97	0.1686
AC	2	2.91	1.46	2.43	0.1027
BC	2	3.63	1.82	3.03	0.0610
ABC	4	4.91	1.23	2.04	0.1089
Error	36	21.61	0.60		
Total	53	63.19			

Impact of Interaction BC

More should be discussed regarding Example 9.3, particularly about dealing with the effect that the interaction between catalyst and washing time is having on the test on the washing time main effect (factor C). Recall our discussion in Section 9.2. Illustrations were given of how the presence of interaction could change the interpretation that we make regarding main effects. In Example 9.3, the BC interaction is significant at approximately the 0.06 level. Consider, however, the two-way table of means in Table 9.10.

It is clear why washing time was found not to be significant. A non-thorough analyst may get the impression that washing time can be eliminated from any future study in which yield is being measured. However, it is obvious how the effect of washing time changes from a negative effect for the first catalyst to what appears to be a positive effect for the third catalyst. If we merely focus on the data for catalyst 1, a simple comparison between the means at the two washing

Table 9.10: Two-Way Table of Means for Example 9.3

Catalyst, B	Washing Time, C	
	15 min	**20 min**
1	12.19	11.29
2	10.86	10.50
3	11.09	11.46
Means	11.38	11.08

times will produce a simple t-statistic:

$$t = \frac{12.19 - 11.29}{\sqrt{0.6(2/9)}} = 2.5,$$

which is significant at a level less than 0.02. Thus, an important negative effect of washing time for catalyst 1 might very well be ignored if the analyst makes the incorrect broad interpretation of the insignificant F-ratio for washing time.

Pooling in Multifactor Models

We have described the three-factor model and its analysis in the most general form by including all possible interactions in the model. Of course, there are many situations where it is known *a priori* that the model should not contain certain interactions. We can then take advantage of this knowledge by combining or pooling the sums of squares corresponding to negligible interactions with the error sum of squares to form a new estimator for σ^2 with a larger number of degrees of freedom. For example, in a metallurgy experiment designed to study the effect on film thickness of three important processing variables, suppose it is known that factor A, acid concentration, does not interact with factors B and C. The sums of squares SSA, SSB, SSC, and $SS(BC)$ are computed using the methods described earlier in this section. The mean squares for the remaining effects will now all independently estimate the error variance σ^2. Therefore, we form our new **mean square error by pooling** $SS(AB)$, $SS(AC)$, $SS(ABC)$, and SSE, along with the corresponding degrees of freedom. The resulting denominator for the significance tests is then the mean square error given by

$$s^2 = \frac{SS(AB) + SS(AC) + SS(ABC) + SSE}{(a-1)(b-1) + (a-1)(c-1) + (a-1)(b-1)(c-1) + abc(n-1)}.$$

Computationally, of course, one obtains the pooled sum of squares and the pooled degrees of freedom by subtraction once SST and the sums of squares for the existing effects are computed. The analysis-of-variance table would then take the form of Table 9.11.

Factorial Experiments in Blocks

In this chapter, we have assumed that the experimental design used is a completely randomized design. By interpreting the levels of factor A in Table 9.11 **as**

Table 9.11: ANOVA with Factor A Noninteracting

Source of Variation	Sum of Squares	Degrees of Freedom	Mean Square	Computed f
Main effect:				
A	SSA	$a-1$	s_1^2	$f_1 = \frac{s_1^2}{s^2}$
B	SSB	$b-1$	s_2^2	$f_2 = \frac{s_2^2}{s^2}$
C	SSC	$c-1$	s_3^2	$f_3 = \frac{s_3^2}{s^2}$
Two-factor interaction:				
BC	$SS(BC)$	$(b-1)(c-1)$	s_4^2	$f_4 = \frac{s_4^2}{s^2}$
Error	SSE	Subtraction	s^2	
Total	SST	$abcn-1$		

different blocks, we then have the analysis-of-variance procedure for a two-factor experiment in a randomized block design. For example, if we interpret the operators in Example 9.3 as blocks and assume no interaction between blocks and the other two factors, the analysis of variance takes the form of Table 9.12 rather than that of Table 9.9. The reader can verify that the mean square error is also

$$s^2 = \frac{4.77 + 2.91 + 4.91 + 21.61}{4 + 2 + 4 + 36} = 0.74,$$

which demonstrates the pooling of the sums of squares for the nonexisting interaction effects. Note that factor B, catalyst, has a significant effect on yield.

Table 9.12: ANOVA for a Two-Factor Experiment in a Randomized Block Design

Source of Variation	Sum of Squares	Degrees of Freedom	Mean Square	Computed f	P-Value
Blocks	13.98	2	6.99		
Main effect:					
B	10.18	2	5.09	6.88	0.0024
C	1.18	1	1.18	1.59	0.2130
Two-factor interaction:					
BC	3.64	2	1.82	2.46	0.0966
Error	34.21	46	0.74		
Total	63.19	53			

Example 9.4: An experiment was conducted to determine the effects of temperature, pressure, and stirring rate on product filtration rate. This was done in a pilot plant. The experiment was run at two levels of each factor. In addition, it was decided that two batches of raw materials should be used, where batches were treated as blocks.

Eight experimental runs were made in random order for each batch of raw materials. It is thought that all two-factor interactions may be of interest. No interactions with batches are assumed to exist. The data appear in Table 9.13. "L" and "H" imply low and high levels, respectively. The filtration rate is in gallons per hour.

(a) Show the complete ANOVA table. Pool all "interactions" with blocks into error.

(b) What interactions appear to be significant?

(c) Create plots to reveal and interpret the significant interactions. Explain what the plot means to the engineer.

Table 9.13: Data for Example 9.4

Batch 1						
	Low Stirring Rate			**High Stirring Rate**		
Temp.	**Pressure L**	**Pressure H**		**Temp.**	**Pressure L**	**Pressure H**
L	43	49		**L**	44	47
H	64	68		**H**	97	102

Batch 2						
	Low Stirring Rate			**High Stirring Rate**		
Temp.	**Pressure L**	**Pressure H**		**Temp.**	**Pressure L**	**Pressure H**
L	49	57		**L**	51	55
H	70	76		**H**	103	106

Solution: (a) The *SAS* printout is given in Figure 9.7.

(b) As seen in Figure 9.7, the temperature by stirring rate (strate) interaction appears to be highly significant. The pressure by stirring rate interaction also appears to be significant. Incidentally, if one were to do further pooling by combining the insignificant interactions with error, the conclusions would remain the same and the *P*-value for the pressure by stirring rate interaction would become stronger, namely 0.0517.

(c) The main effects for both stirring rate and temperature are highly significant, as shown in Figure 9.7. A look at the interaction plot of Figure 9.8(a) shows that the effect of stirring rate is dependent upon the level of temperature. At the low level of temperature the stirring rate effect is negligible, whereas at the high level of temperature stirring rate has a strong positive effect on mean filtration rate. In Figure 9.8(b), the interaction between pressure and stirring rate, though not as pronounced as that of Figure 9.8(a), still shows a slight inconsistency of the stirring rate effect across pressure.

Source	DF	Type III SS	Mean Square	F Value	Pr > F
batch	1	175.562500	175.562500	177.14	<.0001
pressure	1	95.062500	95.062500	95.92	<.0001
temp	1	5292.562500	5292.562500	5340.24	<.0001
pressure*temp	1	0.562500	0.562500	0.57	0.4758
strate	1	1040.062500	1040.062500	1049.43	<.0001
pressure*strate	1	5.062500	5.062500	5.11	0.0583
temp*strate	1	1072.562500	1072.562500	1082.23	<.0001
pressure*temp*strate	1	1.562500	1.562500	1.58	0.2495
Error	7	6.937500	0.991071		
Corrected Total	15	7689.937500			

Figure 9.7: ANOVA for Example 9.4, batch interaction pooled with error.

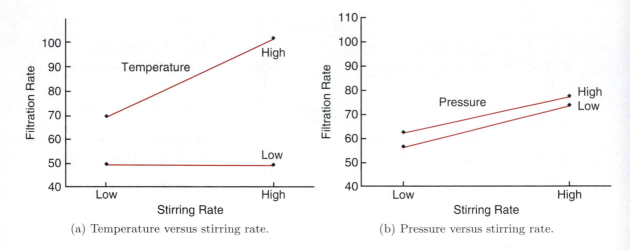

(a) Temperature versus stirring rate.   (b) Pressure versus stirring rate.

Figure 9.8: Interaction plots for Example 9.4.

Exercises

9.16 Consider an experimental situation involving factors A, B, and C, where we assume a three-way fixed effects model of the form $y_{ijkl} = \mu + \alpha_i + \beta_j + \gamma_k + (\beta\gamma)_{jk} + \epsilon_{ijkl}$. All other interactions are considered to be nonexistent or negligible. The data are presented here.

	B_1			B_2		
	C_1	C_2	C_3	C_1	C_2	C_3
A_1	4.0	3.4	3.9	4.4	3.1	3.1
	4.9	4.1	4.3	3.4	3.5	3.7
A_2	3.6	2.8	3.1	2.7	2.9	3.7
	3.9	3.2	3.5	3.0	3.2	4.2
A_3	4.8	3.3	3.6	3.6	2.9	2.9
	3.7	3.8	4.2	3.8	3.3	3.5
A_4	3.6	3.2	3.2	2.2	2.9	3.6
	3.9	2.8	3.4	3.5	3.2	4.3

(a) Perform a test of significance on the BC interaction at the $\alpha = 0.05$ level.

(b) Perform tests of significance on the main effects A, B, and C using a pooled mean square error at the $\alpha = 0.05$ level.

9.17 The data provided are measurements from an experiment conducted using three factors A, B, and C, all fixed effects.

(a) Perform tests of significance on all interactions at the $\alpha = 0.05$ level.

(b) Perform tests of significance on the main effects at the $\alpha = 0.05$ level.

(c) Give an explanation of how a significant interaction has masked the effect of factor C.

	C_1			C_2			C_3		
	B_1	B_2	B_3	B_1	B_2	B_3	B_1	B_2	B_3
A_1	15.0	14.8	15.9	16.8	14.2	13.2	15.8	15.5	19.2
	18.5	13.6	14.8	15.4	12.9	11.6	14.3	13.7	13.5
	22.1	12.2	13.6	14.3	13.0	10.1	13.0	12.6	11.1
A_2	11.3	17.2	16.1	18.9	15.4	12.4	12.7	17.3	7.8
	14.6	15.5	14.7	17.3	17.0	13.6	14.2	15.8	11.5
	18.2	14.2	13.4	16.1	18.6	15.2	15.9	14.6	12.2

9.18 The method of X-ray fluorescence is an important analytical tool for determining the concentration of material in solid missile propellants. In the paper *An X-ray Fluorescence Method for Analyzing Polybutadiene Acrylic Acid (PBAA) Propellants* (Quarterly Report, RK-TR-62-1, Army Ordinance Missile Command, 1962), it is postulated that the propellant mixing process and analysis time have an influence on the homogeneity of the material and hence on the accuracy of X-ray intensity measurements. An experiment was conducted using 3 factors: A, the mixing conditions (4 levels); B, the analysis time (2 levels); and C, the method of loading propellant into sample holders (hot and room temperature). The following data, which represent the weight percent of ammonium perchlorate in a particular propellant, were recorded.

	Hot		Room Temp.	
A	B_1	B_2	B_1	B_2
1	38.62	38.45	39.82	39.82
	37.20	38.64	39.15	40.26
	38.02	38.75	39.78	39.72
2	37.67	37.81	39.53	39.56
	37.57	37.75	39.76	39.25
	37.85	37.91	39.90	39.04
3	37.51	37.21	39.34	39.74
	37.74	37.42	39.60	39.49
	37.58	37.79	39.62	39.45
4	37.52	37.60	40.09	39.36
	37.15	37.55	39.63	39.38
	37.51	37.91	39.67	39.00

Method of Loading, C

(a) Perform an analysis of variance with $\alpha = 0.01$ to test for significant main and interaction effects.

(b) Discuss the influence of the three factors on the weight percent of ammonium perchlorate. Let your discussion involve the role of any significant interaction.

9.19 Corrosion fatigue in metals has been defined as the simultaneous action of cyclic stress and chemical attack on a metal structure. In the study *Effect of Humidity and Several Surface Coatings on the Fatigue Life of 2024-T351 Aluminum Alloy*, conducted by the Department of Mechanical Engineering at Virginia Tech, a technique involving the application of a protective chromate coating was used to minimize corrosion fatigue damage in aluminum. Three factors were used

in the investigation, with 5 replicates for each treatment combination: coating, at 2 levels, and humidity and shear stress, both with 3 levels. The fatigue data, recorded in thousands of cycles to failure, are presented here.

(a) Perform an analysis of variance with $\alpha = 0.01$ to test for significant main and interaction effects.

(b) Make a recommendation for combinations of the three factors that would result in low fatigue damage.

Coating	Humidity	Shear Stress (psi)		
		13,000	17,000	20,000
Uncoated	Low	4580	5252	361
	(20–25% RH)	10,126	897	466
		1341	1465	1069
		6414	2694	469
		3549	1017	937
	Medium	2858	799	314
	(50–60% RH)	8829	3471	244
		10,914	685	261
		4067	810	522
		2595	3409	739
	High	6489	1862	1344
	(86–91% RH)	5248	2710	1027
		6816	2632	663
		5860	2131	1216
		5901	2470	1097
Chromated	Low	5395	4035	130
	(20–25% RH)	2768	2022	841
		1821	914	1595
		3604	2036	1482
		4106	3524	529
	Medium	4833	1847	252
	(50–60% RH)	7414	1684	105
		10,022	3042	847
		7463	4482	874
		21,906	996	755
	High	3287	1319	586
	(86–91% RH)	5200	929	402
		5493	1263	846
		4145	2236	524
		3336	1392	751

9.20 For a study of the hardness of gold dental fillings, five randomly chosen dentists were assigned combinations of three methods of condensation and two types of gold. The hardness was measured. (See Hoaglin, Mosteller, and Tukey, 1991.) Let the dentists play the role of blocks. The data are presented here.

(a) State the appropriate model with the assumptions.

(b) Is there a significant interaction between method of condensation and type of gold filling material?

(c) Is there one method of condensation that seems to be best? Explain.

Type

Dentist	Method	Gold Foil	Goldent
1	1	792	824
	2	772	772
	3	782	803
2	1	803	803
	2	752	772
	3	715	707
3	1	715	724
	2	792	715
	3	762	606
4	1	673	946
	2	657	743
	3	690	245
5	1	634	715
	2	649	724
	3	724	627

9.21 Electronic copiers make copies by gluing black ink on paper, using static electricity. Heating and gluing the ink on the paper comprise the final stage of the copying process. The gluing power during this final process determines the quality of the copy. It is postulated that temperature, surface state of the gluing roller, and hardness of the press roller influence the gluing power of the copier. An experiment is run with treatments consisting of a combination of these three factors at each of three levels. The following data show the gluing power for each treatment combination. Perform an analysis of variance with $\alpha = 0.05$ to test for significant main and interaction effects.

Surface State of Gluing Roller		Hardness of the Press Roller					
		20		**40**		**60**	
Low Temp.	Soft	0.52	0.44	0.54	0.52	0.60	0.55
		0.57	0.53	0.65	0.56	0.78	0.68
	Medium	0.64	0.59	0.79	0.73	0.49	0.48
		0.58	0.64	0.79	0.78	0.74	0.50
	Hard	0.67	0.77	0.58	0.68	0.55	0.65
		0.74	0.65	0.57	0.59	0.57	0.58
Medium Temp.	Soft	0.46	0.40	0.31	0.49	0.56	0.42
		0.58	0.37	0.48	0.66	0.49	0.49
	Medium	0.60	0.43	0.66	0.57	0.64	0.54
		0.62	0.61	0.72	0.56	0.74	0.56
	Hard	0.53	0.65	0.53	0.45	0.56	0.66
		0.66	0.56	0.59	0.47	0.71	0.67
High Temp.	Soft	0.52	0.44	0.54	0.52	0.65	0.49
		0.57	0.53	0.65	0.56	0.65	0.52
	Medium	0.53	0.65	0.53	0.45	0.49	0.48
		0.66	0.56	0.59	0.47	0.74	0.50
	Hard	0.43	0.43	0.48	0.31	0.55	0.65
		0.47	0.44	0.43	0.27	0.57	0.58

9.22 Consider the data set in Exercise 9.21. Construct an interaction plot for any two-factor interaction that is significant.

9.23 Consider combinations of three factors in the removal of dirt from standard loads of laundry. The first factor is the brand of the detergent, X, Y, or Z. The second factor is the type of detergent, liquid or powder. The third factor is the temperature of the water, hot or warm. The experiment was replicated three times. Response is percent dirt removal. The data are as follows:

Brand	Type	Temperature			
X	Powder	Hot	85	88	80
		Warm	82	83	85
	Liquid	Hot	78	75	72
		Warm	75	75	73
Y	Powder	Hot	90	92	92
		Warm	88	86	88
	Liquid	Hot	78	76	70
		Warm	76	77	76
Z	Powder	Hot	85	87	88
		Warm	76	74	78
	Liquid	Hot	60	70	68
		Warm	55	57	54

(a) Are there significant interaction effects at the $\alpha = 0.05$ level?

(b) Are there significant differences between the three brands of detergent?

(c) Which combination of factors would you prefer to use?

9.24 A scientist collects experimental data on the radius of a propellant grain, y, as a function of powder temperature, extrusion rate, and die temperature. Results of the three-factor experiment are as follows:

	Powder Temp			
	150		**190**	
	Die Temp		**Die Temp**	
Rate	**220**	**250**	**220**	**250**
12	82	124	88	129
24	114	157	121	164

Resources are not available to make repeated experimental trials at the eight combinations of factors. It is believed that extrusion rate does not interact with die temperature and that the three-factor interaction should be negligible. Thus, these two interactions may be pooled to produce a 2 d.f. "error" term.

(a) Do an analysis of variance that includes the three main effects and two two-factor interactions. Determine what effects influence the radius of the propellant grain.

(b) Construct interaction plots for the powder temperature by die temperature and powder temperature by extrusion rate interactions.

(c) Comment on the consistency in the appearance of the interaction plots and the tests on the two interactions in the ANOVA.

9.25 In the book *Design of Experiments for the Quality Improvement*, published by the Japanese Standards Association (1989), a study is reported on the extraction of polyethylene by using a solvent and how the amount of gel (proportion) is influenced by three factors: the type of solvent, extraction temperature, and extraction time. A factorial experiment was designed, and the following data were collected on proportion of gel.

		Time					
Solvent Temp.		4		8		16	
Ethanol	120	94.0	94.0	93.8	94.2	91.1	90.5
	80	95.3	95.1	94.9	95.3	92.5	92.4
Toluene	120	94.6	94.5	93.6	94.1	91.1	91.0
	80	95.4	95.4	95.6	96.0	92.1	92.1

(a) Do an analysis of variance and determine what factors and interactions influence the proportion of gel.

(b) Construct an interaction plot for any two-factor interaction that is significant. In addition, explain what conclusion can be drawn from the presence of the interaction.

(c) Do a normal probability plot of residuals and comment.

Review Exercises

9.26 The Statistics Consulting Center at Virginia Tech was involved in analyzing a set of data taken by personnel in the Human Nutrition and Foods Department in which it was of interest to study the effects of flour type and percent sweetener on certain physical attributes of a type of cake. All-purpose flour and cake flour were used, and the percent sweetener was varied at four levels. The following data show information on specific gravity of cake samples. Three cakes were prepared at each of the eight factor combinations.

Sweetener	Flour					
Concentration	All-Purpose			Cake		
0	0.90	0.87	0.90	0.91	0.90	0.80
50	0.86	0.89	0.91	0.88	0.82	0.83
75	0.93	0.88	0.87	0.86	0.85	0.80
100	0.79	0.82	0.80	0.86	0.85	0.85

(a) Treat the analysis as a two-factor analysis of variance. Test for differences between flour type. Test for differences between sweetener concentration.

(b) Discuss the effect of interaction, if any. Give P-values on all tests.

9.27 An experiment was conducted in the Department of Food Science at Virginia Tech in which it was of interest to characterize the texture of certain types of fish in the herring family. The effect of sauce types used in preparing the fish was also studied. The response in the experiment was "texture value," measured with a machine that sliced the fish product. Do the following using the provided data on texture values:

(a) Perform an ANOVA. Determine whether or not there is an interaction between sauce type and fish type.

(b) Based on your results from part (a) and on F-tests on main effects, determine if there is a significant

difference in texture due to sauce types, and determine whether there is a significant difference due to fish types.

	Fish Type					
	Unbleached	Bleached				
Sauce Type	Menhaden	Menhaden		Herring		
Sour Cream	27.6	57.4	64.0	66.9	107.0	83.9
	47.8	71.1	66.5	66.8	110.4	93.4
	53.8		53.8		83.1	
Wine Sauce	49.8	31.0	48.3	62.2	88.0	95.2
	11.8	35.1	54.6	43.6	108.2	86.7
	16.1		41.8		105.2	

9.28 A study was made to determine if humidity conditions have an effect on the force required to pull apart pieces of glued plastic. Three types of plastic were tested using 4 different levels of humidity. The results, in kilograms, are as follows:

	Humidity			
Plastic Type	30%	50%	70%	90%
A	39.0	33.1	33.8	33.0
	42.8	37.8	30.7	32.9
B	36.9	27.2	29.7	28.5
	41.0	26.8	29.1	27.9
C	27.4	29.2	26.7	30.9
	30.3	29.9	32.0	31.5

(a) Assuming a fixed effects experiment, perform an analysis of variance and test the hypothesis of no interaction between humidity and plastic type at the 0.05 level of significance.

(b) Using only plastics A and B and the value of s^2 from part (a), once again test for the presence of interaction at the 0.05 level of significance.

9.29 Personnel in the Materials Engineering Department at Virginia Tech conducted an experiment to study the effects of environmental factors on the stability of a certain type of copper-nickel alloy. The basic response was the fatigue life of the material. The factors are level of stress and environment. The data are as follows:

Environment	Stress Level		
	Low	Medium	High
Dry	11.08	13.12	14.18
Hydrogen	10.98	13.04	14.90
	11.24	13.37	15.10
High	10.75	12.73	14.15
Humidity	10.52	12.87	14.42
(95%)	10.43	12.95	14.25

(a) Do an analysis of variance to test for interaction between the factors. Use $\alpha = 0.05$.

(b) Based on part (a), do an analysis on the two main effects and draw conclusions. Use a P-value approach in drawing conclusions.

9.30 In the experiment of Review Exercise 9.26, cake volume was also used as a response. The units are cubic inches. Test for interaction between factors and discuss main effects. Assume that both factors are fixed effects.

Sweetener Concentration	Flour					
	All-Purpose			Cake		
0	4.48	3.98	4.42	4.12	4.92	5.10
50	3.68	5.04	3.72	5.00	4.26	4.34
75	3.92	3.82	4.06	4.82	4.34	4.40
100	3.26	3.80	3.40	4.32	4.18	4.30

9.31 A control valve needs to be very sensitive to the input voltage, thus generating a good output voltage. An engineer turns the control bolts to change the input voltage. The book *SN-Ratio for the Quality Evaluation*, published by the Japanese Standards Association (1988), described a study on how these three factors (relative position of control bolts, control range of bolts, and input voltage) affect the sensitivity of a control valve. The factors and their levels are shown below. The data show the sensitivity of a control valve.

Factor A, relative position of control bolts:
  center -0.5, center, and center $+0.5$
Factor B, control range of bolts:
  2, 4.5, and 7 (mm)
Factor C, input voltage:
  100, 120, and 150 (V)

Perform an analysis of variance with $\alpha = 0.05$ to test for significant main and interaction effects. Draw conclusions.

		C				
A	B	C_1		C_2		C_3
A_1	B_1	151 135	151 135	151 138
A_1	B_2	178 171	180 173	181 174
A_1	B_3	204 190	205 190	206 192
A_2	B_1	156 148	158 149	158 150
A_2	B_2	183 168	183 170	183 172
A_2	B_3	210 204	211 203	213 204
A_3	B_1	161 145	162 148	163 148
A_3	B_2	189 182	191 184	192 183
A_3	B_3	215 202	216 203	217 205

9.32 Exercise 9.25 on page 419 describes an experiment involving the extraction of polyethylene through use of a solvent.

Solvent	Temp.	Time		
		4	8	16
Ethanol	120	94.0 94.0	93.8 94.2	91.1 90.5
	80	95.3 95.1	94.9 95.3	92.5 92.4
Toluene	120	94.6 94.5	93.6 94.1	91.1 91.0
	80	95.4 95.4	95.6 96.0	92.1 92.1

(a) Do a different sort of analysis on the data. Fit an appropriate regression model with a solvent categorical variable, a temperature term, a time term, a temperature by time interaction, a solvent by temperature interaction, and a solvent by time interaction. Do t-tests on all coefficients and report your findings.

(b) Do your findings suggest that different models are appropriate for ethanol and toluene, or are they equivalent apart from the intercepts? Explain.

(c) Do you find any conclusions here that contradict conclusions drawn in your solution of Exercise 9.25? Explain.

9.33 In the book *SN-Ratio for the Quality Evaluation*, published by the Japanese Standards Association (1988), a study on how tire air pressure affects the maneuverability of an automobile was described. Three different tire air pressures were compared on three different driving surfaces. The three air pressures were both left- and right-side tires inflated to 6 kgf/cm^2, left-side tires inflated to 6 kgf/cm^2 and right-side tires inflated to 3 kgf/cm^2, and both left- and right-side tires inflated to 3 kgf/cm^2. The three driving surfaces were asphalt, dry asphalt, and dry cement. The turning radius of a test vehicle was observed twice for each level of tire pressure on each of the three different driving surfaces. Perform an analysis of variance of the data. Comment on the interpretation of the main and interaction effects.

Driving Surface	Tire Air Pressure		
	1	2	3
Asphalt	44.0 25.5	34.2 37.2	27.4 42.8
Dry Asphalt	31.9 33.7	31.8 27.6	43.7 38.2
Dry Cement	27.3 39.5	46.6 28.1	35.5 34.6

9.34 The manufacturer of a certain brand of freeze-dried coffee hopes to shorten the process time without jeopardizing the integrity of the product. The process engineer wants to use 3 temperatures for the drying chamber and 4 drying times. The current drying time is 3 hours at a temperature of $-15°C$. The flavor response is an average of scores of 4 professional judges. The score is on a scale from 1 to 10, with 10 being the best. The data are as shown in the following table.

	Temperature		
Time	**$-20°C$**	**$-15°C$**	**$-10°C$**
1 hr	9.60 9.63	9.55 9.50	9.40 9.43
1.5 hr	9.75 9.73	9.60 9.61	9.55 9.48
2 hr	9.82 9.93	9.81 9.78	9.50 9.52
3 hr	9.78 9.81	9.80 9.75	9.55 9.58

(a) What type of model should be used? State assumptions.

(b) Analyze the data appropriately.

(c) Write a brief report to the vice-president in charge and make a recommendation for future manufacturing of this product.

9.35 To ascertain the number of tellers needed during peak hours of operation, data were collected by an urban bank. Four tellers were studied during three busy times: (1) weekdays between 10:00 and 11:00 A.M., (2) weekday afternoons between 2:00 and 3:00 P.M., and (3) Saturday mornings between 11:00 A.M. and 12:00 noon. An analyst chose four randomly selected times within each of the three time periods for each of the four teller positions over a period of months, and the numbers of customers serviced were observed. The data are as follows:

	Time Period		
Teller	**1**	**2**	**3**
1	18 24 17 22	25 29 23 32	29 30 21 34
2	16 11 19 14	23 32 25 17	27 29 18 16
3	12 19 11 22	27 33 27 24	25 20 29 15
4	11 9 13 8	10 7 19 8	11 9 17 9

It is assumed that the number of customers served is a Poisson random variable.

(a) Discuss the danger in doing a standard analysis of variance on the data above. What assumptions, if any, would be violated?

(b) Construct a standard ANOVA table that includes F-tests on main effects and interactions. If interactions and main effects are found to be significant, give scientific conclusions. What have we learned? Be sure to interpret any significant interaction. Use your own judgment regarding P-values.

9.5   Potential Misconceptions and Hazards; Relationship to Material in Other Chapters

One of the most confusing issues in the analysis of factorial experiments resides in the interpretation of main effects in the presence of interaction. The presence of a relatively large P-value for a main effect when interactions are clearly present may tempt the analyst to conclude "no significant main effect." However, one must understand that if a main effect is involved in a significant interaction, then the main effect is **influencing the response**. The nature of the effect is inconsistent across levels of other effects.

In light of what is communicated in the preceding paragraph, there is danger of a substantial misuse of statistics when one employs a multiple comparison test on main effects in the clear presence of interaction among the factors.

One must be cautious in the analysis of a factorial experiment when the assumption of a completely randomized design is made when in fact complete randomization is not carried out. For example, it is common to encounter factors that are very **difficult to change**. As a result, factor levels may need to be held without change for long periods of time throughout the experiment. A temperature factor is a common example. Moving temperature up and down in a randomization scheme is a costly plan, and most experimenters will refuse to do it. Experimental designs with *restrictions in randomization* are quite common and are called **split plot designs**. They are beyond the scope of the book, but presentations are found in Montgomery (2008a).

Bibliography

[1] Bartlett, M. S., and Kendall, D. G. (1946). "The Statistical Analysis of Variance Heterogeneity and Logarithmic Transformation," *Journal of the Royal Statistical Society*, Ser. B, **8**, 128–138.

[2] Bowker, A. H., and Lieberman, G. J. (1972). *Engineering Statistics*, 2nd ed. Upper Saddle River, N.J.: Prentice Hall.

[3] Box, G. E. P., Hunter, W. G., and Hunter, J. S. (1978). *Statistics for Experimenters*. New York: John Wiley & Sons.

[4] Brownlee, K. A. (1984). *Statistical Theory and Methodology in Science and Engineering*, 2nd ed. New York: John Wiley & Sons.

[5] Chatterjee, S., Hadi, A. S., and Price, B. (1999). *Regression Analysis by Example*, 3rd ed. New York: John Wiley & Sons.

[6] Cook, R. D., and Weisberg, S. (1982). *Residuals and Influence in Regression*. New York: Chapman and Hall.

[7] Draper, N. R., and Smith, H. (1998). *Applied Regression Analysis*, 3rd ed. New York: John Wiley & Sons.

[8] Dyer, D. D., and Keating, J. P. (1980). "On the Determination of Critical Values for Bartlett's Test," *Journal of the American Statistical Association*, **75**, 313–319.

[9] Geary, R.C. (1947). "Testing for Normality," *Biometrika*, **34**, 209–242.

[10] Gunst, R. F., and Mason, R. L. (1980). *Regression Analysis and Its Application: A Data-Oriented Approach*. New York: Marcel Dekker.

[11] Guttman, I., Wilks, S. S., and Hunter, J. S. (1971). *Introductory Engineering Statistics*. New York: John Wiley & Sons.

[12] Harville, D. A. (1977). "Maximum Likelihood Approaches to Variance Component Estimation and to Related Problems," *Journal of the American Statistical Association*, **72**, 320–338.

[13] Hicks, C. R., and Turner, K. V. (1999). *Fundamental Concepts in the Design of Experiments*, 5th ed. Oxford: Oxford University Press.

[14] Hoaglin, D. C., Mosteller, F., and Tukey, J. W. (1991). *Fundamentals of Exploratory Analysis of Variance*. New York: John Wiley & Sons.

[15] Hocking, R. R. (1976). "The Analysis and Selection of Variables in Linear Regression," *Biometrics*, **32**, 1–49.

[16] Hodges, J. L., and Lehmann, E. L. (2005). *Basic Concepts of Probability and Statistics*, 2nd ed. Philadelphia: Society for Industrial and Applied Mathematics.

[17] Hogg, R. V., and Ledolter, J. (1992). *Applied Statistics for Engineers and Physical Scientists*, 2nd ed. Upper Saddle River, N.J.: Prentice Hall.

[18] Hogg, R. V., McKean, J. W., and Craig, A. (2005). *Introduction to Mathematical Statistics*, 6th ed. Upper Saddle River, N.J.: Prentice Hall.

[19] Johnson, N. L., and Leone, F. C. (1977). *Statistics and Experimental Design in Engineering and the Physical Sciences*, 2nd ed. Vols. I and II. New York: John Wiley & Sons.

[20] Koopmans, L. H. (1987). *An Introduction to Contemporary Statistics*, 2nd ed. Boston: Duxbury Press.

[21] Kutner, M. H., Nachtsheim, C. J., Neter, J., and Li, W. (2004). *Applied Linear Regression Models*, 5th ed. New York: McGraw-Hill/Irwin.

[22] Larsen, R. J., and Morris, M. L. (2000). *An Introduction to Mathematical Statistics and Its Applications*, 3rd ed. Upper Saddle River, N.J.: Prentice Hall.

[23] Lentner, M., and Bishop, T. (1986). *Design and Analysis of Experiments*, 2nd ed. Blacksburg, Va.: Valley Book Co.

[24] Mallows, C. L. (1973). "Some Comments on C_p," *Technometrics*, **15**, 661–675.

[25] McClave, J. T., Dietrich, F. H., and Sincich, T. (1997). *Statistics*, 7th ed. Upper Saddle River, N.J.: Prentice Hall.

[26] Montgomery, D. C. (2008a). *Design and Analysis of Experiments*, 7th ed. New York: John Wiley & Sons.

[27] Montgomery, D. C. (2008b). *Introduction to Statistical Quality Control*, 6th ed. New York: John Wiley & Sons.

[28] Mosteller, F., and Tukey, J. (1977). *Data Analysis and Regression*. Reading, Mass.: Addison-Wesley Publishing Co.

[29] Myers, R. H. (1990). *Classical and Modern Regression with Applications*, 2nd ed. Boston: Duxbury Press.

[30] Myers, R. H., Montgomery, D. C., and Anderson-Cook, C. M. (2009). *Response Surface Methodology: Process and Product Optimization Using Designed Experiments*, 3rd ed. New York: John Wiley & Sons.

[31] Myers, R. H., Montgomery, D. C., Vining, G. G., and Robinson, T. J. (2008). *Generalized Linear Models with Applications in Engineering and the Sciences*, 2nd ed. New York: John Wiley & Sons.

[32] Olkin, I., Gleser, L. J., and Derman, C. (1994). *Probability Models and Applications*, 2nd ed. New York: Prentice Hall.

[33] Ott, R. L., and Longnecker, M. T. (2000). *An Introduction to Statistical Methods and Data Analysis*, 5th ed. Boston: Duxbury Press.

[34] Pacansky, J., England, C. D., and Wattman, R. (1986). "Infrared Spectroscopic Studies of Poly (perfluoropropyleneoxide) on Gold Substrate: A Classical Dispersion Analysis for the Refractive Index," *Applied Spectroscopy*, **40**, 8–16.

[35] Ross, S. M. (2002). *Introduction to Probability Models*, 9th ed. New York: Academic Press, Inc.

[36] Satterthwaite, F. E. (1946). "An Approximate Distribution of Estimates of Variance Components," *Biometrics*, **2**, 110–114.

[37] Snedecor, G. W., and Cochran, W. G. (1989). *Statistical Methods*, 8th ed. Ames, Iowa: The Iowa State University Press.

[38] Steel, R. G. D., Torrie, J. H., and Dickey, D. A. (1996). *Principles and Procedures of Statistics: A Biometrical Approach*, 3rd ed. New York: McGraw-Hill.

[39] Thompson, W. O., and Cady, F. B. (1973). *Proceedings of the University of Kentucky Conference on Regression with a Large Number of Predictor Variables*. Lexington, Ken.: University of Kentucky Press.

[40] Tukey, J. W. (1977). *Exploratory Data Analysis*. Reading, Mass.: Addison-Wesley Publishing Co.

[41] Walpole, R. E., Myers, R. H., Myers, S. L., and Ye, K. (2011). *Probability & Statistics for Engineers & Scientists*, 9th ed. New York: Prentice Hall.

[42] Welch, W. J., Yu, T. K., Kang, S. M., and Sacks, J. (1990). "Computer Experiments for Quality Control by Parameter Design," *Journal of Quality Technology*, **22**, 15–22.

Appendix A
Statistical Tables and Proofs

Table A.1 Binomial Probability Sums $\sum_{x=0}^{r} b(x; n, p)$

							p				
n	r	0.10	0.20	0.25	0.30	0.40	0.50	0.60	0.70	0.80	0.90
1	0	0.9000	0.8000	0.7500	0.7000	0.6000	0.5000	0.4000	0.3000	0.2000	0.1000
	1	1.0000	1.0000	1.0000	1.0000	1.0000	1.0000	1.0000	1.0000	1.0000	1.0000
2	0	0.8100	0.6400	0.5625	0.4900	0.3600	0.2500	0.1600	0.0900	0.0400	0.0100
	1	0.9900	0.9600	0.9375	0.9100	0.8400	0.7500	0.6400	0.5100	0.3600	0.1900
	2	1.0000	1.0000	1.0000	1.0000	1.0000	1.0000	1.0000	1.0000	1.0000	1.0000
3	0	0.7290	0.5120	0.4219	0.3430	0.2160	0.1250	0.0640	0.0270	0.0080	0.0010
	1	0.9720	0.8960	0.8438	0.7840	0.6480	0.5000	0.3520	0.2160	0.1040	0.0280
	2	0.9990	0.9920	0.9844	0.9730	0.9360	0.8750	0.7840	0.6570	0.4880	0.2710
	3	1.0000	1.0000	1.0000	1.0000	1.0000	1.0000	1.0000	1.0000	1.0000	1.0000
4	0	0.6561	0.4096	0.3164	0.2401	0.1296	0.0625	0.0256	0.0081	0.0016	0.0001
	1	0.9477	0.8192	0.7383	0.6517	0.4752	0.3125	0.1792	0.0837	0.0272	0.0037
	2	0.9963	0.9728	0.9492	0.9163	0.8208	0.6875	0.5248	0.3483	0.1808	0.0523
	3	0.9999	0.9984	0.9961	0.9919	0.9744	0.9375	0.8704	0.7599	0.5904	0.3439
	4	1.0000	1.0000	1.0000	1.0000	1.0000	1.0000	1.0000	1.0000	1.0000	1.0000
5	0	0.5905	0.3277	0.2373	0.1681	0.0778	0.0313	0.0102	0.0024	0.0003	0.0000
	1	0.9185	0.7373	0.6328	0.5282	0.3370	0.1875	0.0870	0.0308	0.0067	0.0005
	2	0.9914	0.9421	0.8965	0.8369	0.6826	0.5000	0.3174	0.1631	0.0579	0.0086
	3	0.9995	0.9933	0.9844	0.9692	0.9130	0.8125	0.6630	0.4718	0.2627	0.0815
	4	1.0000	0.9997	0.9990	0.9976	0.9898	0.9688	0.9222	0.8319	0.6723	0.4095
	5	1.0000	1.0000	1.0000	1.0000	1.0000	1.0000	1.0000	1.0000	1.0000	1.0000
6	0	0.5314	0.2621	0.1780	0.1176	0.0467	0.0156	0.0041	0.0007	0.0001	0.0000
	1	0.8857	0.6554	0.5339	0.4202	0.2333	0.1094	0.0410	0.0109	0.0016	0.0001
	2	0.9842	0.9011	0.8306	0.7443	0.5443	0.3438	0.1792	0.0705	0.0170	0.0013
	3	0.9987	0.9830	0.9624	0.9295	0.8208	0.6563	0.4557	0.2557	0.0989	0.0159
	4	0.9999	0.9984	0.9954	0.9891	0.9590	0.8906	0.7667	0.5798	0.3446	0.1143
	5	1.0000	0.9999	0.9998	0.9993	0.9959	0.9844	0.9533	0.8824	0.7379	0.4686
	6	1.0000	1.0000	1.0000	1.0000	1.0000	1.0000	1.0000	1.0000	1.0000	1.0000
7	0	0.4783	0.2097	0.1335	0.0824	0.0280	0.0078	0.0016	0.0002	0.0000	
	1	0.8503	0.5767	0.4449	0.3294	0.1586	0.0625	0.0188	0.0038	0.0004	0.0000
	2	0.9743	0.8520	0.7564	0.6471	0.4199	0.2266	0.0963	0.0288	0.0047	0.0002
	3	0.9973	0.9667	0.9294	0.8740	0.7102	0.5000	0.2898	0.1260	0.0333	0.0027
	4	0.9998	0.9953	0.9871	0.9712	0.9037	0.7734	0.5801	0.3529	0.1480	0.0257
	5	1.0000	0.9996	0.9987	0.9962	0.9812	0.9375	0.8414	0.6706	0.4233	0.1497
	6		1.0000	0.9999	0.9998	0.9984	0.9922	0.9720	0.9176	0.7903	0.5217
	7			1.0000	1.0000	1.0000	1.0000	1.0000	1.0000	1.0000	1.0000

Table A.1 (continued) Binomial Probability Sums $\sum_{x=0}^{r} b(x; n, p)$

						p					
n	r	0.10	0.20	0.25	0.30	0.40	0.50	0.60	0.70	0.80	0.90
8	0	0.4305	0.1678	0.1001	0.0576	0.0168	0.0039	0.0007	0.0001	0.0000	
	1	0.8131	0.5033	0.3671	0.2553	0.1064	0.0352	0.0085	0.0013	0.0001	
	2	0.9619	0.7969	0.6785	0.5518	0.3154	0.1445	0.0498	0.0113	0.0012	0.0000
	3	0.9950	0.9437	0.8862	0.8059	0.5941	0.3633	0.1737	0.0580	0.0104	0.0004
	4	0.9996	0.9896	0.9727	0.9420	0.8263	0.6367	0.4059	0.1941	0.0563	0.0050
	5	1.0000	0.9988	0.9958	0.9887	0.9502	0.8555	0.6846	0.4482	0.2031	0.0381
	6		0.9999	0.9996	0.9987	0.9915	0.9648	0.8936	0.7447	0.4967	0.1869
	7		1.0000	1.0000	0.9999	0.9993	0.9961	0.9832	0.9424	0.8322	0.5695
	8				1.0000	1.0000	1.0000	1.0000	1.0000	1.0000	1.0000
9	0	0.3874	0.1342	0.0751	0.0404	0.0101	0.0020	0.0003	0.0000		
	1	0.7748	0.4362	0.3003	0.1960	0.0705	0.0195	0.0038	0.0004	0.0000	
	2	0.9470	0.7382	0.6007	0.4628	0.2318	0.0898	0.0250	0.0043	0.0003	0.0000
	3	0.9917	0.9144	0.8343	0.7297	0.4826	0.2539	0.0994	0.0253	0.0031	0.0001
	4	0.9991	0.9804	0.9511	0.9012	0.7334	0.5000	0.2666	0.0988	0.0196	0.0009
	5	0.9999	0.9969	0.9900	0.9747	0.9006	0.7461	0.5174	0.2703	0.0856	0.0083
	6	1.0000	0.9997	0.9987	0.9957	0.9750	0.9102	0.7682	0.5372	0.2618	0.0530
	7		1.0000	0.9999	0.9996	0.9962	0.9805	0.9295	0.8040	0.5638	0.2252
	8			1.0000	1.0000	0.9997	0.9980	0.9899	0.9596	0.8658	0.6126
	9					1.0000	1.0000	1.0000	1.0000	1.0000	1.0000
10	0	0.3487	0.1074	0.0563	0.0282	0.0060	0.0010	0.0001	0.0000		
	1	0.7361	0.3758	0.2440	0.1493	0.0464	0.0107	0.0017	0.0001	0.0000	
	2	0.9298	0.6778	0.5256	0.3828	0.1673	0.0547	0.0123	0.0016	0.0001	
	3	0.9872	0.8791	0.7759	0.6496	0.3823	0.1719	0.0548	0.0106	0.0009	0.0000
	4	0.9984	0.9672	0.9219	0.8497	0.6331	0.3770	0.1662	0.0473	0.0064	0.0001
	5	0.9999	0.9936	0.9803	0.9527	0.8338	0.6230	0.3669	0.1503	0.0328	0.0016
	6	1.0000	0.9991	0.9965	0.9894	0.9452	0.8281	0.6177	0.3504	0.1209	0.0128
	7		0.9999	0.9996	0.9984	0.9877	0.9453	0.8327	0.6172	0.3222	0.0702
	8		1.0000	1.0000	0.9999	0.9983	0.9893	0.9536	0.8507	0.6242	0.2639
	9				1.0000	0.9999	0.9990	0.9940	0.9718	0.8926	0.6513
	10					1.0000	1.0000	1.0000	1.0000	1.0000	1.0000
11	0	0.3138	0.0859	0.0422	0.0198	0.0036	0.0005	0.0000			
	1	0.6974	0.3221	0.1971	0.1130	0.0302	0.0059	0.0007	0.0000		
	2	0.9104	0.6174	0.4552	0.3127	0.1189	0.0327	0.0059	0.0006	0.0000	
	3	0.9815	0.8389	0.7133	0.5696	0.2963	0.1133	0.0293	0.0043	0.0002	
	4	0.9972	0.9496	0.8854	0.7897	0.5328	0.2744	0.0994	0.0216	0.0020	0.0000
	5	0.9997	0.9883	0.9657	0.9218	0.7535	0.5000	0.2465	0.0782	0.0117	0.0003
	6	1.0000	0.9980	0.9924	0.9784	0.9006	0.7256	0.4672	0.2103	0.0504	0.0028
	7		0.9998	0.9988	0.9957	0.9707	0.8867	0.7037	0.4304	0.1611	0.0185
	8		1.0000	0.9999	0.9994	0.9941	0.9673	0.8811	0.6873	0.3826	0.0896
	9			1.0000	1.0000	0.9993	0.9941	0.9698	0.8870	0.6779	0.3026
	10					1.0000	0.9995	0.9964	0.9802	0.9141	0.6862
	11						1.0000	1.0000	1.0000	1.0000	1.0000

Table A.1 (continued) Binomial Probability Sums $\sum\limits_{x=0}^{r} b(x; n, p)$

							p				
n	r	0.10	0.20	0.25	0.30	0.40	0.50	0.60	0.70	0.80	0.90
12	0	0.2824	0.0687	0.0317	0.0138	0.0022	0.0002	0.0000			
	1	0.6590	0.2749	0.1584	0.0850	0.0196	0.0032	0.0003	0.0000		
	2	0.8891	0.5583	0.3907	0.2528	0.0834	0.0193	0.0028	0.0002	0.0000	
	3	0.9744	0.7946	0.6488	0.4925	0.2253	0.0730	0.0153	0.0017	0.0001	
	4	0.9957	0.9274	0.8424	0.7237	0.4382	0.1938	0.0573	0.0095	0.0006	0.0000
	5	0.9995	0.9806	0.9456	0.8822	0.6652	0.3872	0.1582	0.0386	0.0039	0.0001
	6	0.9999	0.9961	0.9857	0.9614	0.8418	0.6128	0.3348	0.1178	0.0194	0.0005
	7	1.0000	0.9994	0.9972	0.9905	0.9427	0.8062	0.5618	0.2763	0.0726	0.0043
	8		0.9999	0.9996	0.9983	0.9847	0.9270	0.7747	0.5075	0.2054	0.0256
	9		1.0000	1.0000	0.9998	0.9972	0.9807	0.9166	0.7472	0.4417	0.1109
	10				1.0000	0.9997	0.9968	0.9804	0.9150	0.7251	0.3410
	11					1.0000	0.9998	0.9978	0.9862	0.9313	0.7176
	12						1.0000	1.0000	1.0000	1.0000	1.0000
13	0	0.2542	0.0550	0.0238	0.0097	0.0013	0.0001	0.0000			
	1	0.6213	0.2336	0.1267	0.0637	0.0126	0.0017	0.0001	0.0000		
	2	0.8661	0.5017	0.3326	0.2025	0.0579	0.0112	0.0013	0.0001		
	3	0.9658	0.7473	0.5843	0.4206	0.1686	0.0461	0.0078	0.0007	0.0000	
	4	0.9935	0.9009	0.7940	0.6543	0.3530	0.1334	0.0321	0.0040	0.0002	
	5	0.9991	0.9700	0.9198	0.8346	0.5744	0.2905	0.0977	0.0182	0.0012	0.0000
	6	0.9999	0.9930	0.9757	0.9376	0.7712	0.5000	0.2288	0.0624	0.0070	0.0001
	7	1.0000	0.9988	0.9944	0.9818	0.9023	0.7095	0.4256	0.1654	0.0300	0.0009
	8		0.9998	0.9990	0.9960	0.9679	0.8666	0.6470	0.3457	0.0991	0.0065
	9		1.0000	0.9999	0.9993	0.9922	0.9539	0.8314	0.5794	0.2527	0.0342
	10			1.0000	0.9999	0.9987	0.9888	0.9421	0.7975	0.4983	0.1339
	11				1.0000	0.9999	0.9983	0.9874	0.9363	0.7664	0.3787
	12					1.0000	0.9999	0.9987	0.9903	0.9450	0.7458
	13						1.0000	1.0000	1.0000	1.0000	1.0000
14	0	0.2288	0.0440	0.0178	0.0068	0.0008	0.0001	0.0000			
	1	0.5846	0.1979	0.1010	0.0475	0.0081	0.0009	0.0001			
	2	0.8416	0.4481	0.2811	0.1608	0.0398	0.0065	0.0006	0.0000		
	3	0.9559	0.6982	0.5213	0.3552	0.1243	0.0287	0.0039	0.0002		
	4	0.9908	0.8702	0.7415	0.5842	0.2793	0.0898	0.0175	0.0017	0.0000	
	5	0.9985	0.9561	0.8883	0.7805	0.4859	0.2120	0.0583	0.0083	0.0004	
	6	0.9998	0.9884	0.9617	0.9067	0.6925	0.3953	0.1501	0.0315	0.0024	0.0000
	7	1.0000	0.9976	0.9897	0.9685	0.8499	0.6047	0.3075	0.0933	0.0116	0.0002
	8		0.9996	0.9978	0.9917	0.9417	0.7880	0.5141	0.2195	0.0439	0.0015
	9		1.0000	0.9997	0.9983	0.9825	0.9102	0.7207	0.4158	0.1298	0.0092
	10			1.0000	0.9998	0.9961	0.9713	0.8757	0.6448	0.3018	0.0441
	11				1.0000	0.9994	0.9935	0.9602	0.8392	0.5519	0.1584
	12					0.9999	0.9991	0.9919	0.9525	0.8021	0.4154
	13					1.0000	0.9999	0.9992	0.9932	0.9560	0.7712
	14						1.0000	1.0000	1.0000	1.0000	1.0000

Table A.1 (continued) Binomial Probability Sums $\sum_{x=0}^{r} b(x; n, p)$

n	r	\multicolumn{10}{c}{p}
		0.10	0.20	0.25	0.30	0.40	0.50	0.60	0.70	0.80	0.90
15	0	0.2059	0.0352	0.0134	0.0047	0.0005	0.0000				
	1	0.5490	0.1671	0.0802	0.0353	0.0052	0.0005	0.0000			
	2	0.8159	0.3980	0.2361	0.1268	0.0271	0.0037	0.0003	0.0000		
	3	0.9444	0.6482	0.4613	0.2969	0.0905	0.0176	0.0019	0.0001		
	4	0.9873	0.8358	0.6865	0.5155	0.2173	0.0592	0.0093	0.0007	0.0000	
	5	0.9978	0.9389	0.8516	0.7216	0.4032	0.1509	0.0338	0.0037	0.0001	
	6	0.9997	0.9819	0.9434	0.8689	0.6098	0.3036	0.0950	0.0152	0.0008	
	7	1.0000	0.9958	0.9827	0.9500	0.7869	0.5000	0.2131	0.0500	0.0042	0.0000
	8		0.9992	0.9958	0.9848	0.9050	0.6964	0.3902	0.1311	0.0181	0.0003
	9		0.9999	0.9992	0.9963	0.9662	0.8491	0.5968	0.2784	0.0611	0.0022
	10		1.0000	0.9999	0.9993	0.9907	0.9408	0.7827	0.4845	0.1642	0.0127
	11			1.0000	0.9999	0.9981	0.9824	0.9095	0.7031	0.3518	0.0556
	12				1.0000	0.9997	0.9963	0.9729	0.8732	0.6020	0.1841
	13					1.0000	0.9995	0.9948	0.9647	0.8329	0.4510
	14						1.0000	0.9995	0.9953	0.9648	0.7941
	15							1.0000	1.0000	1.0000	1.0000
16	0	0.1853	0.0281	0.0100	0.0033	0.0003	0.0000				
	1	0.5147	0.1407	0.0635	0.0261	0.0033	0.0003	0.0000			
	2	0.7892	0.3518	0.1971	0.0994	0.0183	0.0021	0.0001			
	3	0.9316	0.5981	0.4050	0.2459	0.0651	0.0106	0.0009	0.0000		
	4	0.9830	0.7982	0.6302	0.4499	0.1666	0.0384	0.0049	0.0003		
	5	0.9967	0.9183	0.8103	0.6598	0.3288	0.1051	0.0191	0.0016	0.0000	
	6	0.9995	0.9733	0.9204	0.8247	0.5272	0.2272	0.0583	0.0071	0.0002	
	7	0.9999	0.9930	0.9729	0.9256	0.7161	0.4018	0.1423	0.0257	0.0015	0.0000
	8	1.0000	0.9985	0.9925	0.9743	0.8577	0.5982	0.2839	0.0744	0.0070	0.0001
	9		0.9998	0.9984	0.9929	0.9417	0.7728	0.4728	0.1753	0.0267	0.0005
	10		1.0000	0.9997	0.9984	0.9809	0.8949	0.6712	0.3402	0.0817	0.0033
	11			1.0000	0.9997	0.9951	0.9616	0.8334	0.5501	0.2018	0.0170
	12				1.0000	0.9991	0.9894	0.9349	0.7541	0.4019	0.0684
	13					0.9999	0.9979	0.9817	0.9006	0.6482	0.2108
	14					1.0000	0.9997	0.9967	0.9739	0.8593	0.4853
	15						1.0000	0.9997	0.9967	0.9719	0.8147
	16							1.0000	1.0000	1.0000	1.0000

Table A.1 (continued) Binomial Probability Sums $\sum\limits_{x=0}^{r} b(x; n, p)$

							p				
n	r	0.10	0.20	0.25	0.30	0.40	0.50	0.60	0.70	0.80	0.90
17	0	0.1668	0.0225	0.0075	0.0023	0.0002	0.0000				
	1	0.4818	0.1182	0.0501	0.0193	0.0021	0.0001	0.0000			
	2	0.7618	0.3096	0.1637	0.0774	0.0123	0.0012	0.0001			
	3	0.9174	0.5489	0.3530	0.2019	0.0464	0.0064	0.0005	0.0000		
	4	0.9779	0.7582	0.5739	0.3887	0.1260	0.0245	0.0025	0.0001		
	5	0.9953	0.8943	0.7653	0.5968	0.2639	0.0717	0.0106	0.0007	0.0000	
	6	0.9992	0.9623	0.8929	0.7752	0.4478	0.1662	0.0348	0.0032	0.0001	
	7	0.9999	0.9891	0.9598	0.8954	0.6405	0.3145	0.0919	0.0127	0.0005	
	8	1.0000	0.9974	0.9876	0.9597	0.8011	0.5000	0.1989	0.0403	0.0026	0.0000
	9		0.9995	0.9969	0.9873	0.9081	0.6855	0.3595	0.1046	0.0109	0.0001
	10		0.9999	0.9994	0.9968	0.9652	0.8338	0.5522	0.2248	0.0377	0.0008
	11		1.0000	0.9999	0.9993	0.9894	0.9283	0.7361	0.4032	0.1057	0.0047
	12			1.0000	0.9999	0.9975	0.9755	0.8740	0.6113	0.2418	0.0221
	13				1.0000	0.9995	0.9936	0.9536	0.7981	0.4511	0.0826
	14					0.9999	0.9988	0.9877	0.9226	0.6904	0.2382
	15					1.0000	0.9999	0.9979	0.9807	0.8818	0.5182
	16						1.0000	0.9998	0.9977	0.9775	0.8332
	17							1.0000	1.0000	1.0000	1.0000
18	0	0.1501	0.0180	0.0056	0.0016	0.0001	0.0000				
	1	0.4503	0.0991	0.0395	0.0142	0.0013	0.0001				
	2	0.7338	0.2713	0.1353	0.0600	0.0082	0.0007	0.0000			
	3	0.9018	0.5010	0.3057	0.1646	0.0328	0.0038	0.0002			
	4	0.9718	0.7164	0.5187	0.3327	0.0942	0.0154	0.0013	0.0000		
	5	0.9936	0.8671	0.7175	0.5344	0.2088	0.0481	0.0058	0.0003		
	6	0.9988	0.9487	0.8610	0.7217	0.3743	0.1189	0.0203	0.0014	0.0000	
	7	0.9998	0.9837	0.9431	0.8593	0.5634	0.2403	0.0576	0.0061	0.0002	
	8	1.0000	0.9957	0.9807	0.9404	0.7368	0.4073	0.1347	0.0210	0.0009	
	9		0.9991	0.9946	0.9790	0.8653	0.5927	0.2632	0.0596	0.0043	0.0000
	10		0.9998	0.9988	0.9939	0.9424	0.7597	0.4366	0.1407	0.0163	0.0002
	11		1.0000	0.9998	0.9986	0.9797	0.8811	0.6257	0.2783	0.0513	0.0012
	12			1.0000	0.9997	0.9942	0.9519	0.7912	0.4656	0.1329	0.0064
	13				1.0000	0.9987	0.9846	0.9058	0.6673	0.2836	0.0282
	14					0.9998	0.9962	0.9672	0.8354	0.4990	0.0982
	15					1.0000	0.9993	0.9918	0.9400	0.7287	0.2662
	16						0.9999	0.9987	0.9858	0.9009	0.5497
	17						1.0000	0.9999	0.9984	0.9820	0.8499
	18							1.0000	1.0000	1.0000	1.0000

Table A.1 (continued) Binomial Probability Sums $\sum_{x=0}^{r} b(x; n, p)$

							p				
n	r	0.10	0.20	0.25	0.30	0.40	0.50	0.60	0.70	0.80	0.90
19	0	0.1351	0.0144	0.0042	0.0011	0.0001					
	1	0.4203	0.0829	0.0310	0.0104	0.0008	0.0000				
	2	0.7054	0.2369	0.1113	0.0462	0.0055	0.0004	0.0000			
	3	0.8850	0.4551	0.2631	0.1332	0.0230	0.0022	0.0001			
	4	0.9648	0.6733	0.4654	0.2822	0.0696	0.0096	0.0006	0.0000		
	5	0.9914	0.8369	0.6678	0.4739	0.1629	0.0318	0.0031	0.0001		
	6	0.9983	0.9324	0.8251	0.6655	0.3081	0.0835	0.0116	0.0006		
	7	0.9997	0.9767	0.9225	0.8180	0.4878	0.1796	0.0352	0.0028	0.0000	
	8	1.0000	0.9933	0.9713	0.9161	0.6675	0.3238	0.0885	0.0105	0.0003	
	9		0.9984	0.9911	0.9674	0.8139	0.5000	0.1861	0.0326	0.0016	
	10		0.9997	0.9977	0.9895	0.9115	0.6762	0.3325	0.0839	0.0067	0.0000
	11		1.0000	0.9995	0.9972	0.9648	0.8204	0.5122	0.1820	0.0233	0.0003
	12			0.9999	0.9994	0.9884	0.9165	0.6919	0.3345	0.0676	0.0017
	13			1.0000	0.9999	0.9969	0.9682	0.8371	0.5261	0.1631	0.0086
	14				1.0000	0.9994	0.9904	0.9304	0.7178	0.3267	0.0352
	15					0.9999	0.9978	0.9770	0.8668	0.5449	0.1150
	16					1.0000	0.9996	0.9945	0.9538	0.7631	0.2946
	17						1.0000	0.9992	0.9896	0.9171	0.5797
	18							0.9999	0.9989	0.9856	0.8649
	19							1.0000	1.0000	1.0000	1.0000
20	0	0.1216	0.0115	0.0032	0.0008	0.0000					
	1	0.3917	0.0692	0.0243	0.0076	0.0005	0.0000				
	2	0.6769	0.2061	0.0913	0.0355	0.0036	0.0002				
	3	0.8670	0.4114	0.2252	0.1071	0.0160	0.0013	0.0000			
	4	0.9568	0.6296	0.4148	0.2375	0.0510	0.0059	0.0003			
	5	0.9887	0.8042	0.6172	0.4164	0.1256	0.0207	0.0016	0.0000		
	6	0.9976	0.9133	0.7858	0.6080	0.2500	0.0577	0.0065	0.0003		
	7	0.9996	0.9679	0.8982	0.7723	0.4159	0.1316	0.0210	0.0013	0.0000	
	8	0.9999	0.9900	0.9591	0.8867	0.5956	0.2517	0.0565	0.0051	0.0001	
	9	1.0000	0.9974	0.9861	0.9520	0.7553	0.4119	0.1275	0.0171	0.0006	
	10		0.9994	0.9961	0.9829	0.8725	0.5881	0.2447	0.0480	0.0026	0.0000
	11		0.9999	0.9991	0.9949	0.9435	0.7483	0.4044	0.1133	0.0100	0.0001
	12		1.0000	0.9998	0.9987	0.9790	0.8684	0.5841	0.2277	0.0321	0.0004
	13			1.0000	0.9997	0.9935	0.9423	0.7500	0.3920	0.0867	0.0024
	14				1.0000	0.9984	0.9793	0.8744	0.5836	0.1958	0.0113
	15					0.9997	0.9941	0.9490	0.7625	0.3704	0.0432
	16					1.0000	0.9987	0.9840	0.8929	0.5886	0.1330
	17						0.9998	0.9964	0.9645	0.7939	0.3231
	18						1.0000	0.9995	0.9924	0.9308	0.6083
	19							1.0000	0.9992	0.9885	0.8784
	20								1.0000	1.0000	1.0000

Table A.2 Poisson Probability Sums $\sum_{x=0}^{r} p(x; \mu)$

					μ				
r	0.1	0.2	0.3	0.4	0.5	0.6	0.7	0.8	0.9
0	0.9048	0.8187	0.7408	0.6703	0.6065	0.5488	0.4966	0.4493	0.4066
1	0.9953	0.9825	0.9631	0.9384	0.9098	0.8781	0.8442	0.8088	0.7725
2	0.9998	0.9989	0.9964	0.9921	0.9856	0.9769	0.9659	0.9526	0.9371
3	1.0000	0.9999	0.9997	0.9992	0.9982	0.9966	0.9942	0.9909	0.9865
4		1.0000	1.0000	0.9999	0.9998	0.9996	0.9992	0.9986	0.9977
5				1.0000	1.0000	1.0000	0.9999	0.9998	0.9997
6							1.0000	1.0000	1.0000

					μ				
r	1.0	1.5	2.0	2.5	3.0	3.5	4.0	4.5	5.0
0	0.3679	0.2231	0.1353	0.0821	0.0498	0.0302	0.0183	0.0111	0.0067
1	0.7358	0.5578	0.4060	0.2873	0.1991	0.1359	0.0916	0.0611	0.0404
2	0.9197	0.8088	0.6767	0.5438	0.4232	0.3208	0.2381	0.1736	0.1247
3	0.9810	0.9344	0.8571	0.7576	0.6472	0.5366	0.4335	0.3423	0.2650
4	0.9963	0.9814	0.9473	0.8912	0.8153	0.7254	0.6288	0.5321	0.4405
5	0.9994	0.9955	0.9834	0.9580	0.9161	0.8576	0.7851	0.7029	0.6160
6	0.9999	0.9991	0.9955	0.9858	0.9665	0.9347	0.8893	0.8311	0.7622
7	1.0000	0.9998	0.9989	0.9958	0.9881	0.9733	0.9489	0.9134	0.8666
8		1.0000	0.9998	0.9989	0.9962	0.9901	0.9786	0.9597	0.9319
9			1.0000	0.9997	0.9989	0.9967	0.9919	0.9829	0.9682
10				0.9999	0.9997	0.9990	0.9972	0.9933	0.9863
11				1.0000	0.9999	0.9997	0.9991	0.9976	0.9945
12					1.0000	0.9999	0.9997	0.9992	0.9980
13						1.0000	0.9999	0.9997	0.9993
14							1.0000	0.9999	0.9998
15								1.0000	0.9999
16									1.0000

Table A.2 (continued) Poisson Probability Sums $\sum_{x=0}^{r} p(x; \mu)$

					μ				
r	5.5	6.0	6.5	7.0	7.5	8.0	8.5	9.0	9.5
0	0.0041	0.0025	0.0015	0.0009	0.0006	0.0003	0.0002	0.0001	0.0001
1	0.0266	0.0174	0.0113	0.0073	0.0047	0.0030	0.0019	0.0012	0.0008
2	0.0884	0.0620	0.0430	0.0296	0.0203	0.0138	0.0093	0.0062	0.0042
3	0.2017	0.1512	0.1118	0.0818	0.0591	0.0424	0.0301	0.0212	0.0149
4	0.3575	0.2851	0.2237	0.1730	0.1321	0.0996	0.0744	0.0550	0.0403
5	0.5289	0.4457	0.3690	0.3007	0.2414	0.1912	0.1496	0.1157	0.0885
6	0.6860	0.6063	0.5265	0.4497	0.3782	0.3134	0.2562	0.2068	0.1649
7	0.8095	0.7440	0.6728	0.5987	0.5246	0.4530	0.3856	0.3239	0.2687
8	0.8944	0.8472	0.7916	0.7291	0.6620	0.5925	0.5231	0.4557	0.3918
9	0.9462	0.9161	0.8774	0.8305	0.7764	0.7166	0.6530	0.5874	0.5218
10	0.9747	0.9574	0.9332	0.9015	0.8622	0.8159	0.7634	0.7060	0.6453
11	0.9890	0.9799	0.9661	0.9467	0.9208	0.8881	0.8487	0.8030	0.7520
12	0.9955	0.9912	0.9840	0.9730	0.9573	0.9362	0.9091	0.8758	0.8364
13	0.9983	0.9964	0.9929	0.9872	0.9784	0.9658	0.9486	0.9261	0.8981
14	0.9994	0.9986	0.9970	0.9943	0.9897	0.9827	0.9726	0.9585	0.9400
15	0.9998	0.9995	0.9988	0.9976	0.9954	0.9918	0.9862	0.9780	0.9665
16	0.9999	0.9998	0.9996	0.9990	0.9980	0.9963	0.9934	0.9889	0.9823
17	1.0000	0.9999	0.9998	0.9996	0.9992	0.9984	0.9970	0.9947	0.9911
18		1.0000	0.9999	0.9999	0.9997	0.9993	0.9987	0.9976	0.9957
19			1.0000	1.0000	0.9999	0.9997	0.9995	0.9989	0.9980
20						0.9999	0.9998	0.9996	0.9991
21						1.0000	0.9999	0.9998	0.9996
22							1.0000	0.9999	0.9999
23								1.0000	0.9999
24									1.0000

Table A.2 (continued) Poisson Probability Sums $\sum_{x=0}^{r} p(x; \mu)$

					μ				
r	10.0	11.0	12.0	13.0	14.0	15.0	16.0	17.0	18.0
0	0.0000	0.0000	0.0000						
1	0.0005	0.0002	0.0001	0.0000	0.0000				
2	0.0028	0.0012	0.0005	0.0002	0.0001	0.0000	0.0000		
3	0.0103	0.0049	0.0023	0.0011	0.0005	0.0002	0.0001	0.0000	0.0000
4	0.0293	0.0151	0.0076	0.0037	0.0018	0.0009	0.0004	0.0002	0.0001
5	0.0671	0.0375	0.0203	0.0107	0.0055	0.0028	0.0014	0.0007	0.0003
6	0.1301	0.0786	0.0458	0.0259	0.0142	0.0076	0.0040	0.0021	0.0010
7	0.2202	0.1432	0.0895	0.0540	0.0316	0.0180	0.0100	0.0054	0.0029
8	0.3328	0.2320	0.1550	0.0998	0.0621	0.0374	0.0220	0.0126	0.0071
9	0.4579	0.3405	0.2424	0.1658	0.1094	0.0699	0.0433	0.0261	0.0154
10	0.5830	0.4599	0.3472	0.2517	0.1757	0.1185	0.0774	0.0491	0.0304
11	0.6968	0.5793	0.4616	0.3532	0.2600	0.1848	0.1270	0.0847	0.0549
12	0.7916	0.6887	0.5760	0.4631	0.3585	0.2676	0.1931	0.1350	0.0917
13	0.8645	0.7813	0.6815	0.5730	0.4644	0.3632	0.2745	0.2009	0.1426
14	0.9165	0.8540	0.7720	0.6751	0.5704	0.4657	0.3675	0.2808	0.2081
15	0.9513	0.9074	0.8444	0.7636	0.6694	0.5681	0.4667	0.3715	0.2867
16	0.9730	0.9441	0.8987	0.8355	0.7559	0.6641	0.5660	0.4677	0.3751
17	0.9857	0.9678	0.9370	0.8905	0.8272	0.7489	0.6593	0.5640	0.4686
18	0.9928	0.9823	0.9626	0.9302	0.8826	0.8195	0.7423	0.6550	0.5622
19	0.9965	0.9907	0.9787	0.9573	0.9235	0.8752	0.8122	0.7363	0.6509
20	0.9984	0.9953	0.9884	0.9750	0.9521	0.9170	0.8682	0.8055	0.7307
21	0.9993	0.9977	0.9939	0.9859	0.9712	0.9469	0.9108	0.8615	0.7991
22	0.9997	0.9990	0.9970	0.9924	0.9833	0.9673	0.9418	0.9047	0.8551
23	0.9999	0.9995	0.9985	0.9960	0.9907	0.9805	0.9633	0.9367	0.8989
24	1.0000	0.9998	0.9993	0.9980	0.9950	0.9888	0.9777	0.9594	0.9317
25		0.9999	0.9997	0.9990	0.9974	0.9938	0.9869	0.9748	0.9554
26		1.0000	0.9999	0.9995	0.9987	0.9967	0.9925	0.9848	0.9718
27			0.9999	0.9998	0.9994	0.9983	0.9959	0.9912	0.9827
28			1.0000	0.9999	0.9997	0.9991	0.9978	0.9950	0.9897
29				1.0000	0.9999	0.9996	0.9989	0.9973	0.9941
30					0.9999	0.9998	0.9994	0.9986	0.9967
31					1.0000	0.9999	0.9997	0.9993	0.9982
32						1.0000	0.9999	0.9996	0.9990
33							0.9999	0.9998	0.9995
34							1.0000	0.9999	0.9998
35								1.0000	0.9999
36									0.9999
37									1.0000

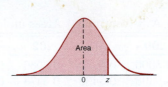

Table A.3 Areas under the Normal Curve

z	.00	.01	.02	.03	.04	.05	.06	.07	.08	.09
−3.4	0.0003	0.0003	0.0003	0.0003	0.0003	0.0003	0.0003	0.0003	0.0003	0.0002
−3.3	0.0005	0.0005	0.0005	0.0004	0.0004	0.0004	0.0004	0.0004	0.0004	0.0003
−3.2	0.0007	0.0007	0.0006	0.0006	0.0006	0.0006	0.0006	0.0005	0.0005	0.0005
−3.1	0.0010	0.0009	0.0009	0.0009	0.0008	0.0008	0.0008	0.0008	0.0007	0.0007
−3.0	0.0013	0.0013	0.0013	0.0012	0.0012	0.0011	0.0011	0.0011	0.0010	0.0010
−2.9	0.0019	0.0018	0.0018	0.0017	0.0016	0.0016	0.0015	0.0015	0.0014	0.0014
−2.8	0.0026	0.0025	0.0024	0.0023	0.0023	0.0022	0.0021	0.0021	0.0020	0.0019
−2.7	0.0035	0.0034	0.0033	0.0032	0.0031	0.0030	0.0029	0.0028	0.0027	0.0026
−2.6	0.0047	0.0045	0.0044	0.0043	0.0041	0.0040	0.0039	0.0038	0.0037	0.0036
−2.5	0.0062	0.0060	0.0059	0.0057	0.0055	0.0054	0.0052	0.0051	0.0049	0.0048
−2.4	0.0082	0.0080	0.0078	0.0075	0.0073	0.0071	0.0069	0.0068	0.0066	0.0064
−2.3	0.0107	0.0104	0.0102	0.0099	0.0096	0.0094	0.0091	0.0089	0.0087	0.0084
−2.2	0.0139	0.0136	0.0132	0.0129	0.0125	0.0122	0.0119	0.0116	0.0113	0.0110
−2.1	0.0179	0.0174	0.0170	0.0166	0.0162	0.0158	0.0154	0.0150	0.0146	0.0143
−2.0	0.0228	0.0222	0.0217	0.0212	0.0207	0.0202	0.0197	0.0192	0.0188	0.0183
−1.9	0.0287	0.0281	0.0274	0.0268	0.0262	0.0256	0.0250	0.0244	0.0239	0.0233
−1.8	0.0359	0.0351	0.0344	0.0336	0.0329	0.0322	0.0314	0.0307	0.0301	0.0294
−1.7	0.0446	0.0436	0.0427	0.0418	0.0409	0.0401	0.0392	0.0384	0.0375	0.0367
−1.6	0.0548	0.0537	0.0526	0.0516	0.0505	0.0495	0.0485	0.0475	0.0465	0.0455
−1.5	0.0668	0.0655	0.0643	0.0630	0.0618	0.0606	0.0594	0.0582	0.0571	0.0559
−1.4	0.0808	0.0793	0.0778	0.0764	0.0749	0.0735	0.0721	0.0708	0.0694	0.0681
−1.3	0.0968	0.0951	0.0934	0.0918	0.0901	0.0885	0.0869	0.0853	0.0838	0.0823
−1.2	0.1151	0.1131	0.1112	0.1093	0.1075	0.1056	0.1038	0.1020	0.1003	0.0985
−1.1	0.1357	0.1335	0.1314	0.1292	0.1271	0.1251	0.1230	0.1210	0.1190	0.1170
−1.0	0.1587	0.1562	0.1539	0.1515	0.1492	0.1469	0.1446	0.1423	0.1401	0.1379
−0.9	0.1841	0.1814	0.1788	0.1762	0.1736	0.1711	0.1685	0.1660	0.1635	0.1611
−0.8	0.2119	0.2090	0.2061	0.2033	0.2005	0.1977	0.1949	0.1922	0.1894	0.1867
−0.7	0.2420	0.2389	0.2358	0.2327	0.2296	0.2266	0.2236	0.2206	0.2177	0.2148
−0.6	0.2743	0.2709	0.2676	0.2643	0.2611	0.2578	0.2546	0.2514	0.2483	0.2451
−0.5	0.3085	0.3050	0.3015	0.2981	0.2946	0.2912	0.2877	0.2843	0.2810	0.2776
−0.4	0.3446	0.3409	0.3372	0.3336	0.3300	0.3264	0.3228	0.3192	0.3156	0.3121
−0.3	0.3821	0.3783	0.3745	0.3707	0.3669	0.3632	0.3594	0.3557	0.3520	0.3483
−0.2	0.4207	0.4168	0.4129	0.4090	0.4052	0.4013	0.3974	0.3936	0.3897	0.3859
−0.1	0.4602	0.4562	0.4522	0.4483	0.4443	0.4404	0.4364	0.4325	0.4286	0.4247
−0.0	0.5000	0.4960	0.4920	0.4880	0.4840	0.4801	0.4761	0.4721	0.4681	0.4641

Table A.3 (continued) Areas under the Normal Curve

z	.00	.01	.02	.03	.04	.05	.06	.07	.08	.09
0.0	0.5000	0.5040	0.5080	0.5120	0.5160	0.5199	0.5239	0.5279	0.5319	0.5359
0.1	0.5398	0.5438	0.5478	0.5517	0.5557	0.5596	0.5636	0.5675	0.5714	0.5753
0.2	0.5793	0.5832	0.5871	0.5910	0.5948	0.5987	0.6026	0.6064	0.6103	0.6141
0.3	0.6179	0.6217	0.6255	0.6293	0.6331	0.6368	0.6406	0.6443	0.6480	0.6517
0.4	0.6554	0.6591	0.6628	0.6664	0.6700	0.6736	0.6772	0.6808	0.6844	0.6879
0.5	0.6915	0.6950	0.6985	0.7019	0.7054	0.7088	0.7123	0.7157	0.7190	0.7224
0.6	0.7257	0.7291	0.7324	0.7357	0.7389	0.7422	0.7454	0.7486	0.7517	0.7549
0.7	0.7580	0.7611	0.7642	0.7673	0.7704	0.7734	0.7764	0.7794	0.7823	0.7852
0.8	0.7881	0.7910	0.7939	0.7967	0.7995	0.8023	0.8051	0.8078	0.8106	0.8133
0.9	0.8159	0.8186	0.8212	0.8238	0.8264	0.8289	0.8315	0.8340	0.8365	0.8389
1.0	0.8413	0.8438	0.8461	0.8485	0.8508	0.8531	0.8554	0.8577	0.8599	0.8621
1.1	0.8643	0.8665	0.8686	0.8708	0.8729	0.8749	0.8770	0.8790	0.8810	0.8830
1.2	0.8849	0.8869	0.8888	0.8907	0.8925	0.8944	0.8962	0.8980	0.8997	0.9015
1.3	0.9032	0.9049	0.9066	0.9082	0.9099	0.9115	0.9131	0.9147	0.9162	0.9177
1.4	0.9192	0.9207	0.9222	0.9236	0.9251	0.9265	0.9279	0.9292	0.9306	0.9319
1.5	0.9332	0.9345	0.9357	0.9370	0.9382	0.9394	0.9406	0.9418	0.9429	0.9441
1.6	0.9452	0.9463	0.9474	0.9484	0.9495	0.9505	0.9515	0.9525	0.9535	0.9545
1.7	0.9554	0.9564	0.9573	0.9582	0.9591	0.9599	0.9608	0.9616	0.9625	0.9633
1.8	0.9641	0.9649	0.9656	0.9664	0.9671	0.9678	0.9686	0.9693	0.9699	0.9706
1.9	0.9713	0.9719	0.9726	0.9732	0.9738	0.9744	0.9750	0.9756	0.9761	0.9767
2.0	0.9772	0.9778	0.9783	0.9788	0.9793	0.9798	0.9803	0.9808	0.9812	0.9817
2.1	0.9821	0.9826	0.9830	0.9834	0.9838	0.9842	0.9846	0.9850	0.9854	0.9857
2.2	0.9861	0.9864	0.9868	0.9871	0.9875	0.9878	0.9881	0.9884	0.9887	0.9890
2.3	0.9893	0.9896	0.9898	0.9901	0.9904	0.9906	0.9909	0.9911	0.9913	0.9916
2.4	0.9918	0.9920	0.9922	0.9925	0.9927	0.9929	0.9931	0.9932	0.9934	0.9936
2.5	0.9938	0.9940	0.9941	0.9943	0.9945	0.9946	0.9948	0.9949	0.9951	0.9952
2.6	0.9953	0.9955	0.9956	0.9957	0.9959	0.9960	0.9961	0.9962	0.9963	0.9964
2.7	0.9965	0.9966	0.9967	0.9968	0.9969	0.9970	0.9971	0.9972	0.9973	0.9974
2.8	0.9974	0.9975	0.9976	0.9977	0.9977	0.9978	0.9979	0.9979	0.9980	0.9981
2.9	0.9981	0.9982	0.9982	0.9983	0.9984	0.9984	0.9985	0.9985	0.9986	0.9986
3.0	0.9987	0.9987	0.9987	0.9988	0.9988	0.9989	0.9989	0.9989	0.9990	0.9990
3.1	0.9990	0.9991	0.9991	0.9991	0.9992	0.9992	0.9992	0.9992	0.9993	0.9993
3.2	0.9993	0.9993	0.9994	0.9994	0.9994	0.9994	0.9994	0.9995	0.9995	0.9995
3.3	0.9995	0.9995	0.9995	0.9996	0.9996	0.9996	0.9996	0.9996	0.9996	0.9997
3.4	0.9997	0.9997	0.9997	0.9997	0.9997	0.9997	0.9997	0.9997	0.9997	0.9998

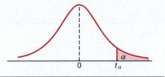

Table A.4 Critical Values of the *t*-Distribution

				α			
v	0.40	0.30	0.20	0.15	0.10	0.05	0.025
1	0.325	0.727	1.376	1.963	3.078	6.314	12.706
2	0.289	0.617	1.061	1.386	1.886	2.920	4.303
3	0.277	0.584	0.978	1.250	1.638	2.353	3.182
4	0.271	0.569	0.941	1.190	1.533	2.132	2.776
5	0.267	0.559	0.920	1.156	1.476	2.015	2.571
6	0.265	0.553	0.906	1.134	1.440	1.943	2.447
7	0.263	0.549	0.896	1.119	1.415	1.895	2.365
8	0.262	0.546	0.889	1.108	1.397	1.860	2.306
9	0.261	0.543	0.883	1.100	1.383	1.833	2.262
10	0.260	0.542	0.879	1.093	1.372	1.812	2.228
11	0.260	0.540	0.876	1.088	1.363	1.796	2.201
12	0.259	0.539	0.873	1.083	1.356	1.782	2.179
13	0.259	0.538	0.870	1.079	1.350	1.771	2.160
14	0.258	0.537	0.868	1.076	1.345	1.761	2.145
15	0.258	0.536	0.866	1.074	1.341	1.753	2.131
16	0.258	0.535	0.865	1.071	1.337	1.746	2.120
17	0.257	0.534	0.863	1.069	1.333	1.740	2.110
18	0.257	0.534	0.862	1.067	1.330	1.734	2.101
19	0.257	0.533	0.861	1.066	1.328	1.729	2.093
20	0.257	0.533	0.860	1.064	1.325	1.725	2.086
21	0.257	0.532	0.859	1.063	1.323	1.721	2.080
22	0.256	0.532	0.858	1.061	1.321	1.717	2.074
23	0.256	0.532	0.858	1.060	1.319	1.714	2.069
24	0.256	0.531	0.857	1.059	1.318	1.711	2.064
25	0.256	0.531	0.856	1.058	1.316	1.708	2.060
26	0.256	0.531	0.856	1.058	1.315	1.706	2.056
27	0.256	0.531	0.855	1.057	1.314	1.703	2.052
28	0.256	0.530	0.855	1.056	1.313	1.701	2.048
29	0.256	0.530	0.854	1.055	1.311	1.699	2.045
30	0.256	0.530	0.854	1.055	1.310	1.697	2.042
40	0.255	0.529	0.851	1.050	1.303	1.684	2.021
60	0.254	0.527	0.848	1.045	1.296	1.671	2.000
120	0.254	0.526	0.845	1.041	1.289	1.658	1.980
∞	0.253	0.524	0.842	1.036	1.282	1.645	1.960

Table A.4 (continued) Critical Values of the *t*-Distribution

				α			
v	0.02	0.015	0.01	0.0075	0.005	0.0025	0.0005
1	15.894	21.205	31.821	42.433	63.656	127.321	636.578
2	4.849	5.643	6.965	8.073	9.925	14.089	31.600
3	3.482	3.896	4.541	5.047	5.841	7.453	12.924
4	2.999	3.298	3.747	4.088	4.604	5.598	8.610
5	2.757	3.003	3.365	3.634	4.032	4.773	6.869
6	2.612	2.829	3.143	3.372	3.707	4.317	5.959
7	2.517	2.715	2.998	3.203	3.499	4.029	5.408
8	2.449	2.634	2.896	3.085	3.355	3.833	5.041
9	2.398	2.574	2.821	2.998	3.250	3.690	4.781
10	2.359	2.527	2.764	2.932	3.169	3.581	4.587
11	2.328	2.491	2.718	2.879	3.106	3.497	4.437
12	2.303	2.461	2.681	2.836	3.055	3.428	4.318
13	2.282	2.436	2.650	2.801	3.012	3.372	4.221
14	2.264	2.415	2.624	2.771	2.977	3.326	4.140
15	2.249	2.397	2.602	2.746	2.947	3.286	4.073
16	2.235	2.382	2.583	2.724	2.921	3.252	4.015
17	2.224	2.368	2.567	2.706	2.898	3.222	3.965
18	2.214	2.356	2.552	2.689	2.878	3.197	3.922
19	2.205	2.346	2.539	2.674	2.861	3.174	3.883
20	2.197	2.336	2.528	2.661	2.845	3.153	3.850
21	2.189	2.328	2.518	2.649	2.831	3.135	3.819
22	2.183	2.320	2.508	2.639	2.819	3.119	3.792
23	2.177	2.313	2.500	2.629	2.807	3.104	3.768
24	2.172	2.307	2.492	2.620	2.797	3.091	3.745
25	2.167	2.301	2.485	2.612	2.787	3.078	3.725
26	2.162	2.296	2.479	2.605	2.779	3.067	3.707
27	2.158	2.291	2.473	2.598	2.771	3.057	3.689
28	2.154	2.286	2.467	2.592	2.763	3.047	3.674
29	2.150	2.282	2.462	2.586	2.756	3.038	3.660
30	2.147	2.278	2.457	2.581	2.750	3.030	3.646
40	2.123	2.250	2.423	2.542	2.704	2.971	3.551
60	2.099	2.223	2.390	2.504	2.660	2.915	3.460
120	2.076	2.196	2.358	2.468	2.617	2.860	3.373
∞	2.054	2.170	2.326	2.432	2.576	2.807	3.290

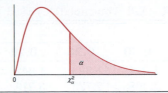

Table A.5 Critical Values of the Chi-Squared Distribution

					α					
v	**0.995**	**0.99**	**0.98**	**0.975**	**0.95**	**0.90**	**0.80**	**0.75**	**0.70**	**0.50**
1	0.0^4393	0.0^3157	0.0^3628	0.0^3982	0.00393	0.0158	0.0642	0.102	0.148	0.455
2	0.0100	0.0201	0.0404	0.0506	0.103	0.211	0.446	0.575	0.713	1.386
3	0.0717	0.115	0.185	0.216	0.352	0.584	1.005	1.213	1.424	2.366
4	0.207	0.297	0.429	0.484	0.711	1.064	1.649	1.923	2.195	3.357
5	0.412	0.554	0.752	0.831	1.145	1.610	2.343	2.675	3.000	4.351
6	0.676	0.872	1.134	1.237	1.635	2.204	3.070	3.455	3.828	5.348
7	0.989	1.239	1.564	1.690	2.167	2.833	3.822	4.255	4.671	6.346
8	1.344	1.647	2.032	2.180	2.733	3.490	4.594	5.071	5.527	7.344
9	1.735	2.088	2.532	2.700	3.325	4.168	5.380	5.899	6.393	8.343
10	2.156	2.558	3.059	3.247	3.940	4.865	6.179	6.737	7.267	9.342
11	2.603	3.053	3.609	3.816	4.575	5.578	6.989	7.584	8.148	10.341
12	3.074	3.571	4.178	4.404	5.226	6.304	7.807	8.438	9.034	11.340
13	3.565	4.107	4.765	5.009	5.892	7.041	8.634	9.299	9.926	12.340
14	4.075	4.660	5.368	5.629	6.571	7.790	9.467	10.165	10.821	13.339
15	4.601	5.229	5.985	6.262	7.261	8.547	10.307	11.037	11.721	14.339
16	5.142	5.812	6.614	6.908	7.962	9.312	11.152	11.912	12.624	15.338
17	5.697	6.408	7.255	7.564	8.672	10.085	12.002	12.792	13.531	16.338
18	6.265	7.015	7.906	8.231	9.390	10.865	12.857	13.675	14.440	17.338
19	6.844	7.633	8.567	8.907	10.117	11.651	13.716	14.562	15.352	18.338
20	7.434	8.260	9.237	9.591	10.851	12.443	14.578	15.452	16.266	19.337
21	8.034	8.897	9.915	10.283	11.591	13.240	15.445	16.344	17.182	20.337
22	8.643	9.542	10.600	10.982	12.338	14.041	16.314	17.240	18.101	21.337
23	9.260	10.196	11.293	11.689	13.091	14.848	17.187	18.137	19.021	22.337
24	9.886	10.856	11.992	12.401	13.848	15.659	18.062	19.037	19.943	23.337
25	10.520	11.524	12.697	13.120	14.611	16.473	18.940	19.939	20.867	24.337
26	11.160	12.198	13.409	13.844	15.379	17.292	19.820	20.843	21.792	25.336
27	11.808	12.878	14.125	14.573	16.151	18.114	20.703	21.749	22.719	26.336
28	12.461	13.565	14.847	15.308	16.928	18.939	21.588	22.657	23.647	27.336
29	13.121	14.256	15.574	16.047	17.708	19.768	22.475	23.567	24.577	28.336
30	13.787	14.953	16.306	16.791	18.493	20.599	23.364	24.478	25.508	29.336
40	20.707	22.164	23.838	24.433	26.509	29.051	32.345	33.66	34.872	39.335
50	27.991	29.707	31.664	32.357	34.764	37.689	41.449	42.942	44.313	49.335
60	35.534	37.485	39.699	40.482	43.188	46.459	50.641	52.294	53.809	59.335

Table A.5 (continued) Critical Values of the Chi-Squared Distribution

v	0.30	0.25	0.20	0.10	0.05	0.025	0.02	0.01	0.005	0.001
1	1.074	1.323	1.642	2.706	3.841	5.024	5.412	6.635	7.879	10.827
2	2.408	2.773	3.219	4.605	5.991	7.378	7.824	9.210	10.597	13.815
3	3.665	4.108	4.642	6.251	7.815	9.348	9.837	11.345	12.838	16.266
4	4.878	5.385	5.989	7.779	9.488	11.143	11.668	13.277	14.860	18.466
5	6.064	6.626	7.289	9.236	11.070	12.832	13.388	15.086	16.750	20.515
6	7.231	7.841	8.558	10.645	12.592	14.449	15.033	16.812	18.548	22.457
7	8.383	9.037	9.803	12.017	14.067	16.013	16.622	18.475	20.278	24.321
8	9.524	10.219	11.030	13.362	15.507	17.535	18.168	20.090	21.955	26.124
9	10.656	11.389	12.242	14.684	16.919	19.023	19.679	21.666	23.589	27.877
10	11.781	12.549	13.442	15.987	18.307	20.483	21.161	23.209	25.188	29.588
11	12.899	13.701	14.631	17.275	19.675	21.920	22.618	24.725	26.757	31.264
12	14.011	14.845	15.812	18.549	21.026	23.337	24.054	26.217	28.300	32.909
13	15.119	15.984	16.985	19.812	22.362	24.736	25.471	27.688	29.819	34.527
14	16.222	17.117	18.151	21.064	23.685	26.119	26.873	29.141	31.319	36.124
15	17.322	18.245	19.311	22.307	24.996	27.488	28.259	30.578	32.801	37.698
16	18.418	19.369	20.465	23.542	26.296	28.845	29.633	32.000	34.267	39.252
17	19.511	20.489	21.615	24.769	27.587	30.191	30.995	33.409	35.718	40.791
18	20.601	21.605	22.760	25.989	28.869	31.526	32.346	34.805	37.156	42.312
19	21.689	22.718	23.900	27.204	30.144	32.852	33.687	36.191	38.582	43.819
20	22.775	23.828	25.038	28.412	31.410	34.170	35.020	37.566	39.997	45.314
21	23.858	24.935	26.171	29.615	32.671	35.479	36.343	38.932	41.401	46.796
22	24.939	26.039	27.301	30.813	33.924	36.781	37.659	40.289	42.796	48.268
23	26.018	27.141	28.429	32.007	35.172	38.076	38.968	41.638	44.181	49.728
24	27.096	28.241	29.553	33.196	36.415	39.364	40.270	42.980	45.558	51.179
25	28.172	29.339	30.675	34.382	37.652	40.646	41.566	44.314	46.928	52.619
26	29.246	30.435	31.795	35.563	38.885	41.923	42.856	45.642	48.290	54.051
27	30.319	31.528	32.912	36.741	40.113	43.195	44.140	46.963	49.645	55.475
28	31.391	32.620	34.027	37.916	41.337	44.461	45.419	48.278	50.994	56.892
29	32.461	33.711	35.139	39.087	42.557	45.722	46.693	49.588	52.335	58.301
30	33.530	34.800	36.250	40.256	43.773	46.979	47.962	50.892	53.672	59.702
40	44.165	45.616	47.269	51.805	55.758	59.342	60.436	63.691	66.766	73.403
50	54.723	56.334	58.164	63.167	67.505	71.420	72.613	76.154	79.490	86.660
60	65.226	66.981	68.972	74.397	79.082	83.298	84.58	88.379	91.952	99.608

Table A.6 Critical Values of the *F*-Distribution

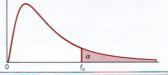

$$f_{0.05}(v_1, v_2)$$

					v_1				
v_2	1	2	3	4	5	6	7	8	9
1	161.448	199.500	215.707	224.583	230.162	233.986	236.768	238.883	240.5433
2	18.513	19.000	19.164	19.247	19.296	19.330	19.353	19.371	19.385
3	10.128	9.552	9.277	9.117	9.013	8.941	8.887	8.845	8.812
4	7.709	6.944	6.591	6.388	6.256	6.163	6.094	6.041	5.999
5	6.608	5.786	5.409	5.192	5.050	4.950	4.876	4.818	4.772
6	5.987	5.143	4.757	4.534	4.387	4.284	4.207	4.147	4.099
7	5.591	4.737	4.347	4.120	3.972	3.866	3.787	3.725	3.686
8	5.318	4.459	4.066	3.838	3.687	3.581	3.500	3.438	3.388
9	5.117	4.256	3.863	3.633	3.482	3.373	3.293	3.230	3.179
10	4.964	4.103	3.708	3.478	3.326	3.217	3.135	3.072	3.020
11	4.844	3.982	3.587	3.357	3.204	3.095	3.012	2.948	2.896
12	4.747	3.885	3.490	3.259	3.106	2.996	2.913	2.849	2.796
13	4.667	3.806	3.411	3.179	3.025	2.915	2.832	2.767	2.714
14	4.600	3.739	3.344	3.112	2.958	2.848	2.764	2.699	2.646
15	4.543	3.682	3.287	3.056	2.901	2.790	2.707	2.641	2.588
16	4.494	3.634	3.239	3.007	2.852	2.741	2.657	2.591	2.538
17	4.451	3.592	3.197	2.965	2.810	2.699	2.614	2.548	2.494
18	4.414	3.555	3.160	2.928	2.773	2.661	2.578	2.510	2.456
19	4.381	3.522	3.127	2.895	2.740	2.628	2.544	2.477	2.423
20	4.351	3.493	3.098	2.866	2.711	2.599	2.514	2.447	2.392
21	4.325	3.467	3.072	2.840	2.685	2.573	2.488	2.420	2.366
22	4.301	3.443	3.049	2.817	2.661	2.549	2.464	2.397	2.342
23	4.279	3.422	3.028	2.795	2.640	2.528	2.442	2.375	2.320
24	4.260	3.403	3.009	2.776	2.621	2.508	2.423	2.355	2.300
25	4.242	3.385	2.991	2.759	2.603	2.490	2.405	2.337	2.282
26	4.225	3.369	2.975	2.743	2.587	2.474	2.388	2.321	2.265
27	4.210	3.354	2.960	2.728	2.572	2.459	2.373	2.305	2.250
28	4.196	3.340	2.947	2.714	2.558	2.445	2.359	2.291	2.236
29	4.183	3.328	2.934	2.701	2.545	2.432	2.356	2.278	2.223
30	4.171	3.316	2.922	2.690	2.534	2.421	2.334	2.226	2.211
40	4.085	3.232	2.839	2.606	2.449	2.339	2.249	2.180	2.124
60	4.001	3.150	2.758	2.525	2.368	2.254	2.167	2.097	2.040
120	3.920	3.072	2.680	2.447	2.290	2.175	2.087	2.016	1.959
∞	3.841	2.996	2.605	2.372	2.214	2.099	2.010	1.938	1.880

Table A.6 (continued) Critical Values of the F-Distribution

$$f_{0.05}(v_1, v_2)$$

v_2	v_1									
	10	12	15	20	24	30	40	60	120	∞
1	241.882	243.906	245.950	248.013	249.052	250.095	251.143	252.196	253.253	254.314
2	19.396	19.413	19.429	19.446	19.454	19.462	19.471	19.479	19.487	19.496
3	8.786	8.745	8.703	8.660	8.639	8.617	8.594	8.572	8.549	8.526
4	5.964	5.912	5.858	5.803	5.774	5.746	5.717	5.688	5.658	5.628
5	4.735	4.678	4.619	4.558	4.527	4.496	4.464	4.431	4.398	4.365
6	4.060	4.000	3.938	3.874	3.841	3.808	3.774	3.740	3.705	3.669
7	3.637	3.575	3.511	3.445	3.410	3.376	3.340	3.304	3.267	3.230
8	3.347	3.284	3.218	3.150	3.115	3.079	3.043	3.005	2.967	2.928
9	3.137	3.073	3.006	2.936	2.900	2.864	2.826	2.787	2.748	2.707
10	2.978	2.913	2.845	2.774	2.737	2.700	2.661	2.621	2.580	2.538
11	2.854	2.788	2.719	2.646	2.609	2.570	2.531	2.490	2.448	2.404
12	2.753	2.687	2.617	2.544	2.505	2.466	2.429	2.384	2.341	2.296
13	2.671	2.602	2.533	2.459	2.420	2.380	2.339	2.297	2.252	2.206
14	2.602	2.534	2.463	2.388	2.349	2.308	2.266	2.223	2.178	2.131
15	2.544	2.475	2.403	2.328	2.288	2.247	2.204	2.160	2.114	2.066
16	2.494	2.425	2.352	2.276	2.235	2.194	2.151	2.106	2.059	2.010
17	2.450	2.381	2.308	2.230	2.190	2.148	2.104	2.058	2.011	1.960
18	2.412	2.342	2.269	2.191	2.150	2.107	2.063	2.017	1.968	1.917
19	2.378	2.308	2.234	2.155	2.114	2.071	2.026	1.980	1.930	1.878
20	2.348	2.278	2.203	2.124	2.082	2.039	1.994	1.946	1.896	1.843
21	2.321	2.250	2.176	2.096	2.054	2.010	1.965	1.916	1.866	1.812
22	2.297	2.229	2.151	2.071	2.028	1.984	1.938	1.889	1.838	1.783
23	2.275	2.204	2.128	2.048	2.005	1.961	1.914	1.865	1.813	1.757
24	2.255	2.183	2.108	2.027	1.984	1.939	1.892	1.842	1.790	1.733
25	2.236	2.165	2.089	2.007	1.964	1.919	1.872	1.822	1.768	1.711
26	2.220	2.148	2.072	1.990	1.946	1.901	1.853	1.803	1.749	1.691
27	2.204	2.132	2.058	1.974	1.930	1.884	1.836	1.785	1.731	1.672
28	2.190	2.118	2.041	1.959	1.915	1.869	1.820	1.769	1.714	1.654
29	2.177	2.104	2.027	1.945	1.901	1.854	1.806	1.754	1.698	1.638
30	2.165	2.092	2.015	1.932	1.887	1.841	1.792	1.740	1.683	1.622
40	2.077	2.003	1.924	1.839	1.793	1.744	1.693	1.637	1.577	1.509
60	1.993	1.917	1.836	1.748	1.700	1.649	1.594	1.534	1.467	1.389
120	1.910	1.834	1.750	1.659	1.608	1.554	1.495	1.429	1.352	1.254
∞	1.831	1.752	1.666	1.570	1.517	1.459	1.394	1.318	1.221	1.000

Table A.6 (continued) Critical Values of the F-Distribution

					$f_{0.01}(v_1, v_2)$				
					v_1				
v_2	1	2	3	4	5	6	7	8	9
1	4052.181	4999.500	5403.352	5624.583	5763.650	5858.986	5928.356	5981.070	6022.473
2	98.503	99.000	99.166	99.249	99.299	99.333	99.356	99.374	99.388
3	34.116	30.817	29.457	28.710	28.237	27.911	27.672	27.489	27.345
4	21.198	18.000	16.694	15.977	15.522	15.207	14.976	14.799	14.659
5	16.258	13.274	12.060	11.392	10.967	10.672	10.455	10.289	10.158
6	13.745	10.925	9.780	9.148	8.746	8.466	8.260	8.102	7.976
7	12.246	9.547	8.451	7.847	7.460	7.191	6.993	6.840	6.719
8	11.259	8.649	7.591	7.006	6.632	6.371	6.178	6.029	5.911
9	10.561	8.022	6.992	6.422	6.057	5.802	5.613	5.467	5.351
10	10.044	7.559	6.552	5.994	5.636	5.386	5.200	5.057	4.942
11	9.646	7.206	6.217	5.668	5.316	5.069	4.886	4.744	4.632
12	9.330	6.927	5.953	5.412	5.064	4.821	4.640	4.499	4.388
13	9.074	6.701	5.739	5.205	4.862	4.620	4.441	4.302	4.191
14	8.862	6.515	5.564	5.035	4.694	4.456	4.278	4.140	4.030
15	8.683	6.359	5.417	4.893	4.556	4.318	4.142	4.004	3.895
16	8.531	6.226	5.292	4.773	4.437	4.202	4.026	3.890	3.780
17	8.400	6.112	5.185	4.669	4.336	4.102	3.927	3.791	3.682
18	8.285	6.013	5.092	4.579	4.248	4.015	3.841	3.705	3.597
19	8.185	5.926	5.010	4.500	4.171	3.939	3.765	3.631	3.523
20	8.096	5.849	4.938	4.431	4.103	3.871	3.699	3.564	3.457
21	8.017	5.780	4.874	4.369	4.042	3.812	3.640	3.506	3.398
22	7.945	5.719	4.817	4.313	3.988	3.758	3.587	3.453	3.346
23	7.881	5.664	4.765	4.264	3.939	3.710	3.539	3.406	3.299
24	7.823	5.614	4.718	4.218	3.895	3.667	3.496	3.363	3.256
25	7.770	5.568	4.675	4.177	3.855	3.627	3.458	3.324	3.217
26	7.721	5.527	4.637	4.140	3.818	3.591	3.421	3.288	3.182
27	7.677	5.488	4.601	4.106	3.785	3.558	3.388	3.256	3.149
28	7.636	5.453	4.568	4.074	3.754	3.528	3.358	3.226	3.120
29	7.598	5.420	4.538	4.045	3.725	3.499	3.330	3.198	3.092
30	7.562	5.390	4.510	4.018	3.699	3.473	3.304	3.173	3.067
40	7.314	5.179	4.313	3.828	3.514	3.291	3.124	2.993	2.888
60	7.077	4.977	4.126	3.649	3.339	3.119	2.953	2.823	2.718
120	6.851	4.787	3.949	3.480	3.174	2.956	2.792	2.663	2.559
∞	6.635	4.605	3.782	3.319	3.017	2.802	2.639	2.511	2.407

Table A.6 (continued) Critical Values of the *F*-Distribution

$$f_{0.01}(v_1, v_2)$$

v_2	v_1									
	10	12	15	20	24	30	40	60	120	∞
1	6055.847	6106.321	6157.285	6208.730	6234.631	6260.649	6286.782	6313.030	6339.391	6365.864
2	99.399	99.416	99.433	99.449	99.458	99.466	99.474	99.483	99.491	99.499
3	27.229	27.052	26.872	26.690	26.598	26.505	26.411	26.316	26.221	26.125
4	14.546	14.374	14.198	14.020	13.929	13.838	13.745	13.652	13.558	13.463
5	10.051	9.888	9.722	9.553	9.466	9.379	9.291	9.202	9.112	9.020
6	7.874	7.718	7.559	7.396	7.313	7.229	7.143	7.057	6.969	6.880
7	6.620	6.469	6.314	6.155	6.074	5.992	5.908	5.824	5.737	5.650
8	5.814	5.667	5.515	5.359	5.279	5.198	5.116	5.032	4.946	4.859
9	5.256	5.111	4.962	4.808	4.729	4.649	4.567	4.483	4.398	4.311
10	4.849	4.706	4.558	4.405	4.327	4.247	4.165	4.082	3.996	3.909
11	4.539	4.397	4.251	4.099	4.021	3.941	3.860	3.776	3.690	3.602
12	4.296	4.155	4.010	3.858	3.780	3.701	3.619	3.535	3.449	3.361
13	4.100	3.960	3.815	3.664	3.587	3.507	3.425	3.341	3.255	3.165
14	3.939	3.800	3.656	3.505	3.427	3.348	3.266	3.181	3.094	3.004
15	3.805	3.666	3.522	3.372	3.294	3.214	3.132	3.047	2.959	2.868
16	3.691	3.553	3.409	3.259	3.181	3.101	3.018	2.933	2.845	2.753
17	3.593	3.455	3.312	3.162	3.084	3.003	2.920	2.835	2.746	2.653
18	3.508	3.371	3.227	3.077	2.999	2.919	2.835	2.749	2.660	2.566
19	3.434	3.297	3.153	3.003	2.925	2.844	2.761	2.674	2.584	2.489
20	3.368	3.231	3.088	2.938	2.859	2.778	2.695	2.608	2.517	2.421
21	3.310	3.173	3.030	2.880	2.801	2.720	2.636	2.548	2.457	2.360
22	3.258	3.121	2.978	2.827	2.749	2.667	2.583	2.495	2.403	2.305
23	3.211	3.074	2.931	2.781	2.702	2.620	2.535	2.447	2.354	2.256
24	3.168	3.031	2.889	2.738	2.659	2.577	2.492	2.403	2.310	2.211
25	3.129	2.993	2.850	2.699	2.620	2.538	2.453	2.363	2.270	2.169
26	3.094	2.958	2.815	2.664	2.585	2.503	2.417	2.327	2.233	2.131
27	3.062	2.926	2.782	2.632	2.552	2.470	2.384	2.294	2.198	2.097
28	3.032	2.896	2.753	2.602	2.522	2.440	2.354	2.263	2.167	2.064
29	3.005	2.868	2.726	2.574	2.495	2.412	2.335	2.234	2.138	2.034
30	2.979	2.843	2.726	2.549	2.469	2.386	2.299	2.208	2.111	2.006
40	2.801	2.665	2.522	2.369	2.288	2.203	2.114	2.019	1.917	1.804
60	2.632	2.496	2.352	2.198	2.115	2.028	1.936	1.836	1.726	1.601
120	2.472	2.336	2.192	2.035	1.950	1.860	1.763	1.656	1.533	1.381
∞	2.321	2.185	2.039	1.878	1.791	1.696	1.592	1.473	1.325	1.000

Table A.7 Tolerance Factors for Normal Distributions

n	Two-Sided Intervals γ=0.05 1−α 0.90	0.95	0.99	Two-Sided Intervals γ=0.01 1−α 0.90	0.95	0.99	One-Sided Intervals γ=0.05 1−α 0.90	0.95	0.99	One-Sided Intervals γ=0.01 1−α 0.90	0.95	0.99
2	32.019	37.674	48.430	160.193	188.491	242.300	20.581	26.260	37.094	103.029	131.426	185.617
3	8.380	9.916	12.861	18.930	22.401	29.055	6.156	7.656	10.553	13.995	17.170	23.896
4	5.369	6.370	8.299	9.398	11.150	14.527	4.162	5.144	7.042	7.380	9.083	12.387
5	4.275	5.079	6.634	6.612	7.855	10.260	3.407	4.203	5.741	5.362	6.578	8.939
6	3.712	4.414	5.775	5.337	6.345	8.301	3.006	3.708	5.062	4.411	5.406	7.335
7	3.369	4.007	5.248	4.613	5.488	7.187	2.756	3.400	4.642	3.859	4.728	6.412
8	3.136	3.732	4.891	4.147	4.936	6.468	2.582	3.187	4.354	3.497	4.285	5.812
9	2.967	3.532	4.631	3.822	4.550	5.966	2.454	3.031	4.143	3.241	3.972	5.389
10	2.839	3.379	4.433	3.582	4.265	5.594	2.355	2.911	3.981	3.048	3.738	5.074
11	2.737	3.259	4.277	3.397	4.045	5.308	2.275	2.815	3.852	2.898	3.556	4.829
12	2.655	3.162	4.150	3.250	3.870	5.079	2.210	2.736	3.747	2.777	3.410	4.633
13	2.587	3.081	4.044	3.130	3.727	4.893	2.155	2.671	3.659	2.677	3.290	4.472
14	2.529	3.012	3.955	3.029	3.608	4.737	2.109	2.615	3.585	2.593	3.189	4.337
15	2.480	2.954	3.878	2.945	3.507	4.605	2.068	2.566	3.520	2.522	3.102	4.222
16	2.437	2.903	3.812	2.872	3.421	4.492	2.033	2.524	3.464	2.460	3.028	4.123
17	2.400	2.858	3.754	2.808	3.345	4.393	2.002	2.486	3.414	2.405	2.963	4.037
18	2.366	2.819	3.702	2.753	3.279	4.307	1.974	2.453	3.370	2.357	2.905	3.960
19	2.337	2.784	3.656	2.703	3.221	4.230	1.949	2.423	3.331	2.314	2.854	3.892
20	2.310	2.752	3.615	2.659	3.168	4.161	1.926	2.396	3.295	2.276	2.808	3.832
25	2.208	2.631	3.457	2.494	2.972	3.904	1.838	2.292	3.158	2.129	2.633	3.601
30	2.140	2.549	3.350	2.385	2.841	3.733	1.777	2.220	3.064	2.030	2.516	3.447
35	2.090	2.490	3.272	2.306	2.748	3.611	1.732	2.167	2.995	1.957	2.430	3.334
40	2.052	2.445	3.213	2.247	2.677	3.518	1.697	2.126	2.941	1.902	2.364	3.249
45	2.021	2.408	3.165	2.200	2.621	3.444	1.669	2.092	2.898	1.857	2.312	3.180
50	1.996	2.379	3.126	2.162	2.576	3.385	1.646	2.065	2.863	1.821	2.269	3.125
60	1.958	2.333	3.066	2.103	2.506	3.293	1.609	2.022	2.807	1.764	2.202	3.038
70	1.929	2.299	3.021	2.060	2.454	3.225	1.581	1.990	2.765	1.722	2.153	2.974
80	1.907	2.272	2.986	2.026	2.414	3.173	1.559	1.965	2.733	1.688	2.114	2.924
90	1.889	2.251	2.958	1.999	2.382	3.130	1.542	1.944	2.706	1.661	2.082	2.883
100	1.874	2.233	2.934	1.977	2.355	3.096	1.527	1.927	2.684	1.639	2.056	2.850
150	1.825	2.175	2.859	1.905	2.270	2.983	1.478	1.870	2.611	1.566	1.971	2.741
200	1.798	2.143	2.816	1.865	2.222	2.921	1.450	1.837	2.570	1.524	1.923	2.679
250	1.780	2.121	2.788	1.839	2.191	2.880	1.431	1.815	2.542	1.496	1.891	2.638
300	1.767	2.106	2.767	1.820	2.169	2.850	1.417	1.800	2.522	1.476	1.868	2.608
∞	1.645	1.960	2.576	1.645	1.960	2.576	1.282	1.645	2.326	1.282	1.645	2.326

Adapted from C. Eisenhart, M. W. Hastay, and W. A. Wallis, *Techniques of Statistical Analysis*, Chapter 2, McGraw-Hill Book Company, New York, 1947.

Table A.8 Critical Values for Bartlett's Test

					$b_k(0.01; n)$				
				Number of Populations, k					
n	2	3	4	5	6	7	8	9	10
3	0.1411	0.1672							
4	0.2843	0.3165	0.3475	0.3729	0.3937	0.4110			
5	0.3984	0.4304	0.4607	0.4850	0.5046	0.5207	0.5343	0.5458	0.5558
6	0.4850	0.5149	0.5430	0.5653	0.5832	O.5978	0.6100	0.6204	0.6293
7	0.5512	0.5787	0.6045	0.6248	0.6410	0.6542	0.6652	0.6744	0.6824
8	0.6031	0.6282	0.6518	0.6704	0.6851	0.6970	0.7069	0.7153	0.7225
9	0.6445	0.6676	0.6892	0.7062	0.7197	0.7305	0.7395	0.7471	0.7536
10	0.6783	0.6996	0.7195	0.7352	0.7475	0.7575	0.7657	0.7726	0.7786
11	0.7063	0.7260	0.7445	0.7590	0.7703	0.7795	0.7871	0.7935	0.7990
12	0.7299	0.7483	0.7654	0.7789	0.7894	0.7980	0.8050	0.8109	0.8160
13	0.7501	0.7672	0.7832	0.7958	0.8056	0.8135	0.8201	0.8256	0.8303
14	0.7674	0.7835	0.7985	0.8103	0.8195	0.8269	0.8330	0.8382	0.8426
15	0.7825	0.7977	0.8118	0.8229	0.8315	0.8385	0.8443	0.8491	0.8532
16	0.7958	0.8101	0.8235	0.8339	0.8421	0.8486	0.8541	0.8586	0.8625
17	0.8076	0.8211	0.8338	0.8436	0.8514	0.8576	0.8627	0.8670	0.8707
18	0.8181	0.8309	0.8429	0.8523	0.8596	0.8655	0.8704	0.8745	0.8780
19	0.8275	0.8397	0.8512	0.8601	0.8670	0.8727	0.8773	0.8811	0.8845
20	0.8360	0.8476	0.8586	0.8671	0.8737	0.8791	0.8835	0.8871	0.8903
21	0.8437	0.8548	0.8653	0.8734	0.8797	0.8848	0.8890	0.8926	0.8956
22	0.8507	0.8614	0.8714	0.8791	0.8852	0.8901	0.8941	0.8975	0.9004
23	0.8571	0.8673	0.8769	0.8844	0.8902	0.8949	0.8988	0.9020	0.9047
24	0.8630	0.8728	0.8820	0.8892	0.8948	0.8993	0.9030	0.9061	0.9087
25	0.8684	0.8779	0.8867	0.8936	0.8990	0.9034	0.9069	0.9099	0.9124
26	0.8734	0.8825	0.8911	0.8977	0.9029	0.9071	0.9105	0.9134	0.9158
27	0.8781	0.8869	0.8951	0.9015	0.9065	0.9105	0.9138	0.9166	0.9190
28	0.8824	0.8909	0.8988	0.9050	0.9099	0.9138	0.9169	0.9196	0.9219
29	0.8864	0.8946	0.9023	0.9083	0.9130	0.9167	0.9198	0.9224	0.9246
30	0.8902	0.8981	0.9056	0.9114	0.9159	0.9195	0.9225	0.9250	0.9271
40	0.9175	0.9235	0.9291	0.9335	0.9370	0.9397	0.9420	0.9439	0.9455
50	0.9339	0.9387	0.9433	0.9468	0.9496	0.9518	0.9536	0.9551	0.9564
60	0.9449	0.9489	0.9527	0.9557	0.9580	0.9599	0.9614	0.9626	0.9637
80	0.9586	0.9617	0.9646	0.9668	0.9685	0.9699	0.9711	0.9720	0.9728
100	0.9669	0.9693	0.9716	0.9734	0.9748	0.9759	0.9769	0.9776	0.9783

Table A.8 (continued) Critical Values for Bartlett's Test

$$b_k(0.05; n)$$

				Number of Populations, k					
n	2	3	4	5	6	7	8	9	10
3	0.3123	0.3058	0.3173	0.3299					
4	0.4780	0.4699	0.4803	0.4921	0.5028	0.5122	0.5204	0.5277	0.5341
5	0.5845	0.5762	0.5850	0.5952	0.6045	0.6126	0.6197	0.6260	0.6315
6	0.6563	0.6483	0.6559	0.6646	0.6727	0.6798	0.6860	0.6914	0.6961
7	0.7075	0.7000	0.7065	0.7142	0.7213	0.7275	0.7329	0.7376	0.7418
8	0.7456	0.7387	0.7444	0.7512	0.7574	0.7629	0.7677	0.7719	0.7757
9	0.7751	0.7686	0.7737	0.7798	0.7854	0.7903	0.7946	0.7984	0.8017
10	0.7984	0.7924	0.7970	0.8025	0.8076	0.8121	0.8160	0.8194	0.8224
11	0.8175	0.8118	0.8160	0.8210	0.8257	0.8298	0.8333	0.8365	0.8392
12	0.8332	0.8280	0.8317	0.8364	0.8407	0.8444	0.8477	0.8506	0.8531
13	0.8465	0.8415	0.8450	0.8493	0.8533	0.8568	0.8598	0.8625	0.8648
14	0.8578	0.8532	0.8564	0.8604	0.8641	0.8673	0.8701	0.8726	0.8748
15	0.8676	0.8632	0.8662	0.8699	0.8734	0.8764	0.8790	0.8814	0.8834
16	0.8761	0.8719	0.8747	0.8782	0.8815	0.8843	0.8868	0.8890	0.8909
17	0.8836	0.8796	0.8823	0.8856	0.8886	0.8913	0.8936	0.8957	0.8975
18	0.8902	0.8865	0.8890	0.8921	0.8949	0.8975	0.8997	0.9016	0.9033
19	0.8961	0.8926	0.8949	0.8979	0.9006	0.9030	0.9051	0.9069	0.9086
20	0.9015	0.8980	0.9003	0.9031	0.9057	0.9080	0.9100	0.9117	0.9132
21	0.9063	0.9030	0.9051	0.9078	0.9103	0.9124	0.9143	0.9160	0.9175
22	0.9106	0.9075	0.9095	0.9120	0.9144	0.9165	0.9183	0.9199	0.9213
23	0.9146	0.9116	0.9135	0.9159	0.9182	0.9202	0.9219	0.9235	0.9248
24	0.9182	0.9153	0.9172	0.9195	0.9217	0.9236	0.9253	0.9267	0.9280
25	0.9216	0.9187	0.9205	0.9228	0.9249	0.9267	0.9283	0.9297	0.9309
26	0.9246	0.9219	0.9236	0.9258	0.9278	0.9296	0.9311	0.9325	0.9336
27	0.9275	0.9249	0.9265	0.9286	0.9305	0.9322	0.9337	0.9350	0.9361
28	0.9301	0.9276	0.9292	0.9312	0.9330	0.9347	0.9361	0.9374	0.9385
29	0.9326	0.9301	0.9316	0.9336	0.9354	0.9370	0.9383	0.9396	0.9406
30	0.9348	0.9325	0.9340	0.9358	0.9376	0.9391	0.9404	0.9416	0.9426
40	0.9513	0.9495	0.9506	0.9520	0.9533	0.9545	0.9555	0.9564	0.9572
50	0.9612	0.9597	0.9606	0.9617	0.9628	0.9637	0.9645	0.9652	0.9658
60	0.9677	0.9665	0.9672	0.9681	0.9690	0.9698	0.9705	0.9710	0.9716
80	0.9758	0.9749	0.9754	0.9761	0.9768	0.9774	0.9779	0.9783	0.9787
100	0.9807	0.9799	0.9804	0.9809	0.9815	0.9819	0.9823	0.9827	0.9830

Table A.9 Critical Values for Cochran's Test

$\alpha = 0.01$

k	n 2	3	4	5	6	7	8	9	10	11	17	37	145	∞
2	0.9999	0.9950	0.9794	0.9586	0.9373	0.9172	0.8988	0.8823	0.8674	0.8539	0.7949	0.7067	0.6062	0.5000
3	0.9933	0.9423	0.8831	0.8335	0.7933	0.7606	0.7335	0.7107	0.6912	0.6743	0.6059	0.5153	0.4230	0.3333
4	0.9676	0.8643	0.7814	0.7212	0.6761	0.6410	0.6129	0.5897	0.5702	0.5536	0.4884	0.4057	0.3251	0.2500
5	0.9279	0.7885	0.6957	0.6329	0.5875	0.5531	0.5259	0.5037	0.4854	0.4697	0.4094	0.3351	0.2644	0.2000
6	0.8828	0.7218	0.6258	0.5635	0.5195	0.4866	0.4608	0.4401	0.4229	0.4084	0.3529	0.2858	0.2229	0.1667
7	0.8376	0.6644	0.5685	0.5080	0.4659	0.4347	0.4105	0.3911	0.3751	0.3616	0.3105	0.2494	0.1929	0.1429
8	0.7945	0.6152	0.5209	0.4627	0.4226	0.3932	0.3704	0.3522	0.3373	0.3248	0.2779	0.2214	0.1700	0.1250
9	0.7544	0.5727	0.4810	0.4251	0.3870	0.3592	0.3378	0.3207	0.3067	0.2950	0.2514	0.1992	0.1521	0.1111
10	0.7175	0.5358	0.4469	0.3934	0.3572	0.3308	0.3106	0.2945	0.2813	0.2704	0.2297	0.1811	0.1376	0.1000
12	0.6528	0.4751	0.3919	0.3428	0.3099	0.2861	0.2680	0.2535	0.2419	0.2320	0.1961	0.1535	0.1157	0.0833
15	0.5747	0.4069	0.3317	0.2882	0.2593	0.2386	0.2228	0.2104	0.2002	0.1918	0.1612	0.1251	0.0934	0.0667
20	0.4799	0.3297	0.2654	0.2288	0.2048	0.1877	0.1748	0.1646	0.1567	0.1501	0.1248	0.0960	0.0709	0.0500
24	0.4247	0.2871	0.2295	0.1970	0.1759	0.1608	0.1495	0.1406	0.1338	0.1283	0.1060	0.0810	0.0595	0.0417
30	0.3632	0.2412	0.1913	0.1635	0.1454	0.1327	0.1232	0.1157	0.1100	0.1054	0.0867	0.0658	0.0480	0.0333
40	0.2940	0.1915	0.1508	0.1281	0.1135	0.1033	0.0957	0.0898	0.0853	0.0816	0.0668	0.0503	0.0363	0.0250
60	0.2151	0.1371	0.1069	0.0902	0.0796	0.0722	0.0668	0.0625	0.0594	0.0567	0.0461	0.0344	0.0245	0.0167
120	0.1225	0.0759	0.0585	0.0489	0.0429	0.0387	0.0357	0.0334	0.0316	0.0302	0.0242	0.0178	0.0125	0.0083
∞	0	0	0	0	0	0	0	0	0	0	0	0	0	0

Reproduced from C. Eisenhart, M. W. Hastay, and W. A. Wallis, *Techniques of Statistical Analysis*, Chapter 15, McGraw-Hill Book Company, New York, 1947.

Table A.9 (continued) Critical Values for Cochran's Test

$\alpha = 0.05$

k	n													
	2	3	4	5	6	7	8	9	10	11	17	37	145	∞
2	0.9985	0.9750	0.9392	0.9057	0.8772	0.8534	0.8332	0.8159	0.8010	0.7880	0.7341	0.6602	0.5813	0.5000
3	0.9669	0.8709	0.7977	0.7457	0.7071	0.6771	0.6530	0.6333	0.6167	0.6025	0.5466	0.4748	0.4031	0.3333
4	0.9065	0.7679	0.6841	0.6287	0.5895	0.5598	0.5365	0.5175	0.5017	0.4884	0.4366	0.3720	0.3093	0.2500
5	0.8412	0.6838	0.5981	0.5441	0.5065	0.4783	0.4564	0.4387	0.4241	0.4118	0.3645	0.3066	0.2513	0.2000
6	0.7808	0.6161	0.5321	0.4803	0.4447	0.4184	0.3980	0.3817	0.3682	0.3568	0.3135	0.2612	0.2119	0.1667
7	0.7271	0.5612	0.4800	0.4307	0.3974	0.3726	0.3535	0.3384	0.3259	0.3154	0.2756	0.2278	0.1833	0.1429
8	0.6798	0.5157	0.4377	0.3910	0.3595	0.3362	0.3185	0.3043	0.2926	0.2829	0.2462	0.2022	0.1616	0.1250
9	0.6385	0.4775	0.4027	0.3584	0.3286	0.3067	0.2901	0.2768	0.2659	0.2568	0.2226	0.1820	0.1446	0.1111
10	0.6020	0.4450	0.3733	0.3311	0.3029	0.2823	0.2666	0.2541	0.2439	0.2353	0.2032	0.1655	0.1308	0.1000
12	0.5410	0.3924	0.3264	0.2880	0.2624	0.2439	0.2299	0.2187	0.2098	0.2020	0.1737	0.1403	0.1100	0.0833
15	0.4709	0.3346	0.2758	0.2419	0.2195	0.2034	0.1911	0.1815	0.1736	0.1671	0.1429	0.1144	0.0889	0.0667
20	0.3894	0.2705	0.2205	0.1921	0.1735	0.1602	0.1501	0.1422	0.1357	0.1303	0.1108	0.0879	0.0675	0.0500
24	0.3434	0.2354	0.1907	0.1656	0.1493	0.1374	0.1286	0.1216	0.1160	0.1113	0.0942	0.0743	0.0567	0.0417
30	0.2929	0.1980	0.1593	0.1377	0.1237	0.1137	0.1061	0.1002	0.0958	0.0921	0.0771	0.0604	0.0457	0.0333
40	0.2370	0.1576	0.1259	0.1082	0.0968	0.0887	0.0827	0.0780	0.0745	0.0713	0.0595	0.0462	0.0347	0.0250
60	0.1737	0.1131	0.0895	0.0765	0.0682	0.0623	0.0583	0.0552	0.0520	0.0497	0.0411	0.0316	0.0234	0.0167
120	0.0998	0.0632	0.0495	0.0419	0.0371	0.0337	0.0312	0.0292	0.0279	0.0266	0.0218	0.0165	0.0120	0.0083
∞	0	0	0	0	0	0	0	0	0	0	0	0	0	0

Table A.10 Upper Percentage Points of the Studentized Range Distribution: $q(0.05; k, v)$

Degrees of Freedom, v	Number of Treatments, k								
	2	3	4	5	6	7	8	9	10
1	18.0	27.0	32.8	37.2	40.5	43.1	15.1	47.1	49.1
2	6.09	8.33	9.80	10.89	11.73	12.43	13.03	13.54	13.99
3	4.50	5.91	6.83	7.51	8.04	8.47	8.85	9.18	9.46
4	3.93	5.04	5.76	6.29	6.71	7.06	7.35	7.60	7.83
5	3.64	4.60	5.22	5.67	6.03	6.33	6.58	6.80	6.99
6	3.46	4.34	4.90	5.31	5.63	5.89	6.12	6.32	6.49
7	3.34	4.16	4.68	5.06	5.35	5.59	5.80	5.99	6.15
8	3.26	4.04	4.53	4.89	5.17	5.40	5.60	5.77	5.92
9	3.20	3.95	4.42	4.76	5.02	5.24	5.43	5.60	5.74
10	3.15	3.88	4.33	4.66	4.91	5.12	5.30	5.46	5.60
11	3.11	3.82	4.26	4.58	4.82	5.03	5.20	5.35	5.49
12	3.08	3.77	4.20	4.51	4.75	4.95	5.12	5.27	5.40
13	3.06	3.73	4.15	4.46	4.69	4.88	5.05	5.19	5.32
14	3.03	3.70	4.11	4.41	4.65	4.83	4.99	5.13	5.25
15	3.01	3.67	4.08	4.37	4.59	4.78	4.94	5.08	5.20
16	3.00	3.65	4.05	4.34	4.56	4.74	4.90	5.03	5.15
17	2.98	3.62	4.02	4.31	4.52	4.70	4.86	4.99	5.11
18	2.97	3.61	4.00	4.28	4.49	4.67	4.83	4.96	5.07
19	2.96	3.59	3.98	4.26	4.47	4.64	4.79	4.92	5.04
20	2.95	3.58	3.96	4.24	4.45	4.62	4.77	4.90	5.01
24	2.92	3.53	3.90	4.17	4.37	4.54	4.68	4.81	4.92
30	2.89	3.48	3.84	4.11	4.30	4.46	4.60	4.72	4.83
40	2.86	3.44	3.79	4.04	4.23	4.39	4.52	4.63	4.74
60	2.83	3.40	3.74	3.98	4.16	4.31	4.44	4.55	4.65
120	2.80	3.36	3.69	3.92	4.10	4.24	4.36	4.47	4.56
∞	2.77	3.32	3.63	3.86	4.03	4.17	4.29	4.39	4.47

Table A.11 The Incomplete Gamma Function: $F(x; \alpha) = \int_0^x \frac{1}{\Gamma(\alpha)} y^{\alpha-1} e^{-y} \, dy$

x	α									
	1	2	3	4	5	6	7	8	9	10
1	0.6320	0.2640	0.0800	0.0190	0.0040	0.0010	0.0000	0.0000	0.0000	0.0000
2	0.8650	0.5940	0.3230	0.1430	0.0530	0.0170	0.0050	0.0010	0.0000	0.0000
3	0.9500	0.8010	0.5770	0.3530	0.1850	0.0840	0.0340	0.0120	0.0040	0.0010
4	0.9820	0.9080	0.7620	0.5670	0.3710	0.2150	0.1110	0.0510	0.0210	0.0080
5	0.9930	0.9600	0.8750	0.7350	0.5600	0.3840	0.2380	0.1330	0.0680	0.0320
6	0.9980	0.9830	0.9380	0.8490	0.7150	0.5540	0.3940	0.2560	0.1530	0.0840
7	0.9990	0.9930	0.9700	0.9180	0.8270	0.6990	0.5500	0.4010	0.2710	0.1700
8	1.0000	0.9970	0.9860	0.9580	0.9000	0.8090	0.6870	0.5470	0.4070	0.2830
9		0.9990	0.9940	0.9790	0.9450	0.8840	0.7930	0.6760	0.5440	0.4130
10		1.0000	0.9970	0.9900	0.9710	0.9330	0.8700	0.7800	0.6670	0.5420
11			0.9990	0.9950	0.9850	0.9620	0.9210	0.8570	0.7680	0.6590
12			1.0000	0.9980	0.9920	0.9800	0.9540	0.9110	0.8450	0.7580
13				0.9990	0.9960	0.9890	0.9740	0.9460	0.9000	0.8340
14				1.0000	0.9980	0.9940	0.9860	0.9680	0.9380	0.8910
15					0.9990	0.9970	0.9920	0.9820	0.9630	0.9300

A.12   Proof of Mean of the Hypergeometric Distribution

To find the mean of the hypergeometric distribution, we write

$$E(X) = \sum_{x=0}^{n} x \frac{\binom{k}{x}\binom{N-k}{n-x}}{\binom{N}{n}} = k \sum_{x=1}^{n} \frac{(k-1)!}{(x-1)!(k-x)!} \cdot \frac{\binom{N-k}{n-x}}{\binom{N}{n}}$$

$$= k \sum_{x=1}^{n} \frac{\binom{k-1}{x-1}\binom{N-k}{n-x}}{\binom{N}{n}}.$$

Since

$$\binom{N-k}{n-1-y} = \binom{(N-1)-(k-1)}{n-1-y} \quad \text{and} \quad \binom{N}{n} = \frac{N!}{n!(N-n)!} = \frac{N}{n}\binom{N-1}{n-1},$$

letting $y = x - 1$, we obtain

$$E(X) = k \sum_{y=0}^{n-1} \frac{\binom{k-1}{y}\binom{N-k}{n-1-y}}{\binom{N}{n}}$$

$$= \frac{nk}{N} \sum_{y=0}^{n-1} \frac{\binom{k-1}{y}\binom{(N-1)-(k-1)}{n-1-y}}{\binom{N-1}{n-1}} = \frac{nk}{N},$$

since the summation represents the total of all probabilities in a hypergeometric experiment when $N-1$ items are selected at random from $N-1$, of which $k-1$ are labeled success.

A.13 Proof of Mean and Variance of the Poisson Distribution

Let $\mu = \lambda t$.

$$E(X) = \sum_{x=0}^{\infty} x \cdot \frac{e^{-\mu}\mu^x}{x!} = \sum_{x=1}^{\infty} x \cdot \frac{e^{-\mu}\mu^x}{x!} = \mu \sum_{x=1}^{\infty} \frac{e^{-\mu}\mu^{x-1}}{(x-1)!}.$$

Since the summation in the last term above is the total probability of a Poisson random variable with mean μ, which can be easily seen by letting $y = x - 1$, it equals 1. Therefore, $E(X) = \mu$. To calculate the variance of X, note that

$$E[X(X-1)] = \sum_{x=0}^{\infty} x(x-1)\frac{e^{-\mu}\mu^x}{x!} = \mu^2 \sum_{x=2}^{\infty} \frac{e^{-\mu}\mu^{x-2}}{(x-2)!}.$$

Again, letting $y = x - 2$, the summation in the last term above is the total probability of a Poisson random variable with mean μ. Hence, we obtain

$$\sigma^2 = E(X^2) - [E(X)]^2 = E[X(X-1)] + E(X) - [E(X)]^2 = \mu^2 + \mu - \mu^2 = \mu = \lambda t.$$

A.14 Proof of Mean and Variance of the Gamma Distribution

To find the mean and variance of the gamma distribution, we first calculate

$$E(X^k) = \frac{1}{\beta^\alpha \Gamma(\alpha)} \int_0^\infty x^{\alpha+k-1} e^{-x/\beta} \, dx = \frac{\beta^{k+\alpha}\Gamma(\alpha+k)}{\beta^\alpha \Gamma(\alpha)} \int_0^\infty \frac{x^{\alpha+k-1}e^{-x/\beta}}{\beta^{k+\alpha}\Gamma(\alpha+k)} \, dx,$$

for $k = 0, 1, 2, \ldots$. Since the integrand in the last term above is a gamma density function with parameters $\alpha + k$ and β, it equals 1. Therefore,

$$E(X^k) = \beta^k \frac{\Gamma(k+\alpha)}{\Gamma(\alpha)}.$$

Using the recursion formula of the gamma function from page 144, we obtain

$$\mu = \beta\frac{\Gamma(\alpha+1)}{\Gamma(\alpha)} = \alpha\beta \quad \text{and} \quad \sigma^2 = E(X^2) - \mu^2 = \beta^2\frac{\Gamma(\alpha+2)}{\Gamma(\alpha)} - \mu^2 = \beta^2\alpha(\alpha+1) - (\alpha\beta)^2 = \alpha\beta^2.$$

Appendix B
Answers to Odd-Numbered
Non-Review Exercises

Chapter 1

1.1   (a) $S = \{8, 16, 24, 32, 40, 48\}$

     (b) $S = \{-5, 1\}$

     (c) $S = \{T, HT, HHT, HHH\}$

     (d) $S =$ {Africa, Antarctica, Asia, Australia, Europe, North America, South America}

     (e) $S = \phi$

1.3 $A = C$

1.5   Using the tree diagram, we obtain

$S = \{1HH,\ 1HT,\ 1TH,\ 1TT,\ 2H,\ 2T,\ 3HH,\ 3HT,\ 3TH,\ 3TT,\ 4H,\ 4T,\ 5HH,\ 5HT,\ 5TH,\ 5TT,\ 6H,\ 6T\}$

1.7   (a) $S = \{M_1M_2, M_1F_1, M_1F_2, M_2M_1, M_2F_1,\ M_2F_2, F_1M_1, F_1M_2, F_1F_2, F_2M_1, F_2M_2,\ F_2F_1\}$

     (b) $A = \{M_1M_2, M_1F_1, M_1F_2, M_2M_1, M_2F_1,\ M_2F_2\}$

     (c) $B = \{M_1F_1, M_1F_2, M_2F_1, M_2F_2, F_1M_1,\ F_1M_2, F_2M_1, F_2M_2\}$

     (d) $C = \{F_1F_2, F_2F_1\}$

     (e) $A \cap B = \{M_1F_1, M_1F_2, M_2F_1, M_2F_2\}$

     (f) $A \cup C = \{M_1M_2, M_1F_1, M_1F_2, M_2M_1,\ M_2F_1, M_2F_2, F_1F_2, F_2F_1\}$

1.11   (a) $\{0, 2, 3, 4, 5, 6, 8\}$

     (b) ϕ, the null set

     (c) $\{0, 1, 6, 7, 8, 9\}$

     (d) $\{1, 3, 5, 6, 7, 9\}$

     (e) $\{0, 1, 6, 7, 8, 9\}$

     (f) $\{2, 4\}$

1.15   (a) The family will experience mechanical problems but will receive no ticket for a traffic violation and will not arrive at a campsite that has no vacancies.

     (b) The family will receive a traffic ticket and arrive at a campsite that has no vacancies but will not experience mechanical problems.

     (c) The family will experience mechanical problems and will arrive at a campsite that has no vacancies.

     (d) The family will receive a traffic ticket but will not arrive at a campsite that has no vacancies.

     (e) The family will not experience mechanical problems.

1.17   18

1.19   8

1.21   48

1.23   210

1.25   72

1.27   362,880

1.29   2880

1.31   (a) 40,320; (b) 336

1.33   360

1.35   24

1.37 $_{365}P_{60}$

1.39   (a) Sum of the probabilities exceeds 1.

     (b) Sum of the probabilities is less than 1.

     (c) A negative probability

     (d) Probability of both a heart and a black card is zero.

1.41 (a) 0.3; (b) 0.2

1.43 $S = \{\$10, \$25, \$100\}$; $P(10) = \frac{11}{20}$, $P(25) = \frac{3}{10}$, $P(100) = \frac{15}{100}$; $\frac{17}{20}$

1.45 (a) 22/25; (b) 3/25; (c) 17/50

1.47 (a) 0.32; (b) 0.68; (c) office or den

1.49 (a) 0.8; (b) 0.45; (c) 0.55

1.51 (a) 0.31; (b) 0.93; (c) 0.31

1.53 (a) 0.009; (b) 0.999; (c) 0.01

1.55 (a) 0.048; (b) \$50,000; (c) \$12,500

1.57 (a) The probability that a convict who sold drugs also committed armed robbery.

(b) The probability that a convict who committed armed robbery did not sell drugs.

(c) The probability that a convict who did not sell drugs also did not commit armed robbery.

1.59 (a) 0.018; (b) 0.614; (c) 0.166; (d) 0.479

1.61 (a) 9/28; (b) 3/4; (c) 0.91

1.63 0.27

1.65 (a) 0.43; (b) 0.12; (c) 0.90

1.67 (a) 0.0016; (b) 0.9984

1.69 (a) 0.75112; (b) 0.2045

1.71 0.588

1.73 0.0960

1.75 0.40625

1.77 0.1124

1.79 0.857

Chapter 2

2.1 Discrete; continuous; continuous; discrete; discrete; continuous

2.3

Sample Space	w
HHH	3
HHT	1
HTH	1
THH	1
HTT	-1
THT	-1
TTH	-1
TTT	-3

2.5 (a) 1/30; (b) 1/10

2.7 (a) 0.68; (b) 0.375

2.9

x	0	1	2
$f(x)$	$\frac{2}{7}$	$\frac{4}{7}$	$\frac{1}{7}$

2.11
$$F(x) = \begin{cases} 0, & \text{for } x < 0, \\ 0.41, & \text{for } 0 \le x < 1, \\ 0.78, & \text{for } 1 \le x < 2, \\ 0.94, & \text{for } 2 \le x < 3, \\ 0.99, & \text{for } 3 \le x < 4, \\ 1, & \text{for } x \ge 4 \end{cases}$$

2.13
$$F(x) = \begin{cases} 0, & \text{for } x < 0, \\ \frac{2}{7}, & \text{for } 0 \le x < 1, \\ \frac{6}{7}, & \text{for } 1 \le x < 2, \\ 1, & \text{for } x \ge 2 \end{cases}$$
(a) 4/7; (b) 5/7

2.15 (a) 3/2; (b) $F(x) = \begin{cases} 0, & x < 0 \\ x^{3/2}, & 0 \le x < 1 \\ 1, & x \ge 1 \end{cases}$; 0.3004

2.17

t	20	25	30
$P(T = t)$	$\frac{1}{5}$	$\frac{3}{5}$	$\frac{1}{5}$

2.19 (a)
$$F(x) = \begin{cases} 0, & x < 0, \\ 1 - \exp(-x/2000), & x \ge 0 \end{cases}$$
(b) 0.6065; (c) 0.6321

2.21 (b)
$$F(x) = \begin{cases} 0, & x < 1, \\ 1 - x^{-3}, & x \ge 1 \end{cases}$$
(c) 0.0156

2.23 (a) 0.2231; (b) 0.2212

2.25 (a) $k = 280$; (b) 0.3633; (c) 0.0563

2.27 (a) 0.1528; (b) 0.0446

2.29 (a) 1/36; (b) 1/15

2.31 (a)

$f(x,y)$		x		
		0	1	2	3
	0	0	3/70	9/70	3/70
y	1	2/70	18/70	18/70	2/70
	2	3/70	9/70	3/70	0

(b) 1/2 (c) 3/10 (d) 3/10, 3/5, 1/10

2.33 (a) 1/16; (b) $g(x) = 12x(1-x)^2$, for $0 \le x \le 1$; (c) 1/4

2.35 (a) 3/64; (b) 1/2

2.37 0.6534

2.39 (a)

x	1	2	3
$g(x)$	0.10	0.35	0.55

(b)

y	1	2	3
$h(y)$	0.20	0.50	0.30

(c) 0.2857

2.41 5/8

2.43 Independent

2.45 (a) 3; (b) 21/512

2.47 Independent

2.49 Dependent

2.51 0.88

2.53 25¢

2.55 $500

2.57 $(\ln 4)/\pi$

2.59 100 hours

2.61 0

2.63 209

2.65 $1855

2.67 (a) 35.2; (b) $\mu_X = 3.20$, $\mu_Y = 3.00$

2.69 2000 hours

2.71 (b) 3/2 micrometers

2.73 (a) 1/6; (b) $(5/6)^5$

2.75 (b) 0.88; (c) 1.62

2.77 0.74

2.79 1/18; in terms of actual profit, the variance is $\frac{1}{18}(5000)^2$

2.81 1/6

2.83 $\mu_Y = 10$; $\sigma_Y^2 = 144$

2.85 −0.0062

2.87 $\sigma_X^2 = 0.8456$, $\sigma_X = 0.9196$

2.89 $-1/\sqrt{5}$

2.91 $0.80

2.93 $\mu = 7/2$, $\sigma^2 = 15/4$

2.95 3/14

2.97 52

2.99 (a) $E(X) = E(Y) = 1/3$ and $\text{Var}(X) = \text{Var}(Y) = 4/9$; (b) $E(Z) = 2/3$ and $\text{Var}(Z) = 8/9$

2.101 (a) 4; (b) 32; 16

Chapter 3

3.1 $\mu = \frac{1}{k} \sum_{i=1}^{k} x_i$, $\sigma^2 = \frac{1}{k} \sum_{i=1}^{k} (x_i - \mu)^2$

3.3 $f(x) = \frac{1}{10}$, for $x = 1, 2, \ldots, 10$, and $f(x) = 0$ elsewhere; 3/10

3.5 (a) 0.0474; (b) 0.0171

3.7 (a) 0.7073; (b) 0.4613; (c) 0.1484

3.9 0.1240

3.11 (a) 0.0778; (b) 0.3370; (c) 0.0870

3.13 $f(x_1, x_2, x_3) = \binom{n}{x_1, x_2, x_3} 0.35^{x_1} 0.05^{x_2} 0.60^{x_3}$

3.15 (a) 0.0749; (b) 0.0023; (c) 0.0782

3.17 0.8670

3.19 (a) 0.2852; (b) 0.9887; (c) 0.6083

3.21 53/65

3.23 0.9517

3.25 0.3222

3.27 (a) 0.6815; (b) 0.1153

3.29 0.2315

3.31 (a) 0.3991; (b) 0.1316

3.33 0.599

3.35 63/64

3.37 (a) 0.3840; (b) 0.0067

3.39 (a) 0.0630; (b) 0.9730

3.41 (a) 0.1429; (b) 0.1353

3.43 0.2657

3.45 $\mu = 6$, $\sigma^2 = 6$

3.47 (a) 0.2650; (b) 0.9596

3.49 (a) 0.8243; (b) 14

3.51 4

3.53 5.53×10^{-4}; $\mu = 7.5$

3.55 (a) 0.0137; (b) 0.0830

3.59 (a) 0.6; (b) 0.7; (c) 0.5

3.61 (a) 0.0823; (b) 0.0250; (c) 0.2424;
(d) 0.9236; (e) 0.8133; (f) 0.6435

3.63 (a) 0.1151; (b) 16.1; (c) 20.275; (d) 0.5403

3.65 (a) 0.0548; (b) 0.4514; (c) 23 cups;
(d) 189.95 milliliters

3.67 (a) 0.8980; (b) 0.0287; (c) 0.6080

3.69 (a) 0.0571; (b) 99.11%; (c) 0.3974;
(d) 27.952 minutes; (e) 0.0092

3.71 6.24 years

3.73 (a) 0.0401; (b) 0.0244

3.75 26 students

3.77 (a) 0.3085; (b) 0.0197

3.79 (a) 0.9514; (b) 0.0668

3.81 (a) 0.8749; (b) 0.0059

3.83 (a) 0.0778; (b) 0.0571; (c) 0.6811

3.85 (a) 0.0228; (b) 0.3974

3.87 (a) 0.01686; (b) 0.0582

3.89 $2.8e^{-1.8} - 3.4e^{-2.4} = 0.1545$

3.91 $\mu = 6$; $\sigma^2 = 18$

3.93 $\sum_{x=4}^{6} \binom{6}{x}(1 - e^{-3/4})^x (e^{-3/4})^{6-x} = 0.3968$

3.95 (a) $\mu = \alpha\beta = 50$; (b) $\sigma^2 = \alpha\beta^2 = 500$;
$\sigma = \sqrt{500}$; (c) 0.815

3.97 (a) 0.1889; (b) 0.0357

3.99 (a) e^{-5}; (b) $\beta = 0.2$

Chapter 4

4.1 (a) Responses of all people in Richmond who
have a telephone;

(b) Outcomes for a large or infinite number of
tosses of a coin;

(c) Length of life of such tennis shoes when
worn on the professional tour;

(d) All possible time intervals for this lawyer
to drive from her home to her office.

4.3 (a) 53.75; (b) 75 and 100

4.5 (a) Range is 10; (b) $s = 3.307$

4.7 (a) 61.15; (b) 61.15

4.9 $s = 0.585$

4.11 (a) 45.9; (b) 5.1

4.13 0.3159

4.15 Yes

4.17 (a) $\mu = 5.3$; $\sigma^2 = 0.81$
(b) $\mu_{\bar{X}} = 5.3$; $\sigma_{\bar{X}}^2 = 0.0225$
(c) 0.9082

4.19 (a) 0.6898; (b) 7.35

4.21 The speculation that the mean amount is 0.20 is
not likely to be true.

4.23 (a) The chance that the difference in mean drying time is larger than 1.0 is 0.0013; (b) 13

4.25 (a) 1/2; (b) 0.3085

4.27 $P(\bar{X} \le 775 \mid \mu = 760) = 0.9332$

4.29 (a) 27.488; (b) 18.475; (c) 36.415

4.31 (a) 0.297; (b) 32.852; (c) 46.928

4.33 (a) 0.05; (b) 0.94

4.35 (a) 0.975; (b) 0.10; (c) 0.875; (d) 0.99

4.37 (a) 2.500; (b) 1.319; (c) 1.714

4.39 The claim is valid.

4.41 (a) 2.71; (b) 3.51; (c) 2.92;
(d) 0.47; (e) 0.34

4.43 The F-ratio is 1.44. The variances are not significantly different.

Chapter 5

5.1 56

5.3 $0.3097 < \mu < 0.3103$

5.5 (a) $22{,}496 < \mu < 24{,}504$; (b) error ≤ 1004

5.7 35

5.9 $10.15 < \mu < 12.45$

5.11 $47.722 < \mu < 49.278$

5.13 $(13{,}075, 33{,}925)$

5.15 323.946 to 326.154

5.17 Upper prediction limit: 9.42;
upper tolerance limit: 11.72

5.19 (a) $(0.9876, 1.0174)$

(b) $(0.9411, 1.0639)$

(c) $(0.9334, 1.0716)$

5.21 Since the manufacturer would be more interested in the mean tensile strength for future products, it is conceivable that prediction interval and tolerance interval may be more interesting than just a confidence interval.

5.23 Yes, the value of 6.9 is outside of the prediction interval.

5.25 $2.80 < \mu_1 - \mu_2 < 3.40$

5.27 $0.69 < \mu_1 - \mu_2 < 7.31$

5.29 $0.70 < \mu_1 - \mu_2 < 3.30$

5.31 $-6536 < \mu_1 - \mu_2 < 2936$

5.33 $-0.74 < \mu_1 - \mu_2 < 6.30$

5.35 $-6.92 < \mu_1 - \mu_2 < 36.70$

5.37 $0.54652 < \mu_B - \mu_A < 1.69348$

5.39 $0.194 < p < 0.262$

5.41 (a) $0.498 < p < 0.642$; (b) error ≤ 0.072

5.43 (a) $0.739 < p < 0.961$; (b) no

5.45 2576

5.47 160

5.49 601

5.51 $-0.0136 < p_F - p_M < 0.0636$

5.53 $0.0011 < p_1 - p_2 < 0.0869$

5.55 $0.293 < \sigma^2 < 6.736$; valid claim

5.57 $3.472 < \sigma^2 < 12.804$

Chapter 6

6.1 (a) Conclude that less than 30% of the public is allergic to some cheese products when, in fact, 30% or more is allergic.

(b) Conclude that at least 30% of the public is allergic to some cheese products when, in fact, less than 30% is allergic.

6.3 (a) The firm is not guilty.

(b) The firm is guilty.

6.5 (a) 0.0559

(b) $\beta = 0.0017$; $\beta = 0.00968$; $\beta = 0.5557$

6.7 (a) 0.1286

(b) $\beta = 0.0901$; $\beta = 0.0708$

(c) The probability of a type I error is somewhat large.

6.9 (a) $\alpha = 0.0850$; (b) $\beta = 0.3410$

6.11 (a) $\alpha = 0.1357$; (b) $\beta = 0.2578$

6.13 $\alpha = 0.0094$; $\beta = 0.0122$

6.15 (a) $\alpha = 0.0718$; (b) $\beta = 0.1151$

6.17 (a) $\alpha = 0.0384$; (b) $\beta = 0.5$; $\beta = 0.2776$

6.19 $z = -2.76$; yes, $\mu < 40$ months;
P-value $= 0.0029$

6.21 $z = -1.64$; P-value $= 0.10$

6.23 $t = 0.77$; fail to reject H_0.

6.25 $z = 8.97$; yes, $\mu > 20,000$ kilometers;
P-value < 0.001

6.27 $t = 12.72$; P-value < 0.0005; reject H_0.

6.29 $t = -1.98$; P-value $= 0.0312$; reject H_0.

6.31 $z = -2.60$; conclude $\mu_A - \mu_B \leq 12$ kilograms.

6.33 $t = 1.50$; there is not sufficient evidence to conclude that the increase in substrate concentration would cause an increase in the mean velocity of more than 0.5 micromole per 30 minutes.

6.35 $t = 0.70$; there is not sufficient evidence to support the conclusion that the serum is effective.

6.37 $t = 2.55$; reject H_0: $\mu_1 - \mu_2 > 4$ kilometers.

6.39 $t' = 0.22$; fail to reject H_0.

6.41 $t' = 2.76$; reject H_0.

6.43 $t = -2.53$; reject H_0; the claim is valid.

6.45 $t = 2.48$; P-value < 0.02; reject H_0.

6.47 $n = 6$

6.49 $n = 78.28 \approx 79$

6.51 $n = 5$

6.53 (a) H_0: $M_{hot} - M_{cold} = 0$,
H_1: $M_{hot} - M_{cold} \neq 0$

(b) paired t, $t = 0.99$; P-value > 0.30; fail to reject H_0.

6.55 P-value $= 0.4044$ (with a one-tailed test); the claim is not refuted.

6.57 $z = 1.44$; fail to reject H_0.

6.59 $z = -5.06$ with P-value ≈ 0; conclude that fewer than one-fifth of the homes are heated by oil.

6.61 $z = 0.93$ with P-value $= P(Z > 0.93) = 0.1762$; there is not sufficient evidence to conclude that the new medicine is effective.

6.63 $z = 2.36$ with P-value $= 0.0182$; yes, the difference is significant.

6.65 $z = 1.10$ with P-value $= 0.1357$; we do not have sufficient evidence to conclude that breast cancer is more prevalent in the urban community.

6.67 $\chi^2 = 10.14$; reject H_0, the ratio is not 5:2:2:1.

6.69 $\chi^2 = 4.47$; there is not sufficient evidence to claim that the die is unbalanced.

6.71 $\chi^2 = 3.125$; do not reject H_0: geometric distribution.

6.73 $\chi^2 = 5.19$; do not reject H_0: normal distribution.

6.75 $\chi^2 = 5.47$; do not reject H_0.

6.77 $\chi^2 = 124.59$; yes, occurrence of these types of crime is dependent on the city district.

6.79 $\chi^2 = 5.92$ with P-value $= 0.4332$; do not reject H_0.

6.81 $\chi^2 = 31.17$ with P-value < 0.0001; attitudes are not homogeneous.

6.83 $\chi^2 = 1.84$; do not reject H_0.

Chapter 7

7.1 (a) $b_0 = 64.529$, $b_1 = 0.561$
   (b) $\hat{y} = 81.4$

7.3 (a) $\hat{y} = 5.8254 + 0.5676x$
   (c) $\hat{y} = 34.205$ at $50°C$

7.5 (a) $\hat{y} = 6.4136 + 1.8091x$
   (b) $\hat{y} = 9.580$ at temperature 1.75

7.7 (b) $\hat{y} = 31.709 + 0.353x$

7.9 (b) $\hat{y} = 343.706 + 3.221x$
   (c) $\hat{y} = \$456$ at advertising costs $= \$35$

7.11 (b) $\hat{y} = -1847.633 + 3.653x$

7.13 (a) $\hat{y} = 153.175 - 6.324x$
   (b) $\hat{y} = 123$ at $x = 4.8$ units

7.15 (a) $s^2 = 176.4$

(b) $t = 2.04$; fail to reject H_0: $\beta_1 = 0$.

7.17 (a) $s^2 = 0.40$
   (b) $4.324 < \beta_0 < 8.503$
   (c) $0.446 < \beta_1 < 3.172$

7.19 (a) $s^2 = 6.626$
   (b) $2.684 < \beta_0 < 8.968$
   (c) $0.498 < \beta_1 < 0.637$

7.21 $t = -2.24$; reject H_0 and conclude $\beta < 6$.

7.23 (a) $24.438 < \mu_{Y|24.5} < 27.106$
   (b) $21.88 < y_0 < 29.66$

7.25 $7.81 < \mu_{Y|1.6} < 10.81$

7.27 (a) 17.1812 mpg
   (b) No, the 95% confidence interval on mean mpg is $(27.95, 29.60)$.
   (c) Miles per gallon will likely exceed 18.

7.29 (b) $\hat{y} = 3.4156x$

7.31 The f-value for testing the lack of fit is 1.58, and the conclusion is that H_0 is not rejected. Hence, the lack-of-fit test is insignificant.

7.33 (a) $\hat{y} = 2.003x$
   (b) $t = 1.40$, fail to reject H_0.

7.35 (a) $b_0 = 10.812$, $b_1 = -0.3437$
   (b) $f = 0.43$; the regression is linear.

7.37 $f = 1.71$ and P-value $= 0.2517$; the regression is linear.

7.39 (a) $\hat{P} = -11.3251 - 0.0449T$
   (b) yes
   (c) $R^2 = 0.9355$
   (d) yes

7.41 (b) $\hat{N} = -175.9025 + 0.0902Y$; $R^2 = 0.3322$

7.43 $r = 0.240$

7.45 (a) $r = -0.979$
   (b) P-value $= 0.0530$; do not reject H_0 at 0.025 level.
   (c) 95.8%

7.47 (a) $r = 0.784$
   (b) Reject H_0 and conclude that $\rho > 0$.
   (c) 61.5%

7.49 $\hat{y} = 0.5800 + 2.7122x_1 + 2.0497x_2$

7.51 (a) $\hat{y} = 27.547 + 0.922x_1 + 0.284x_2$

(b) $\hat{y} = 84$ at $x_1 = 60$ and $x_2 = 4$

7.53 (a) $\hat{y} = -102.7132 + 0.6054x_1 + 8.9236x_2 + 1.4374x_3 + 0.0136x_4$

(b) $\hat{y} = 287.6$

7.55 $\hat{y} = 141.6118 - 0.2819x + 0.0003x^2$

7.57 (a) $\hat{y} = 56.4633 + 0.1525x - 0.00008x^2$

(b) $\hat{y} = 86.7\%$ when temperature is at $225°C$

7.59 $\hat{y} = -6.5122 + 1.9994x_1 - 3.6751x_2 + 2.5245x_3 + 5.1581x_4 + 14.4012x_5$

7.61 (a) $\hat{y} = 350.9943 - 1.2720x_1 - 0.1539x_2$

(b) $\hat{y} = 140.9$

7.63 $\hat{y} = 3.3205 + 0.4210x_1 - 0.2958x_2 + 0.0164x_3 + 0.1247x_4$

7.65 0.1651

7.67 242.72

7.69 (a) $\hat{\sigma}^2_{B_2} = 28.0955$; (b) $\hat{\sigma}_{B_1 B_4} = -0.0096$

7.71 $t = 5.91$ with P-value $= 0.0002$. Reject H_0 and claim that $\beta_1 \neq 0$.

7.73 $0.4516 < \mu_{Y|x_1=900,x_2=1} < 1.2083$ and $-0.1640 < y_0 < 1.8239$

7.75 $263.7879 < \mu_{Y|x_1=75,x_2=24,x_3=90,x_4=98} < 311.3357$ and $243.7175 < y_0 < 331.4062$

7.77 (a) $t = -1.09$ with P-value $= 0.3562$

(b) $t = -1.72$ with P-value $= 0.1841$

(c) Yes; not sufficient evidence to show that x_1 and x_2 are significant.

Chapter 8

8.1 $f = 0.31$; not sufficient evidence to support the hypothesis that there are differences among the 6 machines.

8.3 $f = 14.52$; yes, the difference is significant.

8.5 $f = 8.38$; the average specific activities differ significantly.

8.7 $f = 2.25$; not sufficient evidence to support the hypothesis that the different concentrations of $MgNH_4PO_4$ significantly affect the attained height of chrysanthemums.

8.9 $b = 0.79 > b_4(0.01, 4, 4, 4, 9) = 0.4939$. Do not reject H_0. There is not sufficent evidence to claim that variances are different.

8.11 $b = 0.7822 < b_4(0.05, 9, 8, 15) = 0.8055$. The variances are significantly different.

8.13 (a) P-value < 0.0001, significant

(b) for contrast 1 vs. 2, P-value < 0.0001, significantly different; for contrast 3 vs. 4, P-value $= 0.0648$, not significantly different

8.15 Results of Tukey's tests are given below.

$\bar{y}_{4.}$	$\bar{y}_{3.}$	$\bar{y}_{1.}$	$\bar{y}_{5.}$	$\bar{y}_{2.}$
2.98	4.30	5.44	6.96	7.90

8.17 (a) P-value $= 0.0121$; yes, there is a significant difference.

(b)

Depletion	Modified Hess	Substrate Removal Kicknet	Surber	Kicknet

8.19 $f = 70.27$ with P-value < 0.0001; reject H_0.

$\bar{x}_0$	$\bar{x}_{25}$	$\bar{x}_{100}$	$\bar{x}_{75}$	$\bar{x}_{50}$
55.167	60.167	64.167	70.500	72.833

Temperature is important; both $75°$ and $50°(C)$ yielded batteries with significantly longer activated life.

8.21 The mean absorption is significantly lower for aggregate 4 than for aggregates 3 and 5. However, aggregates 1 and 2 are not significantly different from other three aggregates when we compare them pairwisely.

8.23 $f(\text{fertilizer}) = 6.11$; there is significant difference among the fertilizers.

8.25 $f = 5.99$; percent of foreign additives is not the same for all three brands of jam; brand A

8.27 P-value < 0.0001; significant

8.29 P-value $= 0.0023$; significant

8.31 P-value $= 0.1250$; not significant

8.33 P-value < 0.0001; $f = 122.37$; the amount of dye has an effect on the color density of the fabric.

8.35 (a) $y_{ij} = \mu + A_i + \epsilon_{ij}$, $A_i \sim n(x; 0, \sigma_\alpha)$, $\epsilon_{ij} \sim n(x; 0, \sigma)$

(b) $\hat{\sigma}^2_\alpha = 0$ (the estimated variance component is -0.00027); $\hat{\sigma}^2 = 0.0206$

8.37 (a) $f = 14.9$; operators differ significantly.

(b) $\hat{\sigma}^2_\alpha = 28.91$; $s^2 = 8.32$

Chapter 9

9.1 (a) $f = 8.13$; significant

(b) $f = 5.18$; significant

(c) $f = 1.63$; insignificant

9.3 (a) $f = 14.81$; significant

(b) $f = 9.04$; significant

(c) $f = 0.61$; insignificant

9.5 (a) $f = 34.40$; significant

(b) $f = 26.95$; significant

(c) $f = 20.30$; significant

9.7 Test for effect of temperature: $f_1 = 10.85$ with P-value $= 0.0002$;
Test for effect of amount of catalyst: $f_2 = 46.63$ with P-value < 0.0001;
Test for effect of interaction: $f = 2.06$ with P-value $= 0.074$.

9.9 (a)

Source of Variation	df	Sum of Squares	Mean Squares	f	P
Cutting speed	1	12.000	12.000	1.32	0.2836
Tool geometry	1	675.000	675.000	74.31	< 0.0001
Interaction	1	192.000	192.000	21.14	0.0018
Error	8	72.667	9.083		
Total	11	951.667			

(b) The interaction effect masks the effect of cutting speed.

(c) $f_{\text{tool geometry}=1} = 16.51$ and P-value $= 0.0036$;
$f_{\text{tool geometry}=2} = 5.94$ and P-value $= 0.0407$

9.11 (a)

Source of Variation	df	Sum of Squares	Mean Squares	f	P
Method	1	0.000104	0.000104	6.57	0.0226
Laboratory	6	0.008058	0.001343	84.70	< 0.0001
Interaction	6	0.000198	0.000033	2.08	0.1215
Error	14	0.000222	0.000016		
Total	27	0.008582			

(b) The interaction is not significant.

(c) Both main effects are significant.

(e) $f_{\text{laboratory}=1} = 0.01576$ and P-value $= 0.9019$; no significant difference between the methods in laboratory 1;
$f_{\text{laboratory}=7} = 9.081$ and P-value $= 0.0093$

9.13 (b)

Source of Variation	df	Sum of Squares	Mean Squares	f	P
Time	1	0.060208	0.060208	157.07	< 0.0001
Treatment	1	0.060208	0.060208	157.07	< 0.0001
Interaction	1	0.000008	0.000008	.02	0.8864
Error	8	0.003067	0.000383		
Total	11	0.123492			

(c) Both time and treatment influence the magnesium uptake significantly, although there is no significant interaction between them.

(d) $Y = \mu + \beta_T \text{Time} + \beta_Z Z + \beta_{TZ} \text{Time } Z + \epsilon$, where $Z = 1$ when treatment $= 1$ and $Z = 0$ when treatment $= 2$

(e) $f = 0.02$ with P-value $= 0.8864$; the interaction in the model is insignificant.

9.15 (a) Interaction is significant at a level of 0.05, with P-value of 0.0166.

(b) Both main effects are significant.

9.17 (a) AB: $f = 3.83$; significant;
AC: $f = 3.79$; significant;
BC: $f = 1.31$; not significant;
ABC: $f = 1.63$; not significant

(b) A: $f = 0.54$; not significant;
B: $f = 6.85$; significant;
C: $f = 2.15$; not significant

(c) The differences in the means of the measurements for the three levels of C are not consistent across levels of A.

9.19 (a) Stress: $f = 45.96$ with P-value < 0.0001;
coating: $f = 0.05$ with P-value $= 0.8299$;
humidity: $f = 2.13$ with P-value $= 0.1257$;
coating $\times$ humidity:
$f = 3.41$ with P-value $= 0.0385$;
coating $\times$ stress:
$f = 0.08$ with P-value $= 0.9277$;
humidity $\times$ stress:
$f = 3.15$ with P-value $= 0.0192$;
coating $\times$ humidity $\times$ stress:
$f = 1.93$ with P-value $= 0.1138$

(b) The best combination appears to be uncoated, medium humidity, and a stress level of 20,000 psi.

9.21

Effect	f	P
Temperature	14.22	< 0.0001
Surface	6.70	0.0020
HRC	1.67	0.1954
T $\times$ S	5.50	0.0006
T $\times$ HRC	2.69	0.0369
S $\times$ HRC	5.41	0.0007
T $\times$ S $\times$ HRC	3.02	0.0051

9.23 (a) Yes; brand $\times$ type; brand $\times$ temperature

(b) Yes

(c) Brand Y, powdered detergent, hot temperature

9.25 (a)

Effect	f	P
Time	543.53	< 0.0001
Temp	209.79	< 0.0001
Solvent	4.97	0.0457
Time $\times$ Temp	2.66	0.1103
Time $\times$ Solvent	2.04	0.1723
Temp $\times$ Solvent	0.03	0.8558
Time $\times$ Temp $\times$ Solvent	6.22	0.0140

Although three two-way interactions are shown to be insignificant, they may be masked by the significant three-way interaction.

Index